THE IMMUNE SYSTEM

THE IMMUNE SYSTEM

Peter Parham

Stanford University

The Immune System is derived from
Immunobiology by Charles A. Janeway, Jr., Paul
Travers and Mark Walport; also published by
Garland Publishing and Current Trends.

Garland Publishing
New York and London

Current Trends
London

Distributors:

Inside North America: Garland Publishing, a member of the Taylor & Francis Group, 29 West 35th Street, New York, NY 10001-2299, US.
Outside North America: Garland Publishing, a member of the Taylor & Francis Group, 11, New Fetter Lane, London EC4P 4EE, UK.

ISBN: 0 8153 3043 X   (paperback) Garland
ISBN: 0 8153 3848 1   (paperback) International Student Edition

A catalog record for this book is available from the British Library.

Library of Congress Cataloging-in-Publication Data

Parham, Peter, 1950-
        The immune system / Peter Parham
                p. cm.

        Includes index.
        ISBN 0-8153-3043-X (pbk.)
        1. Immune system.   2. Immunopathology.   I. Title.

        QR181 .P335 2000
        616.07'9 21--dc21

                                        99-042006

This book was produced using QuarkXpress 4.0 and Adobe Illustrator 7.0.

Printed in United States of America.

Published by Current Trends, part of Elsevier Science London, 84 Theobalds Road, London, WC1X 8RR, UK and Garland Publishing, a member of the Taylor & Francis Group, 29 West 35th Street, New York, NY 10001-2299, US.

Preface

In the last century immunology was often taught as a form of epic saga, a progression of elegant experiments, breakthrough discoveries and intrepid investigators. The merit to the approach was its simplicity, as in tracing history using the dates of monarchs, ministers, conflicts and conquests. It also added spice and personality in an era where knowledge was hard won, incomplete and only tenuously related to the human condition. That situation has clearly changed, the direct consequence of technical innovations implemented during the latter part of the twentieth century. Today new information literally pours from the research pipeline, onto the printed page and into the World Wide Web. Elegant experiments, breakthrough discoveries and intrepid investigators are now in such abundance that they in themselves can no longer provide a simplifying principle for explaining immunology. However, as a result of all this industry immunologists now have a simpler and intellectually more satisfying picture of the immune system and how it works, not always successfully, to protect the body from infection. This book is an attempt to sketch that picture without complicating the presentation of essential facts and principles or placing unnecessary burdens on those coming to immunology for the first time.

Throughout the book the emphasis is upon the human immune system and how its successes, failures and compromises affect the lives of each and every one of us. The first eight chapters of the book describe the cells and molecules of the immune system and how they work together in providing defenses against invading microorganisms. This part of the book is built around the biology of the B and T lymphocytes, the cells that make the adaptive immune response and provide long-lasting protective immunity, but also emphasizes the importance of the front-line defenses of innate immunity and the functions of other cells such as macrophages, NK cells and neutrophils. The first chapter introduces basic immunological concepts and the components of the immune system. Antigen recognition, repertoire development, and effector functions of both types of lymphocytes are then dealt with in turn, and this part of the book is completed by a chapter that describes how innate and adaptive immunity work together to battle common types of infection.

The next three chapters focus upon diseases that arise from inadequacies of the human immune system. The first describes situations where infections fail to be controlled, often because the microorganism actively evades, exploits, or subverts the immune response. The following chapter examines conditions in which the immune system overreacts to innocuous substances in the environment and causes chronic inflammatory diseases such as allergies and asthma. The third of these chapters examines the autoimmune diseases, such as insulin-dependent diabetes mellitus, rheumatoid arthritis, and multiple sclerosis, in which the immune system attacks healthy cells and causes tissue damage and loss of function. The last chapter of the book illustrates how the immune system can or might be manipulated in order to improve human health, by consideration of the well-established practices of vaccination and transplantation and the emerging field of cancer immunology.

I would like to thank and acknowledge the authors of *Immunobiology* for giving me license with the text and figures of their book which served as a foundation for this one. I also thank and acknowledge the authors of *Case Studies in Immunology* for similarly permitting me to pick and choose from their book. Miranda Robertson initiated this project and her guidance and encouragement were essential contributions during the drafting phase of the manuscript. By example Frances Brodsky convinced me that a book could indeed be written. Its title was suggested by Vitek Tracz.

I thank the following colleagues who have generously provided their advice and criticism: Julia Albright, Linda Baum, Earl Bloch, Edward Chaperon, Donald Cohen, Neil Greenspan, Anastasia Gregoriades, Edward Hoffmann, Arthur Johnson, Jane McCutcheon, Michael Newlon, Blake Ohlson, Bruce Rabin, Victor Reyes, Ken Rosenthal, Dan Schulze, Peter Sheldon, Michael Siegel, Charles Snow, Gary Splitter, Gerald Stokes and Gill Vince. I thank Nick Rooney for the many excellent photographs he has kindly allowed us to use.

The development and production of the actual book from the draft manuscript was a team effort in which it was a pleasure to be both participant and observer. Eleanor Lawrence not only edited the manuscript and figures but also contributed much more. Her unwavering sympathy for the prospective reader ensured that many pet subjects and old chestnuts never made it to the printed page. Nigel Orme contributed many superb illustrations and Emma Hunt was creative in designing the layout and assiduous in proofing and production. The orchestration of all these efforts has been most ably and good humoredly done by Sarah Gibbs and Richard Woof. I am indebted to Denise Schanck (Garland/Taylor & Francis) and Peter Newmark (Current Biology/Elsevier) who have been consistent both in their encouragement and support of the project. Lastly I would like to thank my sister Joyce and her husband Rowland for living close to Garland's London office and for providing a welcoming place to stay when I am working there.

Peter Parham
Stanford University

Contents

Figure Acknowledgements

Cover image courtesy of Gareth Hobbs.

Chapter frontispieces courtesy of Professor Alexander McPherson; antibody crystals.

Chapter 1

Figure 1.3 panel a © Sinclair Stammers/Science Photo Library; panel b, d, e, g, h, k, l © Eye of Science/Science Photo Library; panel c © A.B. Dowsett/Science Photo Library; panel f © Omikron/Science Photo Library; panel i © Eddy Gray/Science Photo Library; panel j © Philipe Gontier/Science Photo Library.
Figure 1.11 top panel from *Functional Histology: A Text and Colour Atlas*, 2nd edn. p. 174, P.R. Wheater et al. © 1987 Churchill Livingstone.

Chapter 2

Figure 2.5 reprinted by permission from *Nature*, **360**: 369–372. © 1992 Macmillan Magazines Ltd.
Figure 2.10 panel a from *Science*, **248**: 712–719. © 1990 the American Association for the Advancement of Science.
Figure 2.25 from *The European Journal of Immunology*, **18**: 1001–1008. © 1988 Wiley–VCH.

Chapter 3

Figure 3.2 from *Science*, **274**: 209–219. © 1996 the American Association for the Advancement of Science.
Figure 3.17 from *Science*, **274**: 209–219. © 1996 the American Association for the Advancement of Science.

Chapter 4

Figure 4.4 panel b from *The European Journal of Immunology*, **17**: 1473–1484. © 1987 Wiley-VCH.

Chapter 5

Figure 5.6 reprinted by permission from *Nature*, **372**:100–103. © 1993 Macmillan Magazines Ltd.

Chapter 6

Figure 6.23 panel c from *Second International Workshop on Cell Mediated Cytoxicity*, pp. 99–119. Eds. P.A. Henkart, and E. Martz. © 1985 Kluwer Academic/Plenum Publishing.

Chapter 7

Figure 7.4 from *Atlas of Blood Cells: Function and Pathology*, 2nd edn. D. Zucker-Franklin et al. © 1988 Lippincott, Williams and Wilkins.
Figure 7.30 from *FEBS Letters*, **250**: 78–84. © 1989 Elsevier Science.
Figure 7.39 from *Blut*, **60**: 309–318. © 1990 Springer-Verlag.

Chapter 9

Figure 9.6 from *International Reviews in Experimental Pathology*, **28**: 45–78. © 1986 Academic Press.

Chapter 10

Figure 10.1 pollen © E. Gueho/CNRI/Science Photo Library; house dust mite © K.H. Kjeldsen/Science Photo Library; wasp © Claude Nuridsany and Marie Perennon/Science Photo Library; vaccine © Chris Priest and Mark Clarke/Science Photo Library; peanuts © Adrienne Hart-Davis/Science Photo Library; shellfish © Sheila Terry/Science Photo Library; poison ivy © Ken Samuelson/Tony Stone images, Images ® copyright 1999 Photodisc Inc.

Chapter 11

Figure 11.14 from *Cell*, **247**: 59. © 1989 Cell Press.
Figure 11.27 reproduced from *The Journal of Experimental Medicine*, 1992, **176**: 1355–1364. By copyright permission of The Rockefeller University Press.

Chapter 12

Figure 12.40 from *Mechanisms of Cytotoxicity by Natural Killer Cells*, p. 195. Ed. R.B. Herberman. © 1985 Academic Press.

List of Headings

x

Elements of the Immune System and their Roles in Defense

Immunology is the study of the physiological mechanisms that humans and other animals use to defend their bodies from invasion by other organisms. The origins of the subject lie in the practice of medicine and in historical observations that people who survived the ravages of epidemic disease were untouched when faced with that same disease again—they had become **immune** to infection. Infectious diseases are caused by microorganisms, which have the advantage of reproducing and evolving much more rapidly than do their human hosts. During the course of an infection, the microorganism can pit enormous populations of its species against an individual *Homo sapiens*. In response, the human body invests heavily in cells dedicated to defense, which collectively form the **immune system**.

The immune system is crucial to human survival. In the absence of a working immune system, even minor infections can take hold and prove fatal. Without intensive treatment, children born without a functional immune system die in early childhood from the effects of common infections. However, in spite of their immune systems, all humans suffer from infectious disease, especially when young. This is because the immune system takes time to build up a strong response to an invading microorganism, time during which the invader can multiply and cause disease. To provide **protective immunity** for the future, the immune system must first do battle with the microorganism. This places people at highest risk during their first infection with a microorganism and, in the absence of modern medicine, leads to substantial child mortality as witnessed in the developing world. When entire populations face a completely new infection, the outcome can be catastrophic, as experienced by indigenous Americans whose populations were decimated by European diseases to which they were suddenly exposed after 1492.

In medicine the greatest triumph of immunology has been **vaccination**, or **immunization**, a procedure whereby severe disease is prevented by prior exposure to the infectious agent in a form that cannot cause disease. Vaccination provides the opportunity for the immune system to gain the experience needed to make a protective response with little risk to health or life. Vaccination was first used against smallpox, a viral scourge that once ravaged populations and disfigured the survivors. In Asia, small amounts of smallpox virus had been used to induce protective immunity for hundreds of years before 1721, when Lady Mary Wortley Montagu introduced the method into western Europe. Subsequently, in 1796, Edward Jenner, a doctor in rural England, showed how inoculation with cowpox virus offered protection against the related smallpox virus with less risk than the earlier methods. Jenner called his procedure vaccination, after *vaccinia*, the name given to the mild disease produced by cowpox, and he is generally credited with its invention. Since his time, vaccination dramatically reduced the incidence of smallpox worldwide with the last cases being seen by physicians in the 1970s (Figure 1.1).

Effective vaccines have been made from only a fraction of the agents that cause disease and of those some are of limited availability because of their cost. Most of the widely used vaccines were first developed many years ago by processes of trial and error, before very much was known about the workings of the immune system. That approach is no longer so successful for making new vaccines, perhaps because all the easily won vaccines have been obtained. But deeper understanding of the mechanisms of immunity is spawning new ideas for vaccines against infectious diseases and even against other types of disease such as cancer. Much is now known about the molecular and cellular components of the immune system and what they can do in the laboratory. Current research aims at understanding their contributions to fighting infections in the world at large. The new knowledge is also being used to find better ways of manipulating the immune system to prevent the unwanted immune responses that cause allergies, autoimmune diseases, and rejection of organ transplants.

In the first part of this chapter we consider the microorganisms that infect human beings and the defenses they must overcome to start an infection. These include physical barriers, chemical barriers, and the fixed defenses of innate immunity that are ready and waiting to halt infections before they can barely start. Also described are the individual cells and tissues of the immune system and how they integrate their functions with the rest of the human body. The second part of the chapter focuses on the more flexible and forceful defenses of adaptive immunity. These mechanisms are only brought into play if and when an infection is established. The adaptive response is always targeted to the specific problem at hand and is made and refined during the course of the infection. When successful, it clears the infection and provides long-lasting immunity that prevents its recurrence.

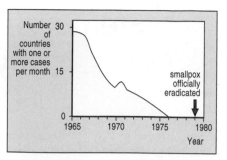

Figure 1.1 The eradication of smallpox by vaccination. In 1979, after 3 years in which no case of smallpox was recorded, the World Health Organization announced that the virus had been eradicated. Since then there has been debate as to whether all laboratory stocks of the virus should be destroyed. One view is that the stocks could be used for nefarious purposes and should therefore be destroyed. The other view is that the virus might not have been eradicated and that if the disease returns, well characterized laboratory stocks will be required to renew study of the disease and its prevention.

Defenses facing invading pathogens

The purpose of the vertebrate immune system is to recognize invading foreign organisms, to prevent their spread, and ultimately to clear them from the body. It consists of billions of cells of various types, which interact with the infectious agent and with each other to fight the infection. With very few exceptions, the cells of the immune system derive originally from the bone marrow, but at some time in their life they leave that site to circulate in the blood, to enter other tissues, and to form part of specialized **lymphoid** tissues. This part of the chapter introduces the cells and tissues of the immune system and some of the molecules through which they carry out their functions. But first we shall consider the organisms and infections that the immune system evolved to protect us against.

1-1 Pathogens are infectious organisms that cause disease

Numerous species of microorganism colonize the human body in large numbers and rarely produce symptoms of disease, for example the benign strains of the bacterium *Escherichia coli* that normally inhabit the gastrointestinal tract. However, for some microorganisms, such as the influenza virus or the typhoid bacillus, infection habitually causes a disease. Any organism with the potential to cause disease is known as a **pathogen.** This definition includes not only microorganisms that generally cause disease if they enter the body, but also ones that can

The immune system protects against four classes of pathogen		
Type of pathogen	Examples	Diseases
Bacteria	*Salmonella enteritidis* *Mycobacterium tuberculosis*	Food poisoning Tuberculosis
Viruses	Variola Influenza HIV	Smallpox Flu AIDS
Fungi	*Epidermophyton floccosum* *Candida albicans*	Ringworm Thrush, systemic candidiasis

Parasites
    protozoa
 	
    worms
	Trypanosoma brucei
	Leishmania donovani
	Plasmodium falciparum
	Ascaris lumbricoides
	Schistosoma mansoni
		Sleeping sickness
		Leishmaniasis
		Malaria
		Ascariasis
		Schistosomiasis

Figure 1.2 The four kinds of pathogen that cause human disease. Examples of the types of pathogen are listed, along with the diseases they cause.

colonize the human body to no ill effect for much of the time but cause illness if the body's defenses are weakened or if the organism gets into the 'wrong' place. The latter kinds of microorganism are known as **opportunistic pathogens**.

Pathogens can be divided into four kinds: **bacteria**, **viruses**, and **fungi**, which are each a group of related microorganisms, and internal **parasites**, a less precise term used to embrace a heterogeneous collection of unicellular protozoa and multicellular invertebrates, mainly worms (Figure 1.2). In this book we consider the functions of the human immune system principally in the context of controlling infections. For some pathogens, this necessitates their complete elimination, but for others it is sufficient to limit the size and location of the pathogen population within the host. Figure 1.3 gives examples of pathogens from the four classes.

Over evolutionary time, the relationship between a pathogen and its human hosts can change, affecting the severity of the disease produced. Most pathogenic organisms have evolved special adaptations that enable them to invade their hosts, replicate in them and be transmitted. However, the rapid death of its host is rarely in a microbe's interest, because this destroys both its home and its source of food. Consequently, those organisms with the potential to cause severe and rapidly fatal disease often tend to evolve towards an accommodation with their hosts. In complementary fashion, host populations have evolved a degree of in-built genetic resistance to common disease-causing organisms, as well as acquiring lifetime immunity to endemic diseases as a result of infection in childhood. Because of the interplay between host and pathogen, the nature and severity of infectious diseases in the human population are always changing.

Influenza is a good example of a common viral disease that, although severe in its symptoms, is usually overcome successfully by the immune system. The fever, aches, and lassitude that accompany infection can be overwhelming, and it is difficult to imagine overcoming foes or predators at the peak of a bout of influenza. However, despite the severity of the symptoms, most strains of influenza do not pose any great danger to healthy people in populations in which influenza is endemic. Warm, well-nourished, and otherwise healthy individuals usually recover in a couple of weeks and take it for granted that their immune systems will accomplish this task. Pathogens new to the human population, in contrast, often cause high mortality—between 60% and 75% in the case of the Ebola virus.

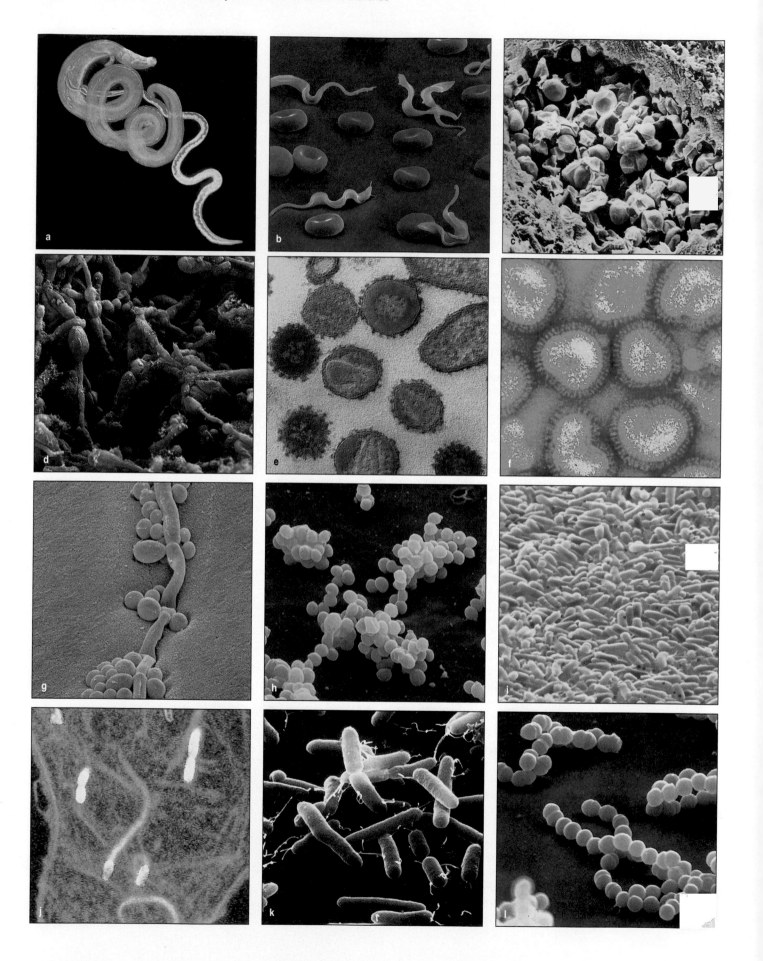

Figure 1.3 (*opposite*) The diversity of human pathogens. Panel a: light micrograph of *Schistosoma mansoni*, the helminth worm that causes schistosomiasis. The adult intestinal blood fluke form is shown: the male is thick and bluish, the female white and thread-like. Magnification × 5. Panel b: false-color scanning electron micrograph of red blood cells and *Trypanosoma brucei*, a protozoan of the type that causes African sleeping sickness. Magnification × 1750. Panel c: false-color scanning electron micrograph of *Pneumocystis carinii*, the fungus that causes opportunistic infections in patients whose immune system is suppressed by disease or drugs. The view is of a lung alveolus from a monkey who is immunosuppressed because of infection with a simian virus that causes an acquired immune deficiency syndrome (AIDS). The fungal cells have been colored green. Magnification × 720. Panel d: scanning electron micrograph of *Epidermophyton floccosum*, the dermatophyte fungus that causes ringworm. Pear-shaped spore-producing structures (macronidia) are seen connected by filaments (hyphae). Magnification × 500. Panel e: false-color transmission electron micrograph of human immunodeficiency virus (HIV), the cause of AIDS. Magnification × 80,000. Panel f: false-color transmission electron micrograph of influenza virus, an orthomyxovirus that causes influenza. Magnification × 40,000. Panel g: false-color scanning electron micrograph of the fungus *Candida albicans*, a normal inhabitant of the human body that occasionally causes thrush and more severe systemic infections. Pseudohyphae are visible in a row. At the junctions of the hyphae, rounded yeast-like cells (blastospores) are budded off into colonies. Magnification × 1400. Panel h: false-color scanning electron micrograph of *Staphylococcus aureus*, a Gram-positive bacterium that colonizes human skin and is the common cause of pimples and boils. Some strains, however, cause food poisoning. The small spherical cells (cocci) typically form grape-like clusters. Magnification × 5000. Panel i: false-color scanning electron micrograph of *Mycobacterium tuberculosis*, the bacterium that causes tuberculosis. Magnification × 15,000. Panel j: false-color transmission electron micrograph of a human cell (colored green) and bacteria of the species *Listeria monocytogenes*, a Gram-positive coccobacillus that can contaminate processed food, causing disease (listeriosis) in immunocompromised individuals and pregnant women. Magnification × 1250. Panel k: false-color scanning electron micrograph of *Salmonella enteritidis*, a Gram-negative, rod-shaped bacterium that is a common cause of food poisoning. The hair-like flagella enable the bacteria to move. Magnification × 6500. Panel l: false-color scanning electron micrograph of *Streptococcus pyogenes*, a Gram-positive bacterium that is the principal cause of tonsilitis and scarlet fever, and can also cause ear infections. It has rounded or spherical cells that sometimes form chains as seen here. Magnification × 6500.

1-2 The skin and mucosal surfaces form physical barriers against infection

The skin is the human body's first defense against infection. It forms a tough impenetrable barrier of epithelium protected by layers of keratinized cells. This barrier can be breached by physical damage, such as wounds, burns, or surgical procedures, which exposes soft tissues and renders them vulnerable to infection. Until the adoption of antiseptic procedures in the nineteenth century, surgery was a very risky business, principally because of the life-threatening infections that the procedures introduced. For the same reason, far more soldiers have died from infection acquired on the battlefield than from the direct effects of enemy action. Ironically, the need to conduct increasingly sophisticated and wide-ranging warfare has been the major force driving improvements in surgery and medicine. As an example from immunology, the burns suffered by fighter pilots during World War II stimulated studies on skin transplantation that led directly to the understanding of the cellular basis of the immune response.

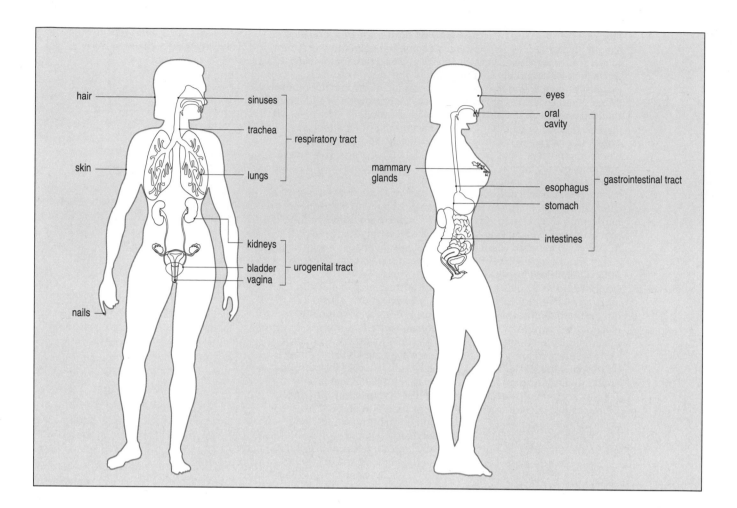

Continuous with the skin are the epithelia lining the respiratory, gastrointestinal, and urogenital tracts (Figure 1.4). On these internal surfaces, the impermeable skin gives way to tissues that are specialized for communication with their environment and are more vulnerable to microbial invasion. Such surfaces are known as **mucosal surfaces** or **mucosa** as they are continually bathed in the **mucus** that they secrete. This thick fluid layer contains glycoproteins, proteoglycans, and enzymes that protect the epithelial cells from damage and help to limit infection. The enzyme lysozyme in tears and saliva is one of a number of anti-bacterial substances in secretions from mucosal surfaces. In the respiratory tract, mucus is continuously removed through the action of epithelial cells bearing beating cilia and is replenished by mucus-secreting goblet cells. The respiratory mucosa are thus continually cleansed of unwanted material, including infectious microorganisms that have been breathed in. Microorganisms are also deterred by the acidic environments of the stomach, the vagina, and the skin. With such defenses, skin and mucosa provide a well-maintained physical and chemical barrier that prevents most pathogens from gaining access to the cells and tissues of the body. When that barrier is breached and pathogens gain entry to the body's soft tissues, the defenses of the immune system are brought into play.

Figure 1.4 The physical barriers that separate the body from its external environment. In these images of a woman, the strong barriers to infection provided by the skin, hair, and nails are colored blue and the more vulnerable mucosal membranes are colored red.

1-3 Immune defenses consist of innate and adaptive immunity

The body has two types of response to invasion by a pathogen—the **innate immune response**, or **innate immunity**, and the **adaptive immune response**. The mechanisms of innate immunity come into play first. They are always present and can rapidly be brought into action, but do not always have the power to

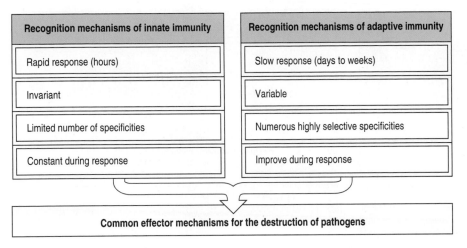

Figure 1.5 The principal characteristics of innate and adaptive immunity.

eliminate the infection. In such circumstances the innate immune response contains the infection while the more powerful forces of the adaptive immune response are martialled.

Innate immunity uses general molecular recognition mechanisms to detect the presence of bacteria and viruses, and it does not lead to long-term immunity to that particular pathogen. The adaptive immune response, in contrast, focuses specifically on the pathogen at hand and leads to a condition of long-lived protection called **adaptive immunity** to that pathogen alone, and no other. Infection with the measles virus, for example, results in immunity to measles but not to mumps, an infection caused by a different virus. However, although the recognition mechanisms of innate and adaptive immune responses differ, the means used to destroy pathogens after their identification are common to both (Figure 1.5). Innate immune defenses are present in vertebrates and invertebrates, but only vertebrates seem to have evolved adaptive immune responses.

The immune system recognizes the presence of a pathogen by the use of so-called 'recognition molecules'. These are proteins that bind to molecules produced by the pathogen or carried on its surface. Some recognition molecules are free in the circulation; others are receptor proteins present on cells of the immune system. Only a few types of recognition molecule are involved in innate immunity; in general, these recognize and distinguish between the major categories of pathogens by binding to distinctive shared features on the pathogen surface. These recognition molecules are always present and can therefore initiate an immediate and rapid response to an invader. In contrast, adaptive immunity has millions of different recognition proteins at its disposal, but requires time for the most useful ones to be selected and mass-produced for use against the particular pathogen.

1-4 White blood cells are responsible for immune responses

The cells responsible for both innate and adaptive immune responses are principally the white blood cells or **leukocytes**, and the tissue cells related to them (Figure 1.6). They all originate in the bone marrow, from where they migrate to develop further and perform their functions. Cells of the immune system are present throughout the body. Some are resident within tissues, where they respond to local trauma and sound the alarm; others circulate in body fluids, from where they are recruited to sites of infection. In defending the body against pathogens, white blood cells cooperate with each other first to recognize the microorganism as an invader and then to destroy it.

The **lymphocytes** provide cells of both innate and adaptive immunity. **Small lymphocytes** are the cells responsible for adaptive immune responses and carry the recognition molecules of adaptive immunity on their surface. Recognition of

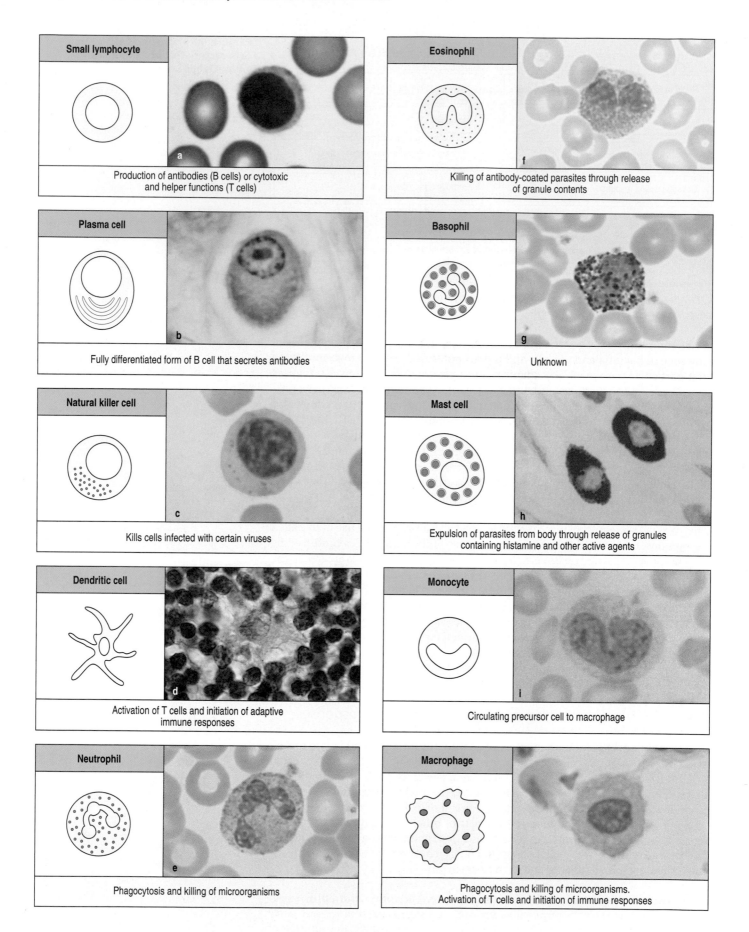

Small lymphocyte

a

Production of antibodies (B cells) or cytotoxic and helper functions (T cells)

Plasma cell

b

Fully differentiated form of B cell that secretes antibodies

Natural killer cell

c

Kills cells infected with certain viruses

Dendritic cell

d

Activation of T cells and initiation of adaptive immune responses

Neutrophil

e

Phagocytosis and killing of microorganisms

Eosinophil

f

Killing of antibody-coated parasites through release of granule contents

Basophil

g

Unknown

Mast cell

h

Expulsion of parasites from body through release of granules containing histamine and other active agents

Monocyte

i

Circulating precursor cell to macrophage

Macrophage

j

Phagocytosis and killing of microorganisms. Activation of T cells and initiation of immune responses

Figure 1.6 (*opposite*) Types of white blood cell. The different types of white blood cell are depicted in schematic diagrams, which indicate their characteristic morphological features, and in accompanying light micrographs. Their main functions are indicated. Red blood cells are also seen in most of the pictures. They are smaller than the white blood cells and have no nucleus. Photographs courtesy of N. Rooney (a, d, e, f, g, j) and D. Friend (b, c, h, i).

a pathogen by small lymphocytes drives a process of lymphocyte selection and differentiation that after 1–2 weeks produces a powerful immune response tailored to the invading organism.

The small lymphocytes, although morphologically indistinguishable from each other, are divided into various classes with respect to their recognition molecules and the functions that they are programmed to perform. The most important difference is between **B lymphocytes** or **B cells**, and **T lymphocytes** or **T cells**. The recognition molecules of B cells are **immunoglobulins** carried on the B-cell surface, whereas those of T cells are known as **T-cell receptors**. On activation by infection, B cells divide and differentiate into **plasma cells**, which make **antibodies**—soluble forms of immunoglobulins that are released into the blood and extracellular fluid. T cells have more diverse functions, which they perform on interaction with other cells of the immune system and cells infected with intracellular pathogens such as viruses. When activated by infection, T cells differentiate into **effector T cells** with various functions. In addition to small lymphocytes, the blood contains large granular lymphocytes called **natural killer cells** or **NK cells**. They function in innate immunity and are important in the defense against viral infections. In the rest of this book the term 'lymphocyte' will be used to denote the small lymphocytes—B cells and T cells.

The principal cell that cooperates with lymphocytes to initiate adaptive immune responses is the **dendritic cell**; this has a distinctive star-shaped morphology and is found in many tissues as well as in the blood.

The job of destroying pathogenic organisms rests largely with two types of phagocytic cell, the **macrophage** and the **neutrophil**. Macrophages are widely distributed in tissues and they derive from circulating white blood cells called **monocytes**. Tissue macrophages are large cells characterized by an extensive cytoplasm with numerous vacuoles, often containing engulfed materials. They are the general scavenger cells of the body, phagocytosing and disposing of dead cells and cell debris. In both innate and adaptive immune responses, one of their roles is to engulf, kill, and break down microorganisms. Macrophages have mechanisms for recognizing and reacting to pathogens, which makes them important cells of innate immunity. They also cooperate with lymphocytes to develop adaptive immune responses.

The neutrophil is smaller than the macrophage and is by far the most abundant white blood cell. Unlike the macrophage, it is not resident in healthy tissues but rapidly migrates to sites of tissue damage, and is at the front line of innate immune defenses, where it is the principal phagocytic and microbicidal cell. Like the macrophage, it possesses general pathogen-recognition mechanisms, and is specialized to engulf and kill microorganisms.

Related to the neutrophil are two other kinds of white blood cell, the **eosinophil** and the **basophil**. Neutrophils, eosinophils, and basophils are collectively called **granulocytes** because of their prominent cytoplasmic granules, which contain reactive substances that kill microorganisms and enhance inflammation. Because granulocytes have irregularly shaped nuclei with two to five lobes they are also called **polymorphonuclear leukocytes**. The eosinophil is the second most abundant granulocyte in the blood and provides defense against parasites such as helminth worms. The least abundant granulocyte is the basophil. They are so rare that little is known of their contribution to the immune response. **Mast cells** are bone-marrow derived cells found in connective tissues. They contain granules containing substances that contribute to inflammation.

1-5 The cells of the immune system derive from precursors in the bone marrow

All the cells that circulate in the blood are derived from a common **progenitor** or precursor cell in the **bone marrow**. This pluripotent precursor is called the

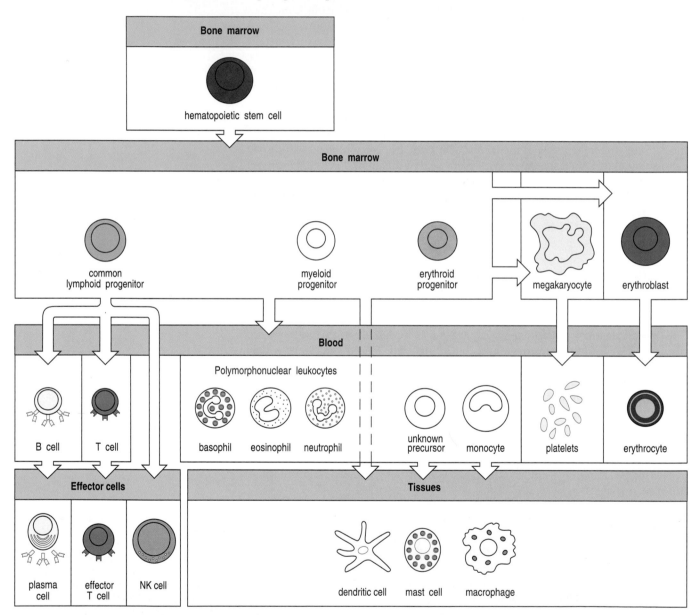

Figure 1.7 Circulating blood cells and certain tissue cells derive from a common hematopoietic stem cell in the bone marrow. The pluripotent stem cell (brown) divides and differentiates into more specialized progenitor cells that give rise to the lymphoid lineage, the myeloid lineage, and the erythrocyte/megakaryocyte lineage. The common lymphoid progenitor divides and differentiates to give B cells (yellow), T cells (blue), and NK cells (purple). On activation by infection, B cells divide and differentiate into plasma cells, whereas T cells differentiate into various types of activated effector T cell. The myeloid progenitor cell divides and differentiates to produce at least six cell types. These are: the three types of granulocyte—the neutrophil, the eosinophil, and the basophil; the mast cell, which takes up residence in connective and mucosal tissues; the circulating monocyte, which gives rise to the macrophages resident in tissues; and the dendritic cell. The erythroid progenitor gives rise to erythrocytes and megakaryocytes.

hematopoietic stem cell. The developmental pathways by which blood cells are produced are outlined in Figure 1.7. As hematopoietic stem cells mature in the bone marrow they give rise to different kinds of stem cells with more limited developmental potential. One kind of stem cell, called the **erythroid progenitor**, gives rise to the erythroid lineage of blood cells—the oxygen-carrying red blood cells (erythrocytes) and the megakaryocytes. The latter remain resident in the bone marrow; platelets are formed from them and released into the blood. A second stem cell gives rise to the monocyte/macrophage lineage, dendritic cells, the granulocytes (neutrophils, eosinophils, and basophils), and mast cells. These cells are known as the **myeloid lineage** of white blood cells and their progenitor stem cell as the **myeloid progenitor**. The third type of stem cell is the **common lymphoid progenitor**. It gives rise to white blood cells of the **lymphoid lineage**, which are the small lymphocytes and the larger NK cells.

1-6 Lymphocytes are found in specialized lymphoid tissues

As well as circulating in the blood, lymphocytes congregate in specialized tissues known as **lymphoid tissues** or **lymphoid organs**. The major lymphoid tissues are bone marrow, thymus, spleen, adenoids, tonsils, appendix, lymph nodes, and Peyer's patches. Lymphoid tissues are also found lining the mucosal surfaces of the respiratory, gastrointestinal, and urogenital tracts. The lymphoid tissues are functionally divided into two types. **Primary** or **central lymphoid tissues** are where lymphocytes develop and mature to the stage at which they are able to respond to a pathogen. The bone marrow and the **thymus** are the primary lymphoid tissues. All other lymphoid tissues are known as **secondary** or **peripheral** lymphoid tissues; they are where mature lymphocytes become stimulated to respond to invading pathogens (Figure 1.8).

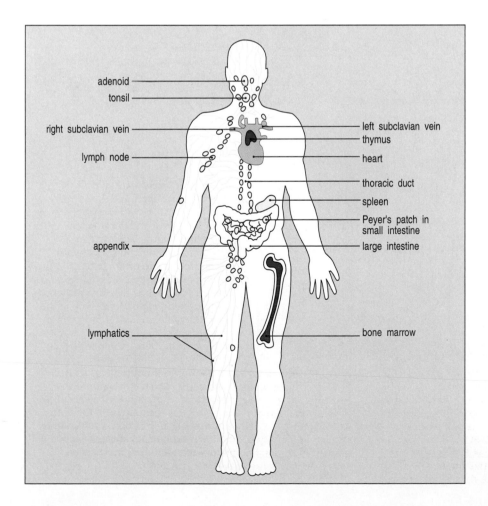

Figure 1.8 The sites of the principal lymphoid tissues within the human body. Lymphocytes arise from stem cells in the bone marrow. B cells complete their maturation in the bone marrow, whereas T cells leave at an immature stage and complete their development in the thymus. The bone marrow and the thymus are the primary lymphoid tissues and are shown in red. The secondary lymphoid tissues and the lymphatics are shown in yellow. Plasma that has leaked from the blood is collected by the lymphatics and returned to the blood supply via the thoracic duct, which empties into the left subclavian vein.

Lymph nodes lie at the junctions of an anastomosing network of **lymphatic vessels** or **lymphatics**, which originate in the connective tissues and collect the plasma that continually leaks out of blood vessels and forms the extracellular fluid. The lymphatics eventually return this fluid to the blood, chiefly via the thoracic duct, which empties into the left subclavian vein in the neck. Mature lymphocytes leave the primary lymphoid tissues and enter the bloodstream, whereupon they travel between the blood, the secondary lymphoid tissues, and the lymphatics, eventually being returned to the blood. The mixture of fluid and cells that flows through the lymphatics is called **lymph**. These movements are called **recirculation** of lymphocytes. An exception to this pattern is the spleen, which has no connections to the lymphatic system. Lymphocytes both enter and leave the spleen in the blood.

Both B and T lymphocytes originate from lymphoid precursors in the bone marrow, but whereas B cells complete their maturation in the bone marrow before entering the circulation, T cells leave the bone marrow at an immature stage and migrate in the blood to the thymus, where they complete their maturation. Lymphocyte maturation is a highly selective process in which the vast majority of immature lymphocytes are destroyed because they fail to develop immunoglobulins or T-cell receptors that are useful to the immune system. The small fraction of immature B and T cells that successfully completes development departs from the primary lymphoid organs and enters the circulation.

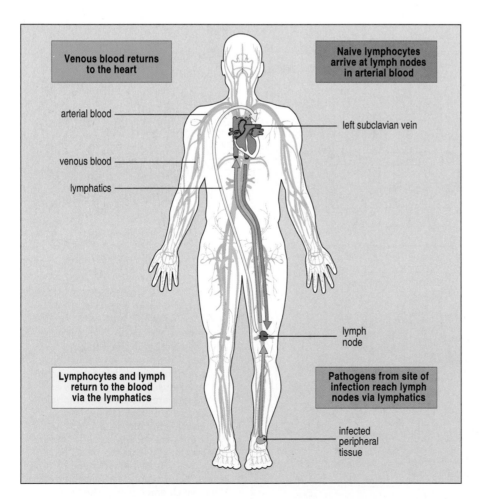

Figure 1.9 Circulating lymphocytes meet lymph-borne pathogens in draining lymph nodes. Lymphocytes leave the blood and enter lymph nodes, where they can be activated by pathogens draining in the afferent lymph from a site of infection. The circulation pertaining to a site of infection in the left foot is shown. When activated by pathogens, lymphocytes stay in the node to divide and differentiate into effector cells. If lymphocytes are not activated, they leave the node in the efferent lymph and are carried by the lymphatics to the thoracic duct (see Figure 1.8), which empties into the blood at the left subclavian vein. Lymphocytes recirculate all the time, irrespective of infection. Every minute 5×10^6 lymphocytes leave the blood and enter secondary lymphoid tissues.

1-7 Lymphocytes are activated in the secondary lymphoid tissues

Secondary lymphoid tissues provide meeting places where lymphocytes circulating in the blood encounter pathogens and their products brought from a site of infection. Precisely how this works depends on where an infection occurs. Frequent sites of infection are the connective tissues, which pathogens penetrate as a result of skin wounds. From such sites, pathogens are carried by the lymphatics to the nearest **lymph node** (Figure 1.9). The lymph node receiving the fluid collected at an infected site is called the **draining lymph node**.

Unlike the blood, the lymph is not driven by a dedicated pump and its flow is comparatively sluggish. One-way valves within lymphatic vessels and the lymph nodes placed at their junctions ensure that net movement of the lymph is always in a direction away from the peripheral tissues and towards the ducts in the upper body where the lymph empties into the blood. The flow of lymph is driven by the continual everyday movements of one part of the body with respect to another. In the absence of such movement, as when a patient is confined to bed for a long time, lymph flow slows and fluid accumulates in tissues, causing the swelling known as **edema**.

Pathogens arrive at a lymph node in **afferent lymphatic vessels**, several of which unite at the node to leave it as a single **efferent lymphatic vessel**. The anatomy of a lymph node is shown in Figure 1.10. As the lymph passes through the node,

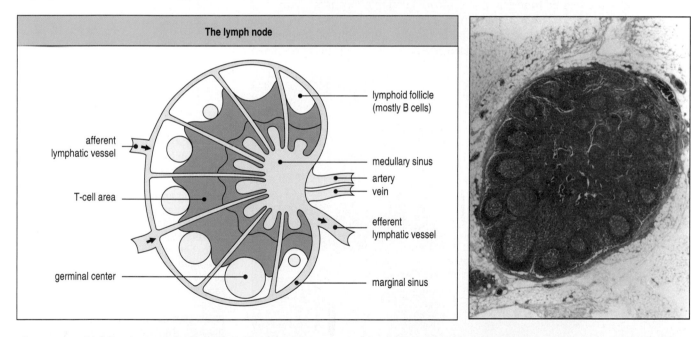

Figure 1.10 Architecture of the lymph node, the site where blood-borne lymphocytes respond to lymph-borne pathogens. Human lymph nodes are small kidney-shaped organs weighing 1 gram or less that form junctions where a number of afferent lymph vessels bringing lymph from the tissues unite to form a single, larger efferent lymph vessel. The lymph node is packed with lymphocytes, macrophages and other cells of the immune system through which the lymph percolates. In a lymph node, pathogens are removed from the lymph by macrophages, then degraded and used to stimulate lymphocytes. Lymphocytes arrive at lymph nodes in the arterial blood. They enter the node by passing between the endothelial cells that line the fine capillaries within the lymph node (not shown). The population of lymphocytes within a node is in a continual state of flux, with new lymphocytes entering from the blood while others leave in the efferent lymph. Within the lymph node there are anatomically discrete areas where B or T cells tend to congregate. A lymph node draining a site of infection increases in size owing to the proliferation of activated lymphocytes, a phenomenon sometimes referred to as 'swollen glands'. The expansion of lymphocyte populations takes place in spherical lymphoid follicles present in the lymph node cortex. As lymphocyte division and differentiation proceeds, the follicle morphology changes and it is then called a germinal center. The photograph shows a section through a lymph node in which there are prominent germinal centers. Photograph (× 7) courtesy of N. Rooney.

pathogens and other extraneous materials are filtered out. This process prevents infectious organisms from reaching the blood and provides a depot of pathogen within the lymph node that is used to activate lymphocytes.

Small lymphocytes recirculate through the body in both the blood and the lymph. When they reach the blood capillaries running through a lymph node, lymphocytes can leave the blood and enter the node. Within the node they can encounter pathogens that interact with their recognition molecules. When this happens, those lymphocytes are retained in the node, where they divide and differentiate into functional **effector cells**. Effector B cells secrete antibodies and are called plasma cells; they either stay in the lymph node or migrate to the bone marrow via lymph and blood. There are various types of effector T cell. **Helper T cells** are so called because they secrete soluble proteins that activate other cells of the immune system, including B cells and macrophages. A second class of T cells is called **cytotoxic T cells** because they kill cells infected with viruses or other intracellular pathogens. Helper T cells that help B cells tend to remain within the lymph node, whereas other helper T cells and cytotoxic T cells migrate to the site of infection, where their functions are needed. Lymphocytes that fail to be activated within the lymph node leave in the efferent lymph to continue recirculation.

Pathogens can enter the blood directly; this can occur when blood-feeding insects transmit disease or when lymph nodes have failed to remove microorganisms from the lymph returned to the blood. The **spleen** is the lymphoid organ that serves as a filter for the blood. One purpose of the filtration is to remove damaged or old red cells; the second function of the spleen is that of a secondary lymphoid tissue—infectious agents are removed and used to activate lymphocytes. In this latter role, the spleen functions in a similar fashion to the lymph node, the only difference being that both pathogens and lymphocytes enter and leave the spleen in the blood (Figure 1.11).

The spleen

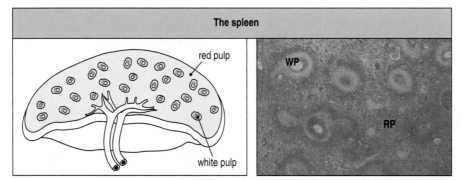

red pulp

white pulp

WP

RP

Transverse section of white pulp of spleen

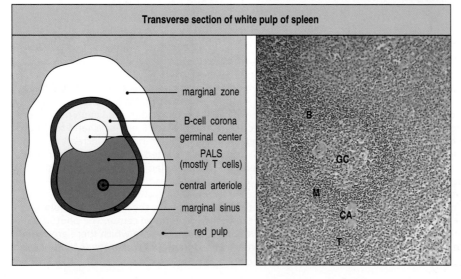

marginal zone

B-cell corona

germinal center

PALS
(mostly T cells)

central arteriole

marginal sinus

red pulp

B

GC

M

CA

T

Figure 1.11 The spleen has aggregations of lymphocytes similar to those in lymph nodes. The human spleen is a large lymphoid organ in the upper left part of the abdomen, weighing around 150 grams. The upper diagram depicts a section of spleen in which nodules of white pulp are scattered within the more extensive red pulp. The red pulp (RP) is where old or damaged red cells are removed from the circulation; the white pulp (WP) is secondary lymphoid tissue, in which lymphocyte responses to blood-borne pathogens are made. The bottom diagram shows a nodule of white pulp in transverse section. It consists of a sheath of lymphocytes surrounding a central arteriole (CA). The sheath is called the periarteriolar lymphoid sheath (PALS). The lymphocytes closest to the arteriole are mostly T cells (T); B cells are placed more peripherally, forming a B-cell corona (B). Lymphoid follicles and germinal centers (GC) lie between the B- and T-cell zones. The marginal zone (M) contains differentiating B cells. Photographs courtesy of H.G. Burkitt and B. Young (top) and N. Rooney (bottom).

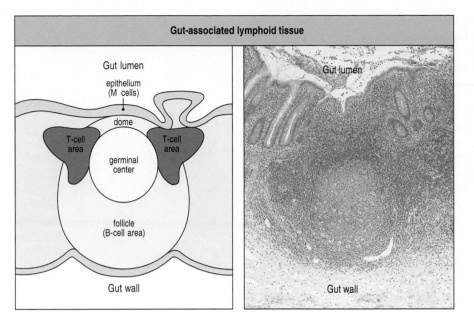

Figure 1.12 A typical region of gut-associated lymphoid tissue. A schematic diagram (left panel) and a light micrograph (right panel) of a typical region of gut-associated lymphoid tissue. M cells of the gut epithelial wall deliver pathogens from the luminal side of the gut mucosa to the lymphoid tissue within the gut wall. These areas are organized similarly to the lymph node and the white pulp of the spleen, with distinctive B- and T-cell zones, lymphoid follicles, and germinal centers. Photograph courtesy of N. Rooney.

The parts of the body that harbor the largest and most diverse populations of microorganisms are the respiratory and gastrointestinal tracts. The most heavily infested site is the oral cavity. The extensive mucosal surfaces of these tissues make them particularly vulnerable to infection and they are therefore heavily invested with secondary lymphoid tissue. The **gut-associated lymphoid tissues (GALT)** include the **tonsils**, **adenoids**, **appendix**, and the **Peyer's patches** which line the small intestine (see Figure 1.8). Similar but less organized aggregates of secondary lymphoid tissue line the respiratory epithelium, where they are called **bronchial-associated lymphoid tissue (BALT)**, and other mucosa, including the gastrointestinal tract. The more diffuse mucosal lymphoid tissues are known generally as **mucosa-associated lymphoid tissue (MALT)**.

Although different from the lymph nodes or spleen in outward appearance, the mucosal lymphoid tissues are similar to them in their microanatomy (Figure 1.12) and in their function of trapping pathogens to activate lymphocytes. The differences are chiefly in the route of pathogen entry and the migration patterns of their lymphocytes. Pathogens arrive at mucosa-associated lymphoid tissues by direct delivery across the mucosa, mediated by specialized cells of the mucosal epithelium called **M cells**. Lymphocytes first enter mucosal lymphoid tissue from the blood and, if not activated, leave via lymphatics that connect the mucosal tissues to draining lymph nodes. Lymphocytes activated in mucosal tissues tend to stay within the mucosal system, moving out from the lymphoid tissue into the lamina propria and the mucosal epithelium, where they perform their effector actions.

1-8 Phagocytic cells that recognize common components of bacterial cell surfaces initiate innate immunity and induce inflammation

The elimination of pathogens is performed principally by cells specialized in the capture, engulfment, and breakdown of microorganisms. These cells, called **phagocytes**, are of two kinds—the macrophages resident in tissues, and the neutrophils, which circulate in the blood and enter tissues but only when they become infected. Macrophages and neutrophils bear cell-surface receptors that bind to common carbohydrate constituents of bacterial surfaces, which are not components of human cells. Such properties are characteristic of the recognition molecules of innate immunity and are analogous to those of antibiotic drugs, in that their targets are biochemical features that distinguish microbes from mammals. The binding of a bacterium to the cell-surface receptors of a neutrophil or macrophage induces engulfment, killing, and degradation of the bacterium (Figure 1.13).

Macrophages resident in the infected tissues are generally the first phagocytic cells to encounter an invading bacterium. As part of their response to the presence of bacteria, macrophages secrete soluble proteins, called **cytokines**, which recruit other cells of the immune system, such as neutrophils, into the infected area. They also secrete a variety of other substances that act as anti-bacterial agents and as inducers of inflammation. Macrophages are versatile and long-lived cells that contribute to both innate and adaptive immune responses and have functions other than phagocytosis.

Neutrophils are specialized to phagocytose and kill pathogens. They begin to act soon after an infection arises and are the principal phagocytic cell in infected and inflamed tissues. They are rapidly mobilized to enter sites of infection and can work in the anaerobic conditions that often prevail in damaged tissue. Neutrophils are short-lived and die at the site of infection, forming **pus**. Collectively, the molecules released by phagocytes when they encounter infection induce a local state of **inflammation**, an ancient concept in medicine that has traditionally been defined by the Latin words *calor, dolor, rubor,* and *tumor;* for heat, pain, redness, and swelling, respectively.

Cytokines induce the local dilation of blood capillaries, which by increasing the blood flow causes the skin to warm and redden. Vascular dilation introduces gaps

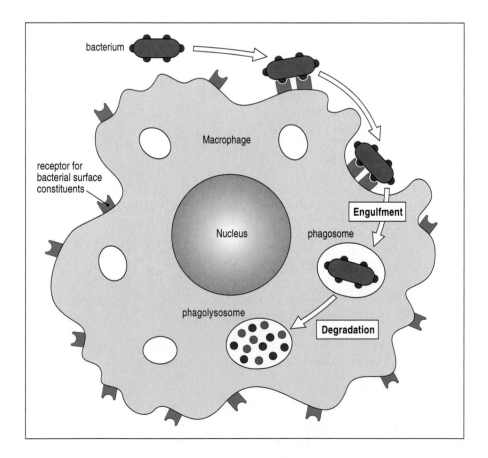

Figure 1.13 Receptor-mediated phagocytosis of bacteria by a macrophage. The bacterium (red) binds to cell-surface receptors of the macrophage (blue), inducing engulfment of the bacterium into an endocytic vesicle called a phagosome within the cytoplasm of the macrophage. Vesicle fusion with lysosomes forms an acidic vesicle called a phagolysosome, which contains toxic small molecules and hydrolytic enzymes that kill and degrade the bacterium.

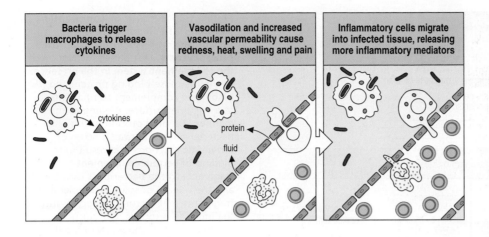

| Bacteria trigger macrophages to release cytokines | Vasodilation and increased vascular permeability cause redness, heat, swelling and pain | Inflammatory cells migrate into infected tissue, releasing more inflammatory mediators |

Figure 1.14 Phagocytosis of infecting bacteria triggers an inflammatory response. When bacteria first penetrate wounded tissue, the first phagocytic cells they encounter are tissue macrophages. In response to the bacteria, the macrophages release cytokines that increase the permeability of local blood vessels, allowing fluid and proteins to pass into the tissue from the blood. The properties of the vessel's endothelial cells change so that inflammatory cells in the blood adhere to the endothelial cell surface and then crawl into the infected tissue. In the figure a monocyte and then a neutrophil are shown leaving the blood in this manner. The accumulation of fluid and cells at the site of infection causes the swelling, redness, heat, and pain that are collectively known as inflammation.

between the endothelial cells, causing more blood plasma to leak into the connective tissue. Such expansion of the local fluid volume causes edema or swelling, putting pressure on nerve endings and causing pain. Cytokines also change the adhesive properties of the vascular endothelium, signaling neutrophils, monocytes/macrophages, and other white blood cells to bind to it and migrate out of the blood into the inflamed tissue (Figure 1.14). White blood cells that are typically found in inflamed tissues and release substances that contribute to the inflammation are called **inflammatory cells**. Infiltration of cells into the inflamed tissue increases the swelling, and some of the molecules they release contribute to the pain. However, the benefit of the discomfort and disfigurement is that inflammation enables cells and molecules of the immune system to be brought rapidly and in large numbers into infected tissues.

1-9 Innate immunity is also mediated by soluble proteins in the blood

The plasma that leaks into infected and inflamed tissue carries with it plasma proteins. Among these are soluble recognition molecules involved in innate immunity. These are made mainly in the liver and travel freely throughout the body. One such recognition molecule, the **mannose-binding protein**, has binding sites for carbohydrates characteristic of bacterial cell surfaces. Once bound to a bacterium, the mannose-binding protein initiates a series of enzymatic reactions involving a set of plasma proteins collectively called **complement**. The cascade of enzymatic reactions triggered by bacterial recognition is described as **complement activation** and leads to the covalent binding of complement proteins to bacterial cell surfaces (Figure 1.15). Complement activation is a general effector mechanism of the immune system and can also be triggered by the presence of bacterial cells alone. Once adaptive immune responses have got under way, complement activation is initiated by antibodies bound to pathogens.

Complement proteins bound to pathogens engage specific complement receptors on the surface of phagocytes, triggering the engulfment and destruction of the complement-coated pathogen. Soluble fragments of complement formed during the cascade of complement reactions recruit additional phagocytes into sites of complement activation, thus enhancing the destruction of the pathogen.

Summary

To restrict the nature, size, and location of microbial infection, animals have evolved a variety of defense mechanisms. The skin and contiguous mucosal surfaces provide physical and chemical barriers which confine microorganisms to the external surfaces of the body. In addition, cells and molecules of the innate

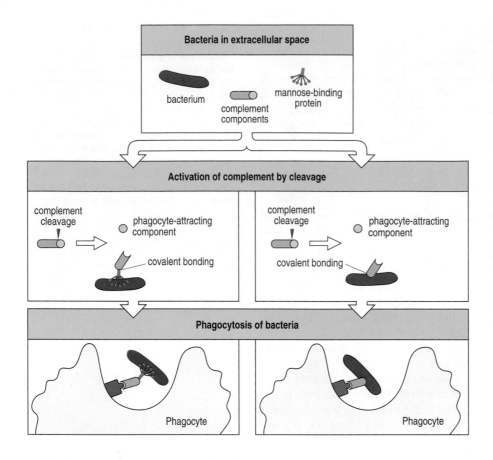

Figure 1.15 Binding of complement to bacterial cell surfaces in the innate immune response facilitates phagocytosis. Left panels: activation of the complement system in the presence of the mannose-binding protein and binding of complement components at a bacterial surface. Binding of the mannose-binding protein to the bacterial surface initates a series of enzymatic reactions that result in the binding of certain complement fragments to the bacterium's surface and the release of soluble complement fragments that recruit phagocytes into the immune response. Cell-surface receptors on the phagocytes bind the complement fixed to the bacterium's surface, initiating the engulfment and destruction of the bacterium. Right panels: complement fixation can occur in the absence of mannose-binding protein.

and adaptive immune systems identify and eliminate microorganisms that have penetrated the physical and chemical barriers and gained entry to the soft tissues. The cells that mediate both innate and adaptive immune responses are principally the various types of leukocyte and allied tissue cells, which derive from the bone marrow. In responding to infection, the immune system uses innate mechanisms that are fast but limited, and adaptive mechanisms that are slow to start but eventually become both powerful and quick to recall. The cells and molecules of innate immunity identify common classes of pathogen and destroy them. Four key elements of innate immunity are: molecules that non-covalently bind to surface macromolecules of pathogens; molecules that covalently bond to pathogen surfaces, forming ligands for phagocyte receptors; phagocytic cells that engulf and kill pathogens; and cytotoxic cells that kill virus-infected cells. Vertebrates have evolved the additional defenses of adaptive immune responses, which involve the T and B lymphocytes. Adaptive immune responses are initiated in lymphoid tissues such as lymph nodes and spleen, in which pathogen-specific lymphocytes encounter and are activated by pathogen antigens.

Principles of adaptive immunity

Innate immunity deals with the diversity of pathogens that a person might encounter in their lifetime by having always available a few types of recognition molecules, each of which recognizes a large number of different pathogens. The adaptive immune response uses a quite different strategy. Millions of different immunoglobulins and T-cell receptors are made by the B and T lymphocytes of the human immune system, and each recognizes a different molecular structure. On infection with a pathogen, only those B and T cells making recognition molecules that can bind to constituents of that pathogen are stimulated to divide, proliferate and differentiate into effector lymphocytes.

1-10 Immunoglobulins and T-cell receptors are the highly variable recognition molecules of adaptive immunity

The molecules and mechanisms of adaptive immunity are restricted to vertebrate animals, suggesting that their evolution became feasible or advantageous only with the degree of cellular and anatomical complexity attained by the vertebrates. Immunoglobulins are expressed on the surface of B cells, where they can bind pathogens. Effector B cells, called plasma cells, secrete soluble forms of these immunoglobulins, which are known as antibodies. In contrast, T-cell receptors are only ever expressed as cell-surface recognition molecules, never as soluble proteins.

Any molecule, macromolecule, virus particle or cell that contains a structure recognized and bound by an immunoglobulin or T-cell receptor is known as its corresponding **antigen**. Surface immunoglobulins and T-cell receptors are thus also referred to as the **antigen receptors** of lymphocytes. The particular part of the antigen bound by the immunoglobulin or T-cell receptor is known as the **antigenic determinant** or **epitope**. Immunoglobulins can bind to a vast variety of different chemical structures, whereas T-cell receptors recognize a more limited range of epitopes. Immunoglobulins and T-cell receptors are said to be **specific**, or to have **specificity**, for the antigens they bind.

Immunoglobulins and T-cell receptors are structurally related molecules whose diversity of antigen-recognition sites is generated by similar genetic mechanisms. Immunoglobulins are formed from two different polypeptides called the heavy and light chains; each Y-shaped immunoglobulin molecule consists of two identical heavy chains and two identical light chains. Both types of polypeptide have an amino-terminal **variable region** that differs in amino-acid sequence from one immunoglobulin to the next, and a **constant region** that is identical in amino-acid sequence from one immunoglobulin to another (Figure 1.16). The variable regions contain the sites that bind antigens. Surface immunoglobulin is anchored in the membrane by transmembrane regions at the carboxyl ends of the heavy chains. Antibodies are a secreted form of immunoglobulin that lack these transmembrane regions but are otherwise identical to surface immunoglobulins (see Figure 1.16).

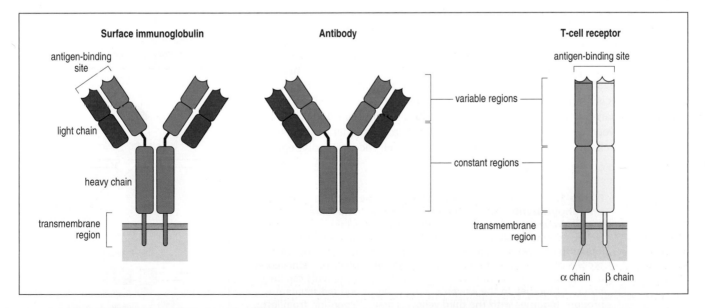

Figure 1.16 Comparison of the basic structures of surface immunoglobulin, antibody, and the T-cell receptor. The heavy chains of surface immunoglobulin and antibody are shown in blue, the light chains in red.

A typical T-cell receptor consists of an α chain and a β chain, both anchored in the T-cell membrane. Like the light chains and heavy chains of immunoglobulins, the α and β chains of T-cell receptors each consist of a variable region and a constant region, with the variable regions forming an antigen-binding site (see Figure 1.16).

The differences in the amino-acid sequences of the variable regions of immunoglobulins and T-cell receptors create a vast variety of binding sites that are specific for different antigens, and thus for different pathogens. This is one of the hallmarks of an adaptive immune response. Antibodies secreted in response to a measles infection, for example, bind to measles virus but not to influenza virus, whereas antibodies specific for influenza virus do not bind to measles virus.

The constant regions of antibodies contain binding sites for cell-surface receptors on phagocytes and inflammatory cells and also for complement proteins. The secreted antibody thus acts as a molecular adaptor or bridge. It binds to pathogens with one type of site and to effector cells and molecules which destroy the pathogen with another. For immunoglobulins, but not T-cell receptors, there are several different types of constant region, known as **isotypes**, which confer different effector functions on the secreted antibody. Every person can make antibodies of all isotypes.

1-11 The diversity of immunoglobulins and T-cell receptors is generated by gene rearrangement

The extensive diversity of variable regions in the antigen receptors of B and T cells is produced by a genetic mechanism unique to the immunoglobulin and T-cell receptor genes. In the genome, the variable regions of immunoglobulin and T-cell receptor chains are encoded in a series of separate gene segments. For a functional gene to be made, the appropriate gene segments need to be brought together by a physical rearrangement of the DNA (Figure 1.17). The immunoglobulin and T-cell receptor genes are quite unlike other human genes in this respect. Rearrangements at the heavy- and light-chain gene loci occur only in B cells, whereas rearrangements at the α- and β-chain gene loci occur only in T cells. Before any rearrangement, the immunoglobulin and T-cell receptor genes are said to be in the **germ-line configuration** because that is how they are present in germ cells: eggs and sperm. The process of gene rearrangement in B and T cells is called **somatic recombination** because it recombines gene segments and occurs in the soma—those cells of the body that are not germ cells.

Many alternative forms of each gene segment are present within the unrearranged DNA and they can be rearranged in numerous different combinations. This is the basis for the diversity of the variable regions. Further diversity is introduced as a result of imprecision in the enzymatic machinery used to cut and splice the DNA during gene rearrangement. Its effect is to introduce additional nucleotides into the joints between gene segments. These processes create sequence diversity in the individual chains of the immunoglobulin and T-cell receptor polypeptides.

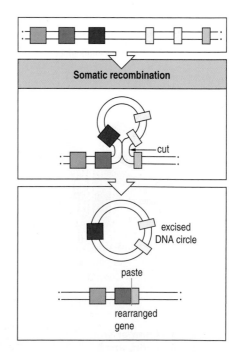

Figure 1.17 Gene rearrangement of the type occurring in immunoglobulin and T-cell receptor genes. In the unrearranged DNA there are three alternative 'red' segments and three alternative 'yellow' segments. A functional gene consists of one red segment joined to one yellow segment. This rearrangement is achieved by a process of 'cut and paste' in which the intervening DNA is removed as a circle. Different combinations of red and yellow segments can be brought together. In the example shown, the second red segment is brought together with the third yellow segment (counting from left to right), but other combinations of a red and a yellow segment would have been equally possible.

Additional diversity in the antigen-binding sites arises from the association of different heavy and light chains in immunoglobulins and of different α and β chains in T-cell receptors.

So far we have considered the mechanisms that produce diversity in lymphocyte receptors during their development in the primary lymphoid tissues. In B cells, but not in T cells, an additional mechanism for diversification is brought into play after activation by pathogens. Activated dividing B cells initiate a process of **somatic hypermutation**, which introduces nucleotide substitutions into the immunoglobulin heavy- and light-chain genes. Some of the variant immunoglobulins produced by somatic hypermutation bind the pathogen more tightly and the cells producing them are preferentially chosen to become antibody-secreting plasma cells.

1-12 Clonal selection of B and T lymphocytes is the guiding principle of the adaptive immune response

A direct consequence of the mechanisms of immunoglobulin and T-cell receptor gene rearrangement is that each lymphocyte expresses immunoglobulin or T-cell receptors of a single specificity. As a population, however, the lymphocytes of the human immune system make millions of different immunoglobulins and T-cell receptors. This hierarchy, whereby individual lymphocytes make a single type of recognition molecule and diversity is expressed at the level of lymphocyte populations, enables the lymphocyte response to be tailored toward particular pathogens. On infection, only the very small proportion of lymphocytes which have receptors that recognize the particular pathogen will be activated to divide and differentiate into effector cells. Consequently, each lymphocyte stimulated by the pathogen gives rise to a clonal population of cells all expressing the identical immunoglobulin or T-cell receptor. The process whereby pathogens select particular clones of lymphocyte for expansion is called **clonal selection** (Figure 1.18). The use of a minute fraction of the total lymphocyte repertoire to respond to each pathogen ensures that the adaptive response is highly specific for the infection at hand.

During the maturation of lymphocytes in the bone marrow and thymus, clonal selection is also used to prevent the emergence of cells that could attack the cells and tissues of the body. Clones of lymphocytes with receptors that bind strongly to the constituents of thymus and bone marrow tissue are signaled to die by programmed cell death, or **apoptosis**. This negative selection of potentially self-reactive lymphocytes helps to render the circulating pool of mature B and T cells non-responsive to, or **tolerant** of, the normal components of the body. The lymphocyte repertoire is thus said to be **self-tolerant**. Apoptosis is a common feature of developmental pathways. Individual cells are induced to commit a form of cell suicide, but there is no generalized tissue damage of the kind that occurs in necrosis. Membrane changes in cells undergoing apoptosis lead to their rapid phagocytosis by macrophages.

Figure 1.18 Clonal selection of lymphocytes by a pathogen. During development each B or T cell is programmed to make a single species of cell-surface antigen receptor. The population of circulating lymphocytes embraces many millions of receptor species, which enables all possible pathogens to be recognized. In the panels, individual clones of lymphocytes are represented by the different colors. On infection by a particular pathogen, the small subsets of B and T cells having receptors that bind to the antigens of the pathogen are stimulated to divide and differentiate, thereby producing a clone of effector cells from each antigen-binding lymphocyte. In the second panel the lymphocyte selected by the pathogen is that colored yellow. After activation and cell division, it gives rise to the clone of cells shown in the third panel.

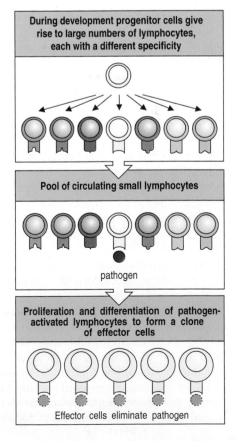

During development progenitor cells give rise to large numbers of lymphocytes, each with a different specificity

Pool of circulating small lymphocytes

pathogen

Proliferation and differentiation of pathogen-activated lymphocytes to form a clone of effector cells

Effector cells eliminate pathogen

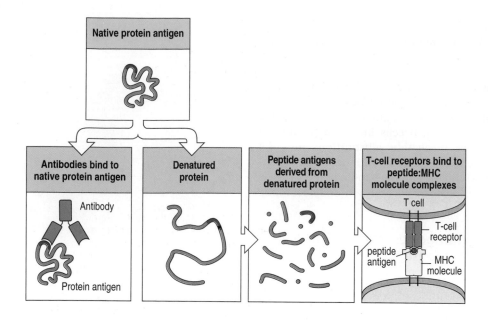

Figure 1.19 B cells recognize native proteins, whereas T cells recognize degraded proteins bound by major histocompatibility complex (MHC) molecules. The surface immunoglobulins of B cells and secreted antibodies bind to the native protein (left panel). To be recognized by a T-cell receptor, the protein must first be processed into peptide antigens by denaturation and proteolytic degradation (two center panels). Secondly, the peptide antigens must be bound to MHC molecules. Only then can the antigen be presented to the T cell (right panel).

1-13 B cells recognize intact pathogens, whereas T cells recognize pathogen-derived peptides bound to proteins of the major histocompatibility complex

Antibodies made by B cells are the soluble recognition molecules of adaptive immunity. By circulating in the body fluids, antibodies can bind to bacterial cells and intact viral particles in extracellular spaces, targeting them for phagocytosis. To fulfill this function, the binding sites of antibodies interact with intact components of the pathogen surface such as glycoproteins and proteoglycans. The epitopes bound by antibodies most commonly include carbohydrate groups, clusters of amino acids on the protein surface, or combinations of the two.

Whereas antibodies bind directly to the native structures of biological macromolecules, T-cell receptors can bind only short peptides that have been assembled into a complex with a membrane glycoprotein called a **major histocompatibility complex (MHC) molecule** (Figure 1.19). T-cell antigens are therefore peptides. They are produced within human cells by the breakdown of pathogens or their protein products, a process called **antigen processing**. Assembly of the peptide:MHC molecule complex takes place in the cell where the antigen was processed and, once formed, the complex is transported to the cell surface. Here it is accessible to T-cell receptors on the surface of T lymphocytes. MHC molecules are therefore said to **present** antigens to T cells, and cells carrying antigen:MHC complexes are known as **antigen-presenting cells**.

There are two classes of MHC molecule (Figure 1.20). **MHC class I molecules** present peptide antigens derived from pathogens that replicate intracellularly, such as viruses and some bacteria, and whose proteins are present in the cytosol of the infected cell (Figure 1.21). **MHC class II molecules** present peptides from pathogens and protein antigens that are present in the extracellular milieu and have been taken up into the endocytic vesicles of phagocytic cells (Figure 1.22). MHC class I molecules are present on almost all cell types, and thus all cell types are able to present viral peptides in case of a viral infection. MHC class II molecules, in contrast, are present on only a few cell types that are specialized in the uptake and processing of pathogens and are present in all tissues. These **professional antigen-presenting cells** include macrophages, dendritic cells, and B cells.

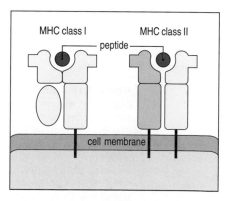

Figure 1.20 There are two types of MHC molecule, MHC class I and MHC class II.

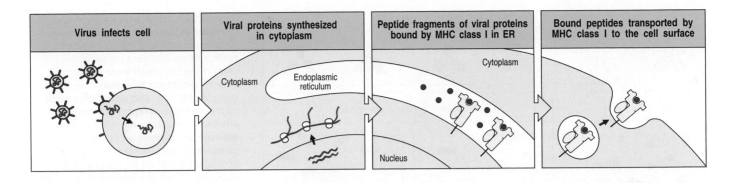

Virus infects cell	Viral proteins synthesized in cytoplasm	Peptide fragments of viral proteins bound by MHC class I in ER	Bound peptides transported by MHC class I to the cell surface

In human populations there are many different genetic variants of MHC molecules; this **polymorphism** is the chief cause of rejection of tissue transplants. When donors and recipients are of different MHC types the immune system of the recipient makes a vigorous immune response against the donor's MHC molecules, which it perceives as 'foreign'. It was this phenomenon that led to the discovery of the MHC and its original naming as a complex of genes that governed the compatibility of tissues (Greek *histo*) on transplantation.

1-14 Extracellular pathogens and their toxins are eliminated by antibodies

Immunoglobulins are divided into five **classes** or isotypes, which differ in their heavy-chain constant regions and have specialized effector functions when secreted as antibodies. The isotypes are called **IgA**, **IgD**, **IgE**, **IgG**, and **IgM**, where Ig stands for immunoglobulin. Cell-surface IgM and IgD are the antigen receptors on circulating B cells that have yet to encounter antigen. IgM is always the first antibody to be secreted in the immune response (IgD antibody is also made but in negligible amounts). As the immune response matures, antibodies of other isotypes emerge. Immunity due to antibodies and their actions is often known as **humoral immunity**. This is because antibodies were first discovered circulating in body fluids (*humors*) such as blood and lymph.

IgM, IgA, and IgG are the main antibodies present in blood, lymph, and connective tissues. Antibodies secreted into the blood from lymph nodes, spleen, and bone marrow circulate in the bloodstream. At sites of infection and inflammation, blood vessels dilate and become permeable, increasing the flow of fluid into the infected tissue and with it the supply of antibodies for binding to extracellular pathogens (bacteria and virus particles) and the protein toxins that many bacterial pathogens secrete. IgA is also made in the lymphoid tissues underlying mucosa and then selectively transported across the mucosal epithelium to bind extracellular pathogens and their toxins on the mucosal surfaces.

Figure 1.21 The MHC class I pathway for presenting antigens derived from intracellular infections. In virus-infected cells, new viral proteins are made on cellular ribosomes in the cytoplasm. Some of these proteins are degraded in the cytoplasm and the resultant peptides are transported into the endoplasmic reticulum (ER). MHC class I molecules bind peptides in the ER and then transport them to the infected cell's surface.

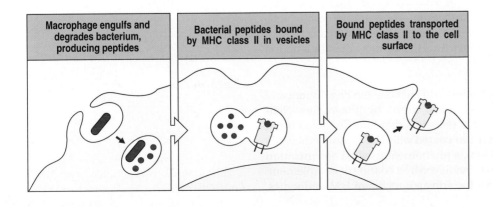

Macrophage engulfs and degrades bacterium, producing peptides	Bacterial peptides bound by MHC class II in vesicles	Bound peptides transported by MHC class II to the cell surface

Figure 1.22 The MHC class II pathway for presenting antigens derived from extracellular infections. Macrophages phagocytose extracellular bacteria and degrade their proteins in endocytic vesicles. MHC class II molecules bind peptides in endocytic vesicles and transport them to the cell surface.

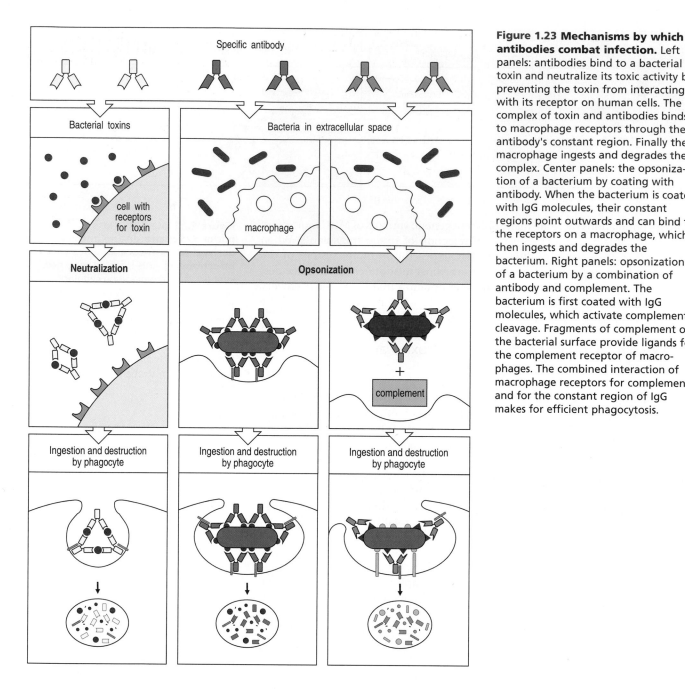

Figure 1.23 Mechanisms by which antibodies combat infection. Left panels: antibodies bind to a bacterial toxin and neutralize its toxic activity by preventing the toxin from interacting with its receptor on human cells. The complex of toxin and antibodies binds to macrophage receptors through the antibody's constant region. Finally the macrophage ingests and degrades the complex. Center panels: the opsonization of a bacterium by coating with antibody. When the bacterium is coated with IgG molecules, their constant regions point outwards and can bind to the receptors on a macrophage, which then ingests and degrades the bacterium. Right panels: opsonization of a bacterium by a combination of antibody and complement. The bacterium is first coated with IgG molecules, which activate complement cleavage. Fragments of complement on the bacterial surface provide ligands for the complement receptor of macrophages. The combined interaction of macrophage receptors for complement and for the constant region of IgG makes for efficient phagocytosis.

One way in which antibodies reduce infection is by binding tightly to a site on a pathogen so as to inhibit pathogen growth, replication, or interaction with human cells. This mechanism is called **neutralization**. For example, certain antibodies when bound to influenza virions prevent the virus from infecting human cells. Similarly, the lethal actions of bacterial toxins can be prevented by a bound antibody (Figure 1.23).

The most important function of IgG antibodies is to facilitate the engulfment and destruction of extracellular microorganisms and toxins by phagocytes. Neutrophils and macrophages have cell-surface receptors that bind to the constant regions of the IgG heavy chains. A bacterium coated with IgG is more efficiently phagocytosed than an uncoated bacterium, a phenomenon called **opsonization** (see Figure 1.23). Opsonization can also be achieved by a coating of complement. There are no receptors on phagocytes for the constant region of IgM antibodies, but IgM bound to a pathogen's surface typically activates the complement system and the pathogen becomes coated with complement. This facilitates its uptake and phagocytosis by macrophages via another set of receptors that bind complement. A combination of IgG antibody and complement produces the greatest stimulation of phagocytosis because macrophage receptors for IgG constant regions and complement both participate in the process (see Figure 1.23).

The constant regions of IgE antibodies bind tightly to receptors on the surface of mast cells; most of the IgE antibody in the body is in this form, rather than circulating freely in blood or lymph like IgM and IgG. In response to worms and other parasitic infestations, the IgE bound to mast cells triggers strong inflammatory reactions that are thought to help expel or destroy the parasites. However, in clinical practice in developed countries, IgE is most often encountered as the antibody involved in unwanted allergic reactions to innocuous substances.

1-15 Three kinds of effector T cell serve different functions in the immune response

The immunity mediated by T cells and the immune system cells with which they interact, such as macrophages, is often called **cellular** or **cell-mediated immunity** to distinguish it from the actions of antibodies. Two main classes of T cell can be distinguished by their expression of the cell-surface glycoproteins CD4 and CD8. T cells that express the CD8 glycoprotein, called **CD8 T cells**, have a cytotoxic function and are also known as cytotoxic T cells. Effector CD8 T cells kill cells infected with viruses or other intracellular pathogens. This prevents the replication of the pathogen and the spread of infection. A CD8 T cell recognizes only cells bearing its corresponding peptide antigen presented by an MHC class I molecule (Figure 1.24). T cells that express the CD4 glycoprotein are called **CD4 T cells**. They secrete cytokines that help activate other types of immune-system cell, and respond to peptide antigens presented by MHC class II antigens.

The CD8 molecule has a binding site for MHC class I molecules; the CD4 molecule has a binding site for MHC class II molecules. These interactions determine the strict specificity of CD8 T cells for antigens presented by MHC class I and of CD4 T cells for antigens presented by MHC class II. T-cell recognition of antigen requires that MHC:peptide complexes bind to both the T-cell receptor and either the CD8 or the CD4 molecule. Because of their roles in T-cell recognition, the CD8 and CD4 molecules are called T-cell **co-receptors**.

The CD4 T cells can be further subdivided according to the cytokines they secrete and the cells they assist. Helper cells that secrete cytokines that mainly activate macrophages are called T_H1 **cells** (where the 'H' stands for helper), whereas helper cells that chiefly help B cells to make antibodies are called T_H2 **cells** (Figure 1.25). Both types of helper cell produce responses that facilitate the elimination of

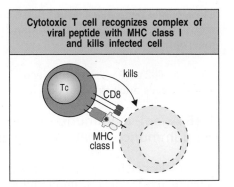
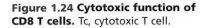

Figure 1.24 Cytotoxic function of CD8 T cells. Tc, cytotoxic T cell.

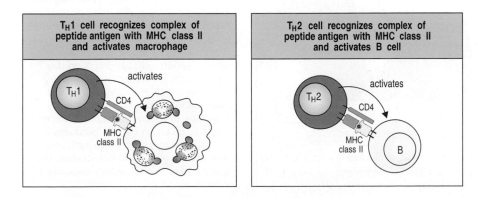

Figure 1.25 Helper functions of CD4 T cells.

extracellular pathogens: B cells make antibodies that coat pathogens, fix complement on their surfaces, and opsonize them for efficient phagocytosis by activated macrophages.

Whereas T_H2 cells work within secondary lymphoid tissues, CD8 cytotoxic T cells and T_H1 cells must travel to the site of infection in order to function.

1-16 Adaptive immune responses generally give rise to long-lived immunological memory and protective immunity

Clonal selection by pathogens is the guiding principle of adaptive immunity and explains the features of immunity that perplexed physicians in the past. The severity of a first encounter with an infectious disease arises because the **primary immune response** is developed from very few lymphocytes; the time taken to expand their number provides an opportunity for the pathogen to establish an infection to the point of causing disease. The clones of lymphocytes produced in a primary response include long-lived **memory cells**, which can respond more quickly and forcefully to subsequent encounters with the same pathogen.

The potency of such **secondary immune responses** can be sufficient to repel the pathogen before there is any detectable symptom of disease. The individual therefore appears immune to that disease. The striking differences between a primary and secondary response are illustrated in Figure 1.26. The immunity due to a secondary immune response is absolutely specific for the pathogen that provoked the primary response. So immunity developed to one disease does not ward off another. Indeed, highly mutable RNA viruses, such as influenza, which change antigenically from year to year, might not be recognized by antibodies and memory cells developed against a previous strain. Thus vaccination against one year's strain of influenza does not always help in the fight against next year's strain.

Vaccination is a way of stimulating protective immunity by administering a pathogen's antigens in a form that will not provoke the disease. It has been successful in preventing many common diseases (Figure 1.27).

Figure 1.26 Comparison of a primary and secondary immune response. This diagram shows how the immune response develops during an experimental immunization of a laboratory animal. The response is measured in terms of the amounts of pathogen-specific antibody that is present in the animal's blood serum, shown on the vertical axis, with time being shown on the horizontal axis. On the first day the animal is immunized with vaccine A. The levels of antibodies against pathogen A are shown in blue. The primary response reaches its maximum level 2 weeks after immunization. After the primary response has subsided, a second immunization with vaccine A on day 60 produces an immediate secondary response which in 5 days is orders of magnitude greater than the primary response. In contrast, vaccine B, which was also given on day 60, produces a typical primary response to pathogen B as shown in yellow, demonstrating the specificity of the secondary response to vaccine A.

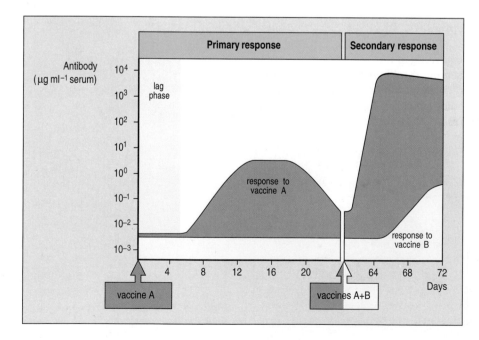

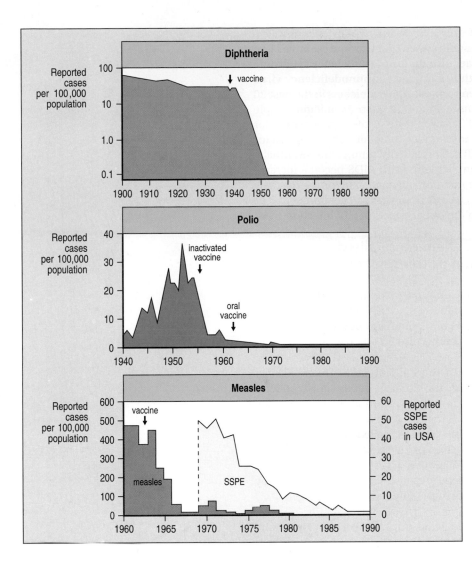

Figure 1.27 Successful vaccination campaigns. Diphtheria, poliomyelitis and measles have been virtually eliminated from the USA, as shown by these three graphs. The arrows indicate when the vaccination campaigns began. Subacute sclerosing panencephalitis (SSPE) is a brain disease that is a late consequence of measles infection for a minority of patients. Reduction of measles was paralleled by a reduction in SSPE 15 years later. Because these diseases have not been eradicated worldwide and the volume of international travel is so high, immunization must be maintained in much of the population to prevent disease recurrence.

1-17 The immune system can be compromised by inherited immunodeficiencies or the actions of certain pathogens

When components of the immune system are either missing or do not work properly, this generally leads to increased susceptibility to microbial infection. One cause of defective immune responses is inherited mutations in genes encoding proteins that contribute to immunity. Most people who carry a mutant gene are healthy because their other, normal, copy of the gene provides adequate function; the small and unfortunate minority who carry two mutant copies of the gene lack the function encoded by that gene. Such deficiencies lead to varying degrees of failure of the immune system and a wide range of **immunodeficiency diseases**. In some of the immunodeficiency diseases, only one aspect of the immune response is affected, leading to susceptibility to particular kinds of infection; in others, adaptive immunity is completely absent, leading to a devastating vulnerability to all infections. These latter gene defects are rare, showing how vital is the protection normally afforded by the immune system. The discovery and study of immunodeficiency diseases has largely been the work of pediatricians, because such conditions usually show up early in life. Before the advent of antibiotics and, more recently, the possibility of replacement therapies, immunodeficiencies would often have caused death in infancy.

Immunodeficiency states are caused not only by non-functional genes, but also by pathogens that subvert the human immune system. An extreme example of an immunodeficiency due to disease is the **acquired immune deficiency syndrome (AIDS)**, which is caused by infection with the **human immunodeficiency virus (HIV)**. Although this disease has only been recognized by clinicians in the past 20 years, it is now at epidemic proportions, with more than 30 million people infected worldwide. HIV infects the CD4 T lymphocyte, a cell type essential for adaptive immunity. During the course of an extended infection, which can last for up to 20 years, the population of CD4 T cells gradually diminishes, eventually leading to collapse of the immune system. Patients with AIDS become increasingly susceptible to a range of infectious microorganisms, many of which rarely trouble uninfected people. Death usually results from the effects of one of these opportunistic infections rather than from the direct effects of HIV infection.

1-18 The actions of the immune system cause allergy, autoimmune disease, and rejection of transplanted tissues

The sensitivity and destructive power of the immune system, which are essential traits in battling infection, can under some circumstances be misdirected towards innocuous materials or normal components of the human body. When this happens, various kinds of chronic and non-infectious disease emerge.

A state of **allergy** arises when antibodies of the IgE isotype are made against innocuous substances in the environment, for example foods, grass pollens, house dust, or dander from pets. Once such antibodies are made they circulate and become bound by their constant regions to mast cells throughout the body's tissues. On repeat exposure to the inducing antigen, called the **allergen**, it is bound by the IgE on mast cells. The combination of IgE and allergen on mast-cell surfaces triggers violent and sometimes life-threatening reactions such as those of asthma and systemic anaphylaxis. The damage and incapacitation caused by allergic reactions is greatly out of proportion to the threat posed by allergen.

Another class of non-infectious immunological disease is caused by chronic immune responses that gradually erode a target tissue. These conditions are called **autoimmune diseases** and result from **autoimmune responses** directed towards normal components of healthy human cells. In insulin-dependent diabetes, for example, the insulin-secreting β cells of the islets of Langerhans in the pancreas are the target of an autoimmune response. Because the human pancreas can produce insulin well in excess of what is normally needed, symptoms of disease do not show themselves until long after the start of the destructive response. Pancreatic β-cell destruction is believed to be due to B and T cells that respond to proteins made only in β cells. Other autoimmune diseases are rheumatoid arthritis, multiple sclerosis, myasthenia gravis, and Graves' disease. Autoimmune responses often seem to be by-products of a protective immune response made against an acute microbial infection. Once activated by pathogen-derived antigens, certain B or T cells cross-react with self components to which they were previously tolerant.

Over the past century, hygiene, vaccination, and anti-microbial drugs have all served to reduce mortality from infectious disease in the developed countries, particularly for children. In these populations the treatment of chronic and malignant diseases has become of growing importance in medicine. For an increasing number of these diseases, the replacement of tissues or organs by **transplantation** is now being used to restore health and prolong life. A major factor limiting the benefits of transplantation are the tissue incompatibilities caused by the extensive polymorphism of MHC class I and class II genes in the human population.

MHC-compatible donors can be found within families, but only for some of the patients needing a transplant. For patients seeking a transplant from an unrelated donor the chances of finding one who is MHC identical are low. In practice, therefore, the majority of clinical transplants are performed across MHC differences which have the potential to stimulate strong immune responses. Currently, the only means of preventing transplant rejection is the prolonged use of drugs that suppress adaptive immunity. Although facilitating graft acceptance, such drugs have toxic effects and also render the transplant patient more vulnerable to infection and cancer.

Summary

The mechanisms of adaptive immunity are ones that improve pathogen recognition rather than pathogen destruction. Adaptive immune responses are due to the lymphocytes of the immune system, which collectively have the ability to recognize a vast array of antigens. Somatic gene rearrangement and somatic mutation in the genes for antigen receptors provide the lymphocyte population with a set of highly diverse antigen receptors—immunoglobulins on B lymphocytes, and T-cell receptors on T lymphocytes. An individual lymphocyte expresses receptors of a single and unique antigen-binding specificity. A pathogen therefore stimulates only the small subset of lymphocytes that express receptors for the pathogen antigens, focusing the adaptive immune response on that pathogen.

One major difference between B and T cells is the type of antigens they recognize. Whereas the immunoglobulin receptors of B cells bind whole molecules and intact pathogens, T-cell receptors recognize only short peptide antigens which are bound to major histocompatibility complex molecules on cell surfaces. A second major difference is that all B cells have the same general function of making immunoglobulins, whereas T cells can have distinctive functions. CD8 cytotoxic T cells kill cells infected with viruses by recognizing peptide antigens presented by MHC class I molecules; CD4 T cells help macrophages and B cells become activated by recognizing peptides presented by MHC class II molecules. The CD4 T cells are further subdivided into T_H1 cells that help macrophages and T_H2 cells that help B cells. Before they can perform these functions all T cells must themselves be activated within a secondary lymphoid tissue through recognition of their specific antigen on a professional antigen-presenting cell.

B cells differentiate into plasma cells that make antibody, a secreted form of the cell-surface immunoglobulin. The production of antibodies by a B cell requires help from a CD4 T_H2 cell that responds to the same antigen and activates the B cell. Antibodies bind to extracellular pathogens and their toxins and deliver them to phagocytes and other effector cells for destruction. Immunity due to antibodies and their actions is known as humoral immunity. Whereas CD4 T_H2 cells work within secondary lymphoid tissues, CD8 cytotoxic T cells and CD4 T_H1 cells must travel to the infected site to perform their functions. Immunity that is principally due to CD8 cytotoxic T cells and/or CD4 T_H1 cells is known as cell-mediated immunity.

When successful, an adaptive immune response terminates infection and provides long-lasting protective immunity against the pathogen that provoked the response. Failures to develop a successful response can arise from inherited deficiencies in the immune system or from the pathogen's ability to escape, avoid, or subvert the immune response. Such failures can lead to debilitating chronic infections or death. Another category of failure of adaptive immunity is the chronic disease caused by a misdirected response. Allergy is the consequence of a strong response to a harmless substance, while autoimmune disease is caused when the destructive potential of the immune system is directed at one or more of the body's own tissues. A further medical context for unwanted adaptive immunity is transplantation. Here the differences in MHC type between donor and recipient trigger an immune response which if left unchecked will reject the transplanted tissue.

Summary to Chapter 1

Throughout their evolutionary history, multicellular animals have been infected by microorganisms. In response, the animals evolved a series of defenses which humans use today. Barriers confine microorgainsims to the outer surfaces of the body and when pathogens manage to breach the barriers they are sought out and destroyed by the immune system. In responding to microbial attack, the immune system starts with innate immunity, whose mechanisms are fast, fixed in their mode of action and effective in stopping a majority of infections at an early stage. The mechanisms of adaptive immunity are slow to start but they eventually become powerful enough to terminate almost all of the infections that eluded innate immunity. The cells and molecules of innate immunity have counterparts in vertebrates and invertebrates. They identify common classes of pathogen and destroy them by using mechanisms that have stood the test of evolutionary time and are continuously useful. Mechanisms of adaptive immunity are found only in vertebrates and are distinguished by elaborate and highly specific ways of recognizing pathogens. They build on the mechanisms of innate immunity to provide a powerful response that is tailored to the pathogen at hand, and can be rapidly reactivated on future challenge with that same pathogen, providing lifelong immunity to many common diseases. Adaptive immunity is an evolving process within a person's lifetime, in which each infection changes the make-up of that individual's lymphocyte population. These changes are neither inherited nor passed on but, during the course of a lifetime, they determine a person's fitness and their susceptibility to disease. The strategy of vaccination aims at circumventing the risk of a first infection and in the twentieth century successful campaigns of vaccination were waged against a number of diseases that were once both familiar and feared. Through the use of vaccination and anti-microbial drugs, as well as better sanitation and nutrition, infectious disease has become a less common cause of death in many countries. In such situations, diseases due to unwanted activities of the immune system take on more importance. These include chronic conditions such as allergy and autoimmunity as well as the rejection of transplanted organs and tissues.

Antibody Structure and the Generation of B-Cell Diversity

2

Antibodies are variable antigen-specific proteins produced by the B lymphocytes of the immune system in response to infection. They circulate as a major component of the plasma in blood and lymph. Their function is to bind to pathogenic microorganisms and their toxins when they are encountered in the extracellular spaces of the body. Binding of antibody to a pathogen can disable the pathogen and also renders it susceptible to destruction by other components of the immune system. Most vaccines provide their protection through stimulating the production of antibodies.

Individual antibodies are **specific**; each antibody can bind to only one or a very small number of different antigens. For this reason, immunity to the measles virus, for example, provides no protection against the influenza virus, nor does immunity to influenza protect from measles. The immune system of an individual has the potential to make antibodies against a vast number of different substances, paralleling the many different foreign antigens that a person is likely to be exposed to during their lifetime. Collectively, therefore, antibodies are diverse in their antigen-binding specificities: the total number of different specific antibodies that can be made by an individual is known as the **antibody repertoire** and it might be as high as 10^{16}.

Antibodies are the secreted form of proteins known more generally as **immunoglobulins** (**Ig**). Before it has encountered antigen, a mature B cell expresses immunoglobulin in a membrane-bound form that serves as the B cell's receptor for antigen. When antigen binds to this receptor, the B cell is stimulated to proliferate and to differentiate into **plasma cells**, which then secrete antibodies of the same specificity as that of the membrane-bound immunoglobulin (Figure 2.1). Antibody production is the single effector function of the B lymphocytes of the immune system.

The first part of this chapter describes the general structure of immunoglobulins, and the second part considers how immunoglobulin diversity is generated in

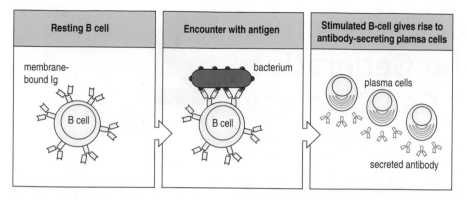

Figure 2.1 Plasma cells secrete antibody of the same antigen specificity as the membrane-bound immunoglobulin expressed by their B-cell precursor. A mature B cell expresses membrane-bound immunoglobulin (Ig) of a single antigen specificity. When a foreign antigen first binds to this immunoglobulin, the B cell is stimulated to proliferate. Its progeny differentiate into plasma cells that secrete antibody of the same specificity as the membrane-bound immunoglobulin.

developing B cells. After a mature B cell has encountered its specific antigen, there are changes in the specificity and effector function of the antibodies that it produces; these are discussed in the final part of the chapter.

The structural basis of antibody diversity

The function of an antibody molecule in host defense is to recognize and bind its corresponding antigen, and to target the bound antigen to other components of the immune system. These then destroy the antigen or clear it from the body. Antigen binding and interaction with other immune system cells and molecules are performed by different parts of the antibody molecule. One part is highly variable in that its amino-acid sequence differs considerably from antibody to antibody. This variable part contains the site of antigen binding and confers specificity on the antibody. The rest of the molecule is far less variable in amino-acid sequence. This constant part of the antibody interacts with other immune system components.

There are five classes of immunoglobulin—IgA, IgD, IgE, IgG, and IgM. They are distinguished on the basis of structural differences in the constant part of the molecule and have different effector functions. We shall first consider the general structure of immunoglobulins and then look more closely at the structural features that confer antigen specificity. IgG, the most abundant antibody in blood and lymph, will be used to illustrate the common structural features of immunoglobulins.

2-1 Antibodies are composed of polypeptides with variable and constant regions

Antibodies are glycoproteins that are built from a basic unit of four polypeptide chains. This unit consists of two identical **heavy chains** (**H chains**) and two identical, smaller, **light chains** (**L chains**), which are assembled into a structure that looks like the letter Y (Figure 2.2, top panel). An IgG molecule has a molecular weight of about 150 kDa, to which each heavy chain contributes approximately

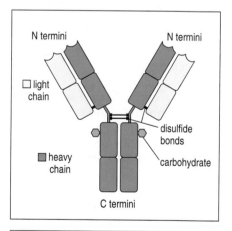

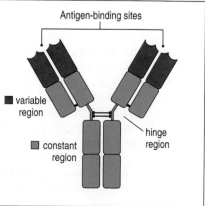

Figure 2.2 The immunoglobulin G (IgG) molecule. As shown in the top panel, each IgG molecule is made up of two identical heavy chains (green) and two identical light chains (yellow). Carbohydrate (turquoise) is attached to the heavy chains. The lower panel shows the location of the variable and constant regions in the IgG molecule. The amino-terminal regions (red) of the heavy and light chains are variable in sequence from one IgG molecule to another; the remaining regions are constant in sequence (blue). The carbohydrate is omitted from this panel and from most subsequent figures for simplicity. In IgG there is a flexible hinge region between the two arms and the stem of the Y.

50 kDa and each light chain approximately 25 kDa. Each arm of the Y is made up of a complete light chain paired with the amino-terminal (N-terminal) part of a heavy chain, covalently linked by a disulfide bond. The stem of the Y consists of the paired carboxy-terminal (C-terminal) portions of the two heavy chains. The two heavy chains are also linked to each other by disulfide bonds.

The polypeptide chains of different antibodies vary greatly in amino-acid sequence, and the sequence differences are concentrated in the amino-terminal region of each type of chain; this is known as the **variable region** or **V region**. This variability is the reason for the great diversity of antigen-binding specificities among antibodies, as the paired V regions of a heavy and a light chain form the **antigen-binding site** (see Figure 2.2, bottom panel). Every Y-shaped antibody molecule therefore has two identical antigen-binding sites, one at the end of each arm. The remaining parts of the light chain and the heavy chain have a much more limited variation in amino-acid sequence between different antibodies and are therefore known as the **constant regions** or **C regions**.

In IgG, a relatively unstructured portion in the middle of the heavy chain forms a flexible hinge region at which the molecule can be cleaved to produce defined antibody fragments. These fragments were instrumental in providing information about antibody structure and function as they can be used for separate analysis of the antigen-binding and effector functions of the molecule. Digestion with the plant protease papain produces three fragments, corresponding to the two arms and the stem (Figure 2.3). The fragments corresponding to the arms are called **Fab** (**F**ragment **a**ntigen **b**inding) because they bind antigen. The fragment

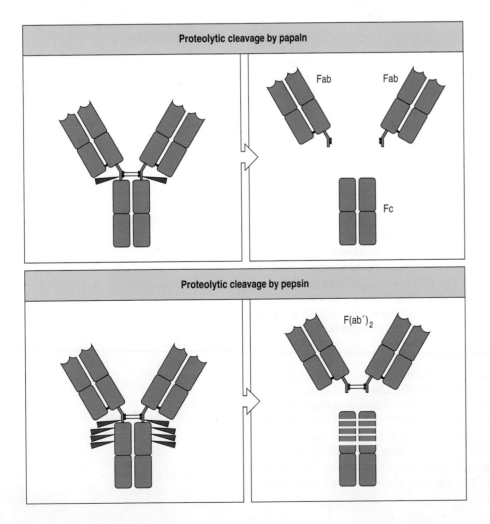

Figure 2.3 The Y-shaped immuno-globulin molecule can be dissected by using proteases. The protease papain cleaves the IgG molecule into three pieces: two Fab fragments and one Fc fragment (upper panels). The protease pepsin cleaves IgG to yield one F(ab')$_2$ fragment. The Fc fragment is not stable to pepsin degradation and is broken into a number of smaller pieces (lower panels). Red arrowheads denote the sites of protease attack.

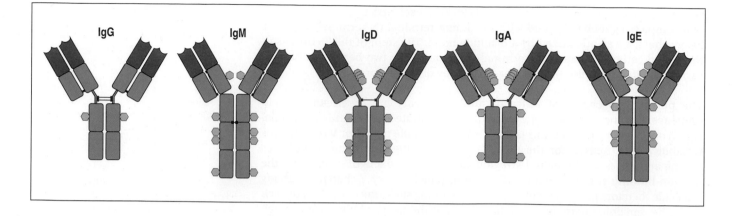

corresponding to the stem is called **Fc** (**F**ragment **c**rystallizable) because it was seen to crystallize in the first experiments of this sort. The stem of a complete antibody molecule is thus often known as the **Fc region** or **Fc piece**, and the arms as Fab. Digestion of IgG with the gut protease pepsin produces a different fragment, **F(ab′)$_2$**, in which the two arms remain linked by disulfide bonds between the heavy chains.

Differences in the heavy-chain constant regions define five main **isotypes** or **classes** of immunoglobulin, which have different functions in the immune response. They are **immunoglobulin A (IgA)**, **immunoglobulin D (IgD)**, **immunoglobulin E (IgE)**, **immunoglobulin G (IgG)**, and **immunoglobulin M (IgM)** (Figure 2.4). Their heavy chains are denoted by the corresponding lower-case Greek letter (α, δ, ε, γ, and μ respectively).

There are only two isotypes or classes of light chain, which are termed **kappa (κ)** and **lambda (λ)**. No functional difference has been found between antibodies carrying κ light chains and those carrying λ light chains; light chains of both isotypes are found associated with all the heavy-chain isotypes. Each antibody, however, contains either κ or λ light chains, not both. The relative abundances of κ and λ light chains vary with the species of animal. In humans, two-thirds of the antibody molecules contain κ chains and one-third have λ chains.

2-2 Immunoglobulin chains are folded into compact and stable protein domains

Antibodies function in extracellular environments in the presence of infection, where they can encounter variations in pH, salt concentration, proteolytic enzymes, and other potentially destabilizing factors. Their structure, however, helps them to withstand such harsh conditions. Heavy and light chains each consist of a series of similar sequence motifs; a single motif is about 100–110 amino acids long and folds up into a compact and exceptionally stable protein domain called an **immunoglobulin domain**. Each immunoglobulin chain is composed of a linear series of these domains.

The variable region at the amino-terminal end of each heavy or light chain is composed of a single **variable domain** (**V domain**): V_H in the heavy chain and V_L in the light chain. A V_H and a V_L domain together form an antigen-binding site. The other domains have little or no sequence diversity within a particular isotype and are termed the **constant domains** (**C domains**). These make up the constant regions. The constant region of a light chain is composed of a single C_L domain, whereas the constant region of a heavy chain is composed of three or four C domains, depending on the isotype. The γ heavy chains of IgG have three

Figure 2.4 The structural organization of the human immunoglobulin isotypes. In particular, note the differences in length of the heavy-chain C regions, the location of the disulfide bonds linking the chains, and the hinge region in IgG, IgA, and IgD, but not in IgM and IgE. The isotypes also differ in the distribution of N-linked carbohydrate groups, as shown in turquoise. All these immunoglobulins occur as monomers in their membrane-bound form. In addition, secreted IgM forms pentamers and secreted IgA is made as monomers and dimers.

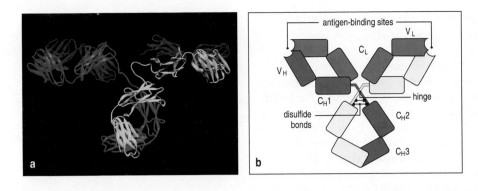

Figure 2.5 The heavy and light chains of an immunoglobulin molecule are made from a series of similar protein domains. Panel a shows a ribbon diagram tracing the course of the polypeptide backbones of chains in an IgG molecule. The heavy chains are shown in yellow and purple, the light chains in red. The molecule is composed of three similarly sized globular portions, which correspond to the Fc stem and the two Fab arms. Each of these globular portions is made up of four immunoglobulin domains. The schematic diagram in panel b shows the domain structure of an IgG molecule. The light chain (red) has one variable domain (V_L) and one constant domain (C_L). The γ heavy chain (one yellow and one purple) has a variable domain (V_H) and three constant domains (C_H1, C_H2, and C_H3). Panel a courtesy of L. Harris.

domains—C_H1, C_H2, and C_H3. Some other isotypes have four C domains (see Figure 2.4). In the complete IgG molecule the pairing of the four polypeptide chains produces three globular regions, corresponding to the two Fab arms and the Fc stem, respectively, each of which is composed of four immunoglobulin domains (Figure 2.5).

The structure of a single immunoglobulin domain can be compared to a bulging sandwich in which two β sheets (the slices of bread) are held together by strong hydrophobic interactions between their constituent amino-acid side chains (the filling). The structure is stabilized by a disulfide bond between the two β sheets. The adjacent strands within the β sheets are connected by loops of the polypeptide chain. This arrangement of β sheets provides the stable structural framework of all immunoglobulin domains, whereas the amino-acid sequence of the loops can be varied to confer different binding properties on the domain. The structural similarities and differences between C and V domains are illustrated for the light chain in Figure 2.6.

Although immunoglobulin domains were first discovered in antibodies, very similar domains, known generally as **immunoglobulin-like domains**, have subsequently been found in many other proteins, and are particularly common in cell-surface and secreted proteins of the immune system. Such proteins collectively form the **immunoglobulin superfamily**.

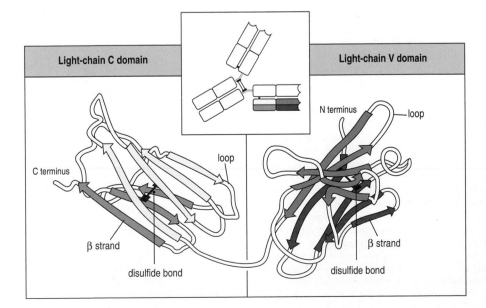

Figure 2.6 The three-dimensional structure of immunoglobulin C and V domains. An individual light chain is folded into a single C domain (left panel) and a single V domain (right panel). The inset shows the location of this light chain in an IgG molecule. The folding of the polypeptide chain backbone is depicted as ribbon diagrams in which the thick arrows correspond to the parts of the chain that form the β strands of the β sheet. The arrows point from the amino terminus of the chain to the carboxy terminus. Adjacent strands in each β sheet run in opposite directions (antiparallel), as shown by the arrows, and are connected by loops. In a C domain there are four β strands in the upper β sheet (yellow) and three in the lower sheet (green). In a V domain there are five strands in the upper sheet (blue) and four in the lower (red).

2-3 An antigen-binding site is formed from the hypervariable regions of a heavy-chain and a light-chain V domain

A comparison of the V domains of heavy and light chains from different antibody molecules shows that the differences in amino-acid sequence are concentrated within particular regions called **hypervariable regions** (**HV**), which are flanked by much less variable **framework regions** (Figure 2.7, top panel). There are three hypervariable regions in each V domain. When mapped onto the folded structure of a V domain, the hypervariable regions locate to three of the loops that are exposed at the end of the domain furthest from the constant region. The framework regions correspond to the β strands and the remaining loops. This relationship is illustrated for the light-chain V domain in Figure 2.7 (center panel). Thus the structure of the immunoglobulin domain provides the capacity for localized variability against a structurally stable background.

The pairing of a heavy and a light chain in an antibody molecule brings together the hypervariable loops from each V domain to create a composite hypervariable surface, which forms the antigen-binding site at the tip of each Fab arm (see Figure 2.7, bottom panel). The differences in the loops between different antibodies create both the specificity of antigen-binding sites and their diversity. The hypervariable loops are also called **complementarity-determining regions** (**CDRs**) because they provide a binding surface that is complementary to that of the antigen.

2-4 Antigen-binding sites vary in shape and physical properties

The function of antibodies is to bind to microorganisms and facilitate their destruction or ejection from the body. The antibodies that are most effective in combating infection are generally those that bind to the exposed and accessible molecules that make up the surface of a pathogen. The part of an antigen to which an antibody binds is called an **antigenic determinant** or an **epitope** (Figure 2.8). In nature these structures are usually either carbohydrate or protein or both, as the surface molecules of pathogens are commonly glycoproteins, polysaccharides, glycolipids, and proteoglycans. Complex macromolecules such as these will usually contain several different epitopes, each of which can be bound by a different antibody. Individual epitopes are typically composed of a cluster of amino acids or a part of a polysaccharide chain. Any antigen that contains more than one epitope, or more than one copy of the same epitope, is known as a **multivalent** antigen (Figure 2.9).

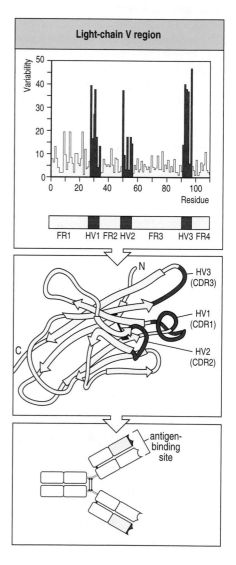

Figure 2.7 The hypervariable regions of antibody V domains lie in discrete loops at one end of the domain structure. The top panel shows the variability plot for the 110 positions within the amino-acid sequence of a light-chain V domain. It is obtained from comparison of many light-chain sequences. Variability is the ratio of the number of different amino acids found at a position and the frequency of the most common amino acid at that position. The maximum value possible for the variability is 400, the square of 20, the number of different amino acids found in antibodies. The minimum value for the variability is 1. Three hypervariable regions (HV1, HV2, and HV3) can be discerned (shown in red) flanked by four framework regions (FR1, FR2, FR3, and FR4) (yellow). The center panel shows the correspondence of the hypervariable regions to three loops at the end of the V domain farthest from the constant region. The location of hypervariable regions in the heavy-chain V domain is similar (not shown). The hypervariable loops contribute much of the antigen specificity of the antigen-binding site located at the tip of each arm of the antibody molecule. Hypervariable regions are also known as complementarity-determining regions: CDR1, CDR2, and CDR3.

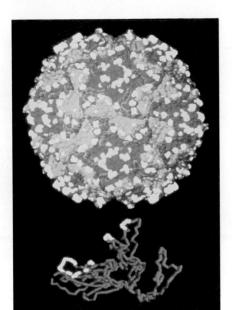

Figure 2.8 Epitopes for antibodies are exposed on the surface of antigens. The coat of a poliovirus is made up of multiple copies of three different proteins (indicated in yellow, blue, and pink). Known epitopes on the surface of the virus are shown as white patches. One of the viral coat proteins, VP1 (blue), is shown separately, but folded as it is in the virus particle. This protein contains several different epitopes (white). They are located at the surface of the protein and are exposed on the surface of the virus particle. Photograph courtesy of D. Filman and J.M. Hogle.

Antibodies can be made against a much wider range of chemical structures than proteins and carbohydrates. Such antibodies are, however, more commonly implicated in allergic reactions and autoimmune diseases than in protective immunity against infection. Antibodies specific for DNA, for example, are characteristic of the autoimmune disease systemic lupus erythematosus (SLE), and allergies can be caused by antibodies against small organic molecules, such as antibiotics and other drugs.

The antigen-binding sites of antibodies vary according to the size and shape of the epitope. Antibodies that bind to the end of a polysaccharide chain, or to a small molecule such as vitamin K_1, use a deep pocket formed between the heavy- and light-chain V domains (Figure 2.10, left panel). In such cases not all the CDRs make contact with the antigen. In contrast, antibodies that bind to a number of adjacent sugars within a polysaccharide, or amino acids within a polypeptide, use longer shallower clefts formed by all the opposing CDRs of the V_H and V_L domains (see Figure 2.10, center panel). Epitopes of this kind, where the antibody binds parts of a molecule that are adjacent in the linear sequence are called **linear epitopes** (Figure 2.11, left panel).

Another kind of epitope is formed by parts of a protein that are separated in the amino-acid sequence but are brought together in the folded protein. These are called **conformational** or **discontinuous epitopes** (see Figure 2.11, right panel). Antibodies that bind to complete proteins often make contacts with a much larger area of the protein surface than can be accommodated within a groove between the CDRs of the V_H and V_L domains. These antibodies might have no discernible binding groove but interact with the antigen by using a surface (in area usually 700–900 Å^2) formed by the CDRs and which can even extend beyond them (see Figure 2.10, right panel).

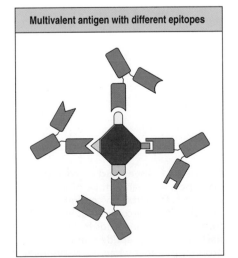

Multivalent antigen with different epitopes

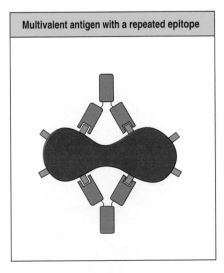

Multivalent antigen with a repeated epitope

Figure 2.9 Two kinds of multivalent antigen.

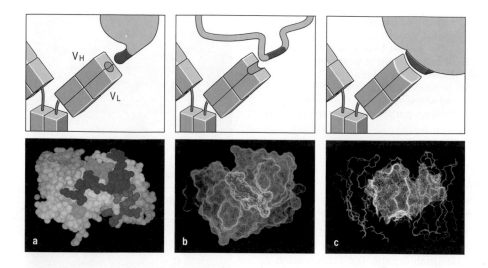

Figure 2.10 Epitopes can bind in pockets or grooves or on extended surfaces in the binding sites of antibodies. The top row shows schematic representations of three different types of antigen-binding site. The antigens are shown in orange and the epitopes in red. Small compact epitopes bind in an antigen-binding pocket between the two V domains, as shown in the left panel. An extended portion of a protein or polysaccharide chain binds in a shallow groove between the two V domains, as shown in the center panel. An antigen with an extended surface binds to a similarly extended surface in an antigen-binding site, as shown in the right panel. Bottom row: panel a shows seven amino-acid residues of a peptide antigen (red) bound in the antigen-binding pocket of a Fab fragment of antibody, viewed looking down into the antigen-binding site. The CDR loops of the antibody are colored. Panel b shows a peptide (red) from the human immunodeficiency virus bound along a groove formed between the two V domains (green) of the antigen-binding site. Panel c shows the protein lysozyme bound to a Fab fragment of an anti-lysozyme antibody. The surface contour of the lysozyme molecule (yellow dots) is superimposed on the antigen-binding site of the antibody. Amino-acid residues in the antibody that make contact with lysozyme are shown in full (red); for the rest of the Fab fragment only the peptide backbone is shown (blue). Photographs courtesy of I.A. Wilson, R.L. Stanfield, and S. Sheriff.

The binding of antigens to antibodies is based solely on non-covalent forces—electrostatic forces, hydrogen bonds, van der Waals forces, and hydrophobic interactions. The antigen-binding sites of antibodies are unusually rich in aromatic amino acids, which can participate in many van der Waals and hydrophobic interactions. The better the general fit between the interacting surfaces of antigen and antibody, the stronger are the bonds formed by these short-range forces. Small differences in shape and chemical properties of the binding site can thus give several antibodies specificity for the same epitope, but they bind to it with different binding strengths, or **affinities**. Binding due to van der Waals forces and hydrophobic interactions is complemented by the formation of electrostatic interactions (salt bridges) and hydrogen bonds between particular chemical groups on the antigen and particular amino acid residues of the antibody.

In the immune response, effective antibodies are those that bind tightly to an antigen and do not let go. This behavior contrasts with that of enzymes, which bind a substrate, chemically change the substrate's structure, and then release it. However, the same set of 20 amino acids is used to make enzymes and antibodies and, after intensive searching, some antibodies have been found that catalyze chemical reactions involving the antigen they bind. The application of such

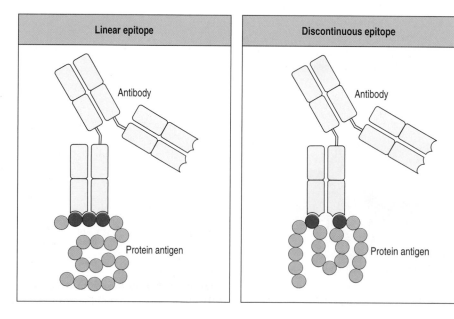

Figure 2.11 Linear and discontinuous epitopes. A linear epitope of a protein antigen is formed from contiguous amino acids. A discontinuous epitope is formed from amino acids from different parts of the polypeptide that are brought together when the chain folds.

catalytic antibodies in medicine is being explored. A possible use for a catalytic antibody would be to convert a toxic chemical within the body into an innocuous product.

2-5 Monoclonal antibodies are produced from a clone of antibody-producing cells

The specificity of an antibody for its antigen makes antibodies useful reagents for detecting and quantifying a particular antigen. They are used in this way in a range of clinical and laboratory tests and in biological research generally. Specific antibodies are also being developed as therapeutic agents. For all such applications, large quantities of identical antibodies are required.

In the past, the only way of generating antibodies in any quantity was by immunizing animals with the appropriate antigen and then preparing **antisera** containing the desired antibodies from the blood. However, the antibodies generated after immunization, or after natural exposure to an antigen in an infection, are a mixture of antibodies of different specificities and affinities. Some of this heterogeneity results from the production of antibodies that bind to different epitopes on the immunizing antigen. However, because of small differences in their binding sites that affect their affinity for antigen, even antibodies directed at a small molecule with a single antigenic determinant can be a mixture of different molecules.

To harness the full potential of antibodies, a way of making an unlimited supply of identical antibody molecules of known specificity was needed. This was achieved through the production of **monoclonal antibodies**. These are antibodies produced by a single clone of antibody-producing cells and thus all have identical antigen-binding sites and are of identical isotype. They are produced on a large scale from hybrid cells derived from a single antibody-producing cell fused with a tumor cell to produce a **hybridoma** cell line that can grow and produce antibodies indefinitely in culture (Figure 2.12).

Monoclonal antibodies are now used in most serological assays, as diagnostic probes and as therapeutic agents. So far, however, only mouse monoclonal antibodies are routinely produced, and efforts to use the same approach to make human monoclonal antibodies have met with limited success. Genetic engineering is being used to overcome this difficulty. Because the CDRs determine the antigen specificity of an antibody and do not affect the overall structure of the V domain it is possible to manipulate the antigen specificity of an antibody molecule solely by replacing the CDRs. This is particularly useful in the development of monoclonal antibodies for use as therapeutics. Mouse monoclonal antibodies of any desired specificity can be made quite easily but are limited in their therapeutic use because of the immune response that can be made by human patients against the constant regions of mouse antibodies. By using genetic engineering techniques, the CDR loops of a human immunoglobulin with irrelevant specificity can be replaced by those of the mouse monoclonal antibody with desired specificity. This confers the antigen specificity of the mouse antibody on the human antibody. This process is called **antibody humanization**.

Summary

IgG antibodies are made of four polypeptide chains—two identical heavy chains and two identical light chains. Each chain has a variable region that contributes to the antigen-binding site and a constant region that, in the heavy chain, determines the antibody isotype and its specialized effector functions. The IgG molecule is shaped like a Y in which the stem and the arms are of comparable sizes and can flex with respect to each other. Each arm contains an antigen-binding site,

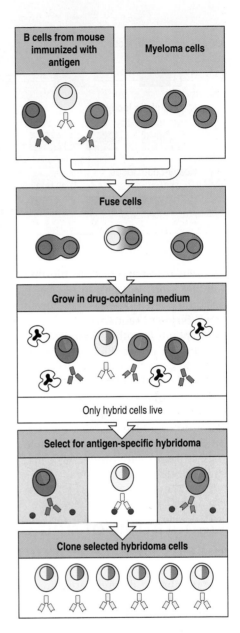

Figure 2.12 Production of a monoclonal antibody. Lymphocytes from a mouse immunized with the antigen are fused with myeloma cells using polyethyleneglycol. The cells are then grown in the presence of drugs which kill myeloma cells but permit the growth of hybridoma cells. Individual cultures of hybridomas are then tested to determine if they make the desired antibody. The cells are then cloned to produce a homogeneous culture of cells making a monoclonal antibody. Myelomas are tumors of plasma cells. Those used to make hybridomas were selected not to express heavy and light chains. Thus hybridomas only express the antibody made by the B-cell fusion partner.

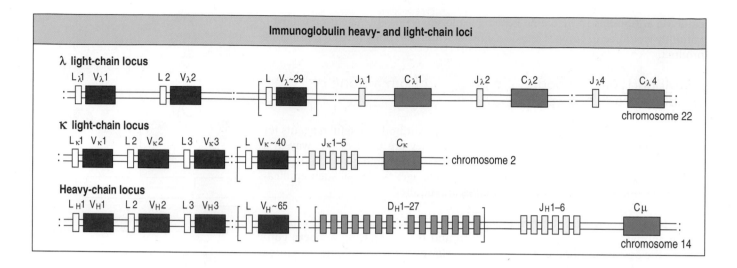

whereas the stem contains binding sites for cells and effector molecules that enable bound antigen to be cleared from the body. Immunoglobulin chains are built from a series of structurally related immunoglobulin domains. Within the V domain, sequence variability is localized to three hypervariable regions corresponding to three loops that are clustered at one end of the domain. In the antibody molecule the hypervariable loops of the heavy and light chains form a variable surface that binds antigen. The type of antigen bound by an antibody depends on the shape of the antigen-binding site: antigens that are small molecules can be bound within deep pockets; linear epitopes from proteins or carbohydrates can be bound within clefts or grooves; and the binding of conformational epitopes of folded proteins takes place over an extended surface. Monoclonal antibodies are antibodies of a single specificity that are derived from a clone of identical antibody-producing cells. They are used in diagnostic tests and as therapeutic agents.

Generation of immunoglobulin diversity in B cells before encounter with antigen

The number of different antibodies that can be produced by the human body seems to be virtually limitless. To achieve this, the immunoglobulin genes are organized differently from other genes. In all cells except B cells, the immunoglobulin genes are in a fragmented form that cannot be expressed; instead of containing a single complete gene, immunoglobulin heavy-chain and light-chain loci consist of families of **gene segments**, sequentially arrayed along the chromosome, with each set of segments containing alternative versions of parts of the immunoglobulin variable region. The immunoglobulin genes are inherited in this form through the germ line (egg and sperm); this arrangement is therefore called the **germline form** or **germline configuration**.

For an immunoglobulin gene to be expressed, individual gene segments must first be rearranged to assemble a functional gene, a process that occurs only in developing B cells. Immunoglobulin-gene rearrangements occur during the development of B cells from B-cell precursors in the bone marrow. When gene rearrangements are complete, heavy and light chains can be produced and membrane-bound immunoglobulin appears at the B-cell surface. The B cell can now recognize and respond to an antigen. Much of the diversity within the mature antibody repertoire is generated during this process of gene rearrangement, as we shall see in this part of the chapter.

Figure 2.13 The germline organization of the human immunoglobulin heavy-chain and light-chain loci. The upper row shows the λ light-chain locus which has about 30 functional V_λ gene segments and four pairs of functional J_λ gene segments and C_λ gene segments. The κ locus (center row) is organized in a similar way, with about 40 functional V_κ gene segments accompanied by a cluster of five J_κ gene segments but with a single C_κ gene segment. In approximately half of the human population, the entire cluster of κ V gene segments is duplicated (not shown for simplicity). The heavy-chain locus (bottom row) has about 65 functional V_H gene segments, a cluster of about 27 D segments and six J_H gene segments. For simplicity only a single C_H gene ($C\mu$) is shown in this diagram. L, leader sequence. This diagram is not to scale: the total length of the heavy-chain locus is over 2 megabases (2 million bases), whereas some of the D segments are only six bases long.

2-6 The DNA sequence encoding a variable region is assembled from two or three gene segments

In humans, the immunoglobulin genes are found at three chromosomal locations: the heavy-chain locus on chromosome 14, the κ light-chain locus on chromosome 2 and the λ light-chain locus on chromosome 22. Different gene segments encode the leader peptide (L), the variable region (V) and the constant region (C) of the heavy and light chains. Gene segments encoding constant regions are commonly called C genes (Figure 2.13). Within the heavy-chain locus are C genes for all the different heavy-chain isotypes.

The gene segments encoding the leader peptide and the constant region consist of exons and introns like those found in other human genes and they are ready to be transcribed. By contrast the V regions are encoded by two (V_L) or three (V_H) gene segments which require rearrangement in order to produce an exon that can be transcribed. The two types of gene segment that encode the light chain V region are called **variable (V)** and **joining (J) gene segments**. The heavy-chain locus includes an additional set of **diversity (D) gene segments** which lie between the arrays of V and J gene segments (see Figure 2.13).

The variable region of a light chain is encoded by the combination of one V and one J segment, whereas the constant region is encoded by a single C gene. The main differences between V gene segments are in the sequences that encode the first and second hypervariable regions of the V domain (see Section 2-3); the third hypervariable region is determined by the junction between the V and J segments. In an individual B cell, only one of the light-chain loci (κ or λ) gives rise to a functional light-chain gene. The variable region of the heavy chain is encoded by one V, one D, and one J segment. The first two hypervariable regions are due to differences in the sequences of the heavy-chain V gene segments; the third hypervariable region is determined by differences in the D gene segments and the junctions they make with the V and J gene segments. The constant region of a heavy chain is encoded by one of the C genes.

2-7 Random recombination of gene segments produces diversity in the antigen-binding sites of immunoglobulins

During the development of B cells the arrays of V, D, and J segments are cut and spliced by DNA recombination. This process is called **somatic recombination** because it occurs in cells of the soma, a term which embraces all of the body's cells except the germ cells. A single gene segment of each type is brought together to form a DNA sequence encoding the variable region of an immunoglobulin chain. For light chains, a single recombination occurs, between a V_L and a J_L segment (Figure 2.14, left panels), whereas for heavy chains, two recombinations are

Figure 2.14 V-region sequences are constructed from gene segments. Light-chain V-region genes are constructed from two segments (left panel). A variable (V) and a joining (J) gene segment in the genomic DNA are joined to form a complete light-chain V-region (V_L) exon. After rearrangement the light-chain gene consists of three exons, encoding the leader peptide, the V region and the C region, that are separated by introns. On transcription of the gene and RNA splicing to remove the introns a functional message encoding a single light chain is produced. Heavy-chain V regions are constructed from three gene segments (right panels). First the diversity (D) and J gene segments join, then the V gene segment joins to the combined DJ sequence, forming a complete heavy-chain V-region (V_H) exon. The heavy-chain C genes (only one, Cμ, shown here for simplicity) are each composed of several exons. After rearrangement the heavy-chain gene is functional and encodes a single heavy chain of the μ isotype.

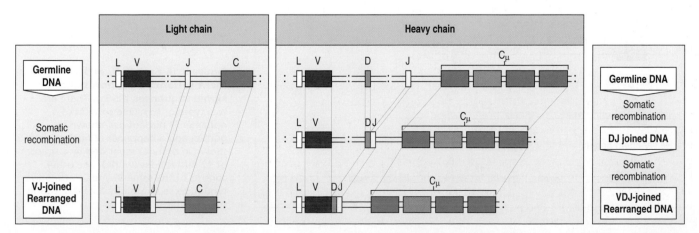

Figure 2.15 The numbers of functional gene segments used to construct the variable regions of human immunoglobulin heavy chains and light chains.

Number of gene segments			
Segment	Light chains		Heavy chain
	κ	λ	H
Variable (V)	40	30	65
Diversity (D)	0	0	27
Joining (J)	5	4	6

needed, the first to join a D and a J_H segment, and the second to join the combined DJ segment to a V_H segment (see Figure 2.14, right panels). In each case, the particular V, D, and J gene segments that are joined together are selected at random. Because of the multiple gene segments of each type, numerous different combinations of V, D, and J gene segments are possible. Thus, the gene rearrangement process generates many different V-region sequences. This is one of the factors contributing to the diversity of immunoglobulin variable regions.

For the human κ light chain, there are approximately 40 V_κ gene segments and 5 J_κ gene segments (Figure 2.15) which can be recombined in 200 (40 × 5) different ways. Similarly, some 30 V_λ gene segments and 4 J_λ gene segments can be recombined in 120 (30 × 4) different ways. At the heavy-chain locus, 65 V_H segments, some 27 D segments and 6 J_H segments can produce 10,530 (65 × 27 × 6) combinations. If recombination between different gene segments were the only mechanism producing V-region diversity, then a maximum of 320 different light chains and 10,530 different heavy chains could be made.

Somatic recombination is performed by enzymes that cut and rejoin the DNA, and exploits some of the mechanisms more ubiquitously used by cells for DNA recombination and repair. The recombination of V, J, and D gene segments is directed by sequences called **recombination signal sequences** (**RSSs**), which flank the 3′ side of the V segment, both sides of the D segment, and the 5′ side of the J segment (Figure 2.16). There are two types of RSS and recombination can occur only between the different types. As well as providing recognition sites for the enzymes that cut and rejoin the DNA, recombination signal sequences ensure that the gene segments are joined in the correct order, as shown for light-chain VJ recombination in Figure 2.17. Because of this strict requirement, recombination in the heavy-chain DNA cannot join a V_H directly to J_H without the involvement of D_H, as V_H and J_H segments are flanked by the same type of recombination signal sequence (see Figure 2.16).

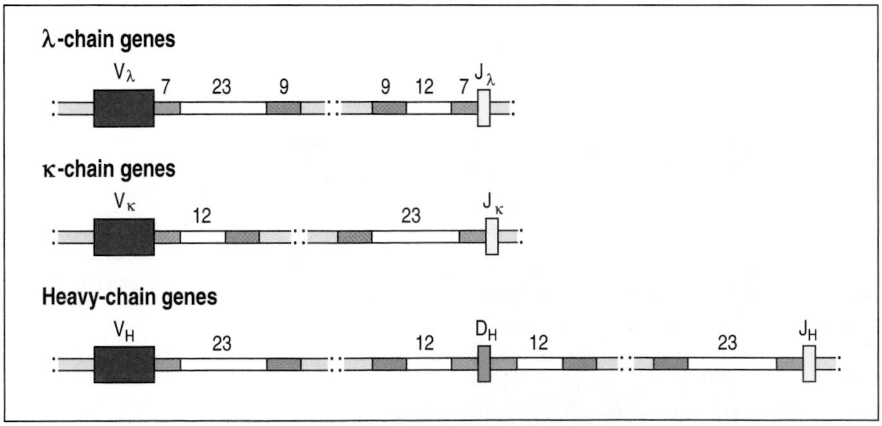

Figure 2.16 Each V, D, or J gene segment is flanked by recombination signal sequences (RSSs). There are two types of RSS. One consists of a nonamer (9 nucleotides, shown in purple) and a heptamer (7 nucleotides, shown in orange) separated by a spacer of 12 nucleotides (white). The other consists of the same 9- and 7-nucleotide sequences separated by a 23-nucleotide spacer (white).

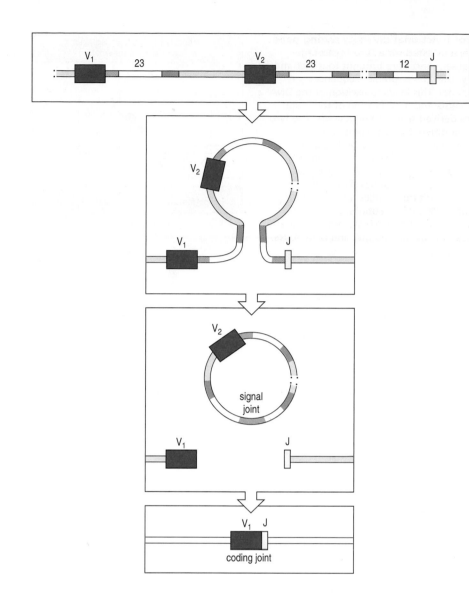

Figure 2.17 Gene segments encoding the variable region are joined by recombination at recombination signal sequences. The recombination between a V (red) and a J (yellow) segment of a light-chain gene is shown here. The recombination signal sequence (RSS) containing a 12-nucleotide spacer is brought together with that containing a 23-nucleotide spacer. This is known as the 12/23 rule and ensures that gene segments are joined in the correct order. The breaking and rejoining of the DNA molecules occurs at the ends of the heptamer sequences. Recombination forms a small circular DNA that contains the DNA lying between the V and J segments that are being joined; this is excised from the chromosome with formation of the signal joint. Within the DNA of the chromosome, the V and J segments are joined to form the coding joint. Nonamers are shown in purple, heptamers in orange, spacers in white.

2-8 Recombination enzymes produce additional diversity in the antigen-binding sites of immunoglobulins

A second factor contributing to the diversity of heavy- and light-chain regions is the enzymatic reactions that cut and splice the DNA during recombination. These recombination and DNA repair reactions introduce additional nucleotides not encoded in the germline DNA (Figure 2.18). Because this diversity is introduced during the formation of the junctions between the cut ends of DNA, it is known as **junctional diversity**. There are two sources of junctional diversity: **P nucleotides** are introduced by a feature of the recombination reaction that generates short palindromic sequences at the cut ends of the DNA strands (a palindromic sequence is identical when read from either end); **N nucleotides** are added at random to the cut ends of the DNA strands by the enzyme **terminal deoxynucleotidyl transferase (TdT)**. N nucleotides are so called because they are non-templated (not encoded) in germline DNA. Junctional diversity is an important source of immunoglobulin variability, adding a factor of 3×10^7 to the overall diversity. The third hypervariable region of the light-chain V domain is encoded by the junction formed between the V and J segments (see Figure 2.7) whereas the third hypervariable region of the heavy-chain V domain is formed by the D segment and its junctions with the rearranged V and J segments.

Figure 2.18 The generation of junctional diversity during gene rearrangement. The process is illustrated for a D to J rearrangement. The recombination signal sequences are brought together and endonuclease cleavage occurs within the heptamer sequences (orange), as shown in the top panel. This leads to excision of the DNA that separates the D and J segments and modification of the D and J segments by extra nucleotides derived from the heptamers. The two DNA strands at the end of the modified D and J segments join together to form hairpins. Further cleavage on one DNA strand at the boundaries between the heptamers and the D and J segments releases the hairpins and generates short palindromic sequences. These extra nucleotides are known as P nucleotides. Terminal deoxynucleotidyl transferase (TdT) adds nucleotides randomly to the ends of the single DNA strands. These nucleotides, which are non-germline, are known as N nucleotides. The single strands pair and through the action of exonuclease, DNA polymerase, and DNA ligase, the double-stranded DNA molecule is repaired to give the coding joint.

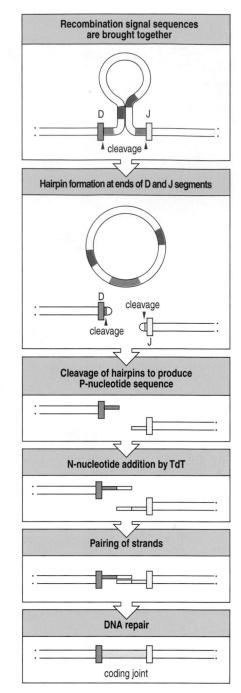

2-9 Naive B cells use alternative mRNA splicing to make both IgM and IgD

After rearrangement of V, D, and J segments the heavy-chain locus can be transcribed. The exon encoding the V region is on the 5′ side of the C gene for the μ chain. The genes encoding the other heavy-chain constant region isotypes are on the 3′ side of the μ gene (Figure 2.19). The exons encoding the leader peptide, the V region and the different domains of the constant region are all separated by introns. As for other human genes, the primary RNA transcripts from the heavy-chain locus include both introns and exons. Functional mRNA is produced by removing the introns from the primary transcripts.

The same assembled V-region gene serves all the heavy-chain C genes. The first immunoglobulins that a B cell expresses on its surface are IgM and IgD. The μ and δ C genes are those nearest to the assembled VDJ sequence (see Figure 2.19); transcription of a complete immunoglobulin heavy-chain gene starts at the V sequence, continues through the μ and δ C genes, and then terminates. This long primary transcript is then processed and spliced to remove the introns and to yield either an IgM mRNA or an IgD mRNA (Figure 2.20). At this stage in the life of a B cell, when it has yet to encounter antigen and is therefore called a **naive B cell**, it expresses both IgM and IgD. These are the only isotypes that B cells produce before they encounter antigen, and they are the only isotypes that can be produced simultaneously by a B cell. After an encounter with their specific antigen, many B cells change the isotype that they produce; they no longer make IgM and IgD but produce IgG, or IgA, or IgE.

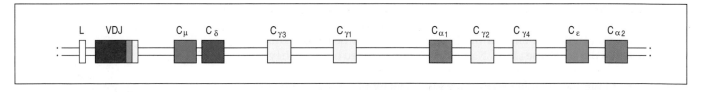

Figure 2.19 Rearrangement of V, D, and J segments produces a functional heavy-chain gene. The assembled VDJ sequence lies some distance from the cluster of C genes. Only functional C genes are shown here. The four different γ genes specify four different subtypes of the γ heavy chain, whereas the two α genes specify two subtypes of the α heavy chain. For simplicity, individual exons in the C genes are not shown. The diagram is not to scale.

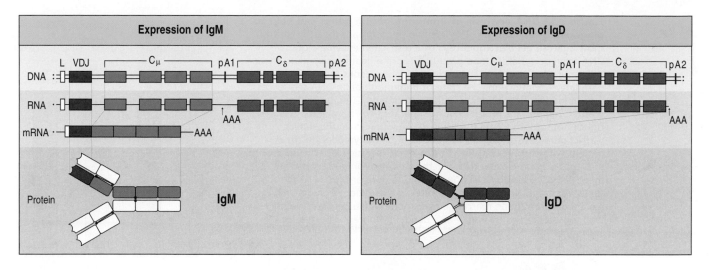

Figure 2.20 Co-expression of IgD and IgM is regulated by RNA processing. In mature B cells, transcription initiated at the V$_H$ promoter extends through both the C$_\mu$ and C$_\delta$ genes. For simplicity we have not shown all the individual C-gene exons but only those of relevance to the differential production of IgM and IgD. The long primary transcript is then processed by cleavage and polyadenylation, and by splicing. Cleavage and polyadenylation at the μ site (pA1) and splicing between C$_\mu$ exons yields an mRNA encoding the μ heavy chain (left panel). Cleavage and polyadenylation at the δ site (pA2) and a different pattern of splicing that removes the C$_\mu$ exons yield mRNA encoding the δ heavy chain (right panel). AAA designates the poly(A) tail.

2-10 Membrane-bound immunoglobulin is complexed with additional proteins to form the functional B-cell receptor for antigen

The immunoglobulin molecules on the surface of B cells bind their specific antigens. However, this interaction alone cannot give the signal that tells the cell's interior when an antigen has bound. The cytoplasmic portions of the immunoglobulin molecule are very short and do not interact with the intracellular proteins that signal cells to divide and differentiate. Antigen–antibody interactions on the outside of the B cell are communicated to the inside of the cell by two transmembrane proteins called **Igα** and **Igβ**, which are associated with immunoglobulin in the B-cell membrane (Figure 2.21). These proteins are invariant in sequence, unlike the immunoglobulins, and have longer cytoplasmic tails, which interact with intracellular signaling molecules. The complex of immunoglobulin, Igα, and Igβ forms the functional **B-cell receptor** for antigen. Like all proteins destined for the cell surface, immunoglobulins enter the endoplasmic reticulum as soon as they are synthesized. Association with Igα and Igβ is necessary for immunoglobulin to be transported out of the endoplasmic reticulum and to appear on the cell surface.

2-11 Each B cell produces immunoglobulin of a single antigen specificity

In a developing B cell, the process of immunoglobulin-gene rearrangement is tightly controlled so that only one heavy chain and one light chain are finally expressed, a phenomenon known as **allelic exclusion**. This ensures that each B

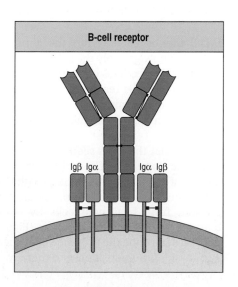

Figure 2.21 Membrane-bound immunoglobulins are associated with two other proteins, Igα and Igβ. Igα and Igβ are disulfide-linked. They have long cytoplasmic tails that can interact with intracellular signaling proteins. The immunoglobulin shown here is IgM, but all isotypes can serve as B-cell receptors.

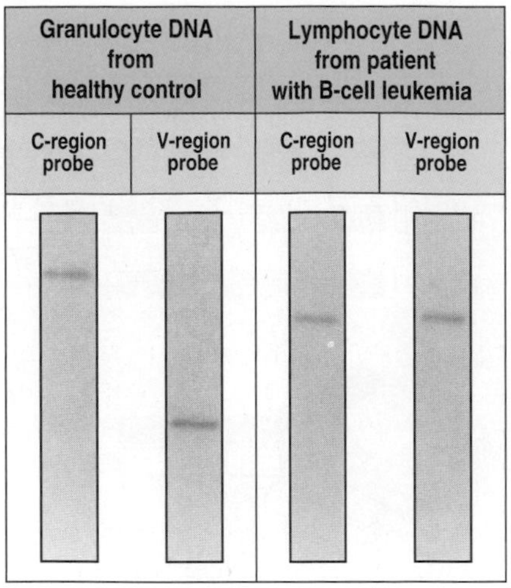

Figure 2.22 DNA analysis of immunoglobulin genes can help to diagnose B-cell tumors. The left two photographs (healthy control) show agarose gel electrophoresis of a restriction enzyme digest of DNA from peripheral blood granulocytes of a healthy person. In these cells, mostly neutrophils, the immunoglobulin genes are in the germline configuration. The state of immunoglobulin DNA sequences is examined by hybridization with V-region and C-region probes. The V and C regions are found in quite separate DNA fragments. The two right photographs (patient with B-cell leukemia) are of a similar restriction digest of DNA from peripheral blood lymphocytes from a patient with chronic lymphocytic leukemia, in which a particular clone of B cells is greatly expanded. For this reason, a unique rearrangement is visible. In this DNA, the V and C regions are found in the same fragment. Normal B lymphocytes in the patient's blood each have a different gene rearrangement, none of which is sufficiently well represented to be visible as a band. Photograph courtesy of S. Wagner and L. Luzzatto.

cell produces immunoglobulin of a single antigen specificity. Thus, although every B cell has two copies or alleles of the heavy-chain locus and two copies of each light-chain locus, only one heavy-chain locus and one light-chain locus are rearranged to produce functional genes.

The combinatorial association of single heavy and light chains within individual B cells is the third factor that produces diversity in the antigen-binding site of immunoglobulins. B cells are free to produce any combination of light and heavy chains and thus the potential number of antibodies that can be made is the product of the total numbers of different heavy and light chains.

Some vertebrate species have little or no germline diversity in the V, D, and J gene segments that are able to rearrange, and gene rearrangement simply produces a population of B cells all expressing a very limited range of antigen specificities. In such species, immunoglobulin diversity is generated after the initial gene rearrangement. In chickens and rabbits, for example, the heavy- and light-chain loci have only a single gene segment of each type that can rearrange to form the initial rearranged gene. The rearranged immunoglobulin genes are further diversified by a recombinational mechanism called gene conversion which uses arrays of 'silent' V gene segments that lie on the 5' side of the rearranging V gene segment. In this mechanism a short segment of the rearranged V gene is replaced with the corresponding part of a non-rearranged silent V gene. By contrast sheep diversify their rearranged immunglobulin genes by a process of somatic hypermutation which introduces many point mutations.

B-cell monospecificity has the consequence that an encounter with a given pathogen engages only the subset of B cells that can make antibodies which bind to that pathogen. This ensures the specificity of the antibody response against an infection. For this reason, vaccination against diphtheria, for example, provides no protection against the influenza virus, nor does vaccination against the influenza virus protect against diphtheria.

The fact that the DNA sequence of the expressed immunoglobulin genes varies from one clone of B cells to the next can be used to detect the large clonal populations of cancer cells in patients with B-cell lymphoma or leukemia. Because the cancer cells are derived from a single clone of B cells, they can be readily distinguished from healthy B cells by comparison of the immunoglobulin-gene rearrangements by DNA analysis (Figure 2.22). In the clinic, such analyses are used to determine the presence of cancer cells in blood samples or tissue biopsies and to monitor the response to therapy.

Summary

In the human genome, the immunoglobulin heavy-chain and light-chain genes are in a form that is incapable of being expressed. In developing B cells, however, the immunoglobulin genes undergo structural rearrangements that permit their expression. The variable domains of immunoglobulin light and heavy chains are encoded in two (V and J) or three (V, D, and J) different kinds of gene segment respectively that are brought into juxtaposition by recombination reactions. One mechanism contributing to the diversity in V-region sequences is the random combination of different V and J segments in light-chain genes and of different V, D, and J segments in rearranged heavy-chain genes. A second mechanism is the introduction of additional nucleotides (P and N nucleotides) at the junctions between gene segments during the process of recombination. A third mechanism that creates diversity in the antigen-binding sites of antibodies is the association of heavy and light chains in different combinations. Gene rearrangement in an individual B cell is strictly controlled so that only one type of heavy chain and one type of light chain are expressed, resulting in an individual B cell and its progeny's expressing only immunoglobulin of a single antigen specificity. A naive B cell that has never encountered antigen expresses membrane-bound immunoglobulin of the IgM and IgD classes. The μ and δ heavy chains are produced from a single transcriptional unit from which RNA processing produces transcripts of different lengths.

Diversification of antibodies after B cells encounter antigen

A mature B cell's first encounter with antigen marks a watershed in its development. Binding of antigen to the surface immunoglobulin of a mature B cell triggers the cell's proliferation and differentiation, and ultimately the secretion of antibodies. As the immune response progresses, antibodies with different properties appear. In this part of the chapter we look at the structural and functional changes that occur in immunoglobulins after antigen has been encountered and how antibodies with different functions are produced.

2-12 Secreted antibodies are produced by an alternative pattern of heavy-chain RNA processing

Gene rearrangement in an immature B cell leads to the expression of functional heavy and light chains and to the production of membrane-bound IgM and IgD on the mature B cell. After encounter with antigen, these isotypes are produced as secreted antibodies. IgM antibodies are produced in large amounts and are important in protective immunity, whereas IgD antibodies are produced only in small amounts and have no known effector function.

All classes of immunoglobulin can be made in two forms: one that is bound to the plasma membrane and serves as the B-cell receptor for antigen, and one, the antibody, that is secreted in order to bind to antigen and aid its destruction. During differentiation to plasma cells, B cells change from making the membrane-bound form to making the secreted form, and plasma cells make only secreted antibody. The difference between the two forms lies at the carboxy terminus of the heavy chain, where membrane-associated immunoglobulin has a hydrophobic anchor sequence that is inserted into the membrane, whereas antibody has a hydrophilic sequence. These two sequences are encoded by different exons; the relative abundances of the two heavy-chain forms are controlled by the use of alternative sites

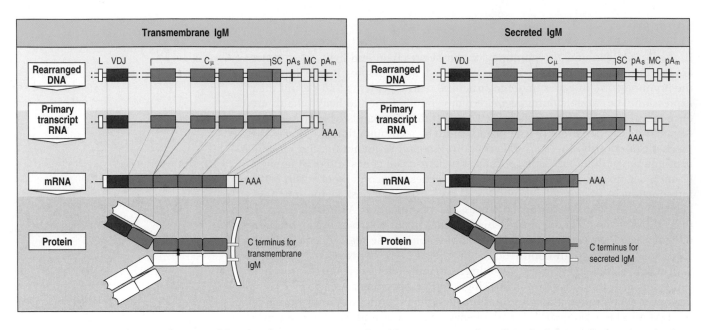

Figure 2.23 The surface and secreted forms of an immunoglobulin are derived from the same heavy-chain gene by alternative RNA processing. Each heavy-chain C gene has two exons (membrane-coding (MC), yellow) encoding the transmembrane region and cytoplasmic tail of the surface form of that isotype, and a secretion-coding (SC) sequence (orange), which encodes the carboxy terminus of the secreted form. The events that dictate whether a heavy-chain RNA will result in a secreted or transmembrane immunoglobulin occur during processing of the initial transcript and are shown here for IgM. Each heavy-chain C gene has two potential polyadenylation sites (shown as pA_s and pA_m). In the left panel, the transcript is cleaved and polyadenylated at the second site (pA_m). Splicing between a site located between the fourth C_μ exon and the SC sequence, and a second site at the 5' end of the MC exons, results in removal of the SC sequence and joining of the MC exons to the fourth C_μ exon. This generates the transmembrane form of the heavy chain. In the right panel, the primary transcript is cleaved and polyadenylated at the first site (pA_s), eliminating the MC exons and giving rise to the secreted form of the heavy chain. AAA designates the poly(A) tail.

of RNA cleavage and polyadenylation to form the 3' end of the RNA, and of alternative sites for RNA splicing. The generation of the membrane-bound and secreted forms of IgM is shown in Figure 2.23.

2-13 Rearranged V-region sequences are further diversified by somatic hypermutation

The diversity generated during gene rearrangement tends to focus sequence diversity in the third CDR of the V_H and V_L domains. Once a B cell has been activated by antigen, however, further diversification of the whole of the V-domain coding sequences occurs through a process of **somatic hypermutation**. This randomly introduces single-nucleotide substitutions (point mutations) at a high rate throughout the rearranged V regions of heavy- and light-chain genes (Figure 2.24). The immunoglobulin constant regions are not affected, and neither are other B-cell genes. The mutations occur at a rate of about one mutation per V-region sequence per cell division, which is more than a million times greater than the ordinary mutation rate for a gene. Mutation occurs at both copies of each immunoglobulin locus, even though one copy is not being expressed. The mechanism of hypermutation and its targeting to immunoglobulin regions are poorly understood, but it appears to act selectively at certain types of DNA sequence motif which are more common than usual within the sequences encoding the V-region hypervariable regions (CDRs). Somatic hypermutation gives rise to B cells bearing mutant immunoglobulin molecules on their surface. Some of these mutant immunoglobulin molecules have a higher affinity for the antigen; they will therefore be more likely to bind the antigen, and so B cells bearing them will

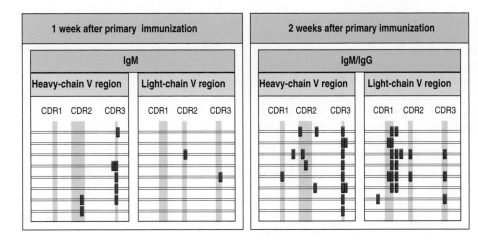

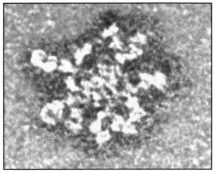

Figure 2.24 Somatic hypermutation introduces diversity into expressed immunoglobulin genes. Antibodies raised by immunization with the same antigen were collected one and two weeks after immunization and their amino-acid sequences were determined. Each line represents one antibody, and the red bars represent amino-acid positions that differ from the prototypic sequence. One week after primary immunization, most of the antibodies were IgM and showed very little sequence variation in the V region. Two weeks after immunization, both IgG and IgM were present and all six CDRs were affected.

be preferentially selected to mature into antibody-secreting cells. As the immune response proceeds, therefore, antibodies of progressively higher affinity for the immunizing antigen are produced. This phenomenon is called **affinity maturation**.

2-14 Isotype switching produces immunoglobulins with different C regions but identical antigen specificities

IgM is the first antibody produced in an immune response. Unlike IgG, IgM is secreted as a circular pentamer of Y-shaped immunoglobulin monomers (Figure 2.25). Because of its ten antigen-binding sites, IgM binds strongly to the surface of pathogens with multiple repetitive epitopes, but it is limited in the effector mechanisms that it uses to clear antigen from the body. Antibodies with other effector functions are produced by the process of **isotype switching** or **class switching**, in which a further DNA recombination event enables the rearranged V-region coding sequence to be used with other heavy-chain C genes.

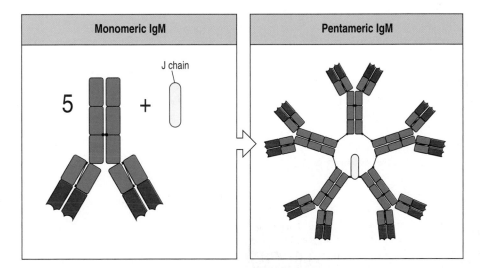

Figure 2.25 IgM is secreted as a pentamer of immuno-globulin monomers. The left two panels show schematic diagrams of the IgM monomer and pentamer. The IgM pentamer is held together by a polypeptide called the J chain, for joining chain (not to be confused with a J segment). The monomers are crosslinked by disulfide bonds to each other and to the J chain. The right panel shows an electron micrograph of an IgM

pentamer, showing the arrangement of the monomers in a flat disc. The lack of a hinge region in the IgM monomer makes the molecule less flexible than, say, IgG, but this is compensated for by the large number of binding sites. Given a pathogen with multiple identical epitopes on its surface, IgM is likely to be able to bind to it with several of its binding sites simultaneously. Photograph ($\times$ 900,000) courtesy of K.H. Roux and J.M. Schiff.

Isotype switching is accomplished by a recombination within the cluster of C genes that excises the previously expressed C gene and brings a different one into juxtaposition with the assembled V-region sequence. Thus the antigen specificity of the antibody remains unchanged, even though its isotype changes. Flanking the 5′ side of each C gene, with the exception of the δ gene, are highly repetitive sequences that mediate recombination. They are called **switch sequences** or **switch regions** (Figure 2.26). In switching the heavy-chain isotype, the switch region that flanks the μ gene interacts with a switch region flanking one of the other C genes. This interaction permits recombination in which the μ, δ, and any other intervening C genes are excised as a circular DNA molecule, bringing the V region into juxtaposition with the new C gene. The mRNA for the new

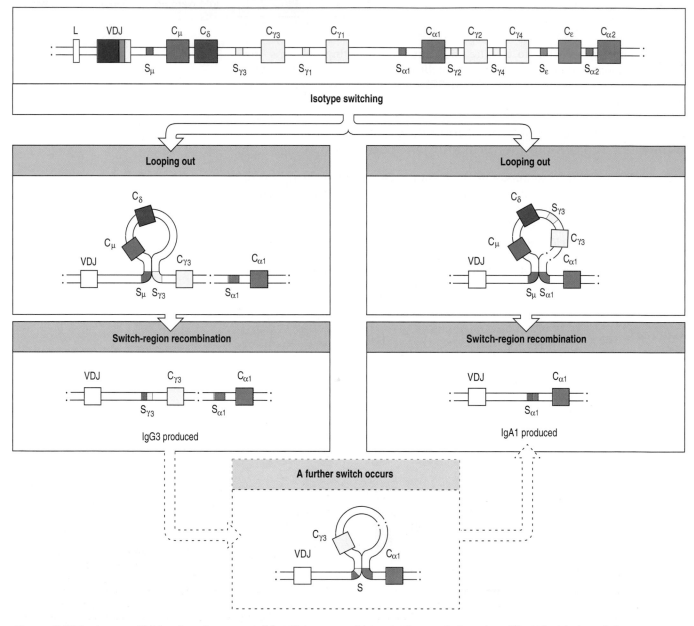

Figure 2.26 Isotype switching involves recombination between specific switch regions. Repetitive DNA sequences are found to the 5′ side of each of the heavy-chain C genes, with the exception of the δ gene. Switching occurs by recombination between these switch regions (S), with deletion of the intervening DNA. The initial switching event takes place from the μ switch region; switching from other isotypes can take place subsequently.

immunoglobulin is then produced as described in Section 2-9. Recombination can take place between the μ switch region and that of any other isotype. Sequential switching can also occur, for example from μ to γ_3 to α_1. Isotype switching only occurs during an active immune response and the patterns of isotype switching are regulated by cytokines secreted by antigen-activated T cells.

2-15 Antibodies with different constant regions have different effector functions

In humans and other mammals the five classes of immunoglobulin are IgA, IgD, IgE, IgG, and IgM (see Figure 2.4). Certain of the classes are divided further into subclasses, which differ in both nomenclature and properties between species. In humans, IgA is divided into two subclasses (IgA1 and IgA2), whereas IgG is divided into four subclasses (IgG1, IgG2, IgG3, and IgG4), which are numbered according to their relative abundance in plasma, IgG1 being the most abundant. The heavy chains of the human IgA subclasses are designated α1 and α2 and the heavy chains of the human IgG subclasses by γ1, γ2, γ3, and γ4. The α, δ, and γ heavy-chain constant regions are made up of three constant domains, whereas the μ and ε heavy chains have four (see Figure 2.4). Each C domain is encoded by a separate exon in the relevant C gene. Additional exons are used to encode the hinge region and the carboxy terminus, depending on the isotype. The physical properties of the human immunoglobulin isotypes are shown in Figure 2.27.

Antibodies aid the clearance of pathogens from the body in various ways. **Neutralizing** antibodies directly inactivate a pathogen or a toxin and prevent it from interacting with human cells. Neutralizing antibodies against viruses, for example, bind to a site on the virus that is normally used to gain entry to cells. Another function of antibodies is **opsonization**, a term used to describe the coating of pathogens with an immune-system protein (see Figure 1.23, p. 24). The common **opsonins** are antibodies and complement proteins. Opsonized pathogens are more efficiently ingested by phagocytes, which have receptors for the Fc region of some antibodies and for certain complement proteins. Complement activation by antibodies bound to a pathogen's surface can also lead to the direct lysis of the bacterium by complement.

IgM is the first antibody produced in an immune response against a pathogen. It is made principally by plasma cells resident in lymph nodes, spleen, and bone marrow and circulates in blood and lymph. On initiation of an immune response, most of the antibodies that bind the antigen will be of low affinity and the multiple antigen-binding sites of IgM are needed if enough antibody is to bind sufficiently strongly to a microorganism to be of any use. When bound to antigen, sites exposed in the constant region of IgM initiate reactions with complement, which

	Immunoglobulin class or subclass								
	IgM	IgD	IgG1	IgG2	IgG3	IgG4	IgA1	IgA2	IgE
Heavy chain	μ	δ	γ_1	γ_2	γ_3	γ_4	α_1	α_2	ε
Molecular weight (kDa)	970	184	146	146	165	146	160	160	188
Serum level (mean adult mg ml^{-1})	1.5	0.03	9	3	1	0.5	2.0	0.5	5×10^{-5}
Half-life in serum (days)	10	3	21	20	7	21	6	6	2

Figure 2.27 The physical properties of the human immunoglobulin isotypes. The molecular weight given for IgM is that of the pentamer (see Figure 2.25) the predominant form in serum. The molecular weight given for IgA is that of the monomer. Large amounts of IgA are also produced in the form of dimers, which is the form found in secretions.

can kill microorganisms directly or facilitate their phagocytosis. Because somatic hypermutation leads to antibodies of increased affinity for the antigen, two antigen-binding sites are then sufficient to produce strong binding; by switching isotype, different effector functions can be brought into play while preserving antigen specificity. Synthesis of IgM then gives way to synthesis of IgG.

IgG is the most abundant antibody in the internal body fluids, including blood and lymph. Like IgM, it is made principally in the lymph nodes, spleen, and bone marrow and circulates in lymph and blood. IgG is smaller and more flexible than IgM, properties that give it easier access to antigens in the extracellular spaces of damaged and infected tissues. The flexibility of the hinge region in IgG enables the two Fab arms to move relative to each other. This enables both the antigen-binding sites to bind to repeated epitopes on the surfaces of pathogens. IgG antibodies can also implement more effector functions than IgM (Figure 2.28). Once they have bound an antigen, the IgG1 and IgG3 subclasses can directly recruit phagocytic cells to ingest the antigen:antibody complex, as well as activating the complement system. During pregnancy, IgG antibodies can be transferred across the placenta providing the fetus with protective antibodies from the mother in advance of possible infection.

Monomeric IgA is made by plasma cells in lymph nodes, spleen and bone marrow and is secreted into the bloodstream. Dimeric IgA is made in the lymphoid tissues underlying mucosal surfaces and is the antibody that is secreted into the lumen of the gut; it is also the principal antibody of other secretions, including milk, saliva, sweat, and tears. The mucosal surface of the gastrointestinal tract provides an extensive surface of contact between the human body and the environment, and the transport processes involved in the uptake of food make it vulnerable to infection. More IgA is made than any other isotype. Some of this is against the resident microorganisms that colonize mucosal surfaces, keeping their population in check. Dimeric IgA includes one copy of the same J chain found in IgM (Figure 2.29).

Function	IgM	IgD	IgG1	IgG2	IgG3	IgG4	IgA	IgE
Neutralization	+	–	++	++	++	++	++	–
Opsonization	–	–	+++	*	++	+	+	–
Sensitization for killing by NK cells	–	–	++	–	++	–	–	–
Sensitization of mast cells	–	–	+	–	+	–	–	+++
Activation of complement system	+++	–	++	+	+++	–	+	–

Property	IgM	IgD	IgG1	IgG2	IgG3	IgG4	IgA	IgE
Transport across epithelium	+	–	–	–	–	–	+++ (dimer)	–
Transport across placenta	–	–	+++	+	++	+/–	–	–
Diffusion into extravascular sites	+/–	–	+++	+++	+++	+++	++ (monomer)	+
Mean serum level (mg ml^{-1})	1.5	0.03	9	3	1	0.5	2.1	5x10^{-5}

Figure 2.28 Each human immunoglobulin isotype has specialized functions and distinct properties. The major effector functions of each isotype (+++) are shaded in dark red; lesser functions (++) are shown in dark pink, and very minor functions (+) in pale pink. Other properties are similarly marked, with mean serum concentrations shown in the bottom row. Opsonization refers to the ability of the antibody itself to facilitate phagocytosis. Antibodies that activate the complement system indirectly cause opsonization via complement. *IgG2 acts as an opsonin in the presence of a genetic variant of its phagocyte Fc receptor, found in about 50% of Caucasians.

The IgE class of antibodies is highly specialized towards the activation of mast cells, which are present in epithelial tissues. IgE bound tightly to mast cells triggers strong inflammatory reactions in the presence of its antigen, and is thought to be involved in the expulsion of worms and other parasites. In medical practice, however, the major impact of IgE is in the allergies that result when it is produced against otherwise harmless antigens.

IgE binds tightly to a high-affinity IgE receptor carried by tissue mast cells and the basophils of the blood. Even though IgE is present in the body at minute concentrations in both normal and allergic individuals, the affinity of this receptor for IgE is so high that it is constantly occupied with IgE. Binding of antigen to receptor-bound IgE induces mast cells to release stored histamine and other activators, which recruit cells and molecules of the immune system to local sites of trauma, causing inflammation.

Summary

The constant region of the heavy chain defines classes of immunoglobulin that serve different functions in immunity. On binding of antigen to the B-cell receptor, the cell receives signals that cause both cell division and proliferation and the initiation of mechanisms that diversify the antigen specificity and effector function of the antibody. Antigen specificity is diversified by somatic hypermutation, which introduces single nucleotide substitutions randomly throughout the V-region sequence. This leads to the selection of B cells that make higher-affinity antibodies. With higher-affinity antibodies, the dependence on pentameric IgM for strong binding to antigen is relaxed and switching of immunoglobulin isotype from IgM/IgD to IgG, IgE, or IgA by further somatic recombination events produces antibodies of different isotype and effector function while preserving antigen specificity. Cell-surface immunoglobulins and secreted antibodies of the same class are produced from the same rearranged heavy-chain gene by different RNA-processing events.

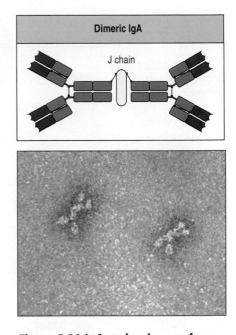

Figure 2.29 IgA molecules can form dimers. In mucosal lymphoid tissue IgA is synthesized as a dimer in association with the same J chain found in pentameric IgM. In dimeric IgA, the monomers have disulfide bonds to the J chain but not to each other. The bottom panel shows an electron micrograph of dimeric IgA. Photograph ($\times$ 900,000) courtesy of K.H. Roux and J.M. Schiff.

Summary of Chapter 2

The principal function of B lymphocytes is to produce antibodies, which are secreted immunoglobulins that bind tightly to infectious agents and tag them for destruction or elimination. Each antibody is highly specific for its corresponding antigen, while the antibody repertoire of each person is enormous, being composed of many millions of different antibodies able to bind a wide variety of different antigens. Antibodies can also be divided into five different effector classes—IgM, IgG, IgD, IgA, and IgE—which have different functions in the immune response. This chapter has provided an overview of the structure and function of the antibody molecule and of the unusual genetic mechanisms that create this diversity in specificity and effector function. Within an antibody molecule, the variable regions that bind antigen are physically separated from the constant region that interacts with effector molecules and cells of the immune system, such as complement, phagocytes, and other leukocytes. Antigen binding is the property of the paired V domains of the heavy and light chains, which can form an almost unlimited number of different binding sites with structural complementarity to a vast range of molecules. The immunoglobulin genes (heavy-chain, κ light chain, and λ light chain) are expressed only in B cells, and their expression involves an unusual process of DNA rearrangement in which somatic DNA recombination assembles a variable-region coding sequence from sets of gene segments that are present in the unrearranged gene. The random selection of gene segments for assembly creates much of the collective diversity of antigen-binding sites. The recombination reactions involved in the rearrangements are imprecise and create

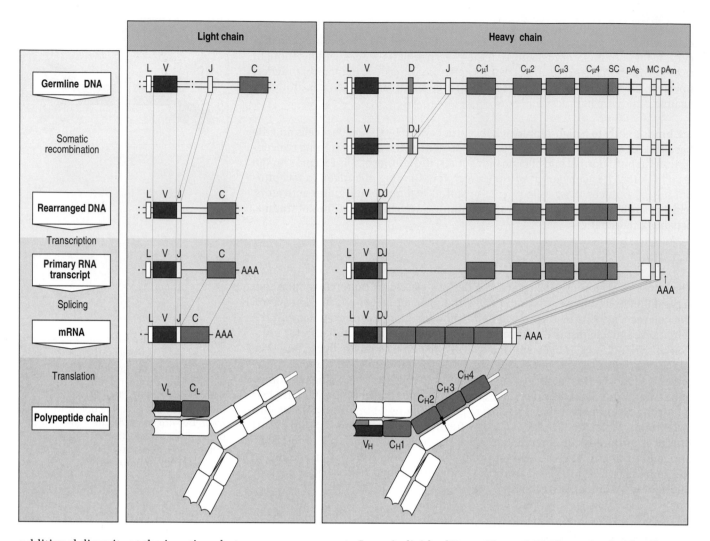

additional diversity at the junctions between gene segments. In an individual B cell, only one rearranged heavy-chain gene and one rearranged light-chain gene become functional, ensuring that each B cell expresses immunoglobulin of a single specificity. On mature B cells, membrane-bound immunoglobulin functions as the specific receptor for antigen; on encounter with antigen the B cell is stimulated to proliferate and differentiate into plasma cells that secrete antibody of the same specificity as the membrane-bound immunoglobulin. This ensures that an immune response is directed only against the invading pathogen or immunizing antigen. On stimulation of a B cell with its specific antigen, a mutational mechanism is also turned on that introduces point mutations into the rearranged variable-region DNA. Thus the diversity of antibodies is in part due to inherited variation that is encoded in the genome and in part to non-inherited diversity that develops in B cells during an individual's lifetime. The first antibody produced after encounter with antigen is always IgM (Figure 2.30). As the immune response proceeds, the process of isotype switching then transfers the antigen specificity onto different constant regions to produce antibody molecules with different effector functions—IgG, IgA, and IgE.

Figure 2.30 Biosynthesis of cell-surface IgM in B cells. Before immunoglobulin light- (center panel) and heavy- (right panel) chain genes can be expressed, rearrangements of gene segments are needed to produce exons encoding the V regions. Once this has been achieved the genes are transcribed to give primary transcripts containing both exons and introns. The latter are spliced out to produce mRNAs that are translated to give κ or λ light chains and μ heavy chains which assemble inside the cell and are expressed as membrane IgM at the cell surface. The main stages in the biosynthesis of the heavy and light chains are shown in the panel on the left.

Antigen Recognition by T Lymphocytes

3

In Chapter 2 we saw that the function of B lymphocytes is to make immunoglobulins, a family of proteins with variable binding sites for antigen. T lymphocytes are a related lineage of antigen-specific cells that perform complementary functions to those of B cells in the adaptive immune response (see Chapter 1). Whereas the sole function of B cells is to produce secreted antibodies, T cells have more diverse roles, all of which involve interactions with other cells. B cells and T cells use similar means of recognizing antigen, but because of the distinct functions of B and T cells there are important differences between them, particularly in the type of antigen that they recognize.

In this chapter we shall consider the antigens that stimulate T cells and the receptors that T cells use to recognize them. In the first part of the chapter we shall describe the structure of the T-cell antigen receptor and the mechanisms that generate its antigen specificity and diversity. The antigen receptor on T cells is commonly referred to simply as the **T-cell receptor**. T-cell receptors have much in common with the immunoglobulins. They have a similar structure, are produced as a result of gene rearrangement, and are highly variable and diverse in their antigen specificity. Like B cells, each clone of T cells expresses a single species of antigen receptor and thus different clones possess different and unique antigen specificities. This clonal distribution of diverse T-cell receptors is produced by genetic mechanisms similar to those that generate immunoglobulins during B-cell development.

Despite the fundamental similarity in immunoglobulin and T-cell receptor structures, the antigens recognized by T cells are quite distinct from those recognized by immunoglobulins. Immunoglobulins bind epitopes on a wide range of intact molecules, such as proteins, carbohydrates, and lipids. These kinds of molecule are present on the surfaces of bacteria, viruses and parasites and also on soluble toxins. T-cell receptors, in contrast, can recognize and bind to only one type of antigen, which must be presented to them on the surface of another human cell.

T-cell receptors recognize peptide antigens that are bound to specialized antigen-presenting glycoproteins called **MHC molecules**, which are expressed on almost all the cells of the body. There are an exceptionally large number of genetically determined variants of the MHC molecules in the human population, and differences between individuals in the MHC molecules that they possess are the primary cause of graft rejection and graft-versus-host disease in clinical transplantation. Because of this clinical effect, these antigen-presenting molecules were first discovered as determinants of tissue incompatibility in transplantation long before their function in antigen presentation was known. The genes that encode them are genetically linked in a chromosomal region called the **major histocompatibility complex** or **MHC**; the glycoproteins have therefore become known as MHC molecules.

In the second part of the chapter we shall consider the complex of peptide antigen and MHC molecule that is recognized by the T-cell receptor and we trace the pathways by which peptides are derived from the proteins of infecting microorganisms and become bound to MHC molecules. The third part of the chapter will deal with the major histocompatibility complex itself and the immunological implications of the exceptional genetic polymorphism of the MHC genes.

T-cell receptor diversity

The T-cell receptor is a membrane-bound glycoprotein that closely resembles a single antigen-binding arm of an immunoglobulin molecule. It is composed of two different polypeptide chains and has one antigen-binding site. T-cell receptors are always membrane bound and there is no secreted form as there is for immunoglobulins. Like immunoglobulins, each chain has a variable region, which binds antigen, and a constant region. During T-cell development, gene rearrangement produces sequence variability in the variable regions of the T-cell receptor by the same mechanisms that are used by B cells to produce the variable regions of immunoglobulins. However, after the T cell is stimulated with antigen, there is no further mutation in the antigen-binding site and there is no switching of constant-region isotype as occurs for immunoglobulins. These differences correlate with the fact that T-cell receptors are used only as receptors to recognize antigen, whereas immunoglobulins serve as both recognition and effector molecules.

3-1 The T-cell receptor resembles a membrane-associated Fab fragment of immunoglobulin

A T-cell receptor consists of two different polypeptide chains, termed the **T-cell receptor α chain (TCRα)** and the **T-cell receptor β chain (TCRβ)**. The genes encoding the α and β chains have similar germline organization to the genes encoding immunoglobulin heavy- and light-chain genes in that they are made up of sets of gene segments that must be rearranged to form a functional gene. As a

Figure 3.1 The T-cell receptor resembles a membrane-bound Fab fragment. Comparison of the T-cell receptor and an IgG antibody molecule. The T-cell receptor is a heterodimer composed of an α chain of 40–50 kDa and a β chain of 35–46 kDa in size. The extracellular portion of each chain consists of two immunoglobulin-like domains: the domain nearest to the membrane is a constant (C) domain (or C region) and the domain furthest from the membrane is a variable (V) domain (or V region). Both the α and β chains span the cell membrane and have very short cytoplasmic tails. The three-dimensional structure formed by the four immunoglobulin-like domains of the T-cell receptor resembles that of the antigen-binding Fab fragment of antibody.

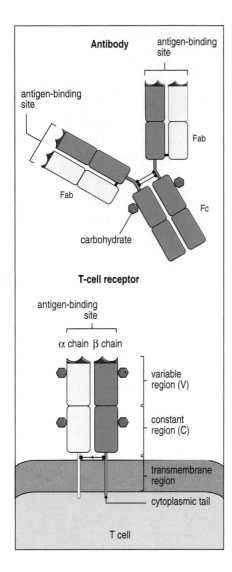

consequence of gene rearrangements that occur as part of T-cell development each mature T cell expresses one functional α chain and one functional β chain, which together define a unique T-cell receptor molecule. Within the population of T cells possessed by a healthy human being are many millions of different T-cell receptors, each of which defines a clone of T cells and a single antigen-binding specificity.

Comparison of the amino-acid sequences of the T-cell receptor α and β chains from different T-cell clones shows that they are organized into variable regions (V regions) and constant regions (C regions) like those found in immunoglobulin chains. The α and β chains are folded into discrete protein domains resembling those in immunoglobulin chains. Each chain consists of an amino-terminal V domain, followed by a C domain, and then a membrane-anchoring domain. The antigen-recognition site of T-cell receptors is formed from the V_α and V_β domains and is the most variable part of the molecule, as in the immunoglobulins. The three-dimensional structure of the four extracellular domains of the T-cell receptor is very similar to that of the antigen-binding Fab fragment of IgG (Figure 3.1).

Comparison of the amino-acid sequences of V domains from different clones of T cells shows that sequence variation in α and β chains is clustered into regions of hypervariability, which correspond to loops of the polypeptide chain at the end of the domain furthest from the T-cell membrane. These loops form the binding site for antigen and are termed complementarity-determining regions (CDRs), as in the immunoglobulins. The T-cell receptor α- and β-chain V domains each have three CDR loops, called CDR1, CDR2, and CDR3 (Figure 3.2).

Immunoglobulins possess two or more binding sites for antigen; this strengthens the interactions of soluble antibody with the repetitive antigens found on the surfaces of microorganisms. T-cell receptors possess a single binding site for antigen and are used only as cell-surface receptors for antigen, never as soluble antigen-binding molecules. Antigen binding to T-cell receptors occurs always in the context of two opposing cell surfaces, where multiple copies of the T-cell receptor bind to multiple copies of the antigen:MHC complex on the opposing cell, thus achieving multipoint attachment.

Figure 3.2 Three-dimensional structure of the T-cell receptor showing the antigen-binding CDR loops. The ribbon diagram shows the α chain (in magenta) and the β chain (in blue). The receptor is viewed from the side as it would sit on a cell surface with the highly variable complementarity-determining region (CDR) loops, which bind the peptide:MHC molecule ligand, arrayed across its relatively flat top surface. The CDR loops are numbered 1–3 for each chain. Courtesy of I.A. Wilson.

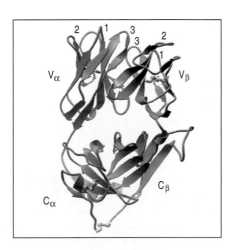

3-2 T-cell receptor diversity is generated by gene rearrangement

In Chapter 2 we divided the mechanisms that generate immunoglobulin diversity into two categories: those operating before the B cell is stimulated with specific antigen and those operating afterwards. In the first category were the gene rearrangements that generate the V-region sequence, whereas in the second category were changes in mRNA splicing that produce a secreted immunoglobulin, C-region DNA rearrangements that switch the heavy-chain isotype, and somatic hypermutation of the V-region gene to give antibodies of higher affinity. In T cells the mechanisms that generate diversity before antigen stimulation are essentially the same as those in B cells, but after antigen stimulation the picture is quite different: whereas immunoglobulin genes continue to diversify, the genes encoding T-cell receptors remain unchanged.

This fundamental difference reflects the fact that the T-cell receptor is used only for the recognition of antigen and not for the generation of effector functions, which is handled by other T-cell molecules. By contrast, the effector functions of B cells are solely dependent on secreted antibodies, whose different C-region isotypes trigger different effector mechanisms. The diversification of an antibody that occurs subsequently to antigen stimulation is directed towards optimizing both its antigen-binding and effector functions.

The human T-cell α-chain locus is on chromosome 14 and the β-chain locus on chromosome 7. The organization of the gene segments encoding T-cell receptor α and β chains is essentially like that of the immunoglobulin gene segments (Figure 3.3). The main difference is the simplicity of the T-cell receptor C region: there is only one C_α gene, and although there are two C_β genes, no functional

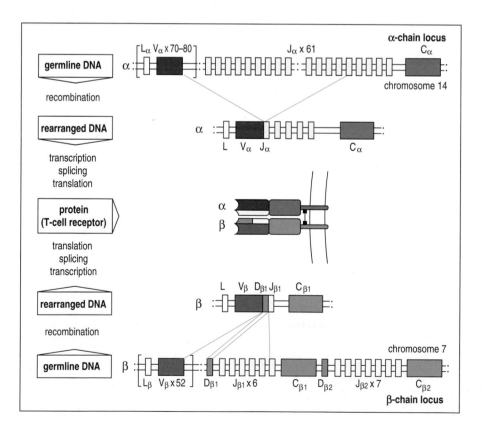

Figure 3.3 Organization and rearrangement of the T-cell receptor genes. The top and bottom rows of the figure show the germline arrangement of the variable (V), diversity (D), joining (J), and constant (C) gene segments at the T-cell receptor α- and β-chain loci. During T-cell development, a V-region sequence for each chain is assembled by DNA recombination. For the α chain (top), a V_α gene segment rearranges to a J_α gene segment to create a functional exon encoding the V domain. For the β chain (bottom), rearrangement of a V_β, a D_β, and a J_β gene segment creates the functional V-domain exon. The assembled genes are transcribed and spliced to produce mRNA (not shown) encoding the α and β chains. Exons encoding the membrane spanning regions are not shown. L, leader sequence.

distinction between them is known. The T-cell receptor α-chain locus is otherwise more similar to an immunoglobulin light-chain locus, containing sets of V and J gene segments only; the β-chain locus is similar to an immunoglobulin heavy-chain locus, containing D gene segments in addition to V and J gene segments. The V domain of the T-cell receptor α chain is thus encoded by a V gene segment and a J gene segment; that of the β chain is encoded by a D gene segment in addition to a V and a J gene segment.

T-cell receptor gene rearrangement occurs during T-cell development in the thymus, and the mechanisms involved are similar to those outlined in Chapter 2 for the immunoglobulin genes. In the α-chain gene, a V gene segment is joined to a J gene segment by somatic DNA recombination to make the V-region sequence; in the β-chain gene, recombination first joins a D and a J gene segment, which are then joined to a V gene segment (see Figure 3.3). The T-cell receptor gene segments are flanked by recombination signal sequences similar to those found in immunoglobulin genes and the same enzymes are involved in the recombination process in both cell types (see Section 2-8, p. 43). During recombination, additional, non-templated P and N nucleotides (see Section 2-8, p. 43) are inserted in the junctions between the V, D, and J gene segments of the T-cell receptor β-chain coding sequence and between the V and J gene segments of the α-chain sequence. These mechanisms contribute junctional diversity to T-cell receptor α and β chains. In people with rare defects in enzymes essential for recombination, the development of functional B and T lymphocytes is equally impaired and a disease called severe combined immunodeficiency results.

After gene rearrangement functional α- and β-chain genes consist of exons encoding the leader peptide, the V region and the C region, as well as the membrane-spanning region. The exons are separated by introns, which in the case of the intron between the V-domain and the C-domain exon may contain unrearranged gene segments (see the α-chain gene in Figure 3.3). Upon transcription, the primary RNA transcript is spliced to remove the introns and is processed to give mRNA. Translation of the α-chain and β-chain mRNA produces α and β chains, respectively. Like all proteins destined for the cell membrane, newly synthesized α and β chains enter the endoplasmic reticulum. There they pair to form the α:β T-cell receptor (see Figure 3.3).

3-3 Expression of the T-cell receptor on the cell surface requires association with additional proteins

T-cell receptors are diverse and specific receptors for antigen. By themselves, however, heterodimers of α and β chains are unable to leave the endoplasmic reticulum and be expressed on the T-cell surface. In this respect the T-cell receptor resembles immunoglobulin. Before leaving the endoplasmic reticulum an α and β chain heterodimer associates with four invariant membrane proteins. Three of the proteins are encoded by closely linked genes of human chromosome 11 and are homologous to each other: these proteins are collectively termed the **CD3 complex** and individually called CD3γ, CD3δ and CD3ε. The fourth protein is known as the ζ chain and is encoded by a gene on human chromosome 1.

At the cell surface the CD3 proteins and the ζ chain remain in stable association with the T-cell receptor and form the functional **T-cell receptor complex** (Figure 3.4). In this complex the CD3 proteins and the ζ chain transduce signals to the cell's interior after antigen has been recognized by the α and β chain heterodimer. Correlating with these functional differences the cytoplasmic domains of the CD3 proteins and the ζ chain contain sequences that associate with intracellular signaling molecules. By contrast the T-cell receptor α and β

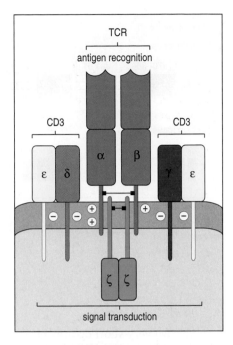

Figure 3.4 Polypeptide composition of the T-cell receptor complex. The functional antigen receptor on the surface of T cells is composed of eight polypeptides and is called the T-cell receptor complex. The α and β chains bind antigen and form the core T-cell receptor (TCR). They associate with one copy each of CD3γ, and CD3δ and two copies each of CD3ε and the ζ chain. These associated invariant polypeptides are necessary for transport of newly synthesized TCR to the cell surface and for transduction of signals to the cell's interior after the TCR has bound antigen. The transmembrane domains of the α and β chains contain positively charged amino acids (+), which form strong electrostatic interactions with negatively charged amino acids (–) in the transmembrane regions of the CD3γ, δ, and ε chains.

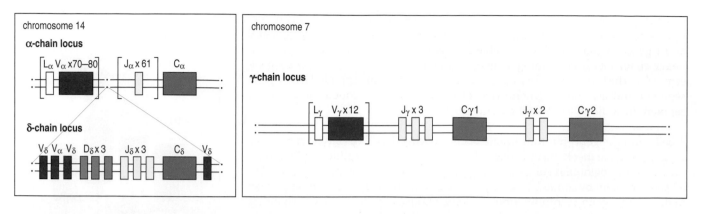

Figure 3.5 The organization of the human T-cell receptor γ- and δ-chain loci. The γ and δ loci, like the α and β loci, contain sets of V, D, J, and C gene segments. The δ locus is located within the α-chain locus on chromosome 14, lying between the clusters of V_α and J_α gene segments. There are at least three V_δ gene segments, three D_δ gene segments, and three J_δ gene segments, and a single C_δ gene segment. V_δ segments are interspersed amongst V_α and other gene segments. The γ locus, on chromosome 7, resembles the β locus, with two C gene segments each with its own set of J segments.

chains have very short cytoplamic tails which lack signaling function, as is also true for immunoglobulin chains. In people lacking functional CD3δ or CD3ε chains, transport of T-cell receptors to the cell surface is inefficient and thus their T cells express abnormally low numbers of receptors. As a consequence of both low T-cell receptor expression and impaired signal transduction these people suffer from immunodeficiency.

3-4 Some T cells bear an alternative form of T-cell receptor composed of γ and δ chains

There is a second type of T-cell receptor which is similar in overall structure to the α:β receptor but is formed of two different protein chains termed γ (not to be confused with CD3γ) and δ. The γ chain resembles the α chain, and the δ chain resembles the β chain. T cells express either α:β receptors or γ:δ receptors but never both. Those expressing α:β T-cell receptors are called **α:β T cells**, whereas those expressing γ:δ receptors are called **γ:δ T cells**. Cells with γ:δ receptors form a small subset of all T cells. Throughout the rest of this book, T cells will refer to α:β T cells and T-cell receptor will refer to the α:β T-cell receptor, unless specified otherwise.

The organization of the γ and δ loci resembles that of the β and α loci, but there are some important differences (Figure 3.5). The δ gene segments are situated within the α-chain locus on chromosome 14, between the V_α and J_α gene segments. This location means that DNA rearrangement within the α-chain locus inevitably results in the deletion and inactivation of the δ-chain locus. The human γ-chain locus is on chromosome 7. The γ- and δ-chain loci contain fewer V gene segments than the α- or β-chain loci, and so can in theory produce less diverse receptors, but for the γ chain this limitation is compensated for by an increase in junctional diversity.

Rearrangement at the γ and δ loci proceeds as for the α and β loci, with the exception that during δ-gene rearrangement two D segments can be incorporated into the final gene sequence. This increases the variability of the δ chain in two ways. First, the potential number of combinations of gene segments is increased. Second, extra N nucleotides can be added at the junction between the two D segments as well as at the VD and DJ junctions.

T cells bearing γ:δ receptors comprise about 1–5% of the T cells found in the circulation, but they can be the dominant T-cell population in epithelial tissue. The immune function of γ:δ T cells is less well defined than that of α:β T cells, as are the antigens to which these cells respond and the ligands that their receptors engage.

Summary

T cells recognize antigen through a cell-surface receptor known as the T-cell receptor, which has structural and functional similarities to the membrane-bound immunoglobulin that serves as the B-cell receptor for antigen. T-cell receptors are heterodimeric glycoproteins in which each polypeptide chain consists of a variable domain and a constant domain, similar to those found in immunoglobulin chains, and a membrane-spanning region. The three-dimensional structure of the extracellular domains of the T-cell receptor resembles that of the antigen-binding Fab fragment of IgG. There are two types of T-cell receptor: one made up of α and β chains and expressed by T cells whose function is understood, and another made up of γ and δ chains and carried by T cells whose function remains elusive. All four types of T-cell receptor chain are encoded by genes that resemble the immunoglobulin genes and require similar DNA rearrangements in order to be expressed. The critical difference between T-cell receptors and immunoglobulins is that T-cell receptors serve only as cell-surface receptors and are not secreted as soluble proteins with effector function; T cells use other molecules for effector function. This explains why the T-cell receptor has only a single binding site for antigen that does not change its affinity on encounter with antigen and a simple constant region that does not switch isotype. Expression of the T-cell receptor at the T-cell surface requires association with proteins of the CD3 complex. These proteins transmit signals to the interior of the cell when the T-cell receptor binds antigen.

Antigen processing and presentation

Unlike the immunoglobulins of B cells, which can recognize a wide range of molecules in their native form, a T-cell receptor can recognize antigen only in the form of a peptide bound to an MHC molecule on a human cell surface. This means that pathogen-derived proteins must be degraded into peptides to be recognized by T cells. This **antigen processing** occurs inside the body's own cells; the peptides are then assembled into peptide:MHC molecule complexes for display on the cell surface, where they can be recognized by T cells. The binding of a peptide antigen by an MHC molecule and its display at the cell surface is termed **antigen presentation** (Figure 3.6).

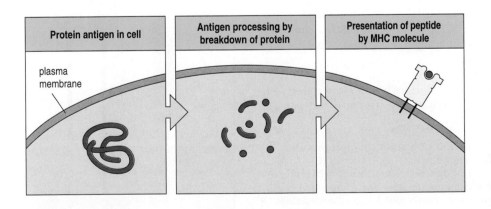

Figure 3.6 Antigen processing and presentation. The antigens recognized by T cells are peptides that arise from the breakdown of macromolecular structures, the unfolding of individual proteins, and their cleavage into short fragments. These events constitute antigen processing. For a T-cell receptor to recognize a peptide antigen, the peptide must be bound by an MHC molecule and displayed at the cell surface, a process called antigen presentation.

The microorganisms that infect the human body can be broadly divided into those that propagate within cells, such as viruses, and those, such as most bacteria, that live in the extracellular spaces. The T-cell population that fights infection is composed of two subpopulations, one of which is specialized to combat intracellular infections, and the other to combat extracellular sources of infection. The latter include, in addition to bacteria that live in extracellular spaces, virus particles free in the extracellular fluid. In this part of the chapter we shall consider how these two sources of antigen are distinguished by the immune system and processed by intracellular pathways, and how the appropriate type of T cell is activated.

3-5 Two classes of T cell are specialized to respond to intracellular and extracellular sources of infection

Circulating T cells fall into one or other of two mutually exclusive classes: one is defined by expression of the **CD4** glycoprotein on the cell surface and the other by expression of the **CD8** glycoprotein (Figure 3.7). These two classes of T cell have different functions and deal with different types of pathogen. **CD8 T cells** are cytotoxic and their main function is to kill cells that have become infected with a virus or some other intracellular pathogen. This response prevents the multiplication of the pathogen and further infection of healthy cells. The general function of **CD4 T cells** is to help other cells of the immune system to respond to extracellular sources of infection. Different aspects of this reponse are carried out by two subclasses of CD4 T cells—T_H1 and T_H2. The H in their names stands for 'helper' and CD4 T cells are often also known as **helper T cells**. **T_H2 cells** are involved mainly in stimulating B cells to make antibodies, which bind to extracellular bacteria and virus particles, whereas an important function of **T_H1 cells** is to activate tissue macrophages to take up extracellular pathogens by phagocytosis and then kill them.

The human immunodeficiency virus (HIV), which causes aquired immunodeficiency syndrome (AIDS), selectively infects CD4 T cells by exploiting the CD4 molecule as its receptor.

3-6 Two classes of MHC molecule present antigen to CD8 and CD4 T cells respectively

The MHC molecules are crucial in ensuring that the appropriate class of T cells is activated in response to a particular source of infection. There are two different classes of MHC molecule—**MHC class I** and **MHC class II**—and each presents peptides from one kind of antigen source to one type of T cell. MHC class I molecules present antigens of intracellular origin to CD8 T cells, whereas MHC class II molecules present antigens of extracellular origin to CD4 T cells. This correlation is achieved by specific molecular interactions between the CD8 glycoprotein and MHC class I molecules and between CD4 and MHC class II molecules. These occur when a T-cell receptor recognizes its specific peptide:MHC molecule ligand. Because of their close involvement in antigen recognition, the CD4 and CD8 molecules are called T-cell **co-receptors**.

Before we consider how the appropriate MHC molecules become associated with peptides from different sources, we shall look at the structure and general peptide-binding properties of MHC molecules.

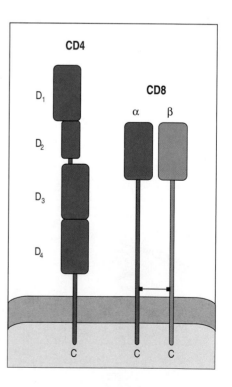

Figure 3.7 The structures of the CD4 and CD8 glycoproteins. CD4 has four extracellular immunoglobulin-like domains (D_1–D_4) with a hinge between the D_2 and D_3 domains. CD8 consists of an α and a β chain, which both have an immunoglobulin-like domain that is connected to the membrane-spanning region by an extended stalk. C denotes the carboxy terminus.

3-7 The two classes of MHC molecule have similar three-dimensional structures

MHC class I and class II molecules are membrane glycoproteins whose function is to bind peptide antigens and present them to T cells. Underlying this common function is a similar three-dimensional structure, which is formed in different ways in the two types of molecule.

The MHC class I molecule is made up of a transmembrane heavy chain, or α chain, which is non-covalently complexed with the small protein **β₂-microglobulin** (Figure 3.8). The heavy chain has three extracellular domains (α₁, α₂, and α₃). The

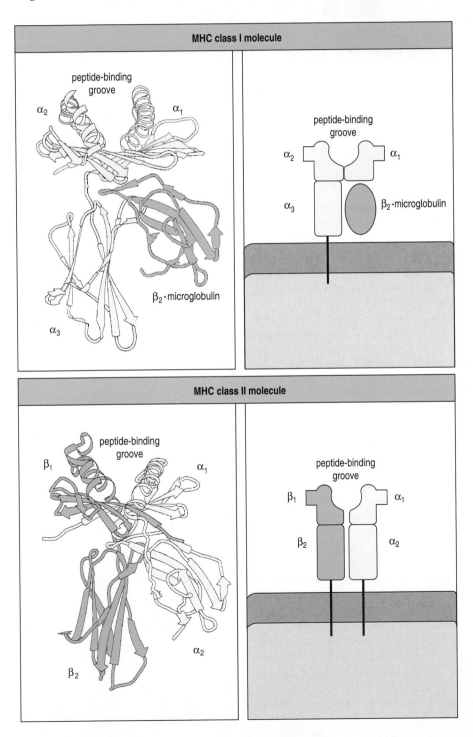

Figure 3.8 The structures of MHC class I and MHC class II molecules are variations on a theme. A class I MHC molecule is composed of one membrane-bound heavy (or α) chain and non-covalently bonded β₂-microglobulin. The heavy chain has three extracellular domains, of which the amino-terminal α₁ and α₂ domains resemble each other in structure and form the peptide-binding site. A class II MHC molecule is composed of two membrane-bound chains, an α chain (which is a different protein from MHC class I α) and a β chain. These have two extracellular domains each, the amino-terminal two (α₁ and β₁) resembling each other in structure and forming the peptide-binding site. The β₂ domain of MHC class II molecules should not be confused with the β₂-microglobulin of MHC class I molecules. The ribbon diagrams in the left panels trace the paths of the polypeptide backbone chains.

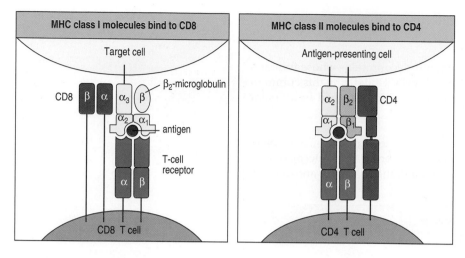

Figure 3.9 MHC class I molecules bind to CD8, and MHC class II molecules bind to CD4. The CD8 co-receptor binds to the α_3 domain of the MHC class I heavy chain, ensuring that MHC class I molecules present peptides only to CD8 T cells (left panel). In a complementary fashion, the CD4 co-receptor binds to the β_2 domain of MHC class II molecules, ensuring that peptides bound by MHC class II stimulate only CD4 T cells (right panel).

peptide-binding site is formed by the folding of α_1 and α_2, the domains farthest from the membrane, and is supported by the α_3 domain and the β_2-microglobulin. The MHC class I heavy chain is encoded by a gene in the MHC, whereas β_2-microglobulin is not.

By contrast, the MHC class II molecule consists of two transmembrane chains (α and β), each of which contributes one domain to the peptide-binding site and one immunoglobulin-like supporting domain (see Figure 3.8). Both of these chains are encoded by genes in the MHC. Thus both classes of MHC molecule have similar three-dimensional structures consisting of two pairs of extracellular domains, with the paired domains farthest from the membrane resembling each other and forming the peptide-binding site. In both types of MHC molecule, the domains supporting the peptide-binding domains are immunoglobulin-like domains: α_3 and β_2-microglobulin in MHC class I molecules, and α_2 and β_2 in MHC class II molecules.

The immunoglobulin-like domains of MHC class I and II molecules are not just a support for the peptide-binding site; they also provide binding sites for the CD4 and CD8 co-receptors. The sites on the MHC molecule that interact with the T-cell receptor and the co-receptor are separated, allowing the simultaneous engagement of both T-cell receptor and co-receptor by an MHC molecule (Figure 3.9).

3-8 MHC molecules bind a wide variety of peptides

MHC molecules have peptide-binding sites that are capable of binding peptides of many different amino-acid sequences. This **degenerate binding specificity** contrasts with the specificity of an immunoglobulin or a T-cell receptor for a single epitope. The peptide-binding site is a deep groove on the surface of the MHC molecule (Figure 3.10), within which a single peptide is held tightly by non-covalent bonds.

There are, however, constraints on the length and amino-acid sequence of peptides that are bound by MHC molecules. These are dictated by the structure of the peptide-binding groove, which differs between MHC class I and MHC class II molecules. The length of the peptides bound by MHC class I molecules is limited because the two ends of the peptide are grasped by pockets situated at the ends of the peptide-binding groove (Figure 3.11). The vast majority of peptides that bind MHC class I molecules are eight, nine, or ten amino acids in length; most are of nine amino acids. The differences in peptide length are accommodated by a slight kinking of the extended conformation of the bound peptide.

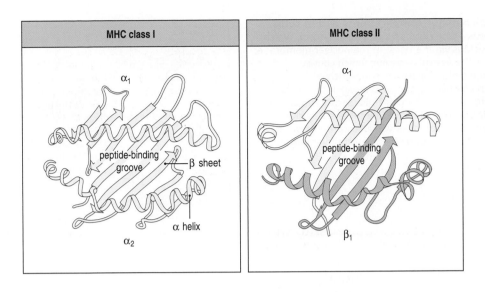

Figure 3.10 The peptide-binding groove of MHC class I and MHC class II molecules. The T-cell receptor's view of the peptide-binding groove is shown. In the MHC class I molecule (left panel) the groove is formed by the α_1 and α_2 domains of the MHC class I heavy chain; in the MHC class II molecule (right panel) it is formed by the α_1 domain of the class II α chain (yellow) and the β_1 domain of the class II β chain (green).

Features that are common to all peptides—the amino terminus, the carboxy terminus and the peptide backbone—interact with binding pockets present in all MHC class I molecules, and these form the basis for all peptide–MHC class I interactions. Most peptides bound by class I MHC molecules also have a hydrophobic or basic residue at the carboxy terminus, which corresponds to a complementary pocket present in the binding groove of MHC class I molecules.

In MHC class II molecules, the two ends of the peptide are not pinned down into pockets at each end of the peptide-binding groove (see Figure 3.11). As a consequence, they can extend out at each end of the groove and so peptides bound by MHC class II molecules are both longer and more variable in length than peptides bound by MHC class I. Peptides that bind to MHC class II molecules are usually 13–25 amino acids in length and some are much longer.

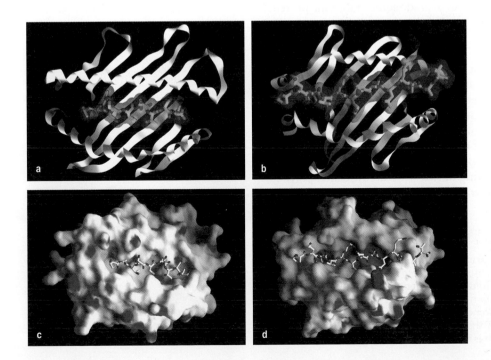

Figure 3.11 Bound peptides are tightly embedded within the MHC structure. Three-dimensional structures of complexes of defined synthetic peptides bound to recombinant MHC molecules. In MHC class I molecules (panels a and c) the peptide is bound in an extended conformation with both ends pinned in pockets at the end of the binding groove. In MHC class II molecules (panels b and d) the peptide is bound in a similarly extended conformation but the ends of the peptide are not pinned and can therefore extend from the ends of the groove. Panels a and b show ribbon diagrams of the polypeptide backbone of the MHC peptide-binding groove (white) in which the peptide is shown in red. In panels c and d a space-filling model of the peptide-binding groove is shown.

Figure 3.12 There are two major compartments within cells, separated by membranes. One compartment is the cytosol, which is contiguous with the nucleus via the pores in the nuclear membrane. The other compartment is the vesicular system, which consists of the endoplasmic reticulum, the Golgi apparatus, endocytic vesicles, lysosomes, and other intracellular vesicles. The vesicular system is effectively contiguous with the extracellular fluid. Secretory vesicles bud off from the endoplasmic reticulum and by successive fusion and budding with the Golgi membranes move vesicular contents out of the cell. By contrast endocytic vesicles take up extracellular material into the vesicular system.

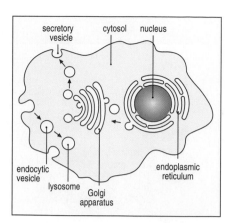

3-9 Peptides generated in the cytosol are transported into the endoplasmic reticulum, where they bind MHC class I molecules

The peptide antigens that are bound and presented by MHC molecules are generated inside cells of the body by the breakdown of larger protein antigens. Proteins derived from 'intracellular' and 'extracellular' antigens are present in different intracellular compartments (Figure 3.12). They are processed into peptides by two intracellular pathways of degradation, and bind to the two classes of MHC molecule in separate intracellular compartments. Peptides derived from the degradation of intracellular pathogens are formed in the cytosol and delivered to the endoplasmic reticulum. This is where MHC class I molecules bind peptides. In contrast, extracellular microorganisms and proteins are taken up by cells via phagocytosis and endocytosis and are degraded in the lysosomes and other vesicles of the endocytic pathways. It is in these cellular compartments that MHC class II molecules bind peptides. In this way, the class of the MHC molecule labels the peptide as being extracellular or intracellular in origin. In this section we shall look at the processing pathway for intracellular antigens.

When viruses infect human cells they exploit the cells' ribosomes to synthesize viral proteins, which are therefore present in the cytosol before being assembled into viral particles. In response, the infected cell uses its normal processes of breakdown and turnover of cellular proteins to degrade some of the viral proteins into peptides that can be bound by MHC class I molecules and presented to CD8 T cells.

Proteins in the cytosol are degraded by a large barrel-shaped protein complex called the **proteasome**, which has several different proteolytic activities. It consists of 28 polypeptide subunits, each of 20–30 kilodaltons molecular weight. Once formed, the antigenic peptides are transported out of the cytosol and into the endoplasmic reticulum (Figure 3.13). Transport of peptides across the endoplasmic reticulum membrane is accomplished by a protein, called the **transporter associated with antigen processing** (**TAP**), embedded in the membrane. TAP is a heterodimer consisting of two structurally related polypeptide chains, TAP-1 and TAP-2. Peptide transport by TAP is dependent on the

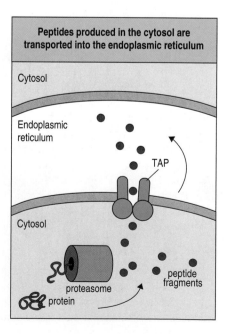

Peptides produced in the cytosol are transported into the endoplasmic reticulum

Figure 3.13 Formation and transport of peptides that bind to MHC class I molecules. In all cells, proteasomes degrade cellular proteins that are poorly folded, damaged, or unwanted. When a cell becomes infected, pathogen-derived proteins in the cytosol are also degraded by the proteasome. Peptides are transported from the cytosol into the lumen of the endoplasmic reticulum by the transporter protein TAP, which is in the endoplasmic reticulum membrane.

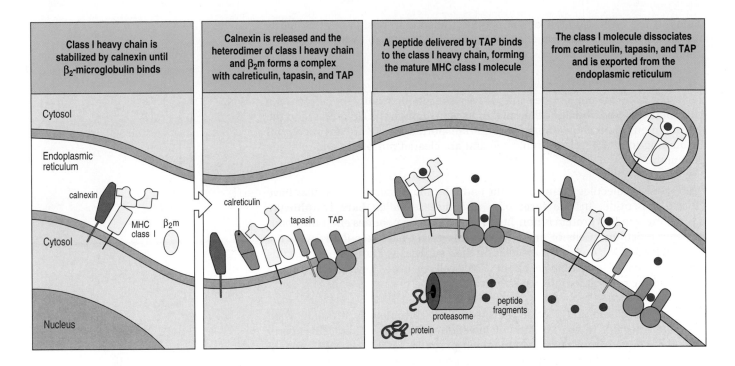

| Class I heavy chain is stabilized by calnexin until β₂-microglobulin binds | Calnexin is released and the heterodimer of class I heavy chain and β₂m forms a complex with calreticulin, tapasin, and TAP | A peptide delivered by TAP binds to the class I heavy chain, forming the mature MHC class I molecule | The class I molecule dissociates from calreticulin, tapasin, and TAP and is exported from the endoplasmic reticulum |

Figure 3.14 Chaperone proteins aid the assembly and peptide loading of MHC class I molecules in the endoplasmic reticulum. MHC class I heavy chains assemble in the endoplasmic reticulum with the membrane-bound protein calnexin. When this complex binds β₂-microglobulin (β₂m) the partly folded MHC class I molecule is released from calnexin and then associates with the TAP-1 subunit of TAP by interacting with the TAP-associated protein tapasin and the chaperone protein calreticulin. The MHC class I molecule is retained in the endoplasmic reticulum until it binds a peptide, which completes the folding of the molecule. The peptide:MHC class I molecule complex is then released from tapasin and calreticulin, leaves the endoplasmic reticulum, and is transported to the cell surface.

binding and hydrolysis of ATP, properties shared with the other transporters in this family. The types of peptide that are preferentially transported by TAP are similar to those that bind MHC class I molecules: they are eight or more amino acids long and have either hydrophobic or basic residues at the carboxy terminus.

Newly synthesized MHC class I heavy chains and β₂-microglobulin are also translocated into the endoplasmic reticulum, where they partially complete their folding, then associate together, and finally bind peptide, at which folding is completed. The correct folding and peptide loading of MHC class I molecules within the endoplasmic reticulum are aided by proteins known as chaperones. These are proteins that help the correct folding and subunit assembly of other proteins while keeping them out of harm's way until they are ready to enter cellular pathways and carry out their functions.

When MHC class I heavy chains first enter the endoplasmic reticulum they bind the membrane protein known as **calnexin**, which retains the partly folded heavy chain in the endoplasmic reticulum (Figure 3.14). Calnexin is a calcium-dependent lectin, a carbohydrate-binding protein, that retains many multi-subunit glycoproteins, including T-cell receptors and immunoglobulins, in the endoplasmic reticulum until they have folded correctly. When the MHC class I heavy chain has bound β₂-microglobulin, calnexin is released from the α:β₂-microglobulin heterodimer to be replaced by a complex of proteins, one of which, **calreticulin**, resembles a soluble form of calnexin and probably has a similar chaperone function. A second component of the complex, called **tapasin**, binds to the TAP-1 subunit of the peptide transporter, positioning the partly folded heterodimer of heavy chain and β_2-microglobulin to await receipt of a suitable peptide from the cytosol.

On binding a peptide, the now completely assembled MHC class I molecule is released from all chaperones and leaves the endoplasmic reticulum in a membrane-bound vesicle. It then makes its way through the Golgi stacks to the plasma membrane. Most of the peptides transported by TAP are not successful in binding to a MHC class I molecule and are cleared out of the endoplasmic reticulum.

MHC class I molecules cannot leave the endoplasmic reticulum unless they have bound a peptide. In one form of a rare disease called **bare lymphocyte syndrome**, the TAP protein is non-functional and so no peptides enter the endoplasmic reticulum. Cells of patients with this defect have less than 1% of the normal level of MHC class I molecules on their surface. As a consequence of this deficiency, patients develop very poor CD8 T-cell responses to viruses and suffer from chronic respiratory infections from a young age.

Protein degradation and peptide transport occur continuously, not only when cells are infected. In the absence of infection, MHC class I molecules carry peptides derived from normal human self-proteins, as also do MHC class II molecules. These do not normally provoke an immune reponse. During T-cell development, maturing T cells that are reactive to these self-peptides are eliminated or rendered inactive. Occasionally, however, this state of T-cell tolerance to self-peptides breaks down, resulting in autoimmunity.

As the CD8 T-cell response to a virus-infected cell kills the cell and shuts down viral replication, viruses have evolved mechanisms that interfere with antigen presentation by MHC class I molecules. The herpes simplex virus, for example, prevents the transport of viral peptides into the endoplasmic reticulum by producing a protein that binds to and inhibits TAP. Adenoviruses make a protein that binds to MHC class I molecules and prevents them from leaving the endoplasmic reticulum. There are many steps in the formation of a peptide:MHC class I complex and they all provide potential targets for disruption by viruses.

3-10 Peptides presented by MHC class II molecules are generated in acidified intracellular vesicles

Proteins of extracellular bacteria, extracellular virus particles, and soluble protein antigens are processed by a different intracellular pathway from that followed by cytosolic proteins and their peptide fragments end up bound to MHC class II molecules. Most cells are continually internalizing extracellular fluid and material bound at their surface by the process of **endocytosis**. In addition cells specialized for **phagocytosis**, such as macrophages, engulf larger objects such as dead cells. These uptake mechanisms produce intracellular vesicles known as **endocytic vesicles** or **phagosomes** (in the case of phagocytosis), in which the vesicular membrane is derived from the plasma membrane and the lumen contains extracellular material.

These membrane-bound vesicles become part of an interconnected vesicle system that carries materials to and from the cell surface. As vesicles travel inwards from the plasma membrane, their interiors become acidified by the action of proton pumps in the vesicle membrane and they fuse with other vesicles, such as lysosomes, that contain proteases and hydrolases that are active in acid conditions. Within the **phagolysosomes** formed by this fusion, enzymes degrade the vesicle contents to produce, among other things, peptides from proteins and glycoproteins.

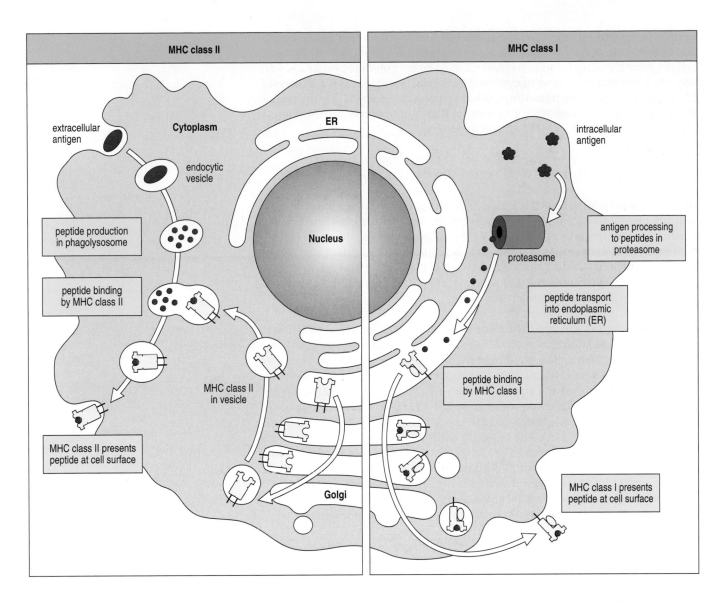

Figure 3.15 Processing of antigens presented by MHC class II and MHC class I molecules occurs in different cellular compartments. The left half of the figure shows the fate of peptides derived from extracellular antigens and pathogens. Extracellular material is taken up by endocytosis and phagocytosis into the vesicular system of the cell, a macrophage. Proteases in these vesicles break down proteins to produce peptides that are bound by MHC class II molecules, which have been transported to the vesicles via the endoplasmic reticulum (ER) and the Golgi apparatus. The peptide:MHC class II complex is transported to the cell surface in outgoing vesicles. The right half of the figure shows the fate of peptides generated in the cytosol as a result of infection with viruses or intracytosolic bacteria. Proteins from such pathogens are broken down in the cytosol by the proteasome to peptides, which enter the endoplasmic reticulum. There the peptides are bound by MHC class I molecules. The peptide:MHC class I complex is transported to the cell surface via the Golgi apparatus.

Microorganisms present in the extracellular environment are taken up by phagocytosis, for example by macrophages, and are then degraded within phagolysosomes. B cells bind specific antigens via their surface immunoglobulin, then internalize these antigens by receptor-mediated endocytosis. These antigens are similarly degraded within the vesicular system. Peptides produced within phagolysosomes become bound to MHC class II molecules within the vesicular system, and the peptide:MHC class II complexes are carried to the cell surface by outward-going vesicles. Thus the MHC class II pathway samples the extracellular environment, complementing the MHC class I pathway, which samples the intracellular environment (Figure 3.15).

Certain pathogens, for example the species of mycobacteria that cause leprosy and tuberculosis, actually exploit the vesicular system as a protected site for intracellular growth and replication. As their proteins do not enter the cytosol, they are not processed and presented to cytotoxic CD8 T cells by MHC class I molecules. They protect themselves from degradation by lysosomal enzymes by preventing fusion of the phagosome with a lysosome. This also prevents presentation of mycobacterial antigens by MHC class II molecules.

3-11 MHC class II molecules are prevented from binding peptides in the endoplasmic reticulum by the invariant chain

Newly synthesized MHC class II α and β chains are translocated from the ribosomes into the membranes of the endoplasmic reticulum. There, an α chain and a β chain associate with a third chain, called the **invariant chain** because it is identical in all individuals (Figure 3.16). One function of the invariant chain is to prevent the peptide-binding site formed by association of an α and a β chain from binding peptides present in the endoplasmic reticulum; these peptides are therefore targeted to MHC class I molecules only.

A second function of the invariant chain is to deliver MHC class II molecules to endocytic vesicles, where they bind peptide. These vesicles, which have been called MIIC, for MHC class II compartment, contain proteases, for example cathepsin L, that selectively attack the invariant chain. A series of cleavages leaves just a small fragment of the invariant chain that covers up the MHC class II peptide-binding site. This fragment is called **CLIP**, for **class II-associated invariant-chain peptide**. Removal of CLIP and binding of peptide is aided by the interaction of the MHC class II molecule with a glycoprotein in the vesicle membrane called **HLA-DM** (see Figure 3.16). HLA-DM resembles an MHC class II molecule in structure but does not bind peptides or appear on the cell surface. After losing CLIP, an MHC class II molecule either quickly binds a peptide or is degraded. Once the MHC class II molecule has lost its invariant chain and has bound peptide, it is carried to the cell surface by outward-going vesicles.

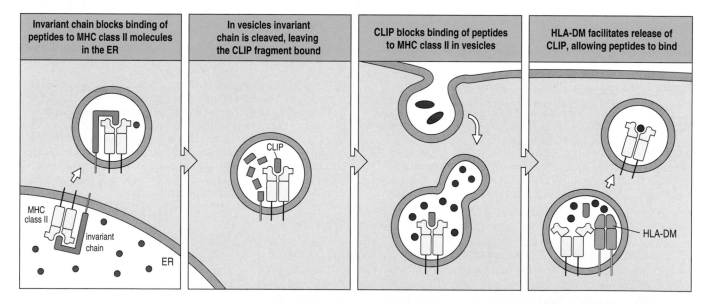

| Invariant chain blocks binding of peptides to MHC class II molecules in the ER | In vesicles invariant chain is cleaved, leaving the CLIP fragment bound | CLIP blocks binding of peptides to MHC class II in vesicles | HLA-DM facilitates release of CLIP, allowing peptides to bind |

Figure 3.16 The invariant chain prevents peptides from binding to a MHC class II molecule until it reaches the site of extracellular protein breakdown. MHC class II α and β chains are assembled with an invariant chain in the endoplasmic reticulum (ER); this complex is transported to the acidified vesicles of the endocytic system. The invariant chain is broken down, leaving just a small fragment called CLIP (for class II-associated invariant-chain peptide) attached in the peptide-binding site. The vesicle membrane protein HLA-DM catalyzes the release of the CLIP fragment and its replacement by a peptide derived from endocytosed antigen that has been degraded within the acidic interior of the vesicles.

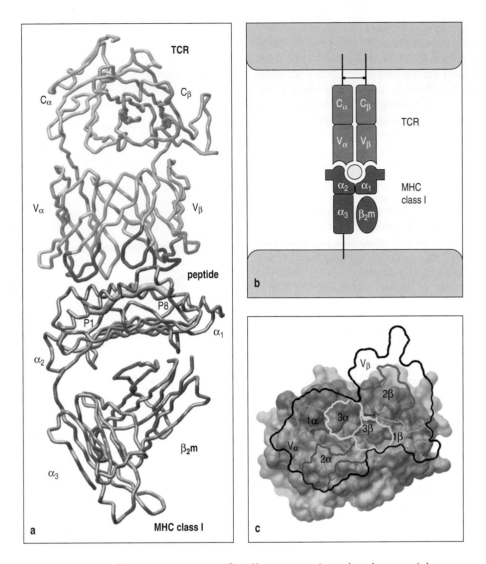

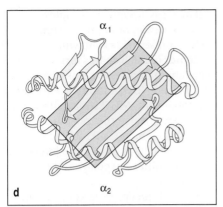

Figure 3.17 The MHC:peptide:T-cell receptor complex. Panel a shows a diagram of the polypeptide backbone of the complex of a T-cell receptor bound to its peptide:MHC class I ligand. Panel b shows a schematic representation of this view of the receptor:ligand complex. In panel a the T-cell receptor's CDRs are colored: the α-chain CDR1 and CDR2 are light and dark blue respectively, while the β-chain CDR1 and CDR2 are light and dark purple respectively. The α chain CDR3 is yellow and the β-chain CDR3 is dark yellow. The eight amino-acid peptide is colored yellow and the positions of the first (P1) and last (P8) amino acids are indicated. Panel c is a view rotated 90 degrees from that of panel a and shows the surface of the peptide:MHC class I ligand and the footprint made upon it by the T-cell receptor (outlined in black). Within this footprint the contributions of the CDRs are outlined in different colors and labeled. In panel d the diagonal orientation of the T-cell receptor with respect to the peptide-binding groove of the MHC class I molecule is shown in a schematic diagram. The T-cell receptor is represented by the black rectangle over the ribbon diagram (yellow) of the α₁ and α₂ domains of the MHC class I molecule. Panels a and c courtesy of I.A. Wilson.

3-12 The T-cell receptor specifically recognizes both peptide and MHC molecule

Once a peptide:MHC complex appears on the cell surface it can be recognized by its corresponding T-cell receptor. When the T-cell receptor binds to a peptide:MHC molecule complex, it makes contacts with both the peptide and the surrounding surface of the MHC molecule. Each peptide:MHC complex thus forms a unique ligand for its T-cell receptor.

The floor of the peptide-binding groove of both classes of MHC molecule is formed by eight strands of antiparallel β-pleated sheet, on which lie two antiparallel α helices (see Figure 3.10). The peptide lies between the helices and parallel to them, such that the top surfaces of the helices and peptide form a roughly planar surface to which the T-cell receptor binds. The residues of the peptide that bind to the MHC molecule lie deep within the peptide-binding groove and are inaccessible to the T-cell receptor; side chains of other peptide amino acids stick out of the binding site and bind to the T-cell receptor.

In its overall organization the T-cell receptor antigen-binding site resembles that of an antibody (see Figure 3.2). The precise interactions between T-cell receptors and the ligands formed from peptides bound to MHC class I molecules have been determined by X-ray crystallography (Figure 3.17). A T-cell receptor binds to an

MHC class I:peptide complex with the long axis of its complementarity-determining region oriented diagonally across the peptide-binding groove of the MHC class I molecule, as shown in panel d of Figure 3.17. The CDR3 loops of the T-cell receptor α and β chains form the central part of the binding site and they grasp the side chain of one of the amino acids in the middle of the peptide. By contrast the CDR1 and CDR2 loops form the periphery of the binding site and contact the α helices of the MHC molecule. The CDR3 loops directly contact peptide antigen and they are also the most variable part of the T-cell receptor antigen-recognition site; the α-chain CDR3 includes the joint between the V and J sequences, and the β chain CDR3 includes the joints between V and D, the whole of the D segment, and the joint between D and J.

The T-cell receptor does not interact symmetrically with the face formed by the peptide and the two α helices of the MHC molecule. Consequently, the CDR1 and CDR2 loops of the α chain make stronger contacts with the peptide:MHC complex than do the CDR1 and CDR2 loops of the β chain. Although crystal structures for T-cell receptor:peptide:MHC class II complexes have yet to be determined, they are predicted to be similar to those of the MHC class I-containing complexes.

3-13 Bacterial superantigens stimulate a massive but ineffective T-cell response

Structural studies of MHC class II molecules have revealed how certain bacteria circumvent the specificity of the T-cell receptor and sabotage the immune system. Such bacteria secrete protein toxins that bind simultaneously to MHC class II molecules and to T-cell receptors in the absence of a specific peptide antigen (Figure 3.18).

By forming a bridge between a CD4 T cell's receptors and the MHC class II molecules of an antigen-presenting cell these toxins mimic specific antigen and cause the T cell to divide and differentiate into effector T cells. Because the toxins bind to sites shared by many different T-cell receptors they stimulate a massive polyclonal response which can involve 2–20% of the total number of circulating CD4 T cells. Because of this property these toxins are called **superantigens**.

The outpouring of cytokines from the CD4 T cells stimulated by a bacterial superantigen has two effects—systemic toxicity and a suppression of the adaptive immune response—both of which further the bacterial infection at the expense of the health of the human host. Prominent among the bacterial superantigens are the **staphylococcal enterotoxins** (**SEs**), which are a common cause of food poisoning, and the **toxic shock syndrome toxin-1** (**TSST-1**), also produced by some strains of staphylococci.

3-14 The two classes of MHC molecule are expressed differentially on cells

Immune responses are guided in an appropriate direction by a differential expression of the two classes of MHC molecules on human cells. Although virtually all the cells of the body express MHC molecules they do not all express both MHC class I and class II. Almost all cells express MHC class I molecules constitutively (Figure 3.19). As all cell types are susceptible to infection by viral pathogens, this enables comprehensive surveillance by CD8 T cells. The erythrocyte is one cell that lacks MHC class I, a property that might facilitate its persistent infection by malarial parasites.

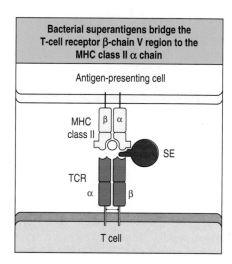

Bacterial superantigens bridge the T-cell receptor β-chain V region to the MHC class II α chain

Antigen-presenting cell

MHC class II

SE

TCR

T cell

Figure 3.18 Binding of bacterial superantigens to MHC class II molecules and T-cell receptors. Superantigens bind simultaneously to MHC class II molecules and to T-cell receptors irrespective of their peptide-binding specificities. Bacterial superantigens such as the staphylococcal enterotoxin (SE) shown here bind to the CDR1 and CDR2 loops of the V$_\beta$ domain of all T-cell receptors (TCR) in which the β chain is encoded by certain V$_\beta$ gene segments, and to the α$_1$ domain of the MHC class II molecule.

Tissue	MHC class I	MHC class II
Lymphoid tissues		
T cells	+++	+*
B cells	+++	+++
Macrophages	+++	++
Other antigen-presenting cells (eg dendritic cells)	+++	+++
Epithelial cells of the thymus	+	+++
Other nucleated cells		
Neutrophils	+++	−
Hepatocytes	+	−
Kidney	+	−
Brain	+	−†
Non-nucleated cells		
Red blood cells	−	−

Figure 3.19 Most human cells express MHC class I, whereas only a few cell types express MHC class II. MHC class I molecules are expressed on almost all nucleated cells, although they are most highly expressed in hemato-poietic cells. MHC class II molecules are normally expressed only by a subset of hematopoietic cells and by stromal cells in the thymus, although they can be produced by other cell types on exposure to the cytokine interferon-γ. *In humans, activated T cells express MHC class II molecules, whereas resting T cells do not. † In the brain, most cell types are MHC class II-negative, but microglia, which are related to macrophages, are MHC class II-positive.

The function of MHC class II molecules, in contrast, is to alert CD4 T cells to the presence of extracellular infections. To be effective, this function does not need to be performed by every cell within a tissue or organ, but only by a subset of cells that guard the extracellular territory. MHC class II molecules are constitutively expressed on only a few cell types, which are cells of the immune system specialized for the uptake, processing, and presentation of antigens from the extracellular environment. These **professional antigen-presenting cells** are: the macrophages, which take up antigens by phagocytosis and endocytosis; the B cells, which efficiently internalize specific antigens bound to their surface immunoglobulin; and the dendritic cells found in lymph nodes and other lymphoid tissues, which are specialized for antigen presentation and the activation of T cells.

During the course of an immune response, cells can increase the synthesis and cell-surface expression of MHC molecules beyond constitutive levels. This upregulation, which enhances antigen presentation and T-cell activation, is induced by several cytokines produced by activated immune-system cells, in particular the interferons. In addition to increasing the levels of constitutively expressed MHC molecules, interferon-γ (IFN-γ) can induce the expression of MHC class II molecules on some cell types that do not normally produce them. In this manner, presentation of antigen to CD4 T cells by MHC class II molecules can be increased in infected or inflamed tissues.

Summary

T cells expressing α:β T-cell receptors recognize peptides presented at cell surfaces by MHC molecules, the third type of antigen-binding molecule in the adaptive immune system. Unlike immunoglobulins and T-cell receptors, however, MHC molecules have degenerate binding sites and each MHC molecule is capable of binding peptides having many different amino-acid sequences. The peptides are produced by the intracellular degradation of proteins of infectious agents and self-proteins. An α:β T-cell either expresses the CD8 co-receptor and recognizes peptides presented by MHC class I molecules or expresses the CD4 co-receptor and recognizes peptides presented by MHC class II molecules. The CD8 co-receptor interacts specifically with MHC class I molecules, and the CD4 co-receptor interacts specifically with MHC class II molecules. Protein antigens from intracellular and extracellular sources are processed into peptides by two different pathways. Peptides generated in the cytosol from viruses and other intracytosolic pathogens enter the endoplasmic reticulum, where they are bound by MHC class I molecules. These peptides are thus recognized by CD8 T cells, which are specialized to fight intracellular infections. As all human cells are susceptible to infection, MHC class I molecules are expressed by most cell types. Extracellular material that has been taken up by endocytosis is degraded into peptides in endocytic vesicles and these peptides are bound by MHC class II molecules within the vesicular system. The resulting complex is recognized by CD4 T cells that are specialized to fight extracellular sources of infection by mobilizing other cells of the immune system such as B cells and macrophages. MHC class II molecules are expressed by a few specialized cell types of the immune system that are able to take up extracellular antigens efficiently and activate CD4 T cells.

The major histocompatibility complex

MHC molecules and other proteins involved in antigen processing and presentation are encoded in a cluster of closely linked genes called the major histocompatibility complex, which in humans is located on chromosome 6. A feature of the MHC class I and class II molecules is the very large number of different genetic variants within the human population. This variability has evolved to help cope with the structural diversity of pathogens and their antigens and to enable the MHC molecules collectively to bind a vast variety of peptide sequences. Although the magnitude of MHC diversity is much smaller than that of immunoglobulins or T-cell receptors, it has major implications for the immune response and the practice of medicine. Differences in MHC molecules between different individuals, for example, are responsible for the problems of graft rejection and graft-versus-host disease encountered in organ transplantation. In this part of the chapter we consider the immunological and medical consequences of MHC diversity within the human population.

3-15 The diversity of MHC molecules in the human population is due to both polygeny and genetic polymorphism

In contrast to immunoglobulins and T-cell receptors, the MHC class I and II molecules are encoded by conventionally stable genes that neither rearrange nor undergo any other developmental or somatic processes of structural change. The inherited diversity of MHC molecules is achieved in two ways. Diversity within an individual is the result of the expression of several different genes encoding MHC class I heavy chains (α chains) and several different genes encoding MHC class II α chains and β chains. This state of affairs is referred to as **polygeny**; the products of the different genes encoding MHC class I or class II chains are called isotypes. A second cause of MHC diversity within the population is **genetic polymorphism**, the existence within human populations of many alternative forms of each MHC class I and II gene. This extensive polymorphism ensures that most individuals inherit different variants of each gene from their mother and father and are therefore heterozygous for all their MHC genes. The different forms of any given gene are called **alleles** and their encoded proteins **allotypes**. When considering the diversity of MHC class I or II molecules that arise from both polygeny and polymorphism, the term **isoform** can be useful to denote any particular MHC protein. The numerous alleles of the MHC class I and II genes, and the many differences that distinguish them, make MHC genes stand out from other polymorphic genes and they are therefore said to be **highly polymorphic**. They are among the most polymorphic human genes yet discovered.

3-16 The MHC class I and class II genes are linked in the major histocompatibility complex

The genes that encode the MHC class I heavy chains and the MHC class II α and β chains are closely linked within a single chromosomal region (Figure 3.20, upper panel). This region is called the major histocompatibility complex because it was first identified as the region containing genetic polymorphisms that are the principal cause of graft rejection when tissues or organs are transplanted between unrelated people or outbred animals. The human MHC is called the

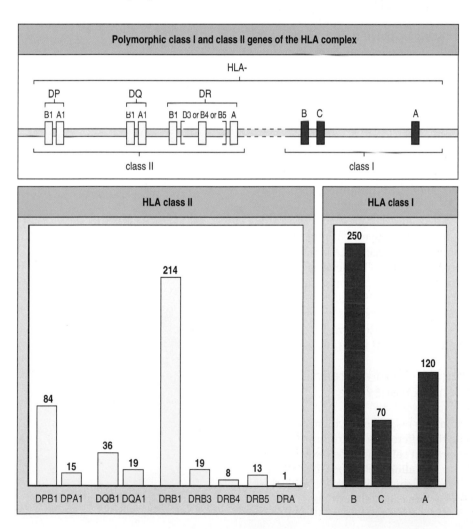

Figure 3.20 Polygeny and polymorphism in the human MHC. The organization of the principal genes involved in antigen presentation within the polygenic HLA complex (the human MHC) is shown in the upper panel. The three HLA class I heavy (α)-chain genes (HLA-A, HLA-B, HLA-C) are shown in red and the HLA class II α- and β-chain genes in yellow. DPB1 denotes the gene for the β chain of HLA-DP, DPA1 the gene for the α chain of HLA-DP, and so on. Note the four different genes for the β chain of HLA-DR. Every chromosome 6 carries the DRB1 gene and for some individuals this is the only DRB gene they carry. Other chromosomes 6 carry, in addition, one of the other functional DRB genes: DRB3, DRB4, or DRB5. The polymorphism of HLA class I and class II genes is illustrated in the lower panels, in which the number of known functional alleles in the human population for each gene is shown. The heights of the bars reflect the numbers of alleles.

HLA complex, where HLA stands for Human Leukocyte Antigen, because the antibodies originally used to identify human MHC molecules react with the white cells of the blood—the leukocytes—but not with the red cells, which lack MHC molecules. This observation distinguished the MHC from the other known systems of polymorphic cell-surface antigens, for example the ABO system which is matched in blood transfusion, which all involved antigens on the surface of red blood cells.

The HLA complex consists of about 4 million base pairs of DNA on the short arm of chromosome 6. The HLA class I heavy-chain genes are clustered together in the **class I region** at the end of the complex furthest from the chromosome's centromere, whereas the HLA class II genes are separately clustered in the **class II region** at the other end of the complex. β_2-Microglobulin and the invariant chain are not polymorphic and are not located in the HLA region: they are encoded by genes on human chromosomes 15 and 5 respectively.

The HLA class I genes are expressed by most cell types. There are three polymorphic HLA class I heavy-chain genes, called **HLA-A**, **HLA-B**, and **HLA-C**. They encode the class I heavy chains that associate with β_2-microglobulin to form complete MHC class I molecules. In humans, 120 HLA-A, 250 HLA-B, and 70 HLA-C alleles have been defined (see Figure 3.20, lower panels). The genes for HLA class II molecules, which consist of an α and a β chain, are not so straightforward. In humans there are three polymorphic MHC class II molecules, called **HLA-DP**, **HLA-DQ**, and **HLA-DR**. The genes for the α and β chains of each molecule cluster together in different subregions within the class II region of the MHC. For HLA-DP and HLA-DQ, both the α and β chains are polymorphic: there are 15 DPA1, 84 DPB1, 19 DQA1 and 36 DQB1 alleles. The α chain of HLA-DR has no polymorphism and is said to be **monomorphic**, whereas there are four different genes within the human population encoding HLA-DR β chains (DRB1, DRB3, DRB4, and DRB5). However, only the DRB1 gene is present on all chromosomes 6 and for some people this is the only DRB gene expressed. Three other types of chromosome 6 have been found: those carrying DRB3, DRB4, or DRB5 in addition to DRB1. Thus, an individual chromosome 6 can encode one or two different HLA-DR molecules. DRB1 is the most polymorphic gene of all the HLA class II genes, with 214 defined alleles; the other three have smaller numbers of alleles. Depending on the chromosomes 6 that a person inherits from their parents they can express between one and four different HLA-DR molecules.

The particular combination of HLA alleles found on a given chromosome 6 is known as a **haplotype**. A person expresses both alleles of each HLA gene and thus a heterozygous person expresses two haplotypes. There is a low level of recombination between the closely linked genes of the MHC and so haplotypes tend to be inherited as single units. The minimum number of HLA isoforms expressed by a person is three class I and three class II isoforms (this person is a homozygote; they do exist and are healthy). The maximum number (for someone heterozygous at all the polymorphic genes) is six class I and eight class II isoforms.

3-17 MHC polymorphism is a primary cause of alloreactions that cause the rejection of transplanted tissues

Transplant rejection is caused by a potent immune response in which B cells and T cells of the transplanted patient respond to the differences in structure between the MHC molecules expressed by the transplanted cells and those expressed by the recipient. Such differences are said to be **allogeneic** and the immune response that they provoke is an **alloreaction**. Antigens such as MHC molecules that vary between members of the same species are called **alloantigens**.

A natural situation in which alloreactions occur is pregnancy, when the mother's immune system can be stimulated by MHC molecules of the fetus that derive from the father and are not found in the mother's cells. This response does not reject the fetus, which is protected, but can lead to antibodies in the mother's circulation that bind to paternal MHC molecules. Before molecular genetic techniques were available, the human MHC and the diversity within its constituent class I and II genes were defined by using **alloantibodies** against MHC molecules, which were obtained from blood donated by women after pregnancy. Alloantibody is the name given to any antibody raised in one member of a species against an allotypic protein from another member of the same species.

HLA types originally defined rather broadly by antibodies are now being refined into individual alleles by DNA sequencing. For example, antibody typing identified several different HLA-DR alloantigens—DR1, DR2, DR3, DR4, DR5, and so on. Now that the DNA encoding the β chains in these DR molecules has been sequenced, each of these alloantigens turns out to consist of a group of related allotypes. The standard nomenclature for HLA alleles is now based on DNA sequence. For example, all alleles of the DRB1 gene start with the letters DRB1. Within the original HLA-DR4 grouping, 22 different alleles of DRB1 have so far been identified. All these are known as DRB1*04. Individual alleles are distinguished by two additional numerals. DRB1*0402, for example, is an allele that is associated with a greater susceptibility to the autoimmune disease pemphigus vulgaris in some ethnic groups.

3-18 The polymorphism of MHC class I and class II molecules affects antigen recognition by T cells

The majority of structural genes in the human genome have a predominant 'wild-type' allele that is present in most individuals. This is not so for the highly polymorphic HLA class I and II genes, for which many alleles are present at relatively even frequencies. The allotypes of a single type of HLA molecule, such as HLA-A, can differ considerably; comparisons of amino-acid sequences show differences at between 1 and 50 positions. The amino-acid substitutions are not randomly distributed within the sequence but are concentrated in the domains that bind peptide and interact with the T-cell receptor. More specifically, they affect the α helices that interact with the T-cell receptor and the amino acids of the binding groove that influence which peptides can bind (Figure 3.21).

The differences between the allotypes of an MHC molecule impart an individuality to the T-cell response in the sense that the T-cell responses that each person makes are determined by their MHC type. A T cell that responds to a peptide presented by one MHC allotype will rarely be able to respond to the combination of that peptide bound to another MHC allotype. The requirement of a T-cell receptor for a defined combination of a specific peptide and an MHC allotype is known as **MHC restriction**, because the antigen-specific T-cell response is

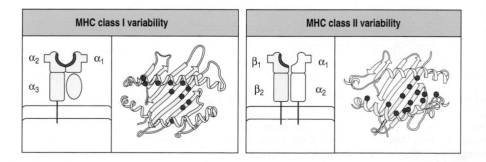

Figure 3.21 Variation between MHC allotypes is concentrated in the sites that bind peptide and T-cell receptor. In the HLA class I molecule (left), allotype variability is clustered in specific sites (shown in red) within the α_1 and α_2 domains. These sites line the peptide-binding groove, lying either in the floor of the groove, where they influence peptide binding, or in the α helices that form the walls, which are also involved in binding the T-cell receptors. In the HLA class II molecule illustrated (right), which is a DR molecule, variability is found only in the β_1 domain because the α chain is monomorphic.

restricted by the MHC type. The first demonstration in the early 1970s of MHC class I restriction of the CD8 T-cell response to a virus infection in mice was a major breakthrough. It revealed the true physiological function of the polymorphic cell-surface MHC glycoproteins that were already well known as the major immunological barrier to clinical transplantation.

3-19 Each MHC isoform binds a characteristic set of peptides

When the amino-acid sequences of peptides that bind to a particular MHC isoform are compared, it turns out that the amino acids at certain positions are identical or very similar in all the peptides. This is because these positions provide important binding contacts between the peptide and that particular MHC molecule. These amino acids are called **anchor residues** because they anchor the peptide to the MHC molecule. The combination of anchor residues that binds to a particular MHC isoform is called its **peptide-binding motif** (Figure 3.22). A change in a single amino acid within the peptide-binding site of an MHC molecule can make it able, or unable, to bind a particular peptide. Variations of this sort probably underlie susceptibility or resistance to some of the diseases that seem to be closely associated with HLA type.

The pattern of sequence variation in MHC molecules demonstrates that the polymorphism of HLA class I and II genes is the result of natural selection for binding different types of peptide. This probably stems from the intense pressures placed on the human immune system by diverse and continually changing pathogens. Most people are heterozygous for the polymorphic HLA class I and II genes. The advantage of being a heterozygote is that two MHC allotypes with different peptide-binding motifs can present distinct sets of pathogen-derived peptides and thus can activate more T-cell clones than one allotype alone. A heterozygote

MHC molecule		Amino acid sequence of peptide-binding motifs and bound peptides	Source of bound peptide
		Position in peptide sequence N — 1 2 3 4 5 6 7 8 9 — C	
Class I	HLA-A*0201	Peptide-binding motif: _ (L/M) _ _ _ (V) _ _ (V/L) Bound peptide: I (L) K E P (V) H G (V)	HIV reverse transcriptase
	HLA-B*2705	Peptide-binding motif: _ (R) _ _ _ _ _ _ (L/F) Bound peptide: R (R) Y P D A V Y (L)	Measles virus F protein
Class II	HLA-DRB1*0401	Self peptide: G V Y F (Y) L Q (W) G R S T (L) V S V S	Igκ light chain
	HLA-DQA1*0501 HLA-DQB1*0301	Self peptide: I P E (L) N K V A R A A A	Transferrin receptor

Figure 3.22 Peptide-binding motifs and the sequences of peptides bound for some MHC isoforms. For the HLA-A and -B isoforms, both the peptide-binding motif for the isoform and the complete amino-acid sequence of one peptide presented by that isoform are given. Blank boxes in the peptide-binding motifs are positions at which the identity of the amino acid can vary. For the HLA-DR and -DQ isoforms, only the sequence of a self peptide that is bound by the isoform is shown. Anchor residues are in green circles. Peptide-binding motifs for MHC class II molecules are not readily defined. The one-letter code for amino acids is used.

can thus make a more effective immune response against the pathogen than can a homozygote. The same argument can also explain the benefit of possessing several different HLA class I and HLA class II genes: each additional gene contributes a different peptide-binding motif for the presentation of antigens to T cells.

A second selective pressure acting on MHC gene polymorphisms is exerted by individual diseases. For example, in West Africa malaria is endemic and the major cause of childhood mortality. Associated with resistance to the severe form of malaria is the HLA-B*5301 allele, which is at high frequency in West African populations, probably as a result of selection by this single disease. At one stage of its life cycle the malaria parasite infects liver cells, and it has been proposed that the protective effect of HLA-B*5301 derives from its ability to bind certain parasite-derived peptides found in infected hepatocytes and thus to present them to CD8 T cells.

3-20 The diversity of MHC molecules in human populations has been generated by meiotic recombination and gene conversion

Comparison of different human populations shows that certain HLA class I and II alleles are widely distributed, whereas others are found only in people from a particular geographical region. Each local allele is closely related to a widely distributed allele, and comparison of the nucleotide sequences of such pairs of alleles has revealed the genetic mechanisms that generate new HLA alleles. A patchwork pattern of nucleotide substitution indicates that new alleles are often formed by genetic recombination between existing alleles during formation of germ cells, such that a segment of one allele is replaced by the corresponding segment of another. This mechanism has been termed **interallelic conversion** or **segmental exchange** (Figure 3.23, left panel). The HLA-B*5301 allele associated with resistance to severe malaria in West Africa is an example of a local allele that was probably formed from a more widespread allele, HLA-B*3501, by interallelic conversion. Some alleles have been formed by similar events involving two different MHC genes, in which case the mechanism is called **gene conversion** (see Figure 3.23, right panel). In comparison with recombination, point mutation appears a less frequent source of new HLA alleles.

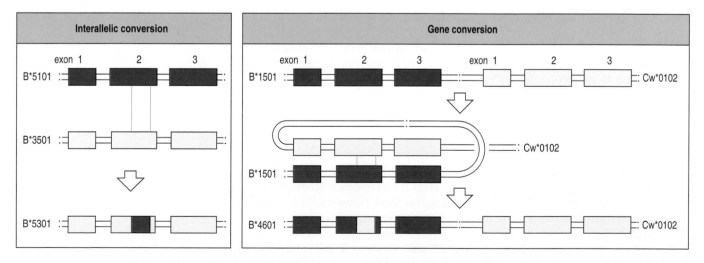

Figure 3.23 New MHC alleles are generated by interallelic conversion or gene conversion. Recombination between alleles of the same gene (HLA-B), as shown on the left, and between alleles of different genes (HLA-B and HLA-C), as shown on the right, can both result in the formation of a new allele in which a small block of DNA sequence has been replaced. Whereas B*5301 is characteristic of African populations, B*4601 is found in southeast Asian populations.

Within the HLA complex, meiotic recombination occurs at a frequency of about 2%. This serves to reassort alleles of polymorphic HLA genes into new haplotypes. Although there are no more than a few hundred alleles for any one HLA gene (see Figure 3.20), they have been recombined into many thousands of different haplotypes. The further random combination of one maternal and one paternal HLA haplotype within each individual produces millions of different combinations of HLA class I and II alleles in the human population.

The combination of HLA class I and II alleles in an individual is referred to as that person's **HLA type**. In medicine, HLA typing is used to inform the selection of donors and recipients for transplantation. The immense diversity of HLA type means that clinical transplants are performed across a varying range of HLA matches and mismatches. In the transplantation of solid organs, such as kidney and heart, considerable HLA mismatch can be overcome by the use of immunosuppressive drugs. In contrast, bone marrow transplantation is more sensitive to HLA mismatch because the healthy tissue being transplanted into a weakened patient is composed of the very cells that form the immune system. In bone marrow transplantation the major problem is not graft rejection by the recipient, but an alloreaction of the graft against the recipient's tissues. This reaction is mounted by lymphocytes in the transplant and is the cause of graft-versus-host disease, which can be fatal. For a patient who needs a bone marrow transplant the first place to look for an HLA-identical donor is among their full siblings. Within a family, four HLA haplotypes segregate, two from the mother and two from the father, and the probability of any one full sibling being HLA-identical to the patient is thus one in four. For situations where there is no HLA-identical sibling, registries consisting of millions of HLA-typed unrelated donors have been established. Even then not every patient can find an HLA-compatible donor.

3-21 Other proteins involved in the immune response are encoded in the MHC

Because of its importance in the immune system and in clinical transplantation the human MHC has been intensively studied as part of the Human Genome Project and its DNA sequence is now completely determined. More than 200 different genes have been identified so far; their functions are highly diverse and relate to many aspects of human biology. In addition to the polymorphic HLA class I and II genes, the MHC contains additional genes that are important for antigen presentation and are situated in the HLA class II region between the HLA-DP and the HLA-DQ genes (Figure 3.24). They include genes encoding the α and β chains of the invariant HLA-DM molecule that facilitates peptide loading of HLA class II molecules, genes encoding the two polypeptides of the TAP peptide transporter, the gene for tapasin, and genes encoding two subunits of the proteasome called LMP2 and LMP7. The class III region contains genes for proteins, such as some complement components, that are involved in other aspects of the immune response.

The expression of the HLA-DM, -DP, -DQ, -DR, and invariant-chain genes, which all contribute to antigen presentation by HLA class II molecules, is co-ordinately regulated by the cytokine IFN-γ. The transcription of these genes involves a transcriptional activator known as **MHC class II transactivator** (**CIITA**), which is itself induced by IFN-γ. Another form of bare lymphocyte syndrome (see Section 3-9) is characterized by an absence of HLA class II molecules on B cells and other professional antigen-presenting cells, which is caused by a lack of functional CIITA. This causes severe immunodeficiency as a result of an inability to activate T cells.

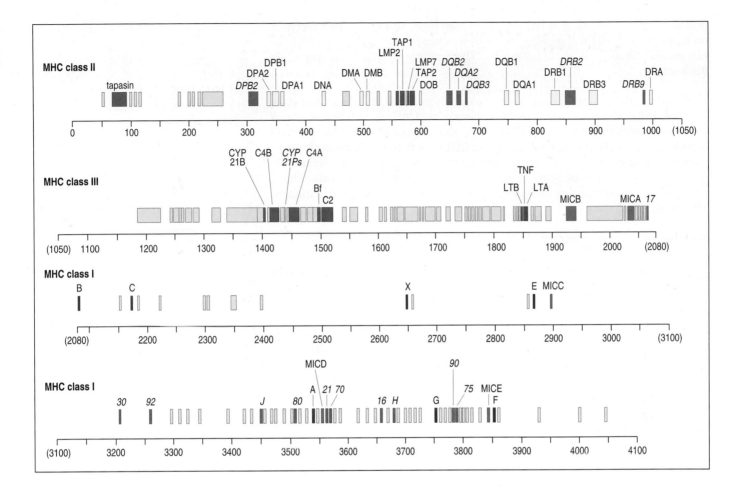

Figure 3.24 Detailed map of the HLA complex. The organization of the class I, class II, and class III regions of the MHC are shown, with approximate distances given in thousands of base pairs (kb). Most of the genes in the class I and class II region are mentioned in the text; the additional class II genes are pseudogenes. The genes shown in the class III region include those for the complement proteins C4 (two genes, shown as C4A and C4B), C2, and factor B (shown as Bf), as well as genes for the cytokines: tumor necrosis factor-α (TNF-α) and TNF-β (lymphotoxin; LTA, LTB). Closely linked to the C4 genes is the gene encoding steroid 21-hydroxylase (shown as CYP 21B), an enzyme involved in the synthesis of steroid hormones in the adrenal cortex. Genes shown in gray and named in italic are pseudogenes. These are DNA sequences that are related to known genes but are not expressed.

The co-ordinate expression of genes involved in the processing and presentation of peptides by HLA class I molecules is also regulated by cytokines. During the early phase of an immune response, IFN-α, -β, and -γ are produced in the area of infection. These cytokines stimulate cells in the vicinity to increase their expression of HLA class I heavy chains, β_2-microglobulin, TAP, and the LMP2 and LMP7 proteasome subunits. In healthy cells, the proteasome degrades cellular proteins and does not use the LMP2 and LMP7 subunits, which are only induced by interferon at sites of infection and inflammation. The effect of replacing the constitutive forms of the proteasome subunits with the inducible forms is to modify the proteasome towards the production of more peptides that are compatible with MHC class I binding requirements.

The HLA class I region also contains genes for the heavy chains of three additional class I molecules, called HLA-E, -F, and -G. In contrast to HLA-A, -B, and -C, the HLA-E, -F, and -G genes have limited polymorphism and are expressed by a more limited range of cell types. HLA-E binds peptides derived from the leader peptides (signal sequences) of HLA-A, -B, -C, and -G heavy chains and forms ligands that interact with an inhibitory receptors on natural killer cells (NK cells). This is thought to be part of a mechanism that prevents NK cells from killing healthy cells expressing MHC class I molecules. HLA-G is of particular interest to reproductive immunologists because its expression is restricted to cells of fetal origin in the placenta that contact the maternal circulation and do not express HLA-A, -B, or -C. It has been suggested that non-polymorphic HLA-G can present antigen in the event of intracellular infection while not inducing alloreactions that could damage fetal tissue. Alternatively, the role of HLA-G could be to provide peptides that bind to HLA-E, which then binds to an inhibitory receptor on maternal NK cells and prevents them from attacking the fetal cells, which do not express HLA-A, -B, or -C molecules. The function of HLA-F has yet to be discovered.

Although the HLA complex contains all the highly polymorphic class I and class II genes, and numerous less polymorphic ones as well, it does not contain all the class I-like genes in the human genome. For example, outside the HLA complex, class I-like heavy chains are encoded by the CD1 gene family on human chromosome 1. These proteins have become specialized to present complex lipid antigens, such as the mycolic acids of mycobacterial membranes, to T cells. The T cells involved in recognizing these atypical T-cell antigens are believed to include γ:δ T cells.

Summary

In humans, the highly polymorphic MHC class I and II genes are closely linked in the HLA region on chromosome 6, which comprises the human major histocompatibility complex (MHC). In contrast to immunoglobulin and T-cell receptor genes, MHC class I and II genes have a conventional organization and do not rearrange. In humans, the MHC class I genes encode the heavy (α) chains of three different class I molecules—HLA-A, HLA-B, and HLA-C—whereas the MHC class II genes encode the α and β chains of three different MHC class II molecules—HLA-DP, HLA-DQ, and HLA-DR. β_2-Microglobulin, the light chain of MHC class I molecules, is encoded outside the MHC, on chromosome 15. Certain of the MHC class I and class II genes are highly polymorphic; some have several hundred alleles. Genes encoding other proteins involved in antigen presentation are located in the MHC and, like the MHC class I and II genes, their expression is regulated by the interferons produced during an immune response. The strategy used by MHC molecules to bind diverse antigens contrasts with that of the T-cell receptors. MHC molecules have highly degenerate binding sites for peptides; an MHC molecule is therefore usually able to present a diversity of peptide antigens to a large number of T-cell receptors with highly specific binding sites. Polygeny and polymorphism of MHC class I heavy-chain genes and MHC class II α- and β-chain genes are secondary strategies that serve to increase the breadth and strength of an individual's T-cell immunity.

Summary to Chapter 3

The antigen receptors of T cells resemble the membrane-bound immunoglobulins of B cells in general structure and are encoded by similarly organized genes that undergo gene rearrangement before they are expressed. As with B cells, this gives rise to a population of T cells each expressing a unique receptor. Differences between immunoglobulins and T-cell receptors reflect the fact that the T-cell receptor is used only as a membrane-bound receptor, whereas immunoglobulins are also used as secreted effector molecules.

T-cell receptors are more limited than immunoglobulins in the antigens that they bind, recognizing only short peptides bound to MHC molecules on a cell surface. The two main classes of T cell—CD8 and CD4—are specialized to respond to intracellular and extracellular pathogens respectively. They are activated in response to an antigen from the appropriate source by specific interactions between the MHC molecules displaying the peptide antigen and the CD4 or CD8 glycoproteins on the T-cell surface. Processing of extracellular and intracellular pathogen antigens into peptides, and their binding to MHC molecules, occurs inside the cells of the infected host. Cytotoxic CD8 T cells are activated by peptides presented by MHC class I molecules, and are directed to destroy the antigen-presenting cell. Most cells express MHC class I molecules and so can present pathogen-derived peptides to CD8 T cells if infected with a virus or other pathogen that penetrates the cytosol. The function of CD4 T cells is the activation of other types of effector cell and their recruitment to sites of infection through cell–cell interactions and the production of cytokines. They are activated by peptides presented by MHC class II molecules, which are usually expressed only on specialized antigen-presenting cells that take up and process material from the extracellular environment and can activate CD4 T cells.

MHC molecules have degenerate peptide-binding sites, which enables the relatively small number of different MHC molecules present in each individual to bind peptides of many different sequences. The diversity of peptides that can be presented by the human population as a whole is further increased by the highly polymorphic nature of MHC class I and II genes. Each person differs from almost all other individuals in some or all of the MHC alleles they possess. The population is thus able to respond to the pathogens it encounters with a wide diversity of individual immune responses. Because of the polymorphism of the MHC, organs or tissues transplanted between individuals of different MHC type provoke strong T-cell responses directed against the 'foreign' MHC molecules.

The Development of B Lymphocytes

4

The B cells of the human immune system have the capacity to make immunoglobulins specific for almost every nuance of chemical structure, which gives a person the potential to make antibodies against all the infectious microorganisms that could possibly be encountered in a lifetime. However, the body does not stockpile all the B cells needed to make an optimal response against all possible pathogens. That would probably mean devoting the vast majority of the body's resources to the immune system, leaving little left for the immune system to protect. Instead, the body carries a less complete inventory of B cells, but one

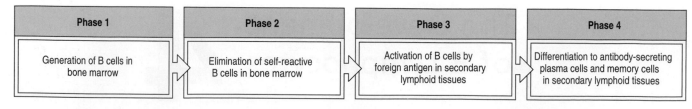

Phase 1	Phase 2	Phase 3	Phase 4
Generation of B cells in bone marrow	Elimination of self-reactive B cells in bone marrow	Activation of B cells by foreign antigen in secondary lymphoid tissues	Differentiation to antibody-secreting plasma cells and memory cells in secondary lymphoid tissues

that expands and contracts its individual clones according to need and circumstance. Fueling this system are stem cells in the bone marrow, which generate tens of billions of new B cells every day of your life.

The life cycle of B cells can be divided into four broad stages (Figure 4.1). The first stage is one of maturation and takes place in the bone marrow, which is therefore designated as a **primary lymphoid tissue**. During this stage, which will be described in the first part of this chapter, developing B cells acquire functional B-cell receptors through the ordered rearrangements of the immunoglobulin genes described in Chapter 2. The second stage involves testing whether a B cell's immunoglobulin receptor binds to normal constituents of the body and therefore has the potential to produce autoreactivity and autoimmune disease. In the third stage, mature naive B cells that survive this selection process leave the bone marrow, enter the blood, and from there move into **secondary lymphoid tissues** (Figure 4.2). If they do not encounter their specific antigens in the lymphoid tissues, they continue to recirculate. If contact is made with antigen in a secondary lymphoid tissue, the fourth stage of a B-cell's life begins. The B cell proliferates and its progeny differentiate either into plasma cells, which synthesize large quantities of antibody, or into long-lived memory B cells, which will respond more quickly than naive B cells on a subsequent encounter with the same antigen. The last three stages of the B-cell life cycle will be dealt with in the second part of this chapter.

Figure 4.1 The development of B cells can be divided into four broad phases.

The development of B cells in the bone marrow

The stages of B-cell development in the bone marrow are marked by successive steps in the rearrangement and expression of the immunoglobulin genes. In this part of the chapter we shall see how gene rearrangement is controlled at each step to produce a mature but naive B cell that makes only one heavy chain and one light chain, and thus expresses immunoglobulin of a single antigen specificity on its surface.

4-1 B-cell development in the bone marrow proceeds through several stages

B cells derive from pluripotential hematopoietic stem cells in the bone marrow. The earliest identifiable cells of the B-cell lineage are called **pro-B cells**. These progenitor cells retain a limited capacity for self-renewal, dividing to produce both more pro-B cells and cells that will go on to develop further. Rearrangement of heavy-chain genes takes place first, and this occurs in pro-B cells: D_H to J_H joining occurs at the **early pro-B cell** stage, followed by V_H to DJ_H joining at the **late pro-B cell** stage (Figure 4.3). A μ heavy chain is the first type of heavy chain to be produced. Once a B cell expresses a μ chain it is known as a **pre-B cell**. The less mature **large pre-B cells** transiently express the μ chain at the cell surface in combination with two proteins that together mimic an immunoglobulin light chain, and are therefore known as the surrogate light chain. The combination of μ chain and the surrogate light chain forms a **pre-B-cell receptor** that, on stimulation by an unknown ligand, induces cell division and the generation of non-dividing **small pre-B cells**.

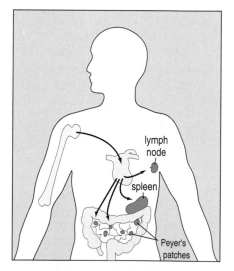

Figure 4.2 B cells develop in bone marrow and then migrate to secondary lymphoid tissues. B cells leaving the bone marrow (yellow) are carried in the blood to lymph nodes, the spleen, Peyer's patches (all shown in green), and other secondary lymphoid tissues such as those lining the respiratory tract (not shown).

	Stem cell	Early pro-B cell	Late pro-B cell	Large pre-B cell	Small pre-B cell	Immature B cell	Mature B cell
H-chain genes	Germline	D–J rearranged	V–DJ rearranged	VDJ rearranged	VDJ rearranged	VDJ rearranged	VDJ rearranged
L-chain genes	Germline	Germline	Germline	Germline	V–J rearrangement	VJ rearranged	VJ rearranged
Surface Ig	Absent	Absent	Absent	μ chain at surface as part of pre-B-cell receptor	μ chain in cytoplasm and at surface as part of pre-B-cell receptor	IgM expressed on cell surface	IgD and IgM made from alternatively spliced H-chain transcripts

In a small pre-B cell most of the μ chains are present inside the cell, whereas a minority form pre-B-cell receptors at the surface. Light-chain gene rearrangements proceed in small pre-B cells. Once light chains have been made, they assemble with μ chains to form IgM molecules, which are transported to the cell surface in the form of a functional B-cell receptor complex (see Section 2-10, p. 45). At this stage the B cell expresses IgM only, and is defined as an **immature B cell**.

Up to this stage, the development of B cells occurs in the bone marrow and does not require interaction with specific antigen. The randomness of the gene rearrangement process leads to a proportion of the B-cell receptors' being reactive against constituents of the body. Unless removed from the repertoire, B cells bearing these receptors have the potential to produce an autoimmune response leading to autoimmune disease. The immature B cells now begin to be selected for tolerance of the normal constituents of the body, and at about this time they also start to enter the peripheral circulation. At this point in development, alternative mRNA splicing of heavy-chain gene transcripts produces IgD, as well as IgM, as membrane-bound immunoglobulin. The cell is now considered to be a **mature B cell**. Such cells are also called **naive B cells**, because they have yet to be exposed to their specific antigen.

B-cell development in the bone marrow is dependent on non-lymphoid **stromal cells**, which provide specialized microenvironments for B cells at various stages of maturation (Figure 4.4). The stromal cells perform two distinct functions. First, they make specific cell-surface contacts with the B cells through the interaction of adhesion molecules and their ligands. Second, they produce growth factors which act on the bound B cells, for example the membrane-bound stem-cell factor (SCF), which is recognized by a receptor called Kit on immature B cells. Another important growth factor for B-cell development is interleukin-7 (IL-7), a cytokine secreted by stromal cells that acts on late pro-B and pre-B cells.

The most immature stem cells lie in a region of the bone marrow called the subendosteum, which is adjacent to the bone's interior surface. As the B cells mature, they move, while maintaining contact with stromal cells, to the central axis of the marrow cavity. Later stages of maturation are less dependent on contact with stromal cells, allowing the B cells eventually to leave the bone marrow. The final stage of development, when immature B cells become mature B cells, can occur either in the bone marrow or in secondary lymphoid organs such as the spleen.

Figure 4.3 The development of B cells proceeds through stages defined by the rearrangement and expression of the immunoglobulin genes. In the stem cell, the immunoglobulin (Ig) genes are in the germline configuration. The first rearrangements are of the heavy-chain (H-chain) genes. Joining D_H to J_H defines the early pro-B cell, which becomes a late pro-B cell on joining V_H to DJ_H. Expression of a functional μ chain and its expression at the cell surface as part of the pre-B receptor defines the large pre-B cell. Large pre-B cells proliferate, producing small pre-B cells in which rearrangement of the light-chain (L-chain) gene occurs. Successful light-chain gene rearrangement and expression of IgM on the cell surface define the immature B cell. The mature B cell is defined by the use of alternative splicing of heavy-chain mRNA to place IgD on the cell surface as well as IgM.

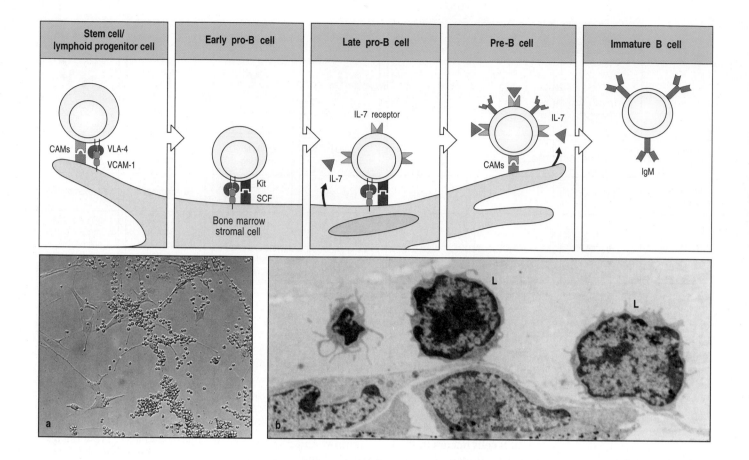

Although B-cell development in humans and mice is quite similar, there are considerable differences in other species. For example, in chickens and sheep, immunoglobulin gene rearrangement itself provides little or no diversity (see Section 2-11, p. 46). After completing heavy- and light-chain gene rearrangement, the B cells in these species migrate from the bone marrow to another anatomical site that provides a specialized microenvironment in which diversification of the immunoglobulin genes occurs. Chicken B cells migrate to the bursa of Fabricius, a lymphoid organ near the cloaca of the gut. B cells were in fact originally named after this organ. In the bursa, B cells are signaled to proliferate and their rearranged heavy- and light-chain genes then diversify through gene conversion (Figure 4.5). This process has similarities to the burst of human or mouse pre-B cell proliferation triggered by expression of the pre-B-cell receptor. The mechanism of gene conversion used to diversify chicken immunoglobulin genes is similar to that used to generate new HLA class I and II alleles in humans (see Section 3-20, p. 79). Sheep, in contrast, diversify their immunoglobulins by using somatic hypermutation, which occurs in an organ named the ileal Peyer's patch.

4-2 The survival of a developing B cell depends on the productive rearrangement of a heavy- and a light-chain gene

Because of imprecision in the gene rearrangement process, including the random addition of N and P nucleotides at joints between gene segments (see Section 2-8, p. 43), not all DNA rearrangements result in a sequence with a reading frame that can be translated into an immunoglobulin chain. Gene rearrangements that cannot be translated into a protein are called **unproductive rearrangements**. Rearrangements that preserve a correct reading frame and give rise to a complete

Figure 4.4 The early stages of B-cell development are dependent on bone marrow stromal cells. The top panels show the interactions of developing B cells with bone marrow stromal cells. Stem cells and early pro-B cells use the integrin VLA-4 to bind to the adhesion molecule VCAM-1 on stromal cells. This and interactions between other cell adhesion molecules (CAMs) promote the binding of the receptor Kit on the B cell to stem-cell factor (SCF) on the stromal cell. Activation of Kit causes the B cell to proliferate. B cells at a later stage of maturation require interleukin-7 (IL-7) to stimulate their growth and proliferation. Panel a is a light micrograph of a tissue culture showing small round B-cell progenitors in intimate contact with stromal cells, which have extended processes fastening them to the plastic dish on which they are grown. Panel b is a high-magnification electron micrograph showing two lymphoid cells (L) adhering to a flattened stromal cell. Photographs courtesy of A. Rolink (a); Paul Kincade and P.L. Witte (b).

Figure 4.5 Diversification of the immunoglobulin genes in chickens is achieved by gene conversion. In chickens the development of mature B cells involves successive modifications of the immunoglobulin genes in two anatomical sites. In the bone marrow, the heavy- and light-chain genes are rearranged. This process produces little immunoglobulin diversity. In the bursa, diversity is introduced by gene conversion between the rearranged genes and a series of V_H and V_λ pseudogenes.

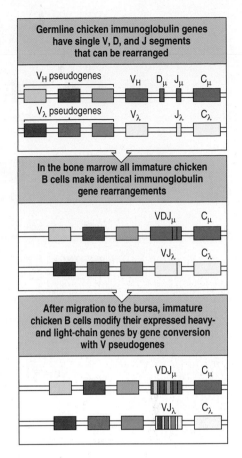

and functional immunoglobulin chain are known as **productive rearrangements**. At each rearrangement event, there is a one in three chance of the correct reading frame being maintained.

Every B cell has two copies of each of the immunoglobulin loci—the heavy-chain locus, the κ light-chain locus, and the λ light-chain locus. The two copies of each locus are on homologous chromosomes; one is inherited from the mother and the other from the father. In the developing B cell, gene rearrangements can be made on both homologous chromosomes. A B cell that has made an unproductive rearrangement on one chromosome therefore still has a chance of producing an immunoglobulin chain if it makes a productive rearrangement at the locus on the other, homologous, chromosome. Developing B cells are allowed to proceed to the next stage only when a productive rearrangement has been made. If all rearrangements are unproductive, the B cell does not produce immunoglobulin and dies in the bone marrow.

The order of gene rearrangements is outlined in Figure 4.6. The immunoglobulin heavy-chain locus rearranges before the light-chain loci. Recombination is heralded

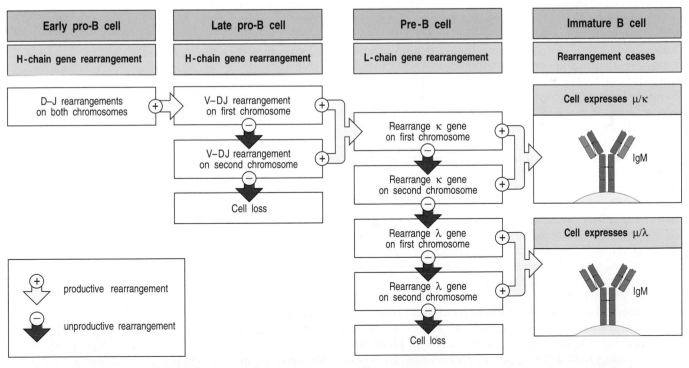

Figure 4.6 The order of gene rearrangements leading to the expression of cell-surface immunoglobulin. The heavy-chain genes are rearranged before the light-chain genes. Developing B cells are allowed to proceed to the next stage only when a productive rearrangement has been made. If an unproductive rearrangement is made on one chromosome of a homologous pair, then rearrangement is attempted on the second chromosome.

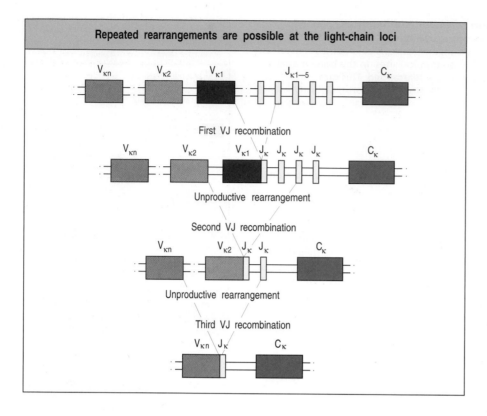

Figure 4.7 Unproductive light-chain gene rearrangements can be replaced by further gene rearrangement. The organization of the light-chain loci allows unproductive rearrangements to be followed by further rearrangement that can result in a functional gene. This type of rescue is shown for the κ light-chain gene. After an unproductive rearrangement of $V_{\kappa 1}$ to a J_κ in which the translational reading frame has been lost, a second rearrangement can be made by $V_{\kappa 2}$, or any other V_κ that is on the 5′ side of the first joint, with a J_κ that is on the 3′ side of the first joint. When the second joint is made, the intervening DNA containing the first joint will be excised. There are five J_κ gene segments and many more V_κ gene segments, so that as many as five successive attempts at productive rearrangement of the κ light-chain gene can be made on a single chromosome.

by expression of the **recombination activation genes *RAG-1*** and ***RAG-2***. The first rearrangement event is the joining of a D_H gene segment to a J_H gene segment, which can occur concurrently in the two copies of the heavy-chain locus. B cells with unproductive D_H to J_H rearrangements on both chromosomes are prevented from developing further because, although they will undergo V_H to DJ_H rearrangement, they cannot produce a heavy chain. In humans, few B cells are lost because of unproductive rearrangements at this stage, because most human D segments can be translated in all three reading frames. In contrast, mouse D segments have a single productive reading frame.

The second event in heavy-chain gene rearrangement is the joining of a V_H gene segment to the rearranged DJ_H sequence. This event first occurs at only one of the heavy-chain loci. When V_H to DJ_H rearrangement on the first chromosome is unproductive, rearrangement then proceeds on the second chromosome. The two-thirds failure rate in maintaining the reading frame, and the independent chances for success offered by having two chromosomes, mean that just over half the total number of pro-B cells make a functional heavy-chain gene. They become large dividing pre-B cells producing μ chains. Those pro-B cells that fail to make μ chains become programmed to die.

The large pre-B cells go through five or six rounds of cell division, after which they become small resting pre-B cells. At this stage, the light-chain genes begin to rearrange. Rearrangements take place at one light-chain locus at a time, with the κ locus tending to be rearranged before the λ locus. In contrast to the heavy-chain locus, several attempts to rearrange the same light-chain gene can be made by using V and J gene segments not involved in previous rearrangements (Figure 4.7). A number of attempts to rearrange the κ light-chain locus on one or other

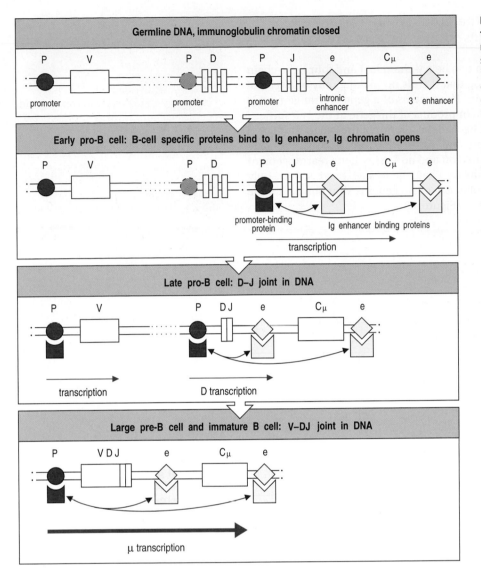

Figure 4.8 Low-level transcription facilitates immunoglobulin-gene rearrangements. First panel: in the stem cell, the immunoglobulin genes are in their germline configuration and the chromatin containing the immunoglobulin genes is tightly packed in a 'closed' configuration. The heavy-chain locus is illustrated here. The positions of promoter (P) and enhancer (e) gene regulatory elements are shown. Second panel: in the early pro-B cell, specific proteins bind to the enhancer elements in the J–C intron and in the region to the 3' side of the C exons. The chromatin is opened and low-level transcription occurs from promoters situated upstream of the J segments. Third panel: the rearrangement of D_H to J_H that follows the initiation of transcription of the J gene segments activates a low level of transcription from promoters situated upstream of the V and D gene segments. Fourth panel: subsequent rearrangement of a V gene segment brings its promoter under the influence of the heavy-chain enhancers, leading to the elevated production of a μ heavy-chain mRNA in large pre-B cells and their progeny.

chromosome can therefore be made before commencing rearrangement of a λ-chain locus on a different chromosome. The possession of four light-chain loci, and the opportunity for making several attempts to rearrange each locus, means that about 85% of the pre-B cell population makes a successful rearrangement of a light-chain gene. Overall, less than half of the B-lineage cells end up producing functional immunoglobulin heavy and light chains.

Before each rearrangement there is a small amount of transcription from the individual gene segments that are about to be joined (Figure 4.8). It is thought that this opens up the chromatin structure and makes the DNA accessible to the enzymes that carry out somatic recombination. It is likely that this low-level transcription is initiated by transcription factors expressed only in B cells, and is the mechanism that directs the recombination enzymes common to both B cells and T cells to the immunoglobulin genes rather than to the T-cell receptor genes (and vice versa in T cells). Also, the gene rearrangement process brings gene regulatory elements within the locus into closer juxtaposition with each other, so that the rearranged immunoglobulin gene can now be transcribed at a high rate.

4-3 Cell-surface expression of the products of rearranged immunoglobulin genes prevents further gene rearrangement

Given that a B cell can rearrange up to two heavy-chain loci and four light-chain loci, there are mechanisms for ensuring that it ends up making only one type of heavy chain and one type of light chain. The success of a gene rearrangement is signaled by the appearance of the protein product of the gene at the cell surface; once that has occurred, a signal is sent back to the interior of the cell to shut down the processes of DNA recombination and repair that are needed for gene rearrangement. The *RAG* genes are turned off and no further gene rearrangement is possible at this time. Once one heavy-chain gene has been successfully rearranged, all further rearrangement of heavy-chain genes is shut down, and the same happens with the light-chain genes (Figure 4.9).

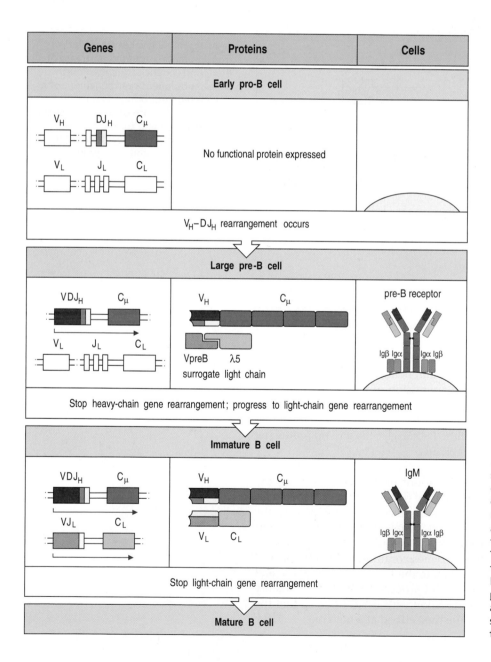

Figure 4.9 Signaling via the functional protein product is used to terminate immunoglobulin gene rearrangement. First panel: in early pro-B cells, no functional μ protein is expressed and the biochemical machinery that performs gene rearrangement is active. Second panel: on completion of a productive heavy-chain gene rearrangement, μ heavy chains are synthesized and assembled into a complex with the λ5 and VpreB polypeptides that make up the surrogate light chain, and with Igα and Igβ. This complex, the pre-B-cell receptor, is transported to the cell surface, where it interacts with an unknown ligand that signals the cell to halt heavy-chain gene rearrangement and to proceed to the large pre-B-cell stage. Proliferation of the large pre-B cell yields small pre-B cells that commence immunoglobulin light-chain rearrangement. Note that λ5 and VpreB are coded by genes that are distinct from those in the immunoglobulin loci. Third panel: on completion of a productive light-chain gene rearrangement a light chain is made and assembles with μ to form IgM. This associates with Igα and Igβ and is transported to the cell surface. Its presence there tells the cell to halt light-chain gene rearrangements.

Once it has formed a functional heavy-chain gene, a pro-B cell begins to synthesize μ chains. In the endoplasmic reticulum of the cell, μ chains assemble into dimers. These cannot, however, assemble with κ or λ light chains to form a regular IgM receptor because the B cell has not yet rearranged its light-chain genes. The receptor that reaches the surface at this stage contains a μ dimer in which each μ chain is bound to an invariant **surrogate light chain**. Each surrogate light chain is formed from two proteins coded by genes away from the immunoglobulin loci; **λ5** substitutes for a light-chain constant region and **VpreB** substitutes for a variable region.

The tetramer of two μ chains and two surrogate light chains associates with the Igα and Igβ polypeptides (see Section 2-10, p. 45) to form a complex that moves to the cell surface to become the pre-B-cell receptor. This receptor is believed to interact with an unknown ligand provided by a self protein, perhaps on the surface of a bone marrow stromal cell. Interaction with this ligand could then signal the pre-B cell to shut down the machinery of somatic recombination and gear up the biosynthetic machinery for the DNA replication and cell division that now follows.

The pre-B-cell receptor is present at the cell surface for only a short period of time, which represents a checkpoint in B-cell development. This checkpoint allows the B cell to check whether a productive rearrangement has occurred and a functional heavy chain has been made. Expression of the pre-B-cell receptor allows the B cell to proceed to the next stage of maturation. After the checkpoint, the surrogate light chain ceases to be synthesized and the pre-B-cell receptor gradually disappears from the cell surface. The μ, Igα, and Igβ chains continue to be synthesized but are now retained in the endoplasmic reticulum.

The *RAG* genes, which were turned off in the dividing pre-B cells, are now turned on again and light-chain gene rearrangement proceeds. When a functional light-chain gene has been formed, light chains are made and assemble with μ to form IgM. This associates with Igα and Igβ to form the mature B-cell receptor, which moves to the cell surface. At this second checkpoint in B-cell maturation, the appearance of a functional B-cell receptor at the cell surface leads to the shutting down of light-chain gene rearrangement.

Cell division initiated by signaling through the pre-B-cell receptor results in a clone of 30–70 small pre-B cells. These all have the same rearranged heavy-chain gene but each has the potential to end up with a different light chain. The combinations of different heavy and light chains generated at this stage contribute to the diversity of the B-cell repertoire.

4-4 The proteins involved in immunoglobulin-gene rearrangement are developmentally controlled

The successive steps in immunoglobulin-gene rearrangement are accompanied by the developmentally regulated expression of the proteins involved (Figure 4.10). Expression of the recombination activation genes *RAG-1* and *RAG-2* is crucial for the rearrangement of both the immunoglobulin genes and the T-cell receptor genes. They are specifically turned on at the stages at which gene rearrangement occurs and encode proteins involved in recognition of the recombination signal sequences (see Section 2-7, p. 42) and making double-strand breaks in the DNA. Addition of N nucleotides at the junctions between rearranging gene segments is due to the enzyme terminal deoxynucleotidyl transferase (TdT). This is expressed in pro-B cells when the heavy-chain genes begin to rearrange, and turned off in small pre-B cells when the light-chain genes begin to rearrange; this explains why N nucleotides are found in all VD and DJ joints of rearranged human heavy-chain genes but in only about half of the VJ joints of rearranged human light-chain genes.

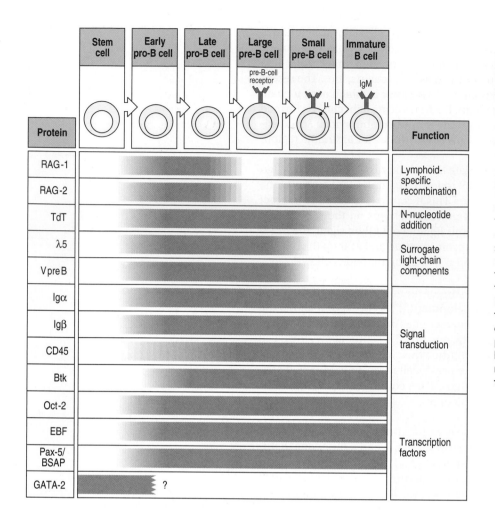

Figure 4.10 The expression of proteins involved in the rearrangement and expression of immunoglobulin genes changes during B-cell development. The rearrangement of immunoglobulin genes and the expression of the pre-B-cell receptor and IgM on the cell surface requires several categories of specialized proteins at different times during B-cell development. Examples of such proteins are listed here, with their expression during B-cell development shown with red shading. Of the proteins not discussed in the text, EBF regulates the transcription of the gene for Igα, and the B-lineage specific-activator protein (BSAP) regulates the expression of several of the other proteins listed. Oct-2 acts at the heavy-chain promoter and GATA-2 is a transcription factor active in several types of hematopoietic cells. CD45 is a cell-surface phosphotyrosine phosphatase that is restricted to hematopoietic cells. It is involved in regulating signal transduction from the pre-B cell and B-cell receptors.

The Igα and Igβ polypeptides are necessary for the cell-surface expression of immunoglobulin and for the transduction of signals from B-cell and pre-B-cell receptors to the interior of the cell. Expression of Igα and Igβ is turned on at the pro-B cell stage and continues throughout the life of a B cell. Only when antigen-stimulated B cells differentiate into antibody-secreting plasma cells, and antigen receptors are no longer required, are the Igα and Igβ genes turned off and immunoglobulin ceases to appear on the cell surface.

A third category of protein contributing to B-cell development is involved in the transduction of signals from cell-surface receptors. One example is the Bruton's tyrosine kinase, Btk, which is encoded by a gene on the X chromosome and is essential for B-cell maturation. Patients who lack a functional Btk gene have almost no circulating antibodies because their B cells are blocked at the pre-B-cell stage. The immune deficiency suffered by these patients is called **X-linked agammaglobulinemia** and leads to recurrent infections from common extracellular bacteria such as *Haemophilus influenzae*, *Streptococcus pneumoniae*, *Streptococcus pyogenes*, and *Staphylococcus aureus*. The infections respond to treatment with antibiotics and can be prevented by regular intravenous infusions of immunoglobulin from pooled blood from healthy donors. Because the Btk gene is located on the X chromosome and the defective gene is recessive, X-linked agammaglobulinemia is seen mostly in boys.

A variety of transcription factors control the genes involved in immunoglobulin gene rearrangement and B-cell development, only some of which are shown in Figure 4.10. Their coordinated activity ensures that the enzymatic machinery for gene rearrangement common to B and T cells is used in B cells to rearrange and express the immunoglobulin genes, and not the T-cell receptor genes.

4-5 Many B-cell tumors carry chromosomal translocations that join immunoglobulin genes to genes regulating cell growth

As B cells cut, splice, and mutate their immunoglobulin genes in the normal course of events, it is perhaps not surprising that this process sometimes goes awry to produce a mutation that helps convert a B cell into a tumor cell. Transformation of a normal cell into a tumor cell involves a series of mutations that release the cell from the normal restraints on its growth. In B-cell tumors, the disruption of regulated growth is often associated with an aberrant immunoglobulin-gene rearrangement that has joined an immunoglobulin gene to a gene on a different chromosome. Events that fuse part of a chromosome with another are called **translocations** and, in B-cell tumors, the immunoglobulin gene has often become joined to a gene involved in the control of cellular growth. Genes that cause cancer when their function or expression is perturbed are collectively called **proto-oncogenes**. Many of them were discovered in the study of RNA tumor viruses which can transform cells directly. The viral genes responsible for transformation were named **oncogenes**, and it was only later realized that they had evolved from cellular genes that control cell growth, division, and differentiation.

The translocations between immunoglobulin genes and proto-oncogenes in B-cell tumors can be seen in metaphase chromosomes examined in the light microscope. Certain translocations define particular types of tumor and are valuable in diagnosis. In Burkitt's lymphoma, for example, the *MYC* proto-oncogene on chromosome 8 is joined by translocation to either an immunoglobulin heavy-chain gene on chromosome 14, a κ light-chain gene on chromosome 2 or a λ light-chain gene on chromosome 22 (Figure 4.11). The Myc protein is normally involved in regulating the cell cycle, but in B cells that carry these translocations its expression is abnormal, removing some of the restraints on cell division. Single genetic changes are rarely sufficient to transform cells malignantly. To produce Burkitt's lymphoma, mutations elsewhere in the genome are required in addition to the translocation between *MYC* and an immunoglobulin gene.

Another translocation found in B-cell tumors is the fusion of an immunoglobulin gene to the proto-oncogene *BCL2*. Normally, the function of the Bcl-2 protein is to prevent premature apoptosis, or programmed cell death, in B-lineage cells. Production of Bcl-2 in the tumors enables B cells to live longer than is normal, during which time they accumulate additional mutations that can lead to malignant transformation. In addition to their value in immunological research, B-cell tumors have provided fundamental advances in knowledge of the proteins involved in the mechanism and regulation of cell division.

4-6 B cells expressing the glycoprotein CD5 express a distinctive repertoire of receptors

Not all B cells conform exactly to the developmental pathway described above. A subset of human B cells that arises early in embryonic development is distinguished from the other B cells by the expression of CD5, a cell-surface glycoprotein that is otherwise considered a marker for the human T-cell lineage. This minority subset of B cells is termed **B-1 cells**, because their development precedes that of the majority subset, whose development we have traced in

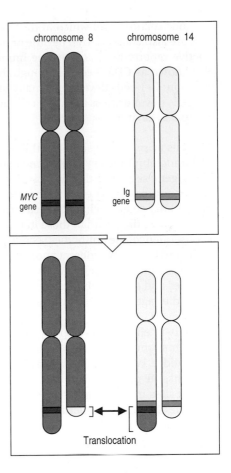

Figure 4.11 Chromosomal rearrangements in Burkitt's lymphoma. In this example from a Burkitt's lymphoma, parts of chromosome 8 and chromosome 14 have been exchanged. The sites of breakage and rejoining are in the proto-oncogene *MYC* on chromosome 8 and the immunoglobulin (Ig) heavy-chain gene on chromosome 14. In these tumors it is usual for the second immunoglobulin gene to be productively rearranged and for the tumor to express cell-surface immunoglobulin. The translocation probably occurred during the first attempt to rearrange a heavy-chain gene. This counted as an unproductive rearrangement and so the other gene was rearranged. In cases where the second rearrangement is also unproductive, the cell dies and thus cannot give rise to a tumor.

previous sections and which in this context are sometimes called **B-2 cells**. B-1 cells are also characterized by little or no cell-surface expression of IgD and a distinctive repertoire of cell-surface immunoglobulin receptors. The B-1 cells are also known as **CD5 B cells**, although the CD5 molecule is not essential for their function: B-1 cells develop normally in mice lacking CD5 glycoprotein, and rat B-1 cells never express CD5. B-1 cells are the dominant B cells in the pleural and peritoneal cavities.

B-1 cells arise from a stem cell that is most active in the prenatal period. Their immunoglobulin heavy-chain gene rearrangements are dominated by the use of the V_H gene segments that lie closest to the D gene segments in the germ line. As TdT is not expressed early in the prenatal period, the rearranged heavy-chain genes of B-1 cells are characterized by a lack of N nucleotides, and their VDJ junctions are less diverse than those in rearranged heavy-chain genes of B-2 cells. As a consequence, the antibodies secreted by B-1 cells tend to be of low affinity and each binds to many different antigens. This property is known as **polyspecificity**. B-1 cells contribute to the antibodies made against common bacterial polysaccharides and other carbohydrate antigens but are of little importance in making antibodies against protein antigens.

Those B-1 cells that develop postnatally use a more diverse repertoire of V gene segments and their rearranged immunoglobulin genes have abundant N nucleotides. With time, B-1 cells are no longer produced by the bone marrow and, in adults, the population of B-1 cells is maintained by the division of existing B-1 cells at sites in the peripheral circulation. This self-renewal is dependent on the cytokine IL-10. The properties of B-1 and B-2 cells are compared in Figure 4.12. In the remainder of this chapter and in the rest of this book, B cell will refer to the population of B-2 cells, unless specified otherwise.

Possibly as a direct result of their capacity for self-renewal, B-1 cells are usually the source for the common B-cell tumors causing chronic lymphocytic leukemia (CLL). A bone-marrow transplant from an HLA-identical sibling is often a successful treatment for CLL. Chemotherapy and/or radiation are used to destroy the dividing cells of the patient's hematopoietic system, which includes the malignant B-1 cells. The donor bone marrow then provides the stem cells that repopulate the immune system with healthy cells.

Property	B-1 cells	Conventional B-2 cells
When first produced	During fetal life	After birth
Mode of renewal	Self-renewing	Replaced from bone marrow
Spontaneous production of immunoglobulin	High	Low
Specificity	Polyreactive	Monospecific, especially after immunization
Isotypes secreted	IgM >> IgG	IgG > IgM
Somatic hypermutation	Low to none	High
Response to carbohydrate antigen	Yes	Maybe
Response to protein antigen	Maybe	Yes

Figure 4.12 Comparison of the properties of B-1 cells and B-2 cells. B-1 cells develop in the omentum as well as in the liver in the fetus, and are produced by the bone marrow for only a short period around the time of birth. A pool of self-renewing B-1 cells is then established, which does not require the microenvironment of the bone marrow for its survival. The limited diversity of the antibodies made by B-1 cells and their tendency to be polyreactive and of low affinity suggests that B-1 cells produce a simpler, less adaptive, immune response than that involving B-2 cells.

Summary

B cells are generated throughout life by development from stem cells in the bone marrow. Different stages in B-cell maturation are correlated with molecular changes that accompany the gene rearrangements required to make functional immunoglobulin heavy and light chains. Many rearrangements are unproductive, an inefficiency compensated for, in part, by the presence of two heavy-chain loci and four light-chain loci in each B cell, all of which can be rearranged in the attempt to obtain a functional heavy chain and light chain. The heavy-chain genes are rearranged first, and only when this produces a functional protein is a B cell allowed to rearrange its light-chain genes. After a successful heavy-chain gene rearrangement, the μ chain is expressed at the cell surface with a surrogate light chain, which signals the cell to halt heavy-chain gene rearrangement and to proceed toward light-chain gene rearrangement. Similarly, after a successful light-chain gene rearrangement, the expression of IgM on the cell surface signals the cell to halt light-chain gene rearrangement. These feedback mechanisms, whereby protein product regulates the progression of gene rearrangement, ensure that each B cell expresses only one heavy chain and one light chain and thus produces immunoglobulin of a single defined antigen specificity. In this manner, the program of gene rearrangement produces the monospecificity essential for the efficient operation of clonal selection. The success rate in obtaining a functional light-chain gene rearrangement is much higher than that for the heavy-chain gene because only two gene segments are involved (compared with three for the heavy-chain gene) and the B cell has four light-chain genes (two κ and two λ) with which to work. Errors in immunoglobulin-gene rearrangement give rise to chromosomal translocations that predispose B cells carrying them to malignant transformation. A minority class of B cells, termed B-1 cells, first develops very early in embryonic life. They produce antibodies that tend to bind to bacterial polysaccharides, but are polyspecific, generally of low affinity, and mainly of IgM isotype. In aggregate, the properties of B-1 cells suggest that they represent a simpler and evolutionarily older lineage of B cell than the majority population.

Selection and further development of the B-cell repertoire

In the previous part of this chapter we saw how the population of immature B cells develops a repertoire of receptors of different antigen specificities. Included in this repertoire are immunoglobulins that bind to normal constituents of the body and have the potential to initiate a damaging immune response. To prevent such responses, mechanisms that either eliminate or inactivate potentially autoreactive B cells now come into play. Once the population of maturing B cells has been purged of self-reactive cells the remaining cells are released to the circulation, where they can engage specific antigen and perform their effector function of producing antibodies.

4-7 Self-reactive immature B cells are eliminated or inactivated by contact with self antigens

When a B cell first expresses IgM on its surface, it is called an immature B cell. Becoming a mature B cell involves emigration from the bone marrow and the use of alternative splicing of heavy-chain mRNA to place IgD, as well as IgM, on the cell surface (see Section 2-9, p. 44). Some immature B cells fail to make this transition because, by chance, their immunoglobulins react with normal components of the body, which in this immunological context are called **self-antigens**. One category of self-antigens consists of the glycoproteins, proteoglycans, and glycolipids that are expressed on the surfaces of many types of cell and are accessible for interaction with the immunoglobulin receptors of B cells. When an immature B cell's surface IgM molecules are crosslinked by a self-antigen on the surface of any cell, the B cell is induced to commit suicide by apoptosis. The apoptotic cell is then removed through ingestion by macrophages.

The elimination of immature B-cell clones with specificity for self-antigens on cell surfaces is called **clonal deletion** (Figure 4.13, left panels), and occurs within the bone marrow or shortly after the B cell enters the peripheral circulation. Some 55 billion B cells per day are lost in the bone marrow because they fail to make a functional immunoglobulin, or are autoreactive and subject to clonal deletion. Cells that survive this selection process go on to express IgD as well as IgM on their surface (see Figure 4.13, right panels). When a B cell expresses only IgM, it undergoes apoptosis upon encounter with its specific antigen on a cell surface; once IgM and IgD are both expressed, however, the B cell has become a mature naive B cell and the identical encounter leads to B-cell activation and an immune response.

A second category of common self-antigens consists of soluble proteins and glycoproteins, some of which are present at very high concentrations in the blood and lymph through which B cells circulate. When the surface IgM of an immature B cell binds to a soluble self-antigen, the B cell is inactivated but it does not die. This state, in which the B cell matures but does not respond to subsequent exposure to antigen, is called **anergy** (see Figure 4.13, center panels). In anergic cells, most of the IgM is retained inside the cell, whereas normal levels of IgD are present at the cell surface. However, binding of antigen to the surface IgD does not activate the B cell.

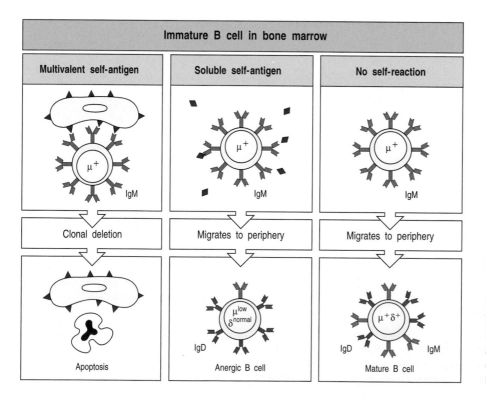

Figure 4.13 Binding to self-antigens in the bone marrow can lead to the deletion or inactivation of immature B cells. When immature B cells express receptors that recognize common cell-surface components of human cells, they are deleted from the repertoire by the induction of apoptosis, as shown in the left panels. Immature B cells that bind soluble self-antigens are rendered unresponsive or anergic to the antigen, as shown in the center panels, and as a consequence express low levels of IgM at the cell surface. They enter the peripheral circulation, where they express IgD, but remain anergic. Immature B cells that do not encounter a stimulatory self-antigen leave the bone marrow and enter the peripheral circulation, expressing both IgM and IgD on their surface, as shown in the right panels.

4-8 Mature naive B cells compete for access to lymphoid follicles

When mature B cells leave the bone marrow, they recirculate between the blood, the secondary lymphoid tissues, such as lymph nodes, spleen, and mucosa-associated lymphoid tissues, and the lymph (see Figure 1.8, p. 11). Within secondary lymphoid tissues, B cells congregate in organized structures called **primary lymphoid follicles** (Figure 4.14). These consist principally of B cells and a specialized stromal cell called the follicular dendritic cell. Mature B cells recirculate, passing from the blood into the primary lymphoid follicles and back through the lymphatics into the blood (Figure 4.15). In the spleen B cells enter and exit via the blood. B cells are particularly enriched in the gut-associated lymphoid tissues, which include the large lymphoid follicles known as Peyer's patches, and also the appendix and tonsils. These tissues in the gastrointestinal tract provide microenvironments in which B cells become committed to the synthesis of dimeric, secretory IgA.

Before it has made contact with its specific antigen, a mature naive B cell must pause periodically in a primary follicle in order to survive and continue its recirculation. In the follicle it receives survival signals that may come directly from the follicular dendritic cells. Throughout a person's life, the bone marrow continues to produce new B cells, which must compete with the existing population of naive B cells in the peripheral circulation. In the bone marrow of a healthy young adult, about 2.5 billion (2.5×10^9) cells per day embark on the program of B-cell development. From these progenitors, some 30 billion B cells leave the bone marrow daily to become circulating naive B cells.

There is continual competition between circulating naive B cells for passage through a limited number of follicular sites; those B cells that fail to gain regular access to a follicle die. Competition is so intense that the majority of mature naive B cells die after only a few days in the peripheral circulation. Naive B cells that gain access to primary follicles are longer lived, but they too disappear from the system with a half-life of 3–8 weeks unless they are stimulated by encounter with their specific antigen.

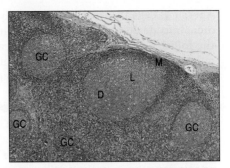

Figure 4.14 Anatomy of lymphoid follicles. This section through a human lymph node shows secondary lymphoid follicles, each of which has developed a germinal center (GC). Three zones can be recognized in a follicle, an outer mantle zone of resting B cells (M), a dark zone of proliferating blasts (D), and a light zone of differentiating cells (L). The areas between the follicles contains T cells, interdigitating cells (antigen-presenting), blood vessels, and sinuses. Courtesy of N. Rooney.

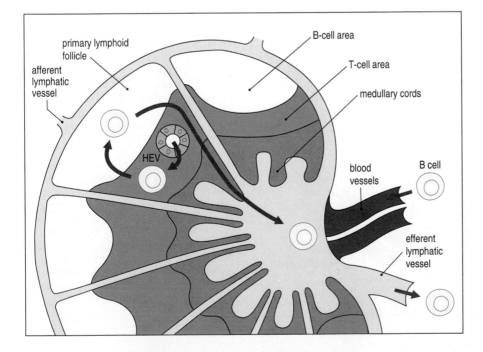

Figure 4.15 The circulation route of mature naive B cells through a lymph node. Having matured in the bone marrow, B cells migrate in the blood to lymph nodes and other secondary lymphoid tissues. B cells leave the blood and enter the cortex of the lymph node through the walls of specialized high endothelial venules (HEV). If they do not encounter their specific antigen, the B cells pass through the primary follicles and leave the node in the efferent lymph, which eventually joins the blood at veins in the neck.

Anergic B cells that reach secondary lymphoid tissue are detained in the T-cell areas outside the primary lymphoid follicles and are thus prevented from entering the follicles. Because anergic B cells cannot be activated, they rapidly undergo apoptosis from lack of stimulation. This ensures that the population of recirculating mature B cells is continuously purged of potentially self-reactive cells.

4-9 Encounter with antigen leads to the differentiation of activated B cells into plasma cells and memory B cells

Secondary lymphoid tissues provide the sites where mature naive B cells encounter specific antigen. When that happens, the antigen-specific B cells are detained in the T-cell areas, where they become activated by antigen-specific CD4 helper T cells. These T cells provide signals that activate the B cells to proliferate and differentiate further. In lymph nodes and spleen, some of the activated B cells proliferate and differentiate straight away into **plasma cells**, which secrete antibody (Figure 4.16). Antibody secretion is effected by a change in the processing of the heavy-chain mRNA that leads to the synthesis of the secreted form of immunoglobulin rather than the membrane-bound form (see Section 2-12, p. 47).

Plasma cells are totally specialized towards the constitutive synthesis and secretion of antibody; the cellular organelles involved in protein synthesis and secretion become highly developed, and between 10 and 20% of the total cell protein made is antibody. As part of this program of terminal differentiation, plasma cells cease to divide, have a limited life-span of about 4 weeks, and no longer express cell-surface immunoglobulin and MHC class II. They thus become unresponsive to antigen and to interaction with T cells.

Other activated B cells migrate to a nearby primary follicle, which changes its morphology to become a **secondary lymphoid follicle** containing a **germinal center** (see Figure 4.16). Here, the activated B cells become large proliferating lymphoblasts called **centroblasts**; these mature into non-dividing B cells called **centrocytes**, which have undergone isotype switching and somatic hypermutation. Those B cells that make surface immunoglobulins with highest affinity for

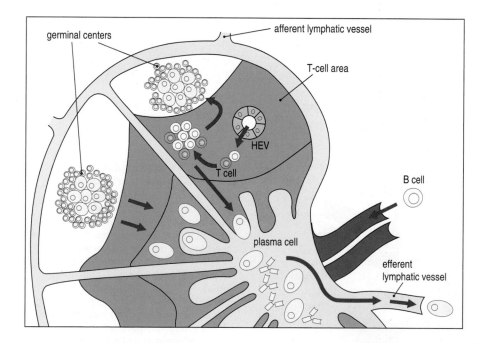

Figure 4.16 B cells encountering antigen in secondary lymphoid tissues form germinal centers and undergo differentiation to plasma cells. A lymph node is illustrated here. A B cell entering the lymph node through a high endothelial venule (HEV) encounters antigen in the lymph node cortex. Antigen was delivered in the afferent lymph that drained from infected tissue. The B cell is activated by CD4 T helper cells (blue) in the T-cell areas to form a primary focus of dividing cells. From this, some B cells migrate directly to the medullary cords and differentiate into antibody-secreting plasma cells. Other B cells migrate into a primary follicle to form a germinal center. B cells continue to divide and differentiate within the germinal center. Activated B cells migrate from the germinal center to the medulla of the lymph node or to the bone marrow to complete their differentiation into plasma cells.

the antigen are selected by the process of affinity maturation (see Section 2-13, p. 48), which occurs in germinal centers. A few weeks after a germinal center is formed, the intense cellular activity, called the germinal center reaction, dies down and the germinal center shrinks in size.

Cells that survive the selection process of affinity maturation undergo further proliferation as lymphoblasts and most migrate from the germinal center to other sites in the secondary lymphoid tissue or to bone marrow, where they complete their differentiation into plasma cells secreting high-affinity, isotype-switched, antibodies. As the primary immune response subsides, germinal center B cells also develop into quiescent resting **memory B cells** possessing high-affinity, isotype-switched, antigen receptors. The production of memory cells after a successful encounter with antigen establishes antigen specificities of proven usefulness permanently in the B-cell repertoire.

Memory B cells persist for long periods of time, and in their recirculation through the body they require only intermittent stimulation in the follicular environment. They are much more easily activated on encounter with antigen than are naive B cells. Their rapid activation and differentiation into plasma cells on a subsequent encounter with antigen enable a secondary antibody response to an antigen to develop more quickly and become stronger than the primary immune response. It also explains why IgG and antibodies of isotype other than IgM predominate in secondary responses.

When B cells become committed to differentiation into plasma cells, they migrate to particular sites in the lymphoid tissues. In lymph nodes these are the medullary cords, in the spleen the red pulp, and in the gut-associated lymphoid tissues prospective plasma cells migrate to the lamina propria, which lies immediately under the gut epithelium. Prospective plasma cells also migrate from lymph nodes and spleen to the bone marrow, which becomes a major site of antibody production. So in one sense, the life of a B-cell both starts and ends in the bone marrow.

4-10 Different types of B-cell tumor reflect B cells at different stages of development

The study of B-cell tumors has provided fundamental insights into both B-cell development and the control of cell growth generally. B-cell tumors arise from both the B-1 and B-2 lineages and from B cells at different stages of maturation and differentiation. The general principle that a tumor represents the uncontrolled growth of a single transformed cell is vividly illustrated by tumors derived from B-lineage cells. In a B-cell tumor, every cell has an identical immunoglobulin-gene rearrangement, proof of their derivation from the same ancestral cell. Although the cells of a patient's B-cell tumor are homogeneous, the B-cell tumors from different patients exhibit a heterogeneity reflective of that seen in the normal B cells of a healthy person.

Tumors retain characteristics of the cell type from which they arose, especially when the tumor is relatively differentiated and slow-growing. This principle is exceptionally well illustrated by the tumors of B cells. Human tumors corresponding to all the stages of B-cell development have been described, from the most immature progenitor to the highly differentiated plasma cell (Figure 4.17). Among the characteristics retained by the tumors is their location at defined sites in the lymphoid tissues. Tumors derived from mature naive B cells grow in the follicles of lymph nodes and form **follicular center cell lymphoma**, whereas plasma-cell tumors, called **myelomas**, propagate in the bone marrow.

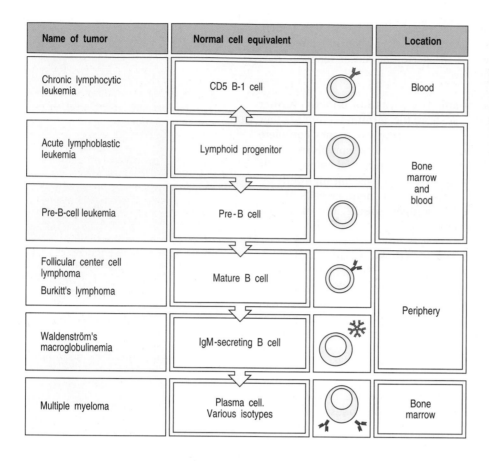

Name of tumor	Normal cell equivalent		Location
Chronic lymphocytic leukemia	CD5 B-1 cell		Blood
Acute lymphoblastic leukemia	Lymphoid progenitor		Bone marrow and blood
Pre-B-cell leukemia	Pre-B cell		
Follicular center cell lymphoma Burkitt's lymphoma	Mature B cell		Periphery
Waldenström's macroglobulinemia	IgM-secreting B cell		
Multiple myeloma	Plasma cell. Various isotypes		Bone marrow

Figure 4.17 The different B-cell tumors reflect the heterogeneity of developmental and differentiation states in the normal B-cell population. Each type of tumor corresponds to a normal state of B-cell development or differentiation. Tumor cells have similar properties to their normal cell equivalent, they migrate to the same sites in the lymphoid tissues, and have similar patterns of expression of cell-surface glycoproteins.

B-cell tumors have proved invaluable in the study of the immune system. What they uniquely provide are cells that largely represent what happens normally but can be had in large quantity, which is highly abnormal. The very first amino-acid sequences of antibody molecules were obtained from patients with plasma-cell tumors, whose bodily fluids are dominated by a single species of antibody. B-cell tumors of mice have also been instrumental in defining the pathways of migration and recirculation of B cells. They provide defined cells that can be grown, manipulated and modified in the laboratory and then tested for their properties *in vivo*.

Summary

B-cell development is inherently wasteful because the vast majority of B cells die without ever contributing to an immune response. Immature B cells in the bone marrow express only IgM at the cell surface. Those cells whose IgM reacts with a self-antigen are removed at this stage by apoptosis or are rendered anergic. B cells that survive this test go on to become mature B cells expressing both IgM and IgD on their surface. Anergic B cells leaving the bone marrow are excluded from the primary lymphoid follicles and have a half-life within the peripheral circulation of only a few days. Even mature B cells that are tolerant to self-antigens and potentially responsive to microorganisms are regularly abandoned, because of the intense competition for access to primary lymphoid follicles. Only after a B cell encounters specific antigen in the secondary lymphoid tissues is it activated to proliferate into a large clone of cells with identical antigen specificity. No sooner has the clone expanded than individual B cells in the clone begin to differentiate and diverge. Some become plasma cells which service the immediate need for antigen-specific antibody, while others become memory cells providing long-term immunity for the future. Over and above these changes, B cells within

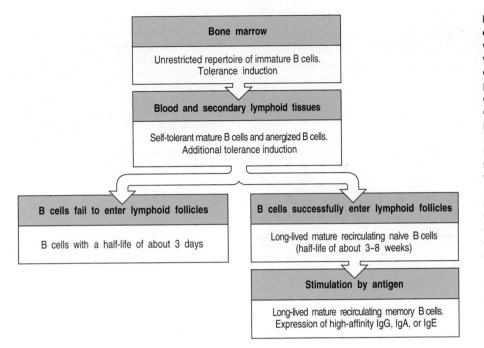

Figure 4.18 Population dynamics of B cells. The B-cell population from which antibody-producing plasma cells will eventually develop is a continually changing and heterogeneous population, consisting of B cells at different stages of development and differentiation. Immature B cells produced in the bone marrow have immunoglobulins specific for the entire range of molecules found in biological systems. Those B cells with receptors that bind to self-antigens present in the bone marrow are either eliminated by clonal deletion or inactivated. Passage through lymphoid follicles is essential for sustaining mature naive B cells in the peripheral circulation; those that fail to enter a follicle soon die. B cells that passage through a follicle will also eventually die unless stimulated by specific antigen, whereupon they proliferate, differentiate, and give rise to memory B cells.

the clone alter their immunoglobulins so that antibodies of higher affinity and more potent effector function can be made (Figure 4.18). The stages of antigen-dependent differentiation correspond to changes in either the structure or the expression of the immunoglobulin genes. A change in the processing of μ heavy-chain mRNA in some cells of the clone leads to the synthesis of secreted IgM antibody. In other cells of the clone, further DNA rearrangement switches the antibody isotype so that cell surface IgG, IgA, and IgE and antibodies can be produced, while somatic hypermutation of rearranged variable regions and selection for higher-affinity immunoglobulin lead to the production of antibodies of increasingly higher affinity for the antigen. Some isotype-switched B cells with high-affinity receptor immunoglobulin become memory cells, which will differentiate into plasma cells on a subsequent encounter with the same antigen. The terminal stage of B-cell development is thus the plasma cell, dedicated entirely to antibody production. B-cell tumors have been instrumental in studying B-cell development because different types of tumor correspond in cell type and location to different stages of normal B-cell development.

Summary to Chapter 4

B lymphocytes are highly specialized cells whose sole function is to recognize foreign antigens by means of cell-surface immunoglobulins and then to differentiate into plasma cells that secrete antibodies of the same antigen specificity. Each B cell expresses immunoglobulin of a single antigen specificity but, as a population, B cells express a diverse repertoire of immunoglobulins. This enables the B-cell response to any antigen to be highly specific. Immunoglobulin diversity is a result of the unusual arrangement and mode of expression of the immunoglobulin genes. In B-cell progenitors the immunoglobulin genes are in the form of arrays of different gene segments that can be rearranged in many different combinations. Gene rearrangement occurs in bone marrow and is independent of B-cell encounter with specific antigen. The gene rearrangements needed for the expression of a functional surface immunoglobulin follow a program; the successive steps define the antigen-independent stages of B-cell development as

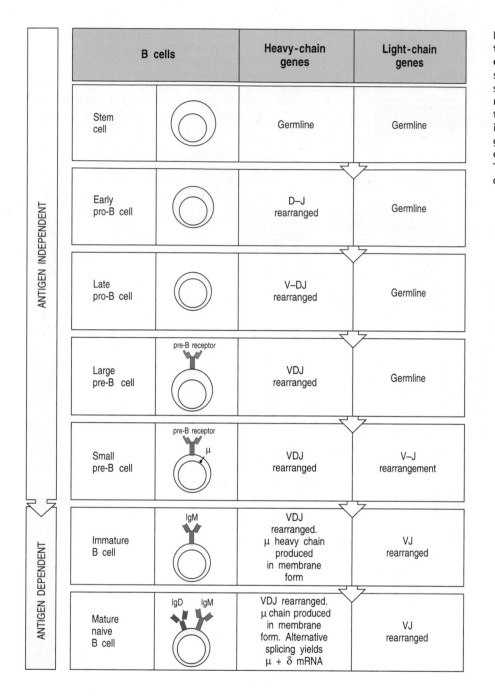

B cells		Heavy-chain genes	Light-chain genes
Stem cell		Germline	Germline
Early pro-B cell		D–J rearranged	Germline
Late pro-B cell		V–DJ rearranged	Germline
Large pre-B cell	pre-B receptor	VDJ rearranged	Germline
Small pre-B cell	pre-B receptor μ	VDJ rearranged	V–J rearrangement
Immature B cell	IgM	VDJ rearranged. μ heavy chain produced in membrane form	VJ rearranged
Mature naive B cell	IgD IgM	VDJ rearranged. μ chain produced in membrane form. Alternative splicing yields μ + δ mRNA	VJ rearranged

ANTIGEN INDEPENDENT

ANTIGEN DEPENDENT

Figure 4.19 A summary of the first two main phases of B-cell development. This diagram shows the stages in B-cell development from the stem cell in bone marrow to the mature naive B cell. The location of B cells at the different stages, the state of the immunoglobulin heavy- and light-chain genes, and the form of immunoglobulin expressed at each stage are indicated. This figure refers only to the development of B-2 cells.

shown in Figure 4.19. The success rate for individual rearrangements is far from optimal, but the use of a stepwise series of reactions allows the quality of the products to be tested at critical checkpoints after heavy- and light-chain gene rearrangement. Failure at any step leads to the death of the B cell. Gene rearrangement is controlled so that only one functional heavy-chain gene and one functional light-chain gene are produced in each cell. Individual B cells thus produce immunoglobulin of a single antigen specificity. For a short period after successful immunoglobulin gene rearrangement, any interaction with specific antigen leads to the elimination or inactivation of the immature B cell, thus rendering the mature B-cell population tolerant of the normal constituents of the body.

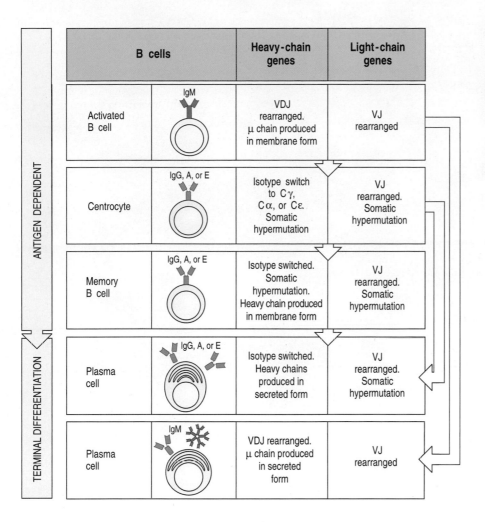

Figure 4.20 A summary of the last two main phases of B-cell development. This diagram shows the stages in B-cell development from the activated mature B cell to the terminally differentiated plasma cell. Plasma cells can differentiate directly from activated B cells, from isotype switched, somatically hypermutated centrocytes, or from memory B cells. This figure refers only to the development of B-2 cells.

Every day the bone marrow pushes out billions of new mature naive B cells into the peripheral circulation. There is, however, limited space for B cells in the secondary lymphoid tissues and unless naive B cells encounter specific antigen they are likely to be dead within weeks. Thus, the B-cell repertoire is never static, and newly generated specificities are continually being tested against the antigens of microorganisms causing infection. Binding of antigen to the cell-surface immunoglobulin of a B cell in secondary lymphoid tissue initiates an antigen-dependent program of cellular proliferation and development and the subsequent terminal differentiation, as shown in Figure 4.20. After encounter with antigen in secondary lymphoid tissues, B cells either differentiate directly into IgM-secreting plasma cells or undergo somatic hypermutation, isotype switching, and affinity maturation in germinal centers before differentiation into plasma cells or into long-lived memory B cells. The end-product of B-cell development is the plasma cell, in which surface immunoglobulin is no longer expressed and all the cell's resources are devoted to antibody secretion.

The Development of T Lymphocytes 5

The paths of development for T and B lymphocytes have much in common: both types of cell derive from bone marrow stem cells and, during development, they must undergo gene rearrangement to produce their antigen receptors. But whereas B cells rearrange their immunoglobulin genes while remaining in the bonc marrow, the precursors of T cells have to leave the bone marrow and enter another primary lymphoid organ—the thymus—before they can rearrange their T-cell receptor genes. The gene rearrangements in developing T cells proceed in a broadly similar fashion to those in B cells. T-cell development has, however, to account for the existence of two distinct lineages of T-cell receptors, one expressing $\alpha{:}\beta$ receptors and the other $\gamma{:}\delta$ receptors.

Where the development of B cells and T cells differ radically is in the selection that the primary repertoire of antigen receptors must undergo. The MHC restriction of $\alpha{:}\beta$ T-cell receptor recognition (see Section 3-18, p. 77) means that any given T-cell receptor recognizes a peptide bound to a particular MHC molecule. To be of use in an immune response, therefore, a person's T cells must be able to interact with ligands formed from peptides bound to one of the MHC molecules expressed by that person, known in this context as **self-MHC**. Screening and selection of immature T cells for this essential property takes place in the thymus.

The population of mature circulating T cells that emerges from the thymus must, however, not interact with self-MHC molecules presenting peptides derived from the constituents of the person's own body, conveniently referred to as **self-peptides**. T cells with receptors that bind too strongly to self-peptide:self-MHC complexes in the thymus are therefore eliminated before they leave the thymus. The mature T-cell population leaving the thymus for the secondary lymphoid organs is thus rendered tolerant of self-antigens, responsive to foreign antigens presented by self-MHC molecules and ready to fight infection. Final differentiation to various types of effector T cells occurs after encounter with antigen.

The first part of this chapter traces the stages in gene rearrangement that produce the primary repertoire of T-cell receptors. The second part of the chapter describes the processes of positive and negative selection that act on this repertoire in the thymus to produce the circulating population of mature naive T cells.

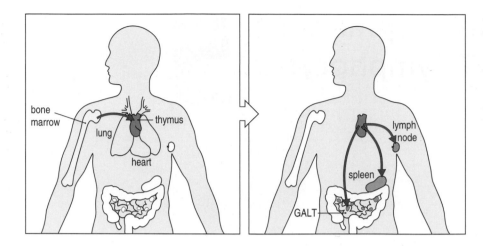

Figure 5.1 T-cell precursors migrate from the bone marrow to the thymus to mature. T cells derive from bone marrow stem cells whose progeny migrate in the blood from the bone marrow to the thymus (left panel), where the development of T cells occurs. Mature T cells leave the thymus in the blood, from where they enter secondary lymphoid tissues (right panel) and then return to the blood in the lymph. In the absence of activation by specific antigen, mature T cells continue to recirculate between the blood, secondary lymphoid tissues, and lymph. GALT, gut-associated lymphoid tissue.

The development of T cells in the thymus

T cells are lymphocytes that originate from bone marrow stem cells but emigrate to mature in the thymus (Figure 5.1). With the discovery of this developmental pathway, these lymphocytes were called **thymus-dependent lymphocytes**, which soon became shortened to **T lymphocytes** or **T cells**. Two lineages of T cells develop in the thymus—the majority α:β T cells and the minority γ:δ T cells. These lineages develop in parallel from a common precursor. While in the thymus, developing thymocytes also start to express other cell-surface proteins related to their eventual effector functions, such as the glycoproteins CD4 and CD8, which are essential for their effective interaction with cells bearing antigen.

5-1 T cells develop in the thymus

The **thymus** is a lymphoid organ in the upper anterior thorax just above the heart. It contains immature T cells, called **thymocytes**, which are embedded in a network of epithelial cells known as the **thymic stroma** (Figure 5.2). Together these elements form an outer close-packed cortex and an inner, less dense, medulla (Figure 5.3). The thymus is designated a primary lymphoid organ because its concern is with the production of useful lymphocytes, not with their application to the problems of infection. Unlike the secondary lymphoid organs, which perform the latter function, the thymus is not involved in lymphocyte recirculation; neither does it receive lymph from other tissues. The blood is the only route by which progenitor cells enter the thymus and by which mature T cells leave.

In the embryonic development of the thymus, the epithelial cells of the cortex arise from ectodermal cells, whereas those of the medulla derive from endodermal cells. Together, these two types of epithelial cell form a rudimentary thymus, called the **thymic anlage**, which subsequently becomes colonized by cells from the bone marrow—the thymocyte progenitors. The progenitor cells also give rise to dendritic cells, which populate the medulla of the thymus. Independently of these progenitors, the thymus is also colonized by bone-marrow derived macrophages, which, although concentrated in the medulla, are also found scattered throughout the cortex. As thymocytes mature, they tend to move progressively from the outer subcapsular region of the cortex radially towards the inner cortex and the medulla.

The importance of the thymus in establishing a functional T-cell repertoire is clearly demonstrated by patients who have complete **DiGeorge's syndrome**. In

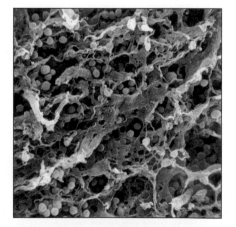

Figure 5.2 The epithelial cells of the thymus form a network surrounding developing thymocytes. In this scanning electron micrograph of the thymus, the developing thymocytes (the spherical cells) occupy the interstices of an extensive network of epithelial cells. Micrograph courtesy of W. van Ewijk.

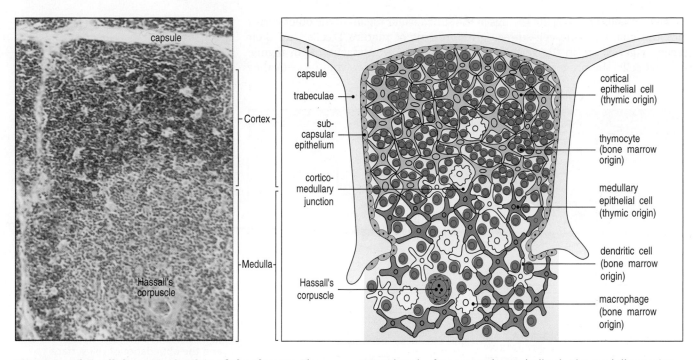

Figure 5.3 The cellular organization of the thymus. The thymus is made up of several lobules. A section through a lobule stained with hematoxylin and eosin and viewed with the light microscope is shown in the left panel. The cells in this view are diagrammed in the right panel. In the left panel, the darker staining of cortex compared with the medulla can be discerned. As shown in the right panel, the cortex consists of immature thymocytes (blue), branched cortical epithelial cells (light orange) and a few macrophages (yellow). The medulla consists of mature thymocytes (blue), medullary epithelial cells (orange), dendritic cells (yellow), and macrophages (yellow). One of the functions of the macrophages in both cortex and medulla is to remove the many thymocytes that fail to mature properly. A characteristic feature of the medulla is Hassall's corpuscles, which are believed to be sites of cell destruction. Photograph courtesy of C.J. Howe.

this genetic disease, the thymus fails to develop, and the resulting susceptibility to a wide range of opportunistic infections resembles that experienced by patients with severe combined immunodeficiency disease.

The thymus is most active in the young and it atrophies markedly with age. The human thymus is fully developed before birth and increases in size until puberty. It then progressively shrinks, with fat gradually claiming the areas once packed with thymocytes. This degeneration, which is almost complete by the age of 30, is called the involution of the thymus. The reduced production of new T cells by the thymus with age does not noticeably impair T-cell immunity; neither does **thymectomy** (removal of the thymus) in adults. Once established, the repertoire of mature peripheral T cells seems to be long-lived and/or self-renewing. In this it seems to differ from the mature B-cell repertoire, which is composed of short-lived cells that are continually being replenished from the bone marrow.

5-2 The two lineages of T cells arise from a common thymocyte progenitor

Maturation of thymocytes into mature T cells occurs in distinct stages. These are defined by changes in the status of the T-cell receptor genes, expression of the T-cell receptor protein, and the production of other T-cell surface glycoproteins essential for the receptor's full function, such as CD4, CD8, and the CD3 complex. Changes in the cell-surface proteins expressed at different developmental stages are used to distinguish between different populations of developing thymocytes and are measured by the technique of flow cytometry (Figure 5.4).

Progenitor cells entering the thymus at the subcapsular region of the outer cortex lack the characteristic cell-surface glycoproteins of mature T cells, and their receptor genes are in the germline configuration. On interaction with thymic stromal cells, the progenitor cells are signaled to proliferate; within a week they

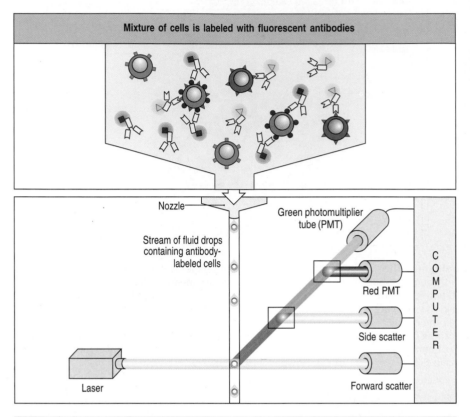

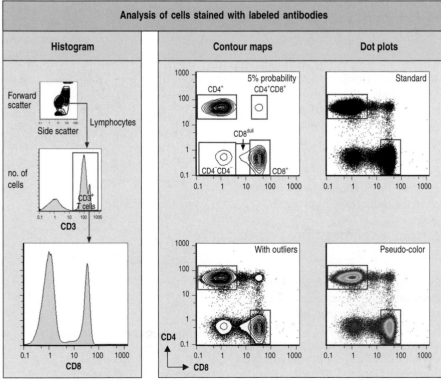

Figure 5.4 The flow cytometer allows individual cells to be identified by their cell-surface antigens. Cells are labeled with fluorescent dyes (top panel) bonded to antibodies specific for cell-surface proteins. They are then forced through a nozzle forming a stream of drops, each containing a single cell that passes through a laser beam (second panel). Photomultiplier tubes (PMTs) detect light scattering, a measure of cell size and granularity, and emissions from the fluorescent dyes. This information is analyzed by computer. Cells with particular characteristics can be counted and the abundance of cell-surface molecules measured. The lower part of the figure shows how this data can be represented, as exemplified by the expression of CD3, CD4, and CD8 on human peripheral blood lymphocytes. The forward and side scatter profiles are used to 'gate' the cell population so that only lymphocytes are analyzed (upper left). Likewise, a gate placed on lymphocytes that express CD3 restricts further analysis to T cells (middle left). The expression of CD4 and CD8 define various T-cell subsets as shown in the right-hand panels. When expression of just one type of molecule is analyzed, the data is usually displayed as a histogram, as shown for the expression of CD3 on lymphocytes and CD8 on T cells. When two or more parameters are measured, various types of two-color plots can be used, as shown in the four right panels. Here the horizontal axis represents CD8 fluorescence intensity and the vertical axis CD4 fluorescence intensity. The standard dot plot (upper right) places a single dot for each cell whose fluorescence is measured. It is good for seeing cells that lie outside the main groups but tends to saturate in areas containing many cells. The pseudo-color dot plot (lower right) uses color density to indicate cell abundance. A contour plot (upper left) draws contours, with 5% of the cells lying between each contour, providing the best monochrome image of regions of high and low density. The lower left plot is a 5% probability contour map which also shows cells lying outside the last contour as dots (outliers).

express certain T-cell specific glycoproteins, for example the CD2 adhesion molecule. They do not yet express any of the proteins of the T-cell receptor complex (see Section 3-3, p. 59) or the T-cell co-receptors CD4 and CD8 (see Section 3-5, p. 62). Because these immature thymocytes express neither CD4 nor CD8, they are called **'double-negative' thymocytes**. As the double-negative thymocytes mature, they first express the adhesion molecule CD44 and then CD25, a component of the receptor for the cytokine interleukin-2 (IL-2). In time, the expression of CD44 decreases and T-cell receptor gene rearrangements commence.

There are two T-cell lineages, which are distinguished by the expression of an α:β or a γ:δ T-cell receptor (see Section 3-4, p. 60). Commitment to one or other lineage does not occur before the T-cell receptor loci start to rearrange, but is the consequence of a race between the different loci to obtain a productive rearrangement. Thymocytes start to rearrange their β-, γ-, and δ-chain genes at about the same time. This is the first major difference from B-cell development, in which each type of immunoglobulin gene is rearranged in turn and in a set order.

If both a productive γ- and δ-chain gene rearrangement are made before a productive β-chain gene rearrangement, then the appearance of the γ:δ receptor on the cell surface signals the cell to stop β-chain rearrrangement and to develop as a γ:δ T cell (Figure 5.5). The more frequent outcome is for the β-chain gene to

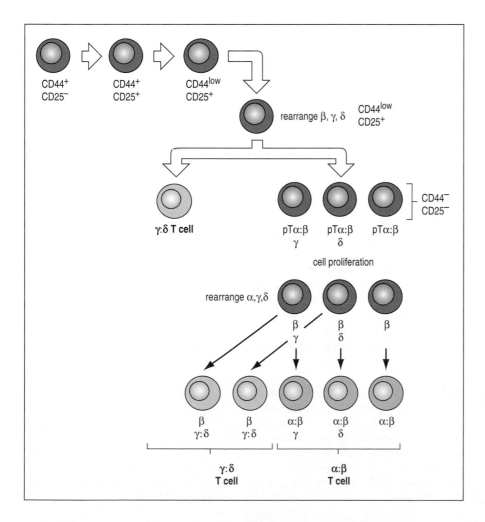

Figure 5.5 Lineage commitment to α:β or γ:δ T cells is a consequence of the order of successful gene rearrangements. T-cell progenitors express no T-cell-specific markers. They first express the CD44 and CD25 cell-surface glycoproteins (top line) and then commence rearrangements at the β-, γ-, and δ-chain loci (second line). If the γ and δ genes are productively rearranged before the β gene, then the cell commits to the γ:δ lineage (orange cell, third line). If the β-chain gene is productively rearranged before either the γ- or δ-chain gene, then the β chain is expressed in combination with pTα to form the pre-T-cell receptor (blue cells, third line). The cell ceases gene rearrangement and proliferates. Its progeny then enter a second phase of gene rearrangement involving the α-chain locus as well as further rearrangements at unproductively rearranged γ and δ loci (fourth line). Lineage commitment depends on whether a functional α:β or γ:δ T-cell receptor is made first (bottom line). Cells committed to one lineage can contain productive gene rearrrangements for the T-cell receptor genes of the other lineage. The exception is that α-chain gene rearrangements are not found in γ:δ cells. Immature thymocytes are colored blue. γ:δ T cells are colored orange; α:β T cells are colored green.

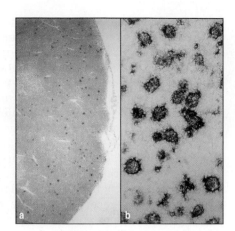

Figure 5.6 Immature T cells that undergo apoptosis are ingested by macrophages in the thymic cortex. Panel a shows a section through the thymic cortex (to the right side of the panel) and part of the medulla, in which cells have been stained for apoptosis with a red dye. Apoptotic cells are scattered throughout the cortex but are rare in the medulla. Panel b shows a section of thymic cortex at higher magnification which has been stained red for apoptotic cells and blue for macrophages. Apoptotic cells can be seen within the macrophages. Magnifications: panel a, × 45; panel b, × 164. Photographs courtesy of J. Sprent and C. Suhr.

rearrange productively before a functional γ:δ receptor can be assembled. In this situation, the β chain assembles with a surrogate α chain called **pTα** and expression of this **pre-T-cell receptor** on the surface signals the cell to halt rearrangement of the β-, γ-, and δ-chain genes and to enter a phase of proliferation.

Once that is over, the recombination machinery is reactivated and becomes targeted to the α-chain locus as well as to the γ and δ loci. In a minority of these cells, successful completion of γ- and δ-chain gene rearrangements before the α-chain gene has rearranged leads to their commitment to the γ:δ lineage. In the majority of these cells, however, productive rearrangement of the α-chain gene occurs first, and leads to expression of an α:β receptor and commitment to differentiation as an α:β T cell (see Figure 5.5). As the δ-chain locus is located in the middle of the α-chain locus, a rearrangement at an α-chain locus also leads to the deletion of the complete δ-chain locus from the chromosome.

Cells that fail to make a productive rearrangement die by apoptosis and are phagocytosed by macrophages in the thymic cortex (Figure 5.6). Apoptosis is the fate of all but a very few thymocytes, and the macrophages of the thymus are continually removing dead and dying cells while not interfering with ongoing thymocyte development.

5-3 Production of a T-cell receptor β chain leads to cessation of β-chain gene rearrangement and to expression of CD4 and CD8

T cells that express α:β receptors are the most abundant type of thymocyte, and we shall consider their development first before returning to that of γ:δ T cells. Rearrangements at the α- and β-chain loci have many features in common with those that take place at the immunoglobulin loci during B-cell development (see Section 4-2, p. 88). Like the immunoglobulin heavy-chain locus, the T-cell receptor β-chain locus contains V, D, and J gene segments and is rearranged first. The T-cell receptor α-chain locus contains no D segments and is rearranged second, like the immunoglobulin light-chain loci. Figure 5.7 shows the sequence of gene rearrangements that lead to α:β T cells, together with an indication of the stage at which the events take place, and the cell-surface molecules expressed at each stage. Gene rearrangement is preceded by small amounts of transcription from the gene segments to be recombined, and by expression of the recombination activation genes *RAG-1* and *RAG-2*, as in B cells.

Once a functional β-chain gene has been produced, a β chain is synthesized and assembles with a surrogate α chain, called preTα, as well as with CD3 proteins and ζ chain (see Section 3-3, p. 59), to form a pre-T-cell receptor, which is transported to the cell surface. The role of the pre-T-cell receptor in T-cell development is analogous to that of the pre-B-cell receptor in B-cell development: its appearance on the cell surface and engagement with some unknown ligand triggers the thymocyte to proliferate and to halt β-chain gene rearrangement. This ensures that a T cell expresses only one type of T-cell receptor β chain.

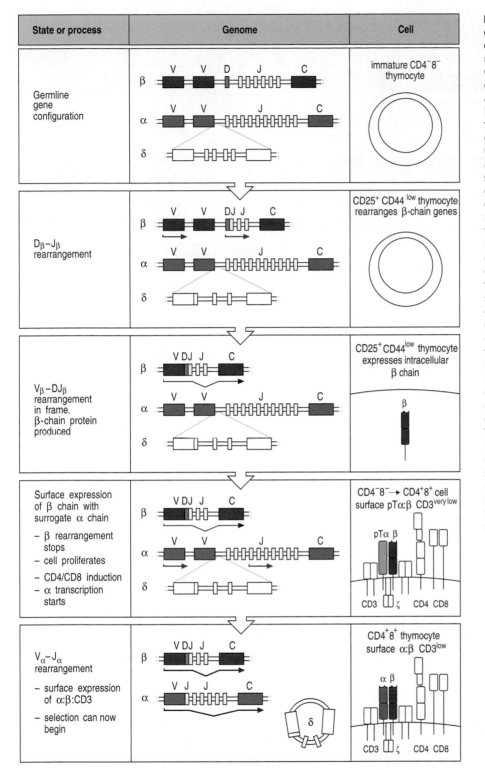

State or process	Genome	Cell

Figure 5.7 Stages of T-cell receptor gene rearrangement in the development of α:β T cells. The T-cell receptor β-chain genes rearrange first in CD4− CD8− (double-negative) thymocytes expressing the cell-surface protein CD25 and low levels of CD44. As with immunoglobulin heavy-chain genes, D to J rearrangement occurs first and a V gene segment then rearranges to DJ (second and third panels). The red arrows in the second panel represent small amounts of transcription from the gene segments to be rearranged, which open up the chromatin. The black arrow represents transcription to produce a functional mRNA. The kink in the arrow represents sequences that are spliced out of the primary transcript. The β chain is expressed within the cell and then appears at low levels on the cell surface in a complex with a surrogate α chain—pTα—and the CD3 chains. This complex is called the pre-T-cell receptor or pTα:β (fourth panel). Expression of the pre-T-cell receptor signals the cell to halt β-chain rearrangement and to undergo cycles of cell division. At the end of proliferation, CD4 and CD8 are expressed at the cell surface and the α chain now rearranges (fifth panel). When a functional α chain is produced, it pairs with the β chain to to form the α:β T-cell receptor. This appears on the surface with the CD3 complex. These double-positive thymocytes express CD4, CD8, and the α:β receptor in association with CD3. They are ready to undergo selection.

If a rearrangement at one β-chain locus is unproductive, the thymocyte attempts a rearrangement at the β-chain locus on the other, homologous, chromosome. An unproductively rearranged β-chain gene can also be rescued by a second rearrangement at the same locus (Figure 5.8). This possibility, which is not available for immunoglobulin heavy-chain genes, exists because two sets of $D_\beta J_\beta$ and C_β gene segments are tandemly associated with the V_β gene segments. The

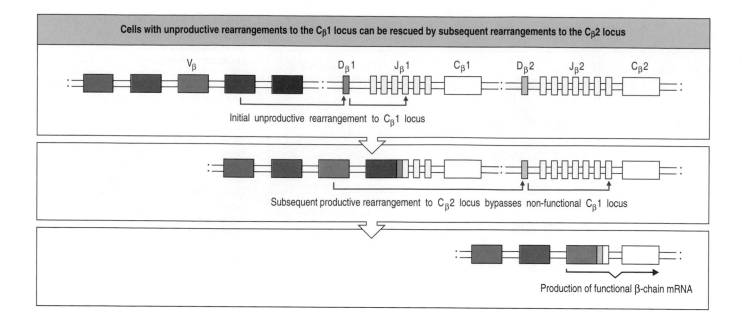

Figure 5.8 Rescue of unproductive β-chain gene rearrangements. Successive rearrangements can rescue an initial unproductive β-chain gene rearrangement, but only if that rearrangement involved D and J gene segments associated with the $C_\beta 1$ gene segment. A second rearrangement is then possible, in which a second V_β gene segment rearranges to a DJ segment associated with the $C_\beta 2$ gene segment, deleting $C_\beta 1$ and the unproductively rearranged gene segments.

potential for trying out up to four gene rearrangements means that 80% of T cells make a successful rearrangement of the β-chain gene, compared with a 55% success rate for productive heavy-chain gene rearrangement by B cells. Successful rearrangement of a β-chain gene induces the expression of the two co-receptors, CD4 and CD8, and cells at this stage are correspondingly called **'double-positive' thymocytes**. These cells are found predominantly in the inner cortex of the thymus, where they interact intimately with the branching network of epithelial cells.

During the spate of cell proliferation initiated by expression of the pre-T-cell receptor, expression of the *RAG-1* and *RAG-2* genes is repressed. Hence, no rearrangement of the α-chain genes can occur until the double-positive cells stop dividing. This ensures that each cell that has made a productive β-chain gene rearrangement gives rise to many daughter cells, each of which has the potential to make and express a different α-chain gene.

5-4 T-cell receptor α-chain genes can undergo several successive rearrangements

Just as an immunoglobulin light-chain locus can undergo several successive gene rearrangements, so can the T-cell receptor α-chain locus. It therefore has a greater chance of achieving a successful rearrangement than the β-chain locus. This difference results from the presence of many V_α and over 50 J_α gene segments, which allows many successive VJ_α rearrangements to be tried (Figure 5.9). As a consequence, productive α-chain gene rearrangements are made in almost every developing T cell. Correlating with this flexibility, the α-chain locus is much larger than the β-chain locus, being spread over some 80 kilobases of DNA.

In T cells, as opposed to B cells, successful rearrangement of one copy of the α-chain gene and cell-surface expression of a functional α:β receptor do not prevent rearrangement at the other copy of the α-chain gene. Many T cells therefore express two α chains and have two different T-cell receptors at the surface at this stage in their development. As the T cell enters the next phase of development, the forces of positive and negative selection can act on either of the two receptors.

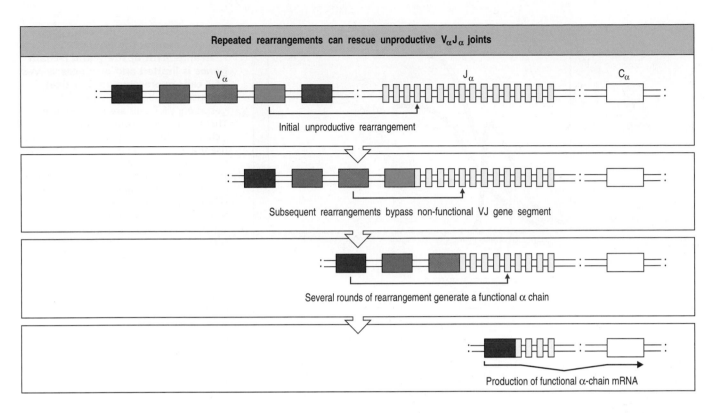

Repeated rearrangements can rescue unproductive V$_\alpha$J$_\alpha$ joints

Initial unproductive rearrangement

Subsequent rearrangements bypass non-functional VJ gene segment

Several rounds of rearrangement generate a functional α chain

Production of functional α-chain mRNA

Figure 5.9 Successive gene rearrangements allow the replacement of one T-cell receptor α chain by another. For the T-cell receptor α-chain genes, the multiplicity of V and J gene segments allows successive rearrangement events to jump over unproductively rearranged VJ segments, deleting the intervening gene segments. This process continues until either a productive rearrangement occurs or the supply of V and J gene segments is exhausted, whereupon the cell dies.

5-5 Cells expressing particular γ:δ receptors arise first in embryonic development

The first T cells to emerge during embryonic development carry γ:δ receptors. In mice, where the development of the immune system can be studied in detail, γ:δ T cells first appear in discrete bursts or waves that migrate to different anatomical sites (Figure 5.10). The first wave of T cells migrates specifically to the skin, where they are called **dendritic epidermal T cells** (**dETC**), whereas the second wave migrates to the epithelial layers of the reproductive tract. Each of these waves is characterized by a distinctive γ:δ receptor which is carried by all the T cells in a wave. In the timing of their emergence, in their restricted use of particular gene segments, and in their lack of N nucleotides, these early γ:δ T cells are analogous to the B-1 cells that emerge from the bone marrow before birth (see Section 4-6, p. 95). Both these minority lineages of lymphocyte are thought to represent older, more primitive, and less specific components of the adaptive immune system.

Later on in embryonic development, T cells are produced continuously, rather than in waves, and α:β cells now dominate the population, making up more than 95% of all T cells. Within the γ:δ subpopulation, the repertoire of receptors has considerably diversified in comparison with the early waves, both in the number of alternative gene segments used and in the abundance of N-nucleotide additions. A further contrast with the early waves of γ:δ T cells is that the later, more diverse, γ:δ cells are found throughout the peripheral lymphoid tissues and not solely in the epithelium of certain non-lymphoid tissues.

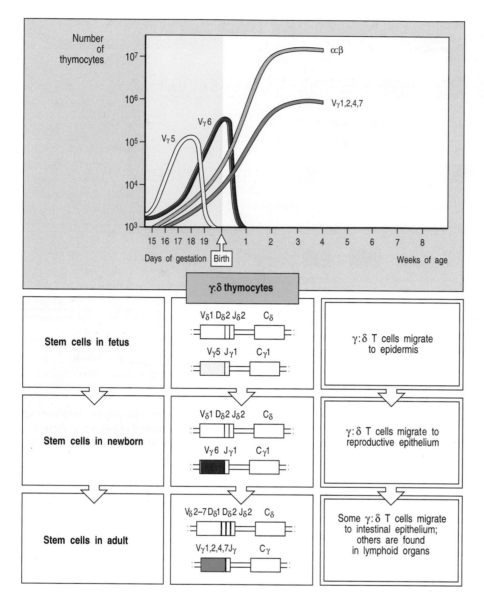

Figure 5.10 Early in the embryonic life of a mouse, the type of rearrangement at the γ- and δ-chain genes is limited and produces waves of T cells expressing an identical receptor. At about two weeks of gestation, γ:δ T cells are first produced. These cells have receptors that contain a γ chain produced from the $C_\gamma1$ locus and use the V gene segment ($V_\gamma5$) nearest to the D gene segments. After a few days, $V_\gamma5$-bearing cells decline (graph in upper panel, yellow line) and are replaced by cells expressing the next most proximal V gene segment, $V_\gamma6$ (red line). Both the $V_\gamma5$ and $V_\gamma6$ chains are associated with the same δ chain, as shown in the lower panels. The antigen specificity of the T cells in these early waves is unknown. The $V_\gamma5$ cells migrate to the epidermis, whereas the $V_\gamma6$ cells migrate to the epithelium of the reproductive tract. After birth, the α:β T-cell lineage becomes dominant and, although γ:δ T cells are still produced (green line), they are more diverse.

It is uncertain to what extent the properties and behavior of mouse γ:δ T cells reflect those of human γ:δ T cells. Although γ:δ T cells are found in the reproductive and gastrointestinal tracts of both species, the dendritic epidermal T cells of mice have no counterpart in human skin. Study of the immune system of the laboratory mouse has been instrumental in working out the basic immunological mechanisms. However, when used as a model for human disease, the mouse can be less informative because of the many details in which its immune system differs from that of humans. For instance, the presence or absence of dendritic epidermal T cells might have important effects on immune responses causing dermatitis or the rejection of a skin allograft.

Summary

The thymus provides a sequestered and organized environment dedicated to T-cell development. Progenitor cells from the bone marrow migrate to the thymus, where they go through phases of division and differentiation associated with the rearrangement of T-cell receptor genes and the expression of other cell-surface glycoproteins involved in T-cell recognition and effector function. Commitment of thymocytes to either the γ:δ or the α:β lineage occurs as a consequence of the gene rearrangements made. The β-, γ-, and δ- chain genes rearrange simultaneously and if a γ:δ receptor is made first, then commitment is made to the γ:δ lineage. If a β-chain gene rearranges productively before this occurs, the β chain forms a pre-T-cell receptor with a surrogate α chain. This stops recombination, initiates cell division, and induces the expression of the CD4 and CD8 co-receptors. Once cellular proliferation stops, the machinery for T-cell receptor gene rearrangement is reactivated and now works on the α-chain gene in addition to the γ- and δ-chain genes. In this second phase of gene rearrangements, lineage commitment is determined by the expression of either γ:δ or α:β receptors on the cell surface. More than 90% of cells that complete successful gene rearrangement commit to the α:β lineage. If cells fail to make successful gene rearrangements and thus do not express a T-cell receptor, they are signaled to die within the thymus.

Positive and negative selection of the T-cell repertoire

In the first phase of T-cell development just described, the role of the thymus is to produce T-cell receptors, irrespective of their antigenic specificity. The second phase of T-cell development involves a critical examination of the receptors produced and selection of those that can work effectively with the individual's own MHC molecules in the recognition of pathogen-derived peptides. These selection processes involve only α:β T cells; γ:δ T cells seem indifferent to peptides presented by MHC molecules and recognize different types of antigen, which remain largely unknown. Once the γ- and δ-chain genes are rearranged productively, the development of γ:δ T cells within the thymus might be complete.

In this part of the chapter we examine this second phase in the development of α:β T cells. We shall see how the population of double-positive thymocytes undergoes two types of screening. In the first screen, positive selection selects T cells that can recognize peptides presented by a self-MHC molecule; in the second, negative selection eliminates potentially autoreactive cells that could be activated by the peptides normally presented by MHC molecules on the surface of healthy cells.

5-6 T cells that can recognize self-MHC molecules are positively selected in the thymus

The primary T-cell receptor repertoire has a bias towards interaction with MHC molecules. This is due to specificities built into the V gene segments. Gene rearrangement thus provides an extensive repertoire of T-cell receptors that could be used with the hundreds of MHC class I and MHC class II isoforms present in the human population. However, the T-cell receptor genes possessed by a given individual are not specifically tailored towards making receptors that interact with the particular forms of MHC molecule expressed by the same individual. Only a small subpopulation of the double-positive thymocytes, at most 1% or 2% of the total, have receptors that can interact with one of the MHC class I or II isoforms expressed by the individual, and will therefore be able to respond to antigens presented by these MHC molecules. **Positive selection** is the name given to the process whereby that small subpopulation is signaled to mature further, leaving the vast majority of double-positive cells to die by apoptosis in the thymic cortex.

Positive selection takes place in the cortex of the thymus (see Figure 5.3). It is mediated by the complexes of self-peptides and self-MHC molecules present on the surface of the cortical epithelial cells. As we saw in Sections 3-9 and 3-10, pp. 66–69, in the absence of infection MHC molecules assemble with self-peptides derived from the normal breakdown of the body's own proteins. The cortical epithelial cells form a web of cell processes that envelop and make contact with the double-positive CD4 CD8 thymocytes. Thymic cortical epithelium expresses both MHC class I and MHC class II molecules. At regions of contact, potential interactions of the α:β receptor of a thymocyte with the self-peptide:self-MHC complexes on the epithelial cell are tested. If a peptide:MHC complex is bound within 3–4 days of the thymocyte's expressing a functional receptor, then a positive signal is delivered to the thymocyte, which continues its maturation. Cells that do not receive such a signal within this period die by apoptosis and are removed by macrophages.

The peptides presented on the surface of thymic epithelial cells are derived from those self-proteins that are present in the thymus. The number of different self-peptides that can be presented by an individual MHC molecule is estimated to be about 10,000, so for someone who is heterozygous for the six major polymorphic HLA genes, around 120,000 self-peptides could be presented by the 12 different MHC molecules that they would possess. As the mature T-cell receptor repertoire is estimated to be of the order of tens of millions or more, a majority of these self-peptide:self-MHC complexes are likely to contribute to positive selection.

Positive selection was first demonstrated by experiments in bone marrow transplantation between mice of different MHC haplotype (Figure 5.11). In recipient mice whose bone marrow has been destroyed by irradiation, thymocyte progenitor cells in the bone-marrow graft migrate to the thymus, where they mature into T cells under the influence of the thymic epithelial cells of the recipient and of the MHC molecules that these cells express. The mature T cells respond only to peptides

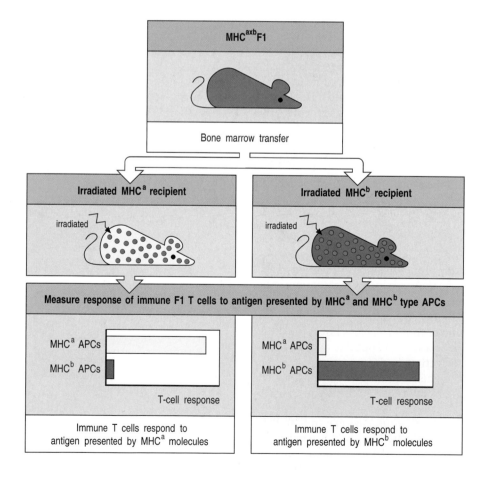

Figure 5.11 Bone marrow transplantation in mice revealed the existence of positive selection. T cells in an F1 hybrid mouse possess both the 'a' and 'b' MHC haplotypes and can respond to antigens presented by both the 'a' and 'b' types of MHC molecule. If bone marrow from the F1 mouse is transplanted into irradiated mice that are homozygous for either the MHC^a or MHC^b haplotype (middle panel), the T cells that mature are restricted to recognizing antigens presented by antigen-presenting cells (APCs) bearing the MHC^a or the MHC^b molecules respectively. T cells from an irradiated MHC^a mouse used as a recipient for hybrid bone marrow respond only to antigen presented by MHC^a molecules, whereas the T cells from a similarly treated MHC^b mouse respond to antigen presented by MHC^b. That the same pool of hybrid progenitor cells gives rise to these two mutually exclusive populations of mature T cells is evidence of positive selection.

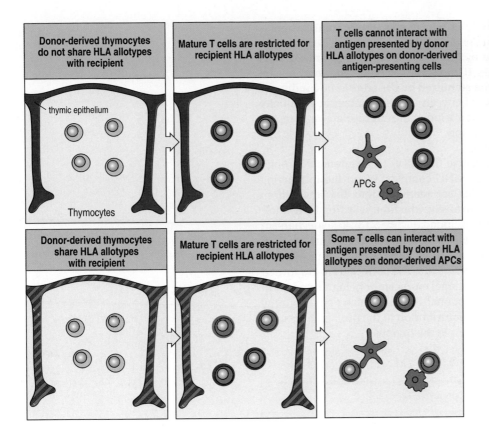

Figure 5.12 The donor and recipient in a bone-marrow transplant must share HLA class I and II molecules in order to reconstitute T-cell function. After bone-marrow transplantation donor-derived thymocytes are positively selected on the recipient's thymic epithelium. The top panels show the situation when none of the recipient's HLA allotypes (red) are the same as the donor's HLA allotypes (blue). The lower panels show the situation when the recipient and donor share the HLA allotypes indicated by blue.

presented by MHC molecules of these types and not to those presented by other MHC molecules possessed by the donor cells themselves. These experiments showed that it was the MHC molecules present in the recipient's thymus that stamped their mark on the T-cell repertoire.

In clinical bone-marrow transplantation a patient's own bone marrow is destroyed and replaced by a graft from a healthy donor which then reconstitutes the entire hematopoietic system. As part of this process, T cells developing from donor-derived stem cells are positively selected in the recipient's thymus for interaction with the recipient's HLA allotypes. To reconstitute T-cell functions the new T cells must be able to respond to antigens presented by the professional antigen-presenting cells (dendritic cells, B cells, and macrophages), which will now all be of donor origin. To satisfy this requirement the donor and recipient must have at least one HLA class I and one HLA class II allotype in common (Figure 5.12). For this reason, and to reduce the deleterious immune reaction caused by HLA differences, hematologists aim to maximize the sharing of HLA allotypes between donors and recipients in bone-marrow transplantation.

5-7 Positive selection controls expression of the CD4 or CD8 co-receptor

Positive selection not only selects a repertoire of cells that can interact with an individual's own MHC allotypes, it is also instrumental in determining whether a double-positive T cell will become a CD4 T cell or a CD8 T cell. As a result of positive selection, double-positive thymocytes mature into cells that express just one or other of the two co-receptors. They are then known as **'single-positive'** **thymocytes.**

As we saw in Chapter 3, CD4 interacts only with MHC class II molecules, whereas CD8 interacts only with MHC class I molecules. CD4 T cells having T-cell receptors that recognize antigens presented by MHC class I molecules, or CD8 cells having receptors that bind MHC class II molecules, would not work, as the receptor:ligand interaction could not be stabilized by the additional binding of the co-receptor. The necessary matching between the T-cell receptor specificity for a particular MHC molecule and the expression of the appropriate co-receptor molecule is achieved during positive selection.

During positive selection, the double-positive CD4 CD8 T cell interacts through its α:β receptor with a particular peptide:MHC complex. When the interacting MHC molecule is class I, CD8 molecules are recruited into the interaction, whereas CD4 molecules are excluded. Conversely, when the selecting MHC molecule is class II, CD4 is recruited and CD8 excluded (Figure 5.13). The mechanism by which this interaction leads to single-positive thymocytes is still unknown. In one model, it is proposed that expression of the non-binding co-receptor is specifically turned off. Another model proposes that the co-receptors are turned off at random. This would result in the functional co-receptor being turned off in half the cells, rendering these cells non-functional. Although such poor economy makes little sense, it would be just another contribution to the profligate wastage of cells that occurs during T-cell development in the thymus.

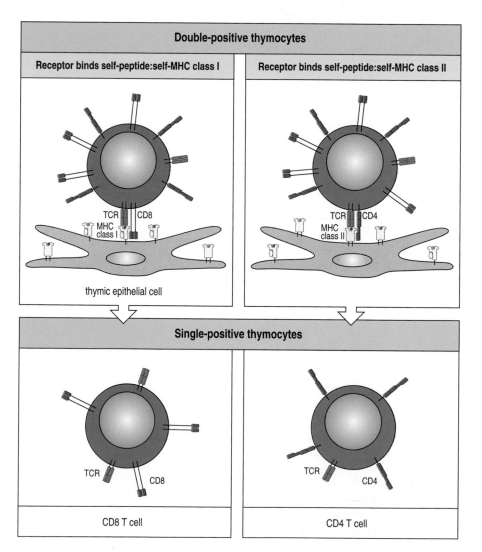

Figure 5.13 **Interaction of a double-positive T cell with a self-peptide:self-MHC complex during positive selection determines whether the T cell will become a CD4 or a CD8 T cell.** The left panels show the selection of a T cell whose T-cell receptor (TCR) interacts with peptide:MHC class I complexes on a thymic epithelial cell. The right panels show the outcome for a cell bearing a receptor that interacts with peptide:MHC class II complexes.

The importance of MHC molecules in co-receptor selection and subsequent T-cell development is demonstrated by human immunodeficiency diseases called **bare lymphocyte syndromes**, which are characterized by a lack of expression of either MHC class I or MHC class II molecules by lymphocytes and thymic epithelial cells. Patients who lack MHC class I expression have CD4 T cells but no CD8 T cells, whereas patients lacking MHC class II have CD8 T cells but only a few, abnormal, CD4 T cells.

5-8 Rearrangement of α-chain genes stops once a cell has been positively selected

Rearrangements at the α-chain locus continue throughout the 3–4-day period of positive selection. Through the use of different gene segments and the production of different α chains, a T cell can change the specificity of the antigen receptor that it expresses. In this way, double-positive thymocytes can successively explore the usefulness of different receptors, thereby improving the chance of their positive selection. Once a T cell has been positively selected, α-chain gene rearrangement stops.

Because of the continued rearrangements, some double-positive cells express two α chains and thus two types of T-cell receptor. Such T cells can be positively selected through the engagement of either one of their receptors. Because the proportion of T-cell receptors that succeed in positive selection is so small, however, it will be a vanishingly rare cell that has two T-cell receptors that can both be activated by peptides presented by self-MHC molecules. Thus, in the vast majority of mature T cells that express two receptors, one receptor will be non-functional. For all practical purposes T cells can be said to have a single working receptor.

5-9 T cells specific for self-antigens are removed in the thymus by negative selection

In Chapter 4, we saw how immature B cells whose immunoglobulin receptors bind to self-antigens on the surface of bone-marrow cells are eliminated from the repertoire by clonal deletion. A similar mechanism, called **negative selection**, serves to delete T cells whose antigen receptors bind too strongly to the complexes of self-peptides and self-MHC molecules presented by cells in the thymus. Such T cells are potentially autoreactive, and if allowed to enter the peripheral circulation could cause tissue damage and autoimmune disease. Whereas positive selection is mediated exclusively by epithelial cells in the cortex of the thymus, negative selection in the thymus can be mediated by several cell types, of which the most important are the bone-marrow derived dendritic cells and macrophages (Figure 5.14). These cell types also serve as the specialized antigen-presenting cells that activate mature T cells in the secondary lymphoid tissues. To prevent their activation of mature self-reactive T cells, such T cells are purged from the repertoire before they can enter the circulation.

Negative selection cannot eliminate T cells whose receptors are specific for self-peptides that are present only in tissues other than the thymus. Such T cells do leave the thymus and enter the peripheral circulation. However, they are soon rendered anergic by mechanisms analogous to those by which B cells specific for soluble self-antigens are inactivated (see Section 4-7, p. 97).

The mechanisms of positive and negative selection used by the thymus both involve the screening of interactions between T-cell receptors and MHC molecules. The population of T cells leaving the thymus has a diverse receptor repertoire, showing that positive selection embraces a much wider range of receptor specificities than does negative selection. A second point to note is that the signals

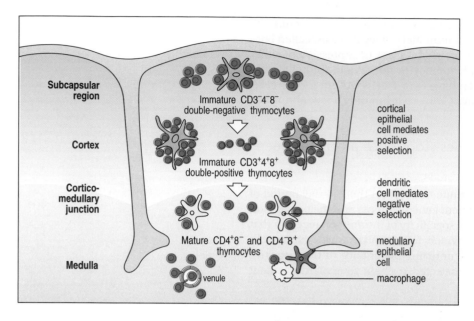

Figure 5.14 Positive and negative selection are mediated by different types of cell in the thymus. As thymocytes mature they move from the subcapsular region deeper into the thymus. Double-positive cells are found in the cortex, where they undergo positive selection on cortical epithelial cells. The positively selected cells encounter dendritic cells and macrophages at the cortico-medullary junction; this is where most negative selection occurs. The surviving mature single-positive T cells leave the thymus and enter the peripheral circulation through venules in the medulla.

given to the thymocyte by similar receptor:peptide:MHC interactions during positive and negative selection must be different, but how this is achieved has yet to be understood. One theory proposes that differences in binding strength of the receptor:peptide:MHC interaction deliver signals that induce either positive or negative selection, whereas another theory proposes that positive and negative selection are induced by different types of ligand.

For two persons of completely different HLA type, the sets of self-peptides presented by their HLA molecules are largely, if not completely, non-overlapping, and the mature T-cell repertoires selected in the two individuals are quite different. When confronted by the same pathogen, the two individuals will make T-cell responses that differ markedly in the T-cell receptors used and the peptides responded to. Because of the diversity of HLA types in the human population, the T-cell component of the immune system becomes highly personalized as a result of positive and negative selection.

5-10 T cells undergo further differentiation in secondary lymphoid tissues after encounter with antigen

Only a small fraction of $\alpha{:}\beta$ T cells survive the obstacle course of positive and negative selection and leave the thymus. Like mature B cells, these mature naive T cells recirculate through the tissues of the body, passing from blood to secondary lymphoid tissues to lymph and then back to the blood. Mature T cells are longer-lived than mature B cells and, in the absence of their specific antigen, continue to circulate through the body for many years.

The T-cell-rich areas of secondary lymphoid tissues provide specialized sites where naive T cells are activated by their specific antigens. Encounter with antigen provokes the final phases of T-cell development and differentiation: the mature T cells divide and differentiate into effector T cells, some of which stay in the lymphoid tissues while others migrate to sites of infection.

Unlike B cells, which have just one terminally differentiated state—the antibody-secreting plasma cell—there are several different types of effector T cells. On activation by antigen, CD8 T cells become activated cytotoxic T cells, whereas CD4 T cells can differentiate under the influence of cytokines into either T_H1 or T_H2 effector T cells. Which type of effector CD4 T cell predominates depends on the nature of the pathogen and the type of immune response required to clear it.

In healthy individuals there are approximately twice as many CD4 T cells in the peripheral circulation as CD8 T cells. The virus that causes acquired immuno-deficiency syndrome (AIDS) selectively infects CD4 cells by exploiting the CD4 molecule as its receptor. In patients with AIDS the number of CD4 T cells declines and this parameter is used by physicians to measure disease progression and to assess the effectiveness of therapy.

5-11 The requirements of thymic selection can limit the number of functional class I and class II genes in the MHC

Each of the mature T cells in a person's circulation owes its existence to interactions between its receptor and a self-peptide:self-MHC complex that led to its positive selection in the thymus. One might think that the greater the number of different MHC molecules that an individual expressed, the larger would be the proportion of the T-cell repertoire that was positively selected. This is indeed true, but increasing the number of MHC molecules beyond a certain point means an unacceptably heavy price to pay when negative selection is initiated.

For each different MHC molecule an individual expresses, about 1% of the positively selected T-cell repertoire is deleted in the thymus. Each MHC molecule is estimated to be able to bind up to 10,000 different self-peptides, and the deleted cells represent receptors that are specific for complexes of that MHC molecule with many different self-peptides. This level of deletion can be demonstrated indirectly by the **mixed lymphocyte reaction**, in which peripheral blood cells from one person are mixed in tissue culture with lethally irradiated cells from a second person of a disparate HLA type. The self-peptide:self-MHC complexes on the second person's cells provoke a strong alloreactive response from the T cells of the first person (Figure 5.15). It has been estimated that about 5% of the T-cell repertoire can be activated by a whole MHC haplotype in such reactions. The magnitude of the alloreactive response to non-self MHC provides an estimate of the relative numbers of T cells that are negatively selected within the thymus of an individual. If the MHC molecules from another person activate a certain fraction of the mature T cells, then a similar stimulation by self-MHC during T-cell development would lead to the deletion of a similar fraction of the positively selected T cells.

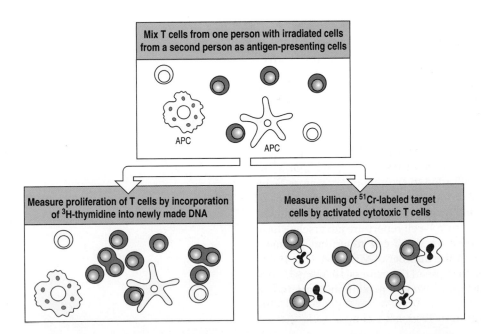

Figure 5.15 The mixed lymphocyte reaction (MLR) can be used to detect HLA differences. Lymphocytes from the two individuals who are to be tested for compatibility are isolated from peripheral blood. The cells from one person (yellow), which include antigen-presenting cells, are either irradiated or treated with mitomycin C to prevent their proliferation; they act as stimulators but cannot respond to antigenic stimulation from the other person's cells. The cells from the two individuals are then mixed (top panel). If the unirradiated lymphocytes (the responders, blue) contain alloreactive T cells, these will be stimulated to proliferate and differentiate into effector cells. Between 3 and 7 days after mixing, the culture is assessed for T-cell proliferation (bottom left panel), which is mainly the result of CD4 T cells recognizing differences in HLA class II molecules, and for the generation of activated cytotoxic T cells (bottom right panel), which respond to differences in HLA class I molecules. The cells used as targets in the cytotoxic assays are lymphoblastoid cells and are therefore larger than the lymphocytes. The target cells are preloaded with radioactive chromate (^{51}Cr) which binds to intracellular proteins and is released into the medium if a cell is killed.

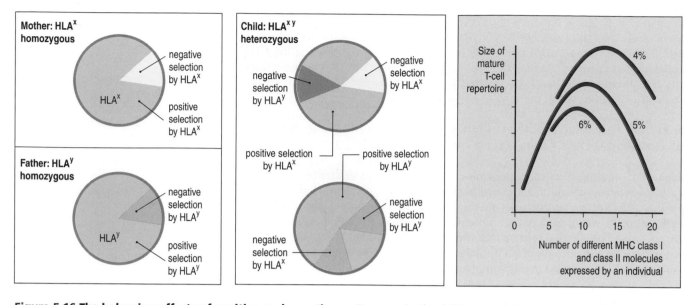

Figure 5.16 The balancing effects of positive and negative selection on T-cell repertoires. The left and center panels show how positive and negative selection have different cumulative effects as the number of MHC molecules (HLA in humans) expressed by an individual increases. In this family, the mother is homozygous for the HLA^x haplotype, whereas the father is homozygous for the disparate HLA^y haplotype. Their child expresses both the x and y haplotypes. In the parents, the x or the y set of HLA molecules positively selects a set of T-cell receptors, which are then reduced by negative selection on the same set of HLA molecules. Whole circles (pie charts) are the populations of thymocytes positively selected by the HLA type indicated; the colored wedges are the proportions of these cells that are subsequently negatively selected by each HLA haplotype. Populations of cells negatively selected by HLA^x are indicated in yellow and those negatively selected by HLA^y in orange. The child can be considered as positively selecting the sum of the T-cell receptors selected by the two parents.

However, in the child, each of these two positively selected repertoires is then negatively selected on both the x and the y sets of HLA molecules. Whereas the number of positively selected subsets of T cells in the child is the sum (2) of those in the two parents (1 + 1), the number of negatively selected subsets of T cells in the child is the square (4) of the number in each parent (2). Thus the effect of negative selection increases disproportionately as the number of different HLA molecules expressed by an individual increases. This effect is shown in the graph on the right, where the relationship between the size of the mature T-cell repertoire and the number of HLA isoforms expressed by an individual is plotted. The three plots vary according to the estimated value (4%, 5%, and 6%) for the percentage of T cells positively selected by one HLA isoform that are negatively selected by a second HLA isoform. A value of 4% gives an optimal number of HLA molecules of 12–13, comparable to the number of different HLA molecules expressed by most human beings.

The effects on positive and negative selection of changing the number of different MHC molecules are illustrated in Figure 5.16. In Chapter 3, we saw that MHC diversity is advantageous to an individual because it broadens the repertoire of pathogen-derived peptides presented. Now we see how a similar expansion in the number of self-peptides presented leads to an equivalent increase in the population of positively selected T cells in the thymus. However, with each additional type of MHC molecule, the proportion of T cells that are negatively selected also goes up, not in an arithmetical manner as with positive selection, but in a geometrical manner that increases with the square of the total number of MHC molecules. The result is that the increase afforded to an individual's mature T-cell repertoire by an additional MHC molecule rapidly declines as the number of MHC molecules increases. Beyond a certain number, the presence of additional MHC molecules will begin to affect the mature T-cell repertoire adversely. These counterbalancing effects of MHC diversity on the T-cell repertoire probably explain why vertebrate species tend to limit the number of MHC class I or II isotypes in the genome to three each or less. Because of the expression of haplotypes from both chromosomes, this results in expression of a maximum of around 12 different MHC molecules.

5-12 T cells are involved in a variety of malignant diseases

Because of the DNA rearrangements that occur during T-cell development, T-cell tumors are not uncommon. Like B-cell tumors (see Section 4-10, p. 101), they correspond to defined stages in development. Whereas the B-cell malignancies cover the whole pathway of B-cell development, tumors of T cells themselves mostly correspond to either early or late stages in T-cell development (Figure 5.17). The absence of tumors corresponding to intermediate stages in T-cell development might be because immature T cells are programmed to die unless rescued within a short period by the next positive signal for maturation. Under these circumstances, there might not be time for thymocytes to accumulate the number of mutations necessary for malignant transformation.

Some tumors represent massive outgrowth of a rare cell type. **Acute lymphoblastic leukemias** (**ALL**), for example, derive from lymphoid progenitor cells. The majority of acute lymphoblastic leukemias resemble immature B cells and are therefore called B-ALL or C-ALL where C stands for common form. A significant minority of acute lymphoblastic leukemias resemble immature T cells or thymocytes and are called T-ALL. Thus, these T-cell tumors provide information about early stages in lymphocyte development. Others help study of the phenotype, migration, and receptor gene status of rare types of normal cells. For example, **cutaneous T-cell lymphoma** (**mycosis fungoides**) arises from a T cell that, on activation by antigen, normally migrates to the skin.

Hodgkin's disease is the most complex of the lymphoid tumors and appears in several forms. The malignant cell seems to be an antigen-presenting cell, and in

Disease	Cell		Characteristic cell-surface markers	Location
	Stem cell		CD34	Bone marrow
Acute lymphoblastic leukemia (B-ALL or C-ALL)	Lymphoid progenitor		CD10 CD19 CD20	
Thymoma	Thymic stromal cell or epithelial cell		Cytokeratins	Thymus
Acute lymphoblastic leukemia (T-ALL)	Thymocyte		CD1	
Adult T-cell leukemia, Mycosis fungoides/ Sézary syndrome, Chronic lymphocytic leukemia (CLL)	T cell		CD3/TCR CD4 or CD8	Peripheral circulation
Hodgkin's disease	Antigen-presenting cell		CD30	

Figure 5.17 T-cell and related tumors of the lymphoid system. Each distinct tumor has a normal equivalent and retains many of the properties of the cell from which it develops. As well as spontaneous tumors of T cells themselves, a virally caused T-cell tumor and two T-cell-related tumors are also included in this figure. Thymomas derive from thymic stromal or epithelial cells, whereas the malignantly transformed cell in Hodgkin's disease is thought to be an antigen-presenting cell. Adult T-cell lymphoma is the result of infection of T cells with the retrovirus HTLV-1. Some characteristic cell-surface markers for each stage are also shown. For example, CD10 is a widely used marker for acute lymphoblastic leukemia. Note that T-cell chronic lymphocytic leukemia (CLL) cells express CD8, whereas the other T-cell tumors shown express CD4.

some patients the disease is dominated by non-malignant T cells that are stimulated by the tumor cells. This form of the disease is called Hodgkin's lymphoma. Other patients have no lymphocytic abnormalities, but present with proliferation of a reticular cell, a condition known as nodular sclerosis. The two manifestations of Hodgkin's disease might reflect real heterogeneity in the transformed cell or, more probably, differences in the T-cell response to the transformed cells. Treatment for Hodgkin's disease is by radiotherapy and chemotherapy and it was one of the first tumors to be successfully treated by radiotherapy. The prognosis for Hodgkin's lymphoma is far better than for nodular sclerosis, suggesting that the non-malignant T cells might actually impede tumor growth and, when helped by clinical intervention, can actually eliminate the tumor.

Like B-cell tumors, T-cell lymphomas have characteristic rearrangements of their T-cell receptor genes, showing that they derive from single transformed cells (Figure 5.18). Analysis of a patient's tissues for T-cell receptor gene rearrangements that are characteristic of the tumor allows tumor growth and dissemination to be monitored with time and the results of therapy to be followed. For some lymphoid tumors, a bone marrow transplant can be used as therapy.

Summary

Once a developing thymocyte expresses an α:β receptor and CD4 and CD8 on its surface, it undergoes two types of selection, both of which involve testing the receptor's interactions with the complexes of self-peptides bound by self-MHC molecules on the surface of thymic cells. Positive selection is the responsibility of epithelial cells in the cortex of the thymus. Double-positive thymocytes whose receptors engage self-peptide:self-MHC complexes on these cells continue their maturation. The vast majority of double-positive thymocytes fail positive selection and die by apoptosis. The class of MHC molecule that drives positive selection determines which co-receptor—CD4 or CD8—is maintained on the single-positive T cell. Negative selection is effected by other cells of the thymus, most importantly the dendritic cells and macrophages, which derive from bone marrow progenitors. Negative selection eliminates autoreactive cells whose receptors bind too strongly to a self-peptide:self-MHC complex, thereby helping to create a mature T-cell repertoire that does not react to the peptide:MHC complexes of normal healthy cells. The requirements of thymic selection place limits on the number of polymorphic MHC molecules encoded in the genome; as the number of molecules increases so does the proportion of the repertoire that is negatively selected. T-cell tumors correspond to some but not all of the stages of T-cell development. Having survived selection the now mature T cells leave the thymus in the blood and circulate through the secondary lymphoid organs, where they encounter foreign antigen. On activation by antigen they differentiate further into effector T cells of various types.

Summary to Chapter 5

In the thymus, three functionally distinct types of T cell develop from a common progenitor that comes from the bone marrow. One type of T cell expresses γ:δ receptors and is not restricted to the recognition of peptide antigens presented by MHC molecules. The other two types of T cell express α:β receptors and are distinguished by the co-receptors that they express—CD4 or CD8—and the class of MHC molecule to which their receptors are restricted. Commitment to either the γ:δ or the α:β T-cell lineage is determined by which pair of T-cell receptor loci successfully rearranges first. Subsequent phases of development in the thymus concern only the α:β T cells. The primary repertoire of T-cell receptors produced by gene rearrangement is acted upon by both positive and negative selection to

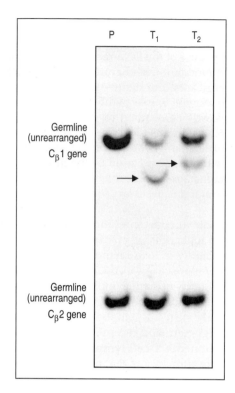

Figure 5.18 The T-cell receptor gene rearrangements that characterize individual clones of T cells can be used to identify tumors of T cells. Because a tumor is an outgrowth of a single transformed cell, the cells of a T-cell tumor all have the same rearranged T-cell receptor genes, which can therefore be detected and characterized. This figure shows a comparison of genomic DNA isolated from placenta (lane P) and two T-cell tumors (lanes T_1 and T_2). The DNAs were digested into fragments with a restriction endonuclease and separated by gel electrophoresis. They were then analyzed by Southern blotting with a radioactive T-cell receptor β-chain C-region cDNA probe. The fragments of genomic DNA that derive from the T-cell receptor β-chain genes show up on the film as dark bands. In placental cells the T-cell receptor genes are in the germline configuration, and bands corresponding to the $C_\beta 1$ and $C_\beta 2$ exons are seen. For the two tumors, bands corresponding to genes in the germline configuration are seen, but there are also bands corresponding to distinct rearrangements of a β-chain gene. Courtesy of T. Diss.

produce a functional repertoire of mature naive T cells whose receptors can be activated by pathogen-derived peptides presented by self-MHC molecules, but not by the self-peptides derived from the body's normal constituents. Positive selection tests the ability of α:β T-cell receptors to interact with self-MHC molecules expressed on the surfaces of cells in the thymus. Thymocytes with receptors that interact with self-peptide:self-MHC complexes are signaled to continue their maturation. Negative selection then eliminates those cells whose receptors interact too strongly with self-peptide:self-MHC complexes. The small fraction of thymocytes that survive both positive and negative selection leave the thymus to become mature circulating α:β T cells. These phases in T-cell development are summarized in Figure 5.19.

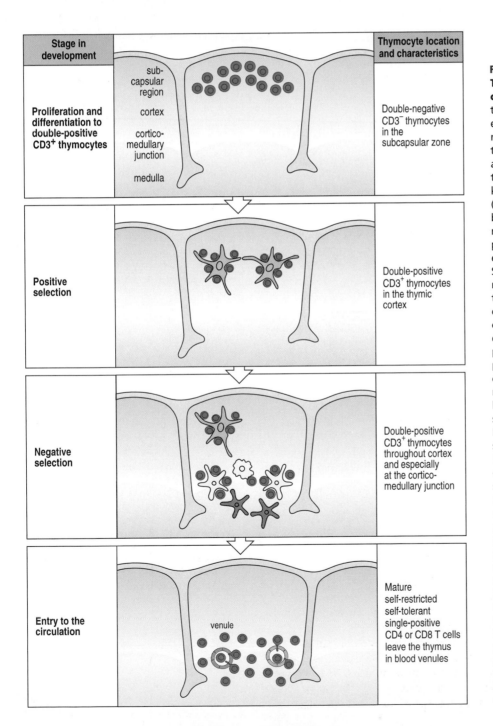

Figure 5.19 The development of T cells in the thymus can be considered as a series of phases. In the first phase, thymocyte progenitors enter the thymus from the blood and migrate to the subcapsular region. At this stage, they do not express the antigen receptor, the CD3 complex, or the CD4 and CD8 co-receptors, and are known as double-negative thymocytes (top panel). These cells proliferate and begin to rearrange their β, γ, and δ T-cell receptor genes, which leads to the production of γ:δ cells and to cells expressing the pre-T-cell receptor. Stimulation through the pre-T-cell receptor leads to cell proliferation and the expression of both CD4 and CD8 co-receptors, producing double-positive cells. As the cells mature, they move deeper into the thymus. In the second phase of development, the double-positive thymocytes rearrange their α-chain genes, express an α:β T-cell receptor and the CD3 complex, and become sensitive to interaction with self-peptide:self-MHC complexes. Double-positive cells undergo positive selection in the thymic cortex through intimate contact with cortical epithelial cells (second panel). During positive selection, matching between the receptor specificity for MHC and the co-receptor molecules starts to take place, which eventually leads to single-positive CD4 or CD8 T cells. In the third phase of development, double-positive cells undergo negative selection for self-reactivity. Negative selection is believed to be most stringent at the cortico-medullary junction, where the nearly mature thymocytes encounter a high density of dendritic cells (third panel). Thymocytes that survive both positive and negative selection leave the thymus in the blood as mature single-positive CD4 or CD8 T cells and enter the circulation (bottom panel).

T-cell development produces several different types of T cell and is therefore more complicated than B-cell development. It also seems to be considerably more wasteful of cells. The vast majority of developing thymocytes die without ever performing a useful task, and only a few percent of thymocytes fulfill the stringent requirements of selection and leave the thymus to enter the circulation. Whereas the bone marrow is continually turning over the B-cell repertoire during the whole of a person's lifetime, the thymus works principally during youth, when it serves to accumulate a repertoire of T cells that can then be used throughout life. This difference might reflect the magnitude of the body's investment in the development of each useful T cell, and the savings to be made by shutting down the thymus before middle age.

T-Cell Mediated Immunity

6

In Chapter 5 we saw how T cells develop in the thymus into a population of mature naive T cells. These T cells now circulate through the peripheral circulation, where they can be activated by specific antigen. Like B cells, naive T cells first meet specific antigen in secondary lymphoid tissues, and it is here that they are activated to undergo clonal expansion and differentiation into effector T cells. Effector T cells can either remain in the lymphoid tissues or migrate to sites of infection, where a subsequent encounter with antigen stimulates them to perform their effector functions.

In the first part of this chapter we consider what happens when a naive T cell encounters its specific antigen for the first time and is stimulated to differentiate into an effector T cell. We shall call this process **T-cell activation**; it is sometimes also referred to as **T-cell priming**. T-cell activation is the first stage of a primary adaptive immune response against most antigens. There are three kinds of effector T cell: **cytotoxic CD8 T cells**, which kill infected cells, and two kinds of **CD4 T cell**

(T_H1 and T_H2) with different functions. The general function of effector CD4 T cells is to secrete cytokines that activate other cells of the immune system. Because the function of CD4 T cells is principally to help other cells achieve their effector function, they are often called **helper T cells**. Once activated, effector T cells interact with their specific antigens which are presented by different types of antigen-presenting cell or **target cell**. This is described later in the chapter. These effector actions, which eventually lead to the removal and destruction of the pathogen, constitute the second stage of the primary immune response. The three kinds of effector T cell enable the human immune system to respond effectively to different types of infection and to different stages of the same infection.

Activation of naive T cells on encounter with antigen

Once an infection begins, the immune system faces the challenge of quickly bringing the minute fraction of naive T cells that are specific for the pathogen into contact with the pathogen's antigens. This is accomplished in the secondary lymphoid tissues, into which antigens brought from outlying tissues by the lymph meet naive T cells brought in by the blood. In this part of the chapter we shall examine the activation of naive T cells to effector T cells by professional antigen-presenting cells within secondary lymphoid tissues. Once activated, antigen-specific CD8 and T_H1 cells are then sent out to the infected sites while T_H2 cells stay in the lymphoid tissues. We shall also see how the interaction of a naive T cell with antigen presented by cells other than professional antigen-presenting cells leads to inactivation rather than activation of the T cell. This mechanism ensures that mature T cells reactive to self-antigens are eliminated before they can become effector cells. This first phase of the primary immune response produces an expanded effector T-cell population that is ready to fight the infecting pathogen but is tolerant of self-antigens.

6-1 Naive T cells first encounter antigen in secondary lymphoid tissues

The immune system does not attempt to initiate adaptive immune responses at the innumerable sites where a pathogen might set up an infection. Instead, it captures some of the pathogen and takes it to the organized secondary lymphoid tissues, whose purpose is the generation of adaptive immune responses.

When bacteria penetrate the skin through a wound, for example, some are carried to the nearest lymph node by the afferent lymph, which originates as extracellular fluid in the damaged connective tissue. Here the bacteria are ingested by antigen-presenting cells such as macrophages, which process the bacterial proteins and display bacterial antigens in the form of peptide:MHC class II complexes on the cell surface as described in Chapter 3. In addition, dendritic cells that reside in the skin can pick up and process bacterial antigens at the site of infection and then carry them to the lymph node by travel in the lymph.

Microorganisms in the blood become trapped in the spleen, whereas those producing infections of the mucosal surfaces of the respiratory or gastrointestinal tracts accumulate, for example, in tonsils, bronchial-associated lymphoid tissues, Peyer's patches, or the appendix. Once within the secondary lymphoid tissues, the trapped antigen is scrutinized by naive circulating T cells.

Naive T cells enter lymphoid tissue through the blood capillaries that provide oxygen and nutrients to the tissue. In lymph nodes, for example, T cells in the

Figure 6.1 Naive T cells encounter antigen during their recirculation through secondary lymphoid organs. Naive T cells (blue and green) recirculate through secondary lymphoid organs, such as the lymph node shown here. They leave the blood at high endothelial venules and enter the lymph node cortex, where they mingle with professional antigen-presenting cells (mainly dendritic cells and macrophages). T cells that do not encounter their specific antigen (green) leave the lymph node in the efferent lymph and eventually rejoin the bloodstream. T cells that encounter antigen (blue) on antigen-presenting cells are activated to proliferate and to differentiate into effector cells. These effector cells can also leave the lymph node in the efferent lymph and enter the circulation.

blood bind to the endothelial cells of the thin-walled high endothelial venules (HEV), squeeze through the vessel wall, and enter the cortical region of the node. The naive T cell then passes through the crowded tissue, constantly encountering antigen-presenting cells and using its antigen receptor to examine the peptide:MHC complexes on their surfaces. When a T cell encounters a peptide:MHC complex to which its T-cell receptor binds, the T cell is retained in the lymph node and activated. It then proliferates and differentiates into a clone of effector T cells (Figure 6.1).

For any given infection, naive T cells specific for the pathogen will represent only one in 10^4 to one in 10^6 of the total pool of circulating T cells. During most passages through a lymph node a T cell does not find its specific antigen and leaves the medulla in the efferent lymph to continue recirculation. In the absence of specific antigen, circulating naive T cells live for many years as small non-dividing cells with condensed chromatin, scanty cytoplasm, and little RNA or protein synthesis. With time, the entire population of circulating T cells will pass through a lymph node. The trapping of pathogens and their antigens in the lymphoid tissue nearest to the site of infection creates a concentrated depot of processed and presented antigens. This enables the small subpopulation of T cells specific for those antigens to be efficiently pulled out of the circulating T-cell pool and activated.

Once an antigen-specific T cell has been trapped in a lymph node by an antigen-presenting cell and activated, it takes several days for the activated T cell to proliferate and for its progeny to differentiate into effector T cells. This accounts for much of the delay between the onset of an infection and the appearance of a primary adaptive immune response. Most effector T cells leave the lymph node in the efferent lymph and on reaching the blood are rapidly carried to the site of infection, where they perform their effector functions.

6-2 Homing of naive T cells to secondary lymphoid tissues is determined by cell adhesion molecules

The behavior of mature T cells before, during, and after their first contact with antigen is controlled entirely by their interactions with other types of cell. Contact between cells is initiated by **cell adhesion molecules** on the T-cell surface, which bind to complementary adhesion molecules on the surfaces of other cells. Although adhesion molecules work independently of antigen, they are sufficiently specific in their interactions with other adhesion molecules to direct the T cell to the appropriate cell type. Thus, T-cell adhesion molecules that bind to high endothelial cells direct naive circulating T cells to leave the blood within a secondary lymphoid tissue, whereas other T-cell adhesion molecules interact with adhesion molecules on antigen-presenting cells to set up the initial cell–cell contacts needed to test the interactions between T-cell receptors and peptide:MHC complexes. Adhesion molecules also stabilize the interaction between an effector T cell and its target cell.

	Name	Tissue distribution	Ligand
Selectins Bind carbohydrates. Initiate leukocyte–endothelial interaction *L-selectin*	L-selectin (MEL-14, CD62L)	Naive and some memory lymphocytes, neutrophils, monocytes, macrophages, eosinophils	Sulfated sialyl Lewis^x, GlyCAM-1, CD34, MAdCAM-1
	P-selectin (PADGEM, CD62P)	Activated endothelium and platelets	Sialyl Lewis^x, PSGL-1
	E-selectin (ELAM-1, CD62E)	Activated endothelium	Sialyl Lewis^x
Mucin-like vascular addressins Bind to L-selectin. Initiate leukocyte–endothelial interaction *CD34*	CD34	Endothelium	L-selectin
	GlyCAM-1	High endothelial venules	L-selectin
	MAdCAM-1	Mucosal lymphoid tissue venules	L-selectin, integrin $\alpha_4\beta_7$
Integrins Bind to cell adhesion molecules and extracellular matrix. Strong adhesion *LFA-1*	$\alpha_L\beta_2$ (LFA-1, CD11a/CD18)	Monocytes, T cells, macrophages, neutrophils, dendritic cells	ICAMs
	$\alpha_M\beta_2$ (Mac-1, CR3, CD11b/CD18)	Neutrophils, monocytes, macrophages	ICAM-1, iC3b, fibrinogen
	$\alpha_X\beta_2$ (CR4, p150.95, CD11c/CD18)	Dendritic cells, macrophages, neutrophils	iC3b
	$\alpha_4\beta_1$ (VLA-4, LPAM-2, CD49d/CD29)	Lymphocytes, monocytes, macrophages	VCAM-1, fibronectin
	$\alpha_5\beta_1$ (VLA-5, CD49d/CD29)	Monocytes, macrophages	Fibronectin
	$\alpha_4\beta_7$ (LPAM-1)	Lymphocytes	MAdCAM-1
	$\alpha_E\beta_7$	Intraepithelial lymphocytes	E-cadherin
Some immunoglobulin superfamily members Various roles in cell adhesion. Target for integrins *CD2*	CD2 (LFA-2)	T cells	LFA-3
	ICAM-1 (CD54)	Activated vessels, lymphocytes, dendritic cells	LFA-1, Mac-1
	ICAM-2 (CD102)	Resting vessels, dendritic cells	LFA-1
	ICAM-3 (CD50)	Lymphocytes	LFA-1
	LFA-3 (CD58)	Lymphocytes, antigen-presenting cells	CD2
	VCAM-1 (CD106)	Activated endothelium	VLA-4

Figure 6.2 Leukocyte adhesion molecules. The four structural classes of adhesion molecule present on white blood cells and the cells with which they interact are: the selectins, which are carbohydrate-binding lectins; the mucin-like vascular addressins, which contain carbohydrate groups to which selectins bind; the integrins; and some proteins in the immunoglobulin superfamily. The figure shows a schematic representation of the structure of one member of each family, and a list of other family members, with alternative names in brackets. The cellular distribution of the adhesion molecules and the ligands to which they bind are also listed. Integrins consist of a large α chain and a smaller β chain. Subfamilies of integrins are defined on the basis of a common β chain that associates with different α chains. The nomenclature for individual adhesion molecules has developed in a rather haphazard fashion, in which many names do not reflect the structural family to which the molecule belongs but are based on assays used to identify cell-surface antigens or adhesion functions. For example, lymphocyte function-associated antigen-1 (LFA-1) is an integrin, whereas LFA-2 and LFA-3 are members of the immunoglobulin superfamily. The CD nomenclature for cell-surface proteins of leukocytes gives each cell-surface protein a unique number but does not reflect its structure or function in any way. iC3b is a complement fragment that becomes bound to pathogen surfaces during an immune response.

Four structural classes of protein are used as adhesion molecules in the immune system: the **selectins**; the mucin-like molecules called **vascular addressins**; the **integrins**; and members of the immunoglobulin superfamily. The different classes of adhesion molecule are described in Figure 6.2.

The entry of recirculating T cells and B cells into secondary lymphoid tissues is directed by selectins on the lymphocyte surface. **L-selectin** expressed by the naive T cell binds to the carbohydrate portions of vascular addressins, which are present on the surface of the endothelial cells of blood vessels. The carbohydrate recognized by L-selectin is sulfated sialyl-Lewis^x, which is structurally related to the carbohydrates that form the Lewis series of blood-group antigens. Two vascular addressins, **CD34** and **GlyCAM-1**, are expressed on the surface of high endothelial venules in lymph nodes and direct naive T cells to leave the blood and enter a node. A third addressin, **MAdCAM-1**, is expressed on the endothelium of capillaries in mucosal tissue, and guides lymphocytes into the lymphoid areas of the gut and other mucosal tissues.

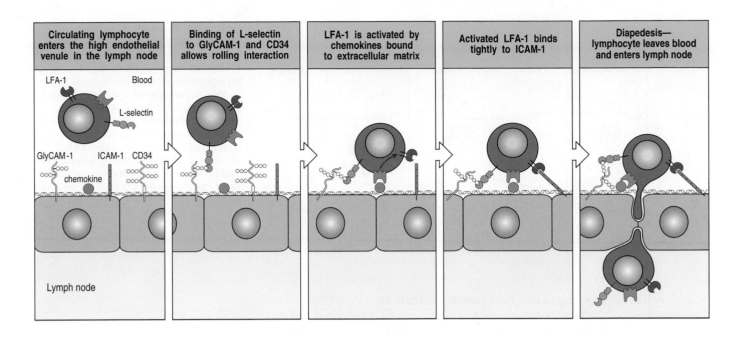

| Circulating lymphocyte enters the high endothelial venule in the lymph node | Binding of L-selectin to GlyCAM-1 and CD34 allows rolling interaction | LFA-1 is activated by chemokines bound to extracellular matrix | Activated LFA-1 binds tightly to ICAM-1 | Diapedesis—lymphocyte leaves blood and enters lymph node |

The interaction between L-selectin and vascular addressin is not in itself sufficient to enable lymphocytes to pass through the endothelium and into lymphoid tissue. This is achieved by strong interactions between integrins on the surface of the lymphocyte and immunoglobulin superfamily members on the surface of the endothelial cells. All T cells express an integrin known as **lymphocyte function-associated antigen-1** (**LFA-1**; $\alpha_L\beta_2$). The principal function of LFA-1 is to bind to the immunoglobulin superfamily **intercellular adhesion molecules** (**ICAMs**). Vascular endothelium expresses **ICAM-1** and **ICAM-2**; binding of these to LFA-1 on a naive T cell enables the T cell to squeeze through the endothelium and enter the cortex of the lymphoid tissue. Small chemoattractant proteins, or **chemokines**, which are also bound to the endothelium, activate the integrin LFA-1 on the lymphocyte surface, enabling it to bind tightly to ICAM-1 on the endothelial cell (Figure 6.3).

As naive T cells negotiate their way through the packed cells of the lymph node cortex, they bind transiently to the antigen-presenting cells that they meet. These interactions involve integrins and members of the immunoglobulin superfamily. Both T cells and professional antigen-presenting cells express the integrin LFA-1. The T cell's LFA-1 binds to ICAM-1 or ICAM-2 on the antigen-presenting cells, whereas the LFA-1 of the antigen-presenting cell binds to a third kind of ICAM—**ICAM-3**—on the T-cell surface. Adhesion is strengthened by interaction between two other members of the immunoglobulin superfamily, **CD2** on the T cell and **LFA-3** on the antigen-presenting cell (Figure 6.4). These transitory cell–cell interactions enable the T-cell receptor to screen the peptide:MHC complexes on the surface of the antigen-presenting cell for ones that engage the receptor and activate the T cell.

When a naive T cell encounters a specific peptide:MHC complex, a signal is delivered through the T-cell receptor. This induces a change in the conformation

Figure 6.3 Naive T and B lymphocytes circulate in the blood and enter lymph nodes by crossing high endothelial venules. L-selectin on the lymphocyte surface binds to sulfated carbohydrates of the vascular addressins GlyCAM-1 and CD34, expressed on the high endothelial cells of venules in the lymph node. Chemokines, which are also bound to the endothelium, activate the integrin LFA-1 on the lymphocyte surface, enabling it to bind tightly to ICAM-1 on the endothelial cell. Establishment of tight binding allows the lymphocyte to squeeze between two endothelial cells, leaving the lumen of the blood vessel and entering the lymph node proper.

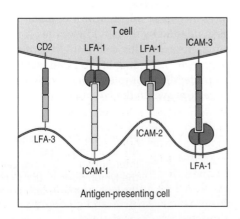

Figure 6.4 Cell-surface molecules of the immunoglobulin superfamily are important in the interaction of lymphocytes with antigen-presenting cells. In the initial encounter of T cells with antigen-presenting cells, CD2 binding to LFA-3 on the antigen-presenting cell synergizes with LFA-1 binding to ICAM-1 and ICAM-2. ICAM-3 expressed on the T cell binds to LFA-1 on the antigen-presenting cell.

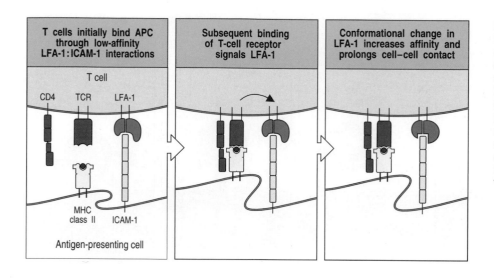

| T cells initially bind APC through low-affinity LFA-1:ICAM-1 interactions | Subsequent binding of T-cell receptor signals LFA-1 | Conformational change in LFA-1 increases affinity and prolongs cell–cell contact |

Figure 6.5 Transient adhesive interactions between T cells and antigen-presenting cells are stabilized by specific antigen recognition. When a T cell binds to its specific peptide:MHC ligand on an antigen-presenting cell (APC), intracellular signaling through the T-cell receptor complex induces a conformational change in LFA-1 that causes it to bind with higher affinity to ICAMs on the antigen-presenting cell. The T cell shown here is a CD4 T cell.

of the T cell's LFA-1 molecules that increases their affinity for ICAMs (Figure 6.5). The interaction of the T cell with the antigen-presenting cell is stabilized and can last for several days, during which time the T cell proliferates, and its progeny, while also remaining in contact with the antigen-presenting cell, differentiate into effector cells.

6-3 Activation of naive T cells requires a co-stimulatory signal delivered by a professional antigen-presenting cell

Ligation of the T-cell receptor alone is insufficient to trigger naive T-cell activation, clonal expansion by cell division, and differentiation of the progeny into effector cells. A second, **co-stimulatory**, signal is required. This can be delivered only by a professional antigen-presenting cell to which the naive T cell is bound by its T-cell receptor. Antigen-specific stimulation and co-stimulation must both be delivered by the same cell.

There are three kinds of professional antigen-presenting cell—**dendritic cells**, **macrophages**, and **B cells**. Dendritic cells are the most potent and they are dedicated entirely to the presentation of antigens to T cells, whereas macrophages and B cells have other effector functions of their own. The three types of antigen-presenting cell are present at different locations within a secondary lymphoid tissue such as a lymph node (Figure 6.6). Macrophages are seen throughout the lymph node, where they actively ingest microorganisms and other particulate material. The dendritic cells are present only in the T-cell areas, whereas B cells are present in the lymphoid follicles (see Section 4-8, p. 99).

Professional antigen-presenting cells are distinguished from other cells by the presence of **co-stimulator molecules** on their surfaces. The co-stimulatory signal is generated by the binding of the co-stimulator molecule **B7** on the professional antigen-presenting cell to the receptor protein **CD28** on the surface of the T cell (Figure 6.7). Binding of the T-cell receptor and the co-receptors CD4 or CD8 to the peptide:MHC complex induces clonal expansion and further differentiation of the T cell only when the co-stimulatory signal is also given.

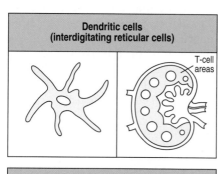

Dendritic cells (interdigitating reticular cells)

T-cell areas

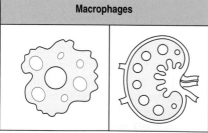

Macrophages

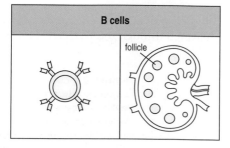

B cells

follicle

Figure 6.6 Professional antigen-presenting cells are distributed differentially in the lymph node. Dendritic cells, also called interdigitating reticular cells, are present throughout the cortex of the lymph node in the T-cell areas. Macrophages are distributed throughout the lymph node. B cells are seen mainly in the follicles.

Figure 6.7 The principal co-stimulatory molecules on professional antigen-presenting cells are B7 molecules, which bind the protein CD28 on the surface of the T cell. Binding of the T-cell receptor (TCR) and its co-receptor CD4 to the peptide:MHC class II complex on the antigen-presenting cell delivers a signal (arrow 1). This signal induces clonal expansion of T cells only when the co-stimulatory signal (arrow 2) is also given by the binding of CD28 to B7. Both CD28 and B7 are members of the immunoglobulin superfamily. There are two forms of B7, called B7.1 (CD80) and B7.2 (CD86), but their functional differences have yet to be understood.

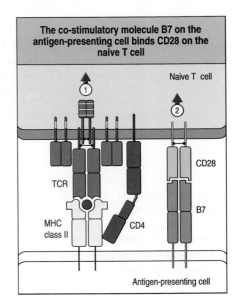

Although CD28 is the only B7 receptor on naive T cells, an additional receptor is expressed once they have been activated. This is **CTLA-4**, which is structurally similar to CD28. CTLA-4 binds B7 twenty times more strongly than does CD28 and it seems to serve an antagonistic function. Whereas B7 binding to CD28 activates a T cell, the engagement of CTLA-4 dampens down activation and limits cell proliferation. This inhibitory role for CTLA-4 is illustrated by the phenotype of mice engineered to lack a functional CTLA-4 gene. In these animals, lymphocytes undergo a massive and uncontrolled proliferation that proves fatal.

6-4 The three kinds of professional antigen-presenting cell present three kinds of antigen to naive T cells

The professional antigen-presenting cells are each specialized to deal with a particular type of antigen. Macrophages specialize in presenting peptides from extracellular, independently replicating microorganisms such as bacteria and yeast, dendritic cells are adept at dealing with viruses, and B cells efficiently present peptides from soluble antigens (Figure 6.8).

	Macrophages	**Dendritic cells**	**B cells**
Antigen uptake	Phagocytosis +++	Phagocytosis by tissue dendritic cells +++ Viral infection ++++	Antigen-specific receptor (Ig) ++++
MHC expression	Inducible by bacteria and cytokines − to +++	Constitutive ++++	Constitutive. Increases on activation +++ to ++++
Co-stimulator delivery	Inducible − to +++	Constitutive in mature non-phagocytic lymphoid dendritic cells ++++	Inducible − to +++
Antigen presented	Particulate antigens. Intracellular and extracellular pathogens	Peptides. Viral antigens. (Allergens?)	Soluble antigens. Toxins. Viruses
Location	Lymphoid tissue. Connective tissue. Body cavities	Lymphoid tissue. Connective tissue. Epithelia	Lymphoid tissue. Peripheral blood

Figure 6.8 The three types of professional antigen-presenting cell. Macrophages, dendritic cells, and B cells are the cells involved in the activation of antigen-specific naive T cells. These three cell types vary in their means of antigen uptake, MHC class II expression, co-stimulator expression, the antigens that they present, and their location within the body. The different professional antigen-presenting cells are adapted to present peptides from different types of pathogen to T cells. Dendritic cells specialize in presenting viral antigens, macrophages in presenting bacterial antigens, and B cells in presenting soluble proteins such as bacterial toxins.

Figure 6.9 Microbial substances induce co-stimulatory activity in macrophages. Phagocytosis of bacteria by macrophages and their breakdown in the phagolysosomes leads to the release of substances such as bacterial lipopolysaccharide, which induce the expression of co-stimulatory B7 molecules on the surface of the macrophage. Peptides derived from the degradation of bacterial proteins in the macrophage vesicular system are bound by MHC class II molecules and presented on the macrophage surface. The combination of B7 binding to CD28 and peptide:MHC complexes binding to the T-cell receptor activates naive T cells.

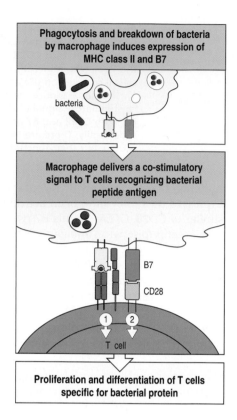

Macrophages are phagocytic cells that take up microorganisms and other particulate material from the extracellular environment and degrade it in phagolysosomes loaded with hydrolytic enzymes. The macrophage surface bears several receptors specific for carbohydrate structures that are unique to bacterial cell walls and capsules and are not present on human cells. These receptors bind bacteria to the macrophage surface, facilitating their phagocytosis by receptor-mediated internalization. The receptors that recognize microbial constituents probably also mediate the induction of co-stimulatory activity, as exposure to a single microbial constituent causes B7 expression by most macrophages. There is evidence that these receptors evolved originally to enable the phagocytic cells in primitive eukaryotic organisms to recognize microorganisms by binding to structures such as bacterial carbohydrates or lipopolysaccharides that are not found in eukaryotes. Macrophage receptors still serve this function in innate immunity as well as having an important role in the initiation of adaptive immune responses.

Before interaction with microbes, resting macrophages do not express the co-stimulator molecule B7 and express little or no MHC class II; in this state they are unable to activate antigen-specific naive T cells. However, as a consequence of receptor stimulation, macrophages are induced to express B7 and MHC class II molecules and thereby become professional antigen-presenting cells (Figure 6.9).

Because macrophages break down pathogens in their endosomes and lysosomes, they most commonly present pathogen-derived peptides on MHC class II molecules (see Section 3-10, p. 68), which then activate naive CD4 T cells. Certain bacteria such as *Listeria monocytogenes* avoid destruction by leaving the endocytic vesicles and exploiting the macrophage's cytosol as a place to live. The infected cell can, however, now be detected by CD8 cytotoxic T cells, which on killing the infected cell release the bacteria into the extracellular space, from which they can be taken up by new macrophages. In normal healthy individuals, symptoms of listerial infection are rare because the combined actions of neutrophils, macrophages and CD8 T cells are sufficient to keep the bacterial population down. For pregnant women, the very young, or the immunosuppressed patient, however, listeriosis can be life-threatening.

The responsibility for initiating T-cell responses against viral infections is borne by the dendritic cells. Dendritic cells are good at presenting virus-derived peptides on MHC class I molecules and it is possible that they have a specialized pathway for processing extracellular virus that produces viral peptides bound to MHC class I molecules. They also ingest extracellular pathogens and present their peptides on MHC class II molecules by the normal processing route for extracellular antigens. Some dendritic cells arise from monocytes, and others from bone marrow progenitors that first migrate via the blood to tissues and organs. These immature dendritic cells are highly phagocytic but they do not have co-stimulatory activity (Figure 6.10, top panel). When they ingest antigens in non-lymphoid tissues they are caused to migrate in the lymph to the nearest lymph node. Here they become mature dendritic cells. They lose their phagocytic ability, but now constitutively express high levels of MHC molecules, co-stimulator molecules, and adhesion

Figure 6.10 Dendritic cells mature through two stages to become potent antigen-presenting cells in lymphoid tissues. Immature dendritic cells (top panel) in non-lymphoid tissues are phagocytic through their use of the receptor DEC 205, but they do not express co-stimulatory molecules. Mature dendritic cells (second panel) found in secondary lymphoid tissues are potent antigen-presenting cells because they constitutively express all the cell-surface molecules required to interact with and activate naive T cells. These include high levels of MHC class I, MHC class II, and the co-stimulator B7, as well as high levels of the adhesion molecules ICAM-1, ICAM-2, LFA-1, and LFA-3. The bottom panel shows an electron micrograph of a mature dendritic cell. Photograph courtesy of R. Steinman.

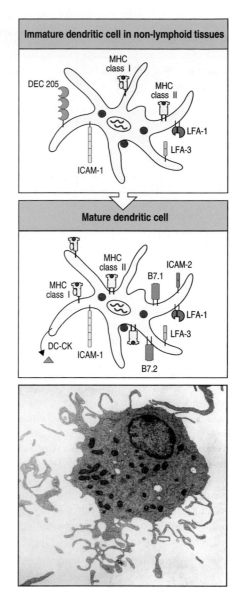

molecules. They also produce various cytokines, including the chemokine DC-CK, which attracts naive T cells (see Figure 6.10, second panel). Mature dendritic cells are thus constantly ready to activate naive T cells. This progression means that a dendritic cell presents only those antigens that it acquires at the site of infection, but it cannot activate naive T cells until it reaches a lymphoid organ. By carrying pathogen antigens from their site of origin into the lymphoid tissues, dendritic cells provide a rapid means for initiation of both the CD4 and CD8 T-cell responses.

Immature dendritic cells are present in the epithelial tissues that confront the external environment, particularly those lining the skin, gut, and respiratory tract. Of these immature cells, the **Langerhans' cells** of the skin are the best characterized. Once Langerhans' cells at a site of infection have taken up antigen, they migrate in the lymph to the nearest lymph node. Here they home to the T-cell areas and mature into dendritic cells with co-stimulatory activity, which are also known as **interdigitating reticular cells** (Figure 6.11).

The third type of professional antigen-presenting cell, the B cell, binds soluble protein antigens from the extracellular environment by means of its surface immunoglobulin. Receptor-mediated endocytosis causes selective uptake of the protein antigen, followed by its processing into peptides that bind to MHC class II molecules (Figure 6.12). B cells therefore present peptides to CD4 T cells. Like macrophages, B cells do not express B7 co-stimulator molecules constitutively but are induced to do so by microbial constituents such as bacterial lipopolysaccharide. Because B7 expression needs to be induced, B cells activate naive T cells to respond to soluble protein antigens only in the context of an infection.

Immunologists appreciated that antigen alone was usually insufficient to induce an immune response long before the reasons for this effect were understood. Much research in immunology involved immunizing laboratory animals with purified protein antigens, but immunization with the protein alone rarely led to an immune response. However, when the protein antigen was mixed with certain bacteria or their breakdown products, a strong response to the protein was obtained. It was subsequently discovered that the microbial components, known in this context as **adjuvants**, were inducing co-stimulatory activity in B cells and macrophages. This explains why whole microorganisms are usually more effective vaccines than are highly purified antigenic macromolecules. The induction of co-stimulatory activity by common microbial constituents is believed to allow the immune system to distinguish antigens borne by infectious agents from antigens associated with innocuous proteins, including self-proteins.

Figure 6.11 Electron micrograph of an interdigitating reticular cell in the T-cell area of a human lymph node. The interdigitating cell has a complex folded nucleus (N). Its cytoplasm (C) makes a complex mesh around the surrounding T lymphocytes. Magnification × 10,000. Photograph courtesy of N. Rooney.

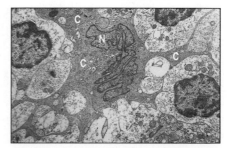

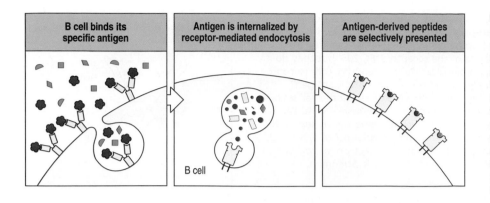

B cell binds its specific antigen	Antigen is internalized by receptor-mediated endocytosis	Antigen-derived peptides are selectively presented

B cell

Figure 6.12 B cells use their surface immunoglobulin to internalize specific antigen efficiently by receptor-mediated endocytosis. B cells selectively bind specific antigen from the extracellular milieu. The antigen:immunoglobulin complexes are then internalized by receptor-mediated endocytosis. They are delivered to endocytic vesicles, where they are degraded into peptides which are bound by MHC class II molecules. The peptide:MHC class II complexes are transported to the cell surface.

6-5 When T cells are activated by antigen, signals from T-cell receptors and co-receptors alter the pattern of gene transcription

When a T cell binds the peptide:MHC complexes on an antigen-presenting cell, the receptor–ligand interactions occur at localized and apposed areas of the two cell membranes. Within these areas, the specific MHC:peptide complexes on the antigen-presenting cell and the T-cell receptors and co-receptors cluster together, with cell adhesion molecules forming a tight seal around the area. The signal that antigen has bound to the T-cell receptor is transmitted to the interior of the T cell by the cytoplasmic tails of the CD3 proteins, which are associated with the antigen-binding α and β chains of the receptor (see Figure 3.4, p. 59). The cytoplasmic tails of all the CD3 proteins contain sequences called **immunoreceptor tyrosine-based activation motifs** (**ITAMs**), which associate with cytoplasmic **protein tyrosine kinases**. These kinases are activated by receptor clustering and phosphorylate tyrosine residues in the ITAMs. Enzymes and other signaling molecules bind to the phosphorylated tyrosine residues and become activated in their turn. In this way, pathways of intracellular signaling are initiated that end with alterations in gene expression.

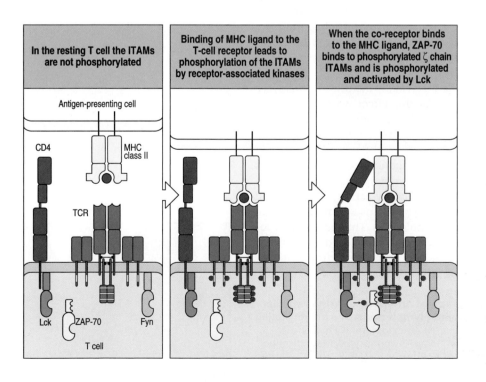

In the resting T cell the ITAMs are not phosphorylated	Binding of MHC ligand to the T-cell receptor leads to phosphorylation of the ITAMs by receptor-associated kinases	When the co-receptor binds to the MHC ligand, ZAP-70 binds to phosphorylated ζ chain ITAMs and is phosphorylated and activated by Lck

Antigen-presenting cell

CD4   MHC class II

TCR

Lck   ZAP-70   Fyn

T cell

Figure 6.13 Clustering of the T-cell receptor and a co-receptor initiates signaling within the T cell. When T-cell receptors become clustered on binding MHC:peptide complexes on the surface of an antigen-presenting cell, activation of receptor-associated kinases such as Fyn leads to phosphorylation of the CD3γ, δ, and ε ITAMs (yellow, with phosphorylated tyrosines shown as small pink circles) as well as those on the ζ chain. The tyrosine kinase ZAP-70 binds to the phosphorylated ITAMs of the ζ chain but is not activated until the co-receptor binds to the MHC molecule on the antigen-presenting cell (here shown as CD4 binding to an MHC class II molecule), which brings the kinase Lck into the complex. This phosphorylates and activates ZAP-70.

Signals from both the T-cell receptor and the CD4 or CD8 co-receptor combine to stimulate the T cell (Figure 6.13). The cytoplasmic tails of both CD4 and CD8 are associated with a protein tyrosine kinase called **Lck**. On formation of the T-cell receptor:MHC:co-receptor complex this kinase activates a cytoplasmic protein tyrosine kinase called **ZAP-70**, which binds to the phosphorylated tyrosines on the ζ chain of the T-cell receptor complex. ZAP-70 is instrumental in initiating the intracellular signaling pathway. The importance of ZAP-70 for all subsequent signaling events was revealed by a patient with immunodeficiency due to the absence of functional ZAP-70. Although the patient had normal numbers of T cells, these were unable to develop intracellular signals on engagement of their antigen receptors.

The participation of the co-receptor is essential for effective T-cell stimulation. A target cell minimally requires about 100 specific peptide:MHC complexes to trigger a naive T cell. Human cells express between 10,000 and 100,000 MHC molecules per cell, so that 0.1–1% of the MHC molecules on a target cell must bind the same peptide for the cell to activate a T cell. In the absence of the correct co-receptor—CD8 for peptides presented by MHC class I, and CD4 for peptides presented by MHC class II—stimulation of the T cell becomes highly inefficient, requiring about 10,000 specific peptide:MHC complexes on the target cell, a number almost never reached *in vivo*.

Once activated, the T-cell-specific kinase ZAP-70 triggers three signaling pathways that are common to many types of cell (Figure 6.14). In naive T cells, they lead to changes in gene expression produced by the transcriptional activator **NFAT** (**Nuclear Factor of Activated T cells**) in combination with other transcription factors. One pathway initiated by ZAP-70 leads via the second messenger inositol trisphosphate to the activation of NFAT. A second pathway leads to the activation of protein kinase C, which results in the induction of the transcription factor **NFκB**. The third signaling pathway initiated by ZAP-70 involves the activation of Ras, a GTP-binding protein, and leads to the activation of a nuclear protein called Fos, which is a component of the transcription factor **AP-1**. The co-stimulatory signals delivered through CD28, lead, among other things, to activation of the Jun protein, which with Fos forms the AP-1 transcription factor.

The combined actions of NFAT, AP-1, and NFκB turn on the transcription of genes that direct T-cell proliferation and the development of effector function. One of the most important of these genes is that for the cytokine interleukin-2 (IL-2).

6-6 Proliferation and differentiation of activated T cells are driven by the cytokine interleukin-2

Activation by a professional antigen-presenting cell initiates a program of differentiation in the T cell that starts with a burst of cell division and then leads to the acquisition of effector function. This program is under the control of a cytokine called **interleukin-2** (**IL-2**), which is synthesized and secreted by the activated T cell itself. IL-2 binds to IL-2 receptors at the T-cell surface to drive clonal expansion of the activated cell. IL-2 is one of a number of cytokines produced by activated and effector T cells that control the development and differentiation of cells in the immune response.

The production of IL-2 requires both the signal delivered through the T-cell receptor:co-receptor complex and the co-stimulatory signal delivered through CD28 (see Section 6-3). Signals delivered through the T-cell receptor complex activate the transcription factor NFAT, which activates transcription of the IL-2 gene. IL-2 and other cytokines have powerful effects on cells of the immune system, and their production is precisely controlled in both time and space. To avoid overproduction,

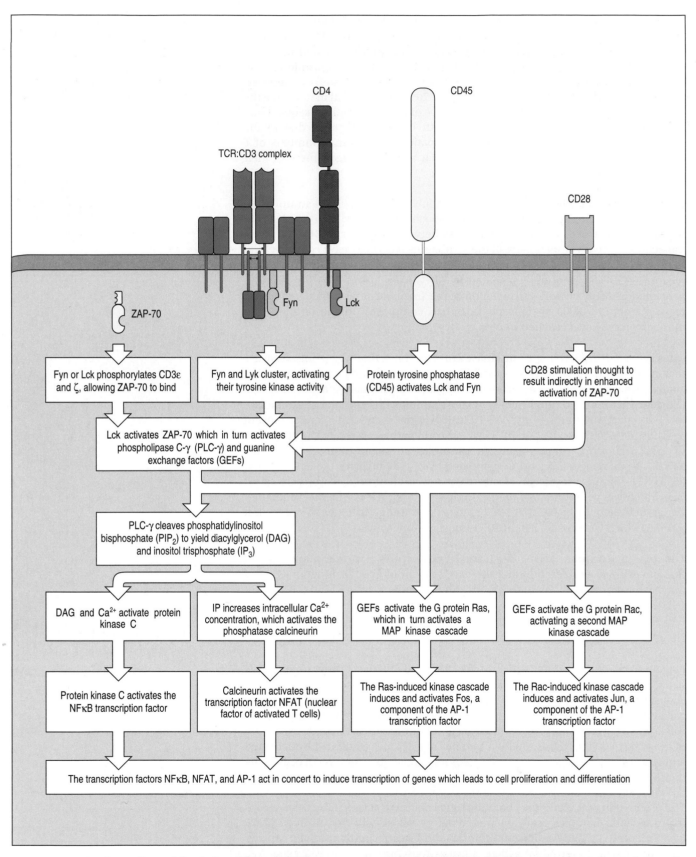

Figure 6.14 Simple outline of the intracellular signaling pathways initiated by the T-cell receptor complex, its CD4 co-receptor, and CD28. Similar pathways operate in CD8 T cells, as CD8 interacts with Lck like CD4.

Figure 6.15 Activated T cells secrete and respond to interleukin-2 (IL-2). Naive T cells express the low-affinity receptor for IL-2, which consists of β and γ chains. Activation of a naive T cell by the recognition of a peptide:MHC complex accompanied by co-stimulation induces the synthesis and secretion of IL-2 and the synthesis of the IL-2 receptor α chain (yellow). The cell also enters the first phase (G1) of the cell-division cycle. The α chain combines with the β and γ chains to make a high-affinity receptor for IL-2. IL-2 binds to the IL-2 receptor, producing an intracellular signal that promotes T-cell proliferation.

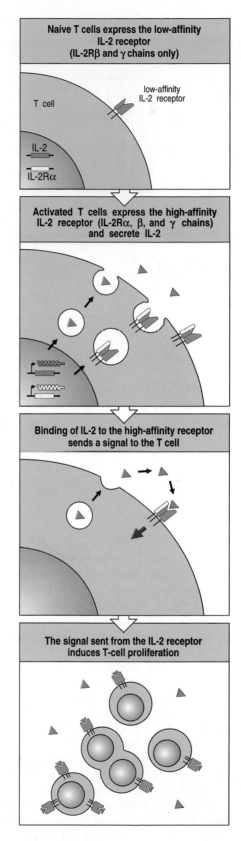

cytokine mRNA is inherently unstable, and sustained production of IL-2 requires the stabilization of the mRNA. Stabilization is one of the functions of the co-stimulatory signal, and it causes a 20- to 30-fold increase in IL-2 production by the T cell. A second effect of co-stimulation is the activation of additional transcription factors that increase the rate of transcription of the IL-2 gene threefold. The principal effect of co-stimulation is therefore to increase the synthesis of IL-2 by the T cell by some 100-fold.

IL-2 binds to a high-affinity IL-2 receptor whose expression is also induced by T-cell activation (Figure 6.15). On binding IL-2, this receptor triggers the T cell to progress through cell division. T cells activated in this way can divide two to three times a day for about a week, enabling a single activated T cell to produce thousands of daughter cells. This proliferative phase is of crucial importance to the immune response because it produces large numbers of antigen-specific effector cells from rare antigen-specific naive T cells. In the response to certain viruses, nearly 50% of the CD8 T cells present at the peak of the response are specific for a single viral peptide:MHC complex.

The importance of IL-2 in the activation of the adaptive immune response is reflected in the mode of action of the immunosuppressive drugs cyclosporin A (cyclosporine), tacrolimus (also called FK506), and rapamycin (also called sirolimus), which are used to prevent the rejection of organ transplants. Cyclosporin A and tacrolimus inhibit IL-2 production by disrupting signals from the T-cell receptor, whereas rapamycin inhibits signaling from the IL-2 receptor. These drugs therefore suppress the activation and differentiation of naive T cells and all immune responses that require activated T cells.

6-7 Antigen recognition by a naive T cell in the absence of co-stimulation leads to the T cell becoming non-responsive

Among the mature naive T cells entering the peripheral circulation from the thymus are some that are specific for self-proteins expressed by cells not encountered in the thymus. However, these T cells are unlikely to be activated, because the cells presenting these self-antigens will not express co-stimulatory molecules. When the T-cell receptor on a mature naive T cell binds to a peptide:MHC complex on a cell that does not express the co-stimulatory molecule B7, the T cell becomes non-responsive, a state described as anergy. In this state, the T cell cannot be activated even if it subsequently encounters its specific antigen presented by a professional antigen-presenting cell (Figure 6.16). The principal characteristic of anergic T cells is their inability to make IL-2; they are therefore unable to stimulate their own proliferation and differentiation. Induction of tolerance in the mature T-cell repertoire thus seems to be based on the requirement that antigen-specific stimulation and co-stimulation of T cells are both delivered by the same antigen-presenting cell.

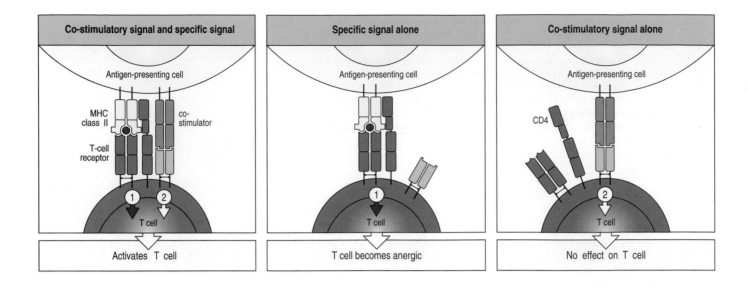

6-8 Activated CD4 T cells can differentiate in ways that favor either a humoral or a cell-mediated immune response

Towards the end of the proliferative phase, activated T cells acquire the capacity to synthesize proteins that are needed to perform the specialized functions of effector T cells. Mature naive CD8 T cells are already committed to becoming cytotoxic effector cells should they be activated by antigen. The effector options for CD4 T cells are, however, more varied, and commitment to a particular function is made only when the naive CD4 T cell is first stimulated by antigen.

CD4 T cells can differentiate along two pathways, giving rise to either CD4 T_H1 **cells** or CD4 T_H2 **cells** (Figure 6.17). Most immune responses involve contributions from both T_H1 and T_H2 cells. Which differentiation pathway is taken in any particular interaction depends on factors that are not completely understood. These include the cytokines that are already present as a result of preceding innate immune responses, cytokines produced by the antigen-presenting cell, the abundance of the antigen and thus the density of specific peptide:MHC complexes on the surface of the antigen-presenting cell, and the affinity of the peptide:MHC complexes for the T-cell receptor. The cytokines produced by effector T_H1 and T_H2 cells also tend to suppress each other's differentiation. So once a CD4 T-cell response has been pointed in one direction, this bias becomes reinforced.

Effector CD4 T cells are helper cells that secrete cytokines which activate and recruit other immune system cells into the immune response. The cytokines secreted by T_H1 cells bias the immune response toward macrophage activation, which leads to inflammation and a **cell-mediated immune response**. This means a response dominated by cytotoxic CD8 T cells and/or CD4 T_H1 cells, macrophages, and other effector cells of the immune system. The cytokines secreted by T_H2 cells, in contrast, mainly induce B-cell differentiation and the production of antibodies. The response that involves antibody production is known as the **humoral immune response**, because antibodies are present in the body fluids, which were historically called humors. This division of labor is not absolute, however, because T_H1 cells have some influence on antibody production. Thus both T_H1 and T_H2 cells contribute to the induction of isotype switching in B cells.

Figure 6.16 T-cell tolerance to antigens expressed on non-professional antigen-presenting cells results from antigen recognition in the absence of co-stimulation. A naive T cell can be activated only by an antigen-presenting cell carrying both a specific peptide:MHC complex and a co-stimulatory molecule on its surface. This combination results in the naive T cell's receipt of signal 1 from the T-cell receptor and signal 2 from the co-stimulator (left panel). When the antigen-presenting cell has the specific peptide:MHC complex to deliver signal 1, but no co-stimulator to deliver signal 2, the T cell enters a non-responsive state called anergy (center panel). When the antigen-presenting cell has a co-stimulator to deliver signal 2, but no specific peptide:MHC complex to deliver signal 1, the naive T cell neither responds nor becomes anergic (right panel).

Certain infectious agents can elicit either a T_H1 or a T_H2 response, rather than a mixture of the two, and the choice made by the immune system can have profound consequences for the human host. An impressive example is leprosy, which is caused by infection with *Mycobacterium leprae*, a bacterium that grows within the vesicular system of macrophages. The most effective response to this microorganism is given by T_H1 cells. They secrete cytokines that activate macrophages, which then destroy the bacteria that they contain. When the host's response consists mainly of T_H1 cells, bacterial populations are kept low and, although skin and peripheral nerves are damaged by the inflammatory response associated with macrophage activation, the disease progresses slowly and the patient usually survives. However, if the CD4 T-cell response consists mainly of T_H2 cells, then bacterial populations expand because the antibodies made by the host cannot reach the bacteria inside the macrophages. Unchecked bacterial growth within macrophages causes gross tissue destruction, which is eventually fatal. The visible symptoms of disease in patients making either a T_H1 or a T_H2 response to *Mycobacterium leprae* are so dissimilar that the conditions are given different names: tuberculoid leprosy and lepromatous leprosy, respectively.

6-9 Naive CD8 T cells can be activated in different ways to become cytotoxic effector cells

The activation of naive CD8 T cells to cytotoxic effector cells generally requires stronger co-stimulatory activity than is needed to activate CD4 T cells. Only dendritic cells, the most potent of the antigen-presenting cells, can provide sufficient co-stimulation. When activated by antigen and co-stimulatory molecules on a dendritic cell, CD8 T cells are stimulated to synthesize both the cytokine IL-2 and its high-affinity receptor, which together induce the proliferation and differentiation of the CD8 T cells (Figure 6.18, left panels).

In circumstances where the antigen-presenting cell offers sub-optimal co-stimulation, CD4 T cells can help to activate naive CD8 T cells. To do this, the naive CD8 cell and the CD4 T cell recognize their specific antigens on the same antigen-presenting cell. When the CD4 T cell is already an effector cell, recognition of antigen causes it to secrete cytokines that then induce the antigen-presenting cell to increase its level of co-stimulators (see Figure 6.18, center panels). The activated antigen-presenting cell then activates the naive CD8 T cell. In this mechanism the CD4 T cell and the CD8 T cell can either interact simultaneously (as shown in Figure 6.18) or successively with the antigen-presenting cell. When the CD4 T cell is a naive cell, it is activated by the antigen-presenting cell to produce IL-2, which can then drive the proliferation and differentiation of the CD8 T cell (see Figure 6.18, right panels). This latter mechanism works because engagement of the T-cell receptor is sufficient to induce CD8 T cells to express the high-affinity IL-2 receptor, although not to make their own IL-2. This mechanism requires the two T cells to interact simultaneously with the antigen-presenting cell; the close juxtaposition of the two T cells on the surface of the antigen-presenting cell is needed to ensure that enough IL-2 is captured by the CD8 T cell to induce its activation.

The more stringent requirements for the activation of naive CD8 T cells means that they are only activated when the evidence of infection is unambiguous. Cytotoxic T cells inflict damage on any target tissue to which they are directed, and their actions will only be of benefit to the host if a pathogen is eliminated in the process. Even then the actions of cytotoxic T cells can have deleterious effects. For example, in fighting viral infections of the airways, cytotoxic T cells prevent viral replication by destroying the epithelial layer, but this makes the underlying tissue vulnerable to secondary bacterial infection.

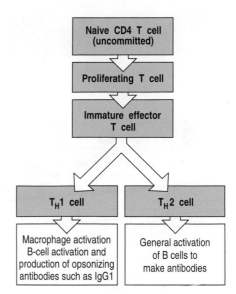

Figure 6.17 The stages of activation of CD4 T cells. Naive CD4 T cells first respond to peptide:MHC class II complexes by synthesis of IL-2 and proliferation. The progeny cells have the potential to become either T_H1 or T_H2 cells.

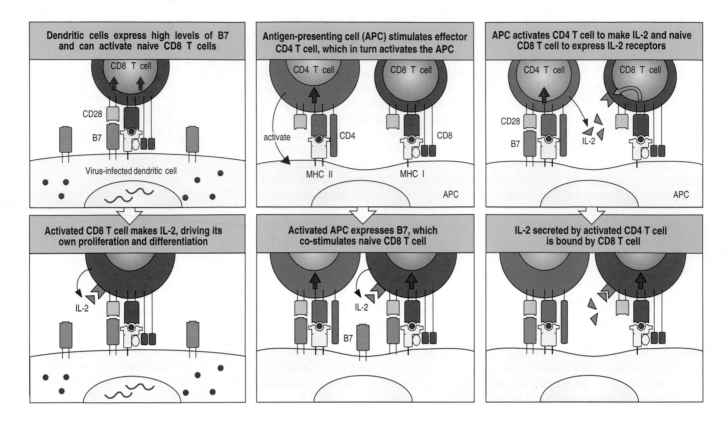

Figure 6.18 Three ways of activating a naive CD8 T cell. The left panels show how a naive CD8 T cell can be activated directly by a virus-infected dendritic cell. The center and right panels show two ways in which antigen-presenting cells (APC) that offer sub-optimal co-stimulation can interact with a CD4 T cell to stimulate a naive CD8 T cell. One way is for cytokines secreted by the CD4 T cell to improve the co-stimulation of the antigen-presenting cell, for example by the induction of B7 expression (center panels). A second way is for cytokines secreted by the CD4 T cell, for example IL-2, to act directly on a neighboring CD8 T cell (right panels).

Summary

Almost all types of adaptive immune response are started by the activation of antigen-specific naive T cells. Their proliferation and differentiation to form large clones of antigen-specific effector T cells form the first stage of a primary immune response. T-cell activation takes place in the secondary lymphoid tissues and requires antigen to be presented to the naive T cell by a professional antigen-presenting cell. Professional antigen-presenting cells express co-stimulatory molecules that engage ligands on the T-cell surface. This interaction induces the production of the cytokine IL-2 by the T cell, which is required for clonal expansion and differentiation. Activation requires signals from the T-cell receptor, the co-receptor, and co-stimulatory molecules. When a naive T cell engages specific antigen on a non-professional antigen-presenting cell, the lack of co-stimulation induces anergy, a state of tolerance. The whole course of T-cell activation occurs in the immediate environment of the professional antigen-presenting cell. Throughout the process, the naive T cell and its numerous progeny maintain contact with the same antigen-presenting cell.

The three major types of professional antigen-presenting cell are the dendritic cell, the macrophage, and the B cell. These three cells specialize in the presentation of different types of antigen and in the induction of effector T cells with distinct functions. The same mechanisms are used to activate CD4 T cells and CD8 T cells, although CD8 T cells require stronger co-stimulation. Whereas CD8 T cells are all destined to have cytotoxic effector function, CD4 T cells can differentiate along alternative pathways to produce effector cells that secrete different cytokines and drive the immune response in different directions. CD4 T_H1 cells secrete mainly cytokines that favor cell-mediated immune responses, whereas CD4 T_H2 cells secrete mainly cytokines that stimulate B cells to produce antibodies.

The properties and functions of effector T cells

After differentiating in the secondary lymphoid tissues, effector T cells detach themselves from the antigen-presenting cell that nursed their differentiation. CD8 cytotoxic T cells and most CD4 T_H1 cells leave the lymphoid tissues and enter the blood to seek out the sites of infection, whereas CD4 T_H2 cells mostly remain in the secondary lymphoid tissues. T-cell effector function is turned on when the T-cell receptors bind to peptide:MHC complexes on a target cell. This stimulates the release of effector molecules by the T cell that act on the target cell. As well as possessing specialized functions, effector T cells differ from naive T cells in ways that enable them to act more effectively as effector cells. We shall consider this first before discussing the specialized functions of CD8 T cells and CD4 T_H1 and T_H2 cells.

6-10 Effector T cells can be stimulated by antigen in the absence of co-stimulatory signals

One of the major changes that occurs in effector T cells enables them to respond to their specific antigen without the need for co-stimulation via B7–CD28 interaction. This means they can respond to antigen on cells other than professional antigen-presenting cells. The benefit gained from these relaxed activation requirements is most easily understood for cytotoxic CD8 T cells. They must be able to recognize and kill all manner of cell types that become infected with viruses, even though only a small minority of cell types express co-stimulators. Effector CD4 T cells, which interact mainly with B cells in lymphoid tissues and with macrophages at sites of infection, also benefit from this change. Because macrophages and B cells express varying levels of co-stimulatory activity, relaxation of the requirement for co-stimulation increases the number of antigen-presenting cells that can stimulate effector CD4 T cells, and thus strengthens the overall immune response.

Effector T cells also express two to four times as much of the cell adhesion molecules CD2 and LFA-1 as naive T cells, and so are able to interact with target cells expressing lower levels of ICAM-1 and LFA-3 than those found on the professional antigen-presenting cells. The interaction between an effector T cell and its target is short-lived unless the T-cell receptor is engaged by specific antigen. When this happens, a conformational change in LFA-1 strengthens the adhesion between the two cells (Figure 6.19). Changes in the adhesion molecules expressed by effector T cells also cause their pattern of migration to differ from that of naive T cells. Effector T cells no longer express L-selectin and thus do not recirculate through lymph nodes by leaving the blood at high endothelial venules. Instead they express the integrin **VLA-4** (see Figure 6.2). This enables them to bind to adhesion molecules expressed on the endothelial cells of blood vessels in infected and inflamed tissues, and thus to enter tissues in which their effector functions are needed.

6-11 Effector T-cell functions are performed by cytokines and cytotoxins

The molecules that perform the functions of effector T cells fall into two broad classes: **cytokines**, which alter the behavior of their target cells, and secreted cytotoxic proteins or **cytotoxins**, which are used to kill target cells. All effector T cells produce cytokines, but of different types and in different combinations. Cytotoxins, in contrast, are the specialized products of cytotoxic CD8 T cells.

Cytokines are small secreted proteins and related membrane-bound proteins that act through cell-surface receptors and generally induce changes in gene expression within their target cell. Secreted cytokines can act on the cell that made them

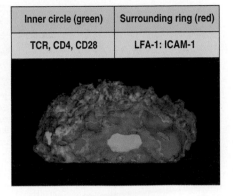

Inner circle (green)	Surrounding ring (red)
TCR, CD4, CD28	LFA-1: ICAM-1

Figure 6.19 Tight junctions are formed between effector T cells and their targets. Confocal fluorescence micrograph of the area of contact between a T cell and a B cell (as viewed down through one of the cells). The central green area includes T-cell receptor complexes, the co-receptor CD4, and CD28. Surrounding this is a red ring made up of LFA-1 on the T cell and its counter-receptors on the target cell. The outer, less intense green ring represents staining of receptors elsewhere on the cell surface. Photograph courtesy of A. Kupfer.

Cytokine	T-cell source	Effects on					Effect of gene knock-out in mice
		B cells	T cells	Macrophages	Hematopoietic cells	Other somatic cells	
Hematopoietins							
Interleukin-2 (IL-2)	T_H0, T_H1, some CTL	Stimulates growth and J-chain synthesis	Growth	–	Stimulates NK cell growth	–	↓ T-cell responses IBD
Interleukin-3 (IL-3)	T_H1, T_H2, some CTL	–	–	–	Growth factor for progenitor hematopoietic cells (multi-CSF)	–	–
Interleukin-4 (IL-4)	T_H2	Activation, growth IgG1, IgE ↑MHC class II induction	Growth, survival	Inhibits macrophage activation	↑Growth of mast cells	–	No T_H2
Interleukin-5 (IL-5)	T_H2	Differentiation IgA synthesis	–	–	↑ Eosinophil growth and differentiation	–	–
Granulocyte-macrophage colony-stimulating factor (GM-CSF)	T_H1, some T_H2, some CTL	Differentiation	Inhibits growth	Activation. Differentiation to dendritic cells	↑ Production of granulocytes and macrophages (myelopoiesis) and dendritic cells	–	–
Interferons							
Interferon-γ (IFN-γ)	T_H1, CTL	Differentiation IgG2a synthesis	Inhibits T_H2 cell growth	Activation, ↑ MHC class I and class II	Activates NK cells	Antiviral ↑ MHC class I and class II	Susceptibility to mycobacteria
Tumor necrosis factor family							
Tumor necrosis factor-α (TNF-α)	T_H1, some T_H2, some CTL	–	–	Activates, induces NO production	–	Activates microvascular endothelium	Resistance to Gram −ve sepsis
Lymphotoxin (LT, TNF-β)	T_H1, some CTL	Inhibits	Kills	Activates, induces NO production	Activates neutrophils	Kills fibroblasts and tumor cells	Absence of lymph nodes. Disorganized spleen
Fas ligand (FasL)	Activated CTL, CD4 T cells	Induces apoptosis in cells bearing Fas	Induces apoptosis in cells bearing Fas	Induces apoptosis in cells bearing Fas	?	?	Lymphoproliferative disease and autoimmunity
CD40 ligand (CD40L)	Activated CD4 T cells	Growth differentiation and isotype switching	–	Activation and cytokine production	Cytokine production by dendritic cells	–	Hyper-IgM syndrome
Others							
Interleukin-10 (IL-10)	T_H2	↑ MHC class II	Inhibits T_H1	Inhibits cytokine release	Co-stimulates mast cell growth	–	IBD
Transforming growth factor-β (TGF-β)	CD4 T cells	Inhibits growth IgA switch factor	–	Inhibits activation	Activates neutrophils	Inhibits/ stimulates cell growth	Death at ~10 weeks
Macrophage chemoattractant protein (MCP)	T_H1	–	–	Attracts into tissues	–	–	–

Figure 6.20 (*opposite page*) The properties of cytokines produced by T cells. Each of the cytokines is named and their cells of origin and action are listed. Cytokines can be grouped into families on the basis of their structures. Those shown here mostly fall into the hematopoietin, interferon, and tumor necrosis factor (TNF) families. The cell-surface receptors to which cytokines bind can similarly be grouped into structural families that correlate with the families of ligands. Cytokine functions are shown in the boxes, with the principal activities of effector cytokines highlighted in red. [↑], increase; [↓], decrease; CSF, colony-stimulating factor; CTL, cytotoxic CD8 T cell; IBD, inflammatory bowel disease; MHC, major histocompatibility complex; NK cell, natural killer cell; NO, nitric oxide.

(**autocrine** action), as with IL-2, or they can act locally on another type of cell (**paracrine** action). Many of the cytokines made by T cells are called **interleukins** and were assigned numbers according to the order of their discovery, for example interleukin-2 (IL-2). Cytokines made by lymphocytes are often called **lymphokines**. In this book, we shall use the general term cytokine for all these molecules. The cytokines that are referred to in this chapter are listed in Figure 6.20.

Secreted cytokines generally work locally and over a short period. Membrane-bound cytokines, such as **tumor necrosis factor-α (TNF-α)**, **CD40 ligand**, and **Fas ligand**, can have an effect only on the target cell in the localized area where the T cell is bound. The secretion of soluble cytokines is similarly focused on the target cell by polarization of the T cell's intracellular secretory apparatus, which occurs on binding of the T-cell receptor to the target cell.

The cytoplasmic tails of most cytokine receptors are associated with protein kinases known as **Janus kinases** (**JAKs**) (Figure 6.21). Cytokine binding causes

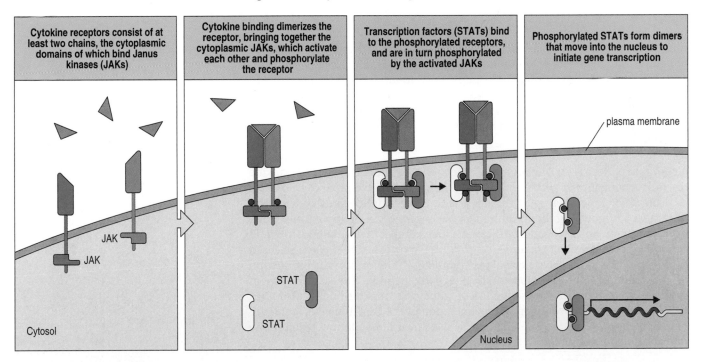

Figure 6.21 Many cytokine receptors signal by a rapid pathway in which receptor-associated kinases directly activate transcription factors. These receptors consist of at least two chains, each associated with a specific Janus kinase (JAK) (first panel). Ligand binding and dimerization of the receptor chains brings together the JAKs, which can transactivate each other, subsequently phosphorylating tyrosines in the receptor tails (second panel). Members of the STAT family of proteins bind to the phosphorylated receptors and are themselves phosphorylated by the JAKs (third panel). On phosphorylation, STAT proteins dimerize and go rapidly to the nucleus, where they activate transcription from a variety of genes important for adaptive immunity (fourth panel).

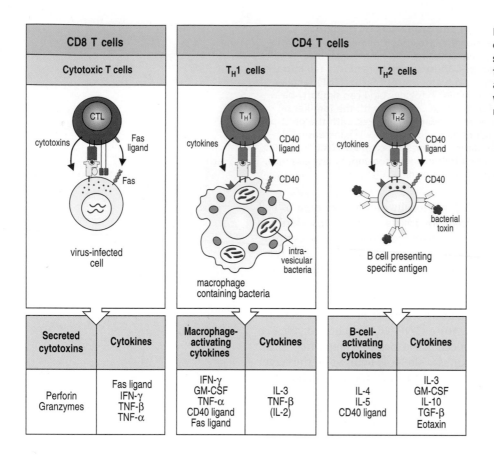

Figure 6.22 The three types of effector T cell produce distinct sets of effector molecules. In this figure the three types of effector T cell are shown, the types of target cell with which they interact, and the effector molecules that they make.

dimerization of the cytokine receptors, which in turn activates the kinases to phosphorylate members of a protein family called **STATs** (for **S**ignal **T**ransducers and **A**ctivators of **T**ranscription). On phosphorylation, two STATs dimerize and move from the cytoplasm to the nucleus. Here they activate specific genes, which differ according to the individual cytokine receptor–JAK–STAT pathway. These are short direct intracellular signaling pathways that enable cells to respond rapidly to cytokine stimulation.

A common pattern is for a membrane-associated cytokine and a secreted cytokine to work synergistically within a local area. Because of these properties, clinical trials in which individual cytokines such as IL-2 were administered systemically as a means of boosting the immune response were often disappointing. However, some cytokines do work at a distance, for example **interleukin-3 (IL-3)** and **granulocyte–macrophage colony-stimulating factor (GM-CSF)**. When released by effector CD4 T cells, these cytokines stimulate **myelopoiesis**—the production of macrophages and granulocytes in the bone marrow. The clinical use of cytokines that stimulate the production of blood cells from the bone marrow has been successful in speeding up the regeneration of the hematopoietic system in patients who have undergone ablation of their own bone marrow before bone marrow transplantation.

T_H1 and T_H2 CD4 T cells are distinguished by the sets of cytokines that they make and the effects that they have on the immune response. T_H1 cells work principally with macrophages in developing a cell-mediated immune response, whereas T_H2 cells work principally with B cells in developing an antibody-mediated immune response. CD8 T cells act mainly through the cytotoxins that they produce but also produce cytokines, which can have effects on other cells of the immune system (Figure 6.22).

6-12 Cytotoxic CD8 T cells are selective and serial killers of target cells at sites of infection

Once inside a cell, a pathogen becomes inaccessible to antibodies and other immune-system proteins. It can be eliminated either through the efforts of the infected cell itself or by direct attack on the infected cell by the immune system. The function of cytotoxic CD8 T cells is to kill cells that have become overwhelmed by intracellular infection. Overall, the sacrifice of the infected cells serves to prevent the spread of infection to healthy cells. The importance of cytotoxic T cells in combating viral infection is seen in humans and mice that lack functional cytotoxic T cells and who suffer from persistent viral infections.

Effector cytotoxic T cells contain stored **lytic granules**, which are modified lysosomes containing a mixture of specialized proteins called cytotoxins. CD8 T cells start to synthesize cytotoxins in inactive forms and to package them into lytic granules as soon as the T cells are activated by specific antigen in secondary lymphoid tissues. Effector CD8 T cells then migrate to sites of infection, where they recognize specific peptide:MHC class I complexes presented by infected cells. Binding via the T-cell receptor tells a T cell to secrete its cargo of lytic granules, which is delivered directly onto the surface of the infected target cell.

At sites of infection, cytotoxic CD8 T cells and the infected target cells are surrounded by healthy cells and cells of the immune system that have infiltrated the infected tissue. Because of their antigen specificity, cytotoxic T cells pick out only infected cells for attack and leave healthy cells alone. The T cell focuses granule secretion on the small localized area of the target cell where it is attached to the T cell (Figure 6.23). In this way, cytotoxic granules neither attack healthy neighbors

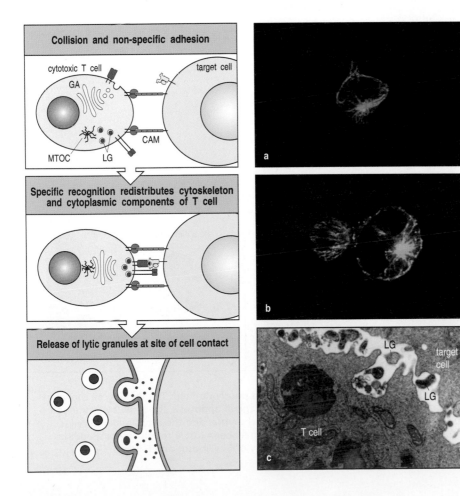

Figure 6.23 The polarization of cytotoxic T cells during recognition of specific antigen allows the delivery of cytotoxins to be focused on the antigen-bearing target cell. As shown in the panels on the left, initial binding to a target cell through adhesion molecules has no effect on the location of the lytic granules (LG) (top panel). Engagement of the T-cell receptor causes the T cell to become polarized: the cortical actin cytoskeleton at the site of contact reorganizes, enabling the microtubule-organizing center (MTOC), the Golgi apparatus (GA), and the lytic granules to align towards the target cell (center panel). Proteins stored in lytic granules are then directed onto the target cell (bottom panel). The photomicrograph in panel a shows an unbound, isolated cytotoxic T cell. The microtubules are stained green and the lytic granules red. Note how the lytic granules are dispersed throughout the T cell. Panel b depicts a cytotoxic T cell bound to a (larger) target cell. The lytic granules are now clustered at the site of cell–cell contact in the bound T cell. The electron micrograph in panel c shows the release of granules from a cytotoxic T cell. Panels a and b courtesy of G. Griffiths. Panel c courtesy of E.R. Podack.

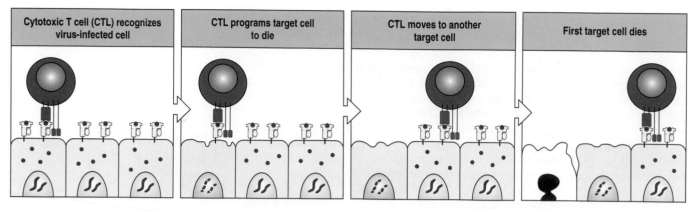

| Cytotoxic T cell (CTL) recognizes virus-infected cell | CTL programs target cell to die | CTL moves to another target cell | First target cell dies |

Figure 6.24 Cytotoxic CD8 T cells selectively kill infected cells. Specific recognition of peptide:MHC complexes on an infected cell by a cytotoxic CD8 T cell (CTL) programs the infected cell to die. The T cell detaches from its target cell and synthesizes a new set of lytic granules. The cytotoxic T cell then seeks out and kills another target.

of an infected cell nor kill the T cell itself. As the target cell starts to die, the cytotoxic T cell is released from the target cell and starts to make new granules. Once new granules have been made, the cytotoxic T cell is able to kill another target cell. In this manner, one cytotoxic T cell can kill many infected cells in succession (Figure 6.24).

Besides their cytotoxic action, CD8 T cells also contribute to the immune response by secreting cytokines. Secretion of **interferon-γ** (**IFN-γ**) inhibits the replication of viruses in the infected cells and increases the processing and presentation of viral antigens by MHC class I molecules. Another effect of IFN-γ is to activate macrophages in the vicinity of the cytotoxic T cells. These macrophages get rid of the dying infected cells, thereby allowing the T cells more room for maneuver and also helping the damaged tissue to heal and regenerate.

6-13 Cytotoxic T cells kill their target cells by inducing apoptosis

Cells killed by cytotoxic CD8 T cells do not lyse or disintegrate, like cells undergoing **necrosis** due to physical or chemical injury (Figure 6.25). Instead, the cells targeted by cytotoxic T cells shrivel and shrink. This type of cell death prevents not only pathogen replication but also the release of infectious bacteria or virus particles. This cell suicide is called **apoptosis** or **programmed cell death**, and is induced in the target cell by the cytotoxins released by the cytotoxic T cell.

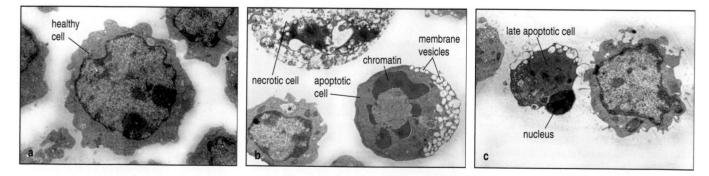

Figure 6.25 Apoptosis. Panel a shows an electron micrograph of a healthy cell with a normal nucleus. In the bottom right of panel b is an apoptotic cell at an early stage. The chromatin in the nucleus has become condensed (shown in red); the plasma membrane is well defined and is shedding vesicles. In contrast the plasma membrane of the necrotic cell shown in the upper left part of panel b is poorly defined. The middle cell shown in panel c is at a late stage in apoptosis. It has a very condensed nucleus and no mitochondria, and the cytoplasm and cell membranes have largely been lost through vesicle shedding. Photographs courtesy of R. Windsor and E. Hirst.

Figure 6.26 Time course of programmed cell death. The four panels show time-lapse photographs of a cytotoxic T cell killing a target cell. The lytic granules of the T cell are labeled with a red fluorescent dye. In the top panel, the T cell has just made contact with the target cell and this event is designated as the Start. At this time, the T-cell granules are distant from the point of contact with the target cell. After 1 minute, as shown in the second panel, the granules have begun to move toward the point of target-cell attachment, a move that is essentially completed after 4 minutes. After 40 minutes, as shown in the bottom panel, the granules have been secreted and are seen in between the T cell and the target cell. The target cell has now begun to undergo apoptosis as shown by the fragmented nucleus. The T cell is now ready to detach from the apoptotic cell and seek out a further target cell. Photographs courtesy of G. Griffiths.

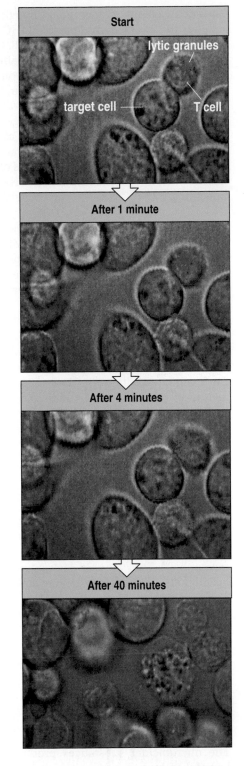

Soon after contact with a cytotoxic T cell, the target cell's DNA starts to be fragmented by the cell's own nucleases. These cleave between the nucleosomes, to give DNA fragments that are multiples of 200 base pairs in length and are characteristic of apoptosis. Eventually the nucleus becomes disrupted, and there is loss of membrane integrity and of normal cell morphology. The cell destroys itself from within. It shrinks by the shedding of membrane-bound vesicles (see Figure 6.25, center panel) and the degradation of the cell contents until little is left. The changes that occur in the plasma membrane during apoptosis are recognized by phagocytes, which speed the dying cell on its way by ingesting and digesting it. The apoptotic processes that degrade the infected human cell also act upon the infecting pathogen. In particular, the breakdown of viral nucleic acids prevents the assembly of infectious virus particles that might cause further infection were they to escape from the dying cell. A five-minute contact between a cytotoxic T cell and its target cell is all it takes for the target cell to be programmed to die, even though visible evidence of death takes longer to become obvious (Figure 6.26).

Cytotoxic T cells can induce apoptosis by two different pathways. The first is initiated by cytotoxins that they release. These are **perforin**, a protein that polymerizes to form transmembrane pores of 160 Å diameter in cell membranes, and the **granzymes**, a family of three serine proteases related to the pancreatic protease trypsin. A current model for the killing mechanism is that perforin makes pores in the target cell membrane through which the granzymes can enter. Once inside, the granzymes cleave certain cell proteins, leading to the activation of nucleases and other enzymes that initiate apoptosis.

A second way of inducing apoptosis is by interactions between cell-surface molecules on the cytotoxic T cell and the target cell. Activated cytotoxic T cells express the cell-surface cytokine Fas ligand, which binds to **Fas** molecules on the target-cell surface. This interaction sends signals to the target cell to undergo apoptosis. Fas–Fas ligand interactions are also likely to be involved in the removal of self-reactive lymphocytes, as patients with defective Fas genes suffer a lymphoproliferative disease associated with autoimmunity.

6-14 T$_H$1 CD4 cells induce macrophages to become activated

Macrophages have receptors that bind to bacteria and other microorganisms and facilitate their phagocytosis, destruction, and intracellular degradation. As a consequence of these processes, pathogen-derived peptides are presented by MHC class II molecules on the macrophage surface, where they are able to activate naive CD4 T cells to become T$_H$1 effector cells. Some species of microorganism have adapted to the macrophage: they interfere with macrophage function by living and replicating inside the phagosome. A principal function of T$_H$1 cells is to act back on macrophages, increasing their phagocytic ability and their capacity to kill ingested microorganisms.

The enhancement of macrophage function is called **macrophage activation** and requires the interaction of peptide:MHC class II complexes on the macrophage with the T-cell receptor of the T_H1 cell. One effect of macrophage activation is to cause the phagosomes which contain captured microorganisms to be more efficiently fused with lysosomes, the source of hydrolytic degradative enzymes. Another is to increase the synthesis by macrophages of highly reactive and microbicidal molecules such as oxygen radicals, nitric oxide (NO), and proteases, which together kill the engulfed pathogens. In patients with AIDS, the number of CD4 T cells decreases progressively, as does the activation of macrophages. Under these circumstances, microorganisms such as *Pneumocystis carinii* and mycobacteria, which live in the vesicles of macrophages and are normally kept in check by macrophage activation, flourish as opportunistic and sometimes fatal infections.

Other changes that occur on macrophage activation help to amplify the immune response. Increased expression of MHC class II molecules and B7 on the macrophage surface increases antigen presentation to naive T cells, thus recruiting more T cells into the immune response. This in turn enhances macrophage activation and maintains increased numbers of macrophages in the activated state.

Macrophages require two signals for activation, both of which can be delivered by effector T_H1 cells. The primary signal is provided by IFN-γ, which is a characteristic cytokine produced by T_H1 cells, whereas the second signal, which makes a macrophage responsive to IFN-γ, is delivered by CD40 ligand on the surface of the T cell interacting with CD40 on the macrophage (Figure 6.27). Macrophage activation results in the increased expression of CD40 and of TNF receptors, which raises the sensitivity of the macrophage to CD40 ligand and TNF-α. TNF-α, produced by the activated macrophage itself, synergizes with IFN-γ to raise the level of activation.

CD4 T cells make their effector molecules only on demand, unlike CD8 T cells. After encounter with antigen on a macrophage, an effector T_H1 cell takes several hours to synthesize the requisite effector cytokines and cell-surface molecules. During this time the T cell must maintain contact with its target cell. Newly synthesized cytokines are translocated into the endoplasmic reticulum of the T cell and delivered by secretory vesicles to the site of contact between T cell and macrophage. They are thus focused on the target cell. Newly synthesized CD40 ligand is also selectively expressed at the region of contact with the macrophage. Together, this localized production of cytokines ensures the selective activation of those macrophages carrying the specific peptide:MHC complexes recognized by the T cell.

CD8 T cells are also an important source of IFN-γ and because of this they can activate macrophages. Macrophage sensitization to IFN-γ need not require the action of CD40 ligand; small amounts of bacterial polysaccharide have a similar effect and can be of particular importance when CD8 T cells, which do not express CD40 ligand, are the principal source of IFN-γ.

The microbicidal substances produced by activated macrophages are also harmful to human tissues, which inevitably suffer damage from macrophage activity. For this reason, the activation of macrophages by CD4 T_H1 cells is under strict control. Cytokines secreted by CD4 T_H2 cells, which include **transforming growth factor-β (TGF-β)**, **IL-4**, **IL-10**, and **IL-13**, inhibit macrophage activation, an example of how cytokines secreted by T_H2 cells can control the T_H1 response. That T_H1 cells will stop production of IFN-γ if their antigen receptors lose contact with the peptide:MHC complexes on a macrophage further controls the T_H1 response.

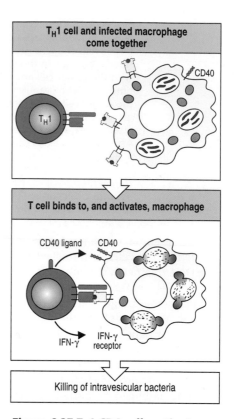

Figure 6.27 T_H1 CD4 cells activate macrophages to become highly microbicidal. When a T_H1 cell specific for a bacterial peptide contacts a macrophage that presents the peptide, the T_H1 cell is induced to secrete the macrophage-activating cytokine IFN-γ and also to express CD40 ligand at its surface. Together these newly synthesized proteins activate the macrophage to kill the bacteria living inside its vesicles.

6-15 T_H1 cells coordinate the host response to intravesicular pathogens

Certain microorganisms, including the mycobacteria that cause tuberculosis and leprosy, are intracellular pathogens that enjoy a protected life in the vesicular system of macrophages. Such microorganisms subvert the destructive mission of the macrophage for their own purpose. By sequestering themselves in this cellular compartment they cannot be reached by antibodies; neither are their peptides presented by MHC class I molecules, thus preventing the infected macrophage from being attacked by cytotoxic T cells. Mycobacteria avoid digestion by lysosomal enzymes by preventing the acidification of the phagolysosome that is required to activate the lysosomal hydrolases. Infections of this type are fought by T_H1 CD4 T cells, which help the macrophage to become activated to the point at which the intracellular pathogens are eliminated or killed.

The activation of macrophages by IFN-γ and CD40 ligand is central to the immune response against pathogens that proliferate in macrophage vesicles. In mice lacking functional IFN-γ or CD40 ligand, the ability of macrophages to kill intravesicular bacteria is impaired, and doses of *Mycobacterium* or *Leishmania* that normal mice can withstand prove fatal. Although IFN-γ and CD40 ligand are probably the most important effector molecules of T_H1 cells, other cytokines secreted by these cells help to coordinate responses to intravesicular bacteria (Figure 6.28). Macrophages chronically infected with intravesicular bacteria can lose the capacity to be activated. Such cells can be killed by an effector T_H1 cell that uses Fas ligand or TNF to engage Fas or a TNF receptor on the surface of the macrophage, inducing the macrophage to undergo apoptosis. This releases the bacteria, which are then taken up and killed by fresh macrophages.

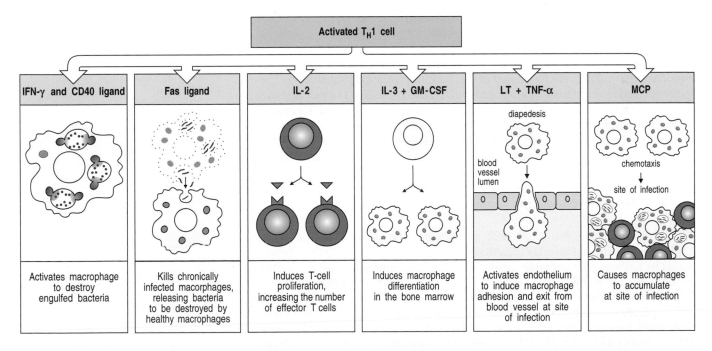

Figure 6.28 The immune response to intravesicular bacteria is coordinated by activated T_H1 cells. The activation of T_H1 cells by infected macrophages results in the synthesis of cytokines that activate the macrophage and coordinate the immune response to intravesicular pathogens. The six panels show the effects of different cytokines secreted by T_H1 cells. LT, lymphotoxin (TNF-β); MCP, macrophage chemoattractant protein.

Figure 6.29 Granulomas form when an intracellular pathogen or its constituents resists elimination. In some circumstances mycobacteria (red) resist the killing effects of macrophage activation (top panel). A characteristic localized inflammatory response called a granuloma develops (second panel). The granuloma consists of a central core of infected macrophages, which can include multinucleated giant cells formed by macrophage fusions, surrounded by large single macrophages often called epithelioid cells. Mycobacteria can persist in the cells of the granuloma. The central core is surrounded by T cells, many of which are CD4 T cells. The photomicrograph (bottom panel) shows a granuloma from the lung. Photograph courtesy of J. Orrell.

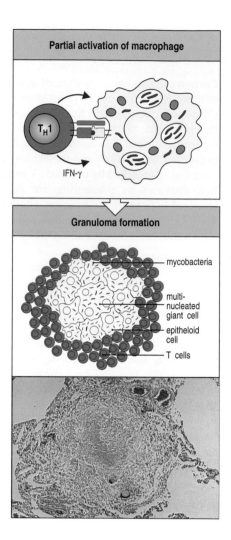

The IL-2 produced by T_H1 cells induces T-cell proliferation and potentiates the production and release of other cytokines. Some of these recruit phagocytes—macrophages and neutrophils—to sites of infection. First, T_H1 cells secrete IL-3 and GM-CSF, which stimulate the increased production of macrophages and neutrophils in the bone marrow. Second, the secretion of TNF-α and **TNF-β** (also called **lymphotoxin** or **LT**) by T_H1 cells induces vascular endothelial cells at sites of infection to change the adhesion molecules they express so phagocytes circulating in the blood can bind to them. At this point, **macrophage chemoattractant protein** (**MCP**), which is also produced by the T_H1 cells, guides the phagocytes between the endothelial cells and into the infected area. In total, the CD4 T_H1 cell orchestrates a multifaceted macrophage response that focus on the destruction of pathogens taken up by macrophages.

When microbes successfully resist the microbicidal effects of activated macrophages, a chronic infection with inflammation can develop. Such areas of tissue often have a characteristic morphology, called a **granuloma**, in which a central area containing infected macrophages is surrounded by activated T cells. Giant cells resulting from the fusion of macrophages are present at the center of a granuloma; they contain the resistant pathogens. Large single macrophages, sometimes called epithelioid cells, form an epithelium-like layer around the center (Figure 6.29).

In tuberculosis, the centers of large granulomas can become cut off from the blood supply and the cells in the center die, probably from a combination of oxygen deprivation and the cytotoxic effects of the macrophages. The resemblance of the dead tissue to cheese led to this process being called **caseation necrosis**. It provides a vivid example of how CD4 T_H1 cells can produce a local pathology. Their absence, however, leads to death from disseminated infection, which is commonly seen in AIDS patients infected with opportunistic mycobacteria.

6-16 CD4 T_H2 cells activate only those B cells that recognize the same antigen as they do

During an infection, the T-cell zone of secondary lymphoid tissues contains pathogen-specific T_H2 effector cells that are the progeny of naive CD4 T cells activated by antigen-presenting dendritic cells. The main function of these T cells is to help B cells to mount an antibody response against the infectious agent. Mature naive B cells passing through the lymphoid tissue pick up and present their specific antigens. As the circulating B cells pass through the T-cell zones, they make transient interactions with the T_H2 cells, whose T-cell receptors screen the peptides presented by the MHC class II molecules on the surface of the B cell. When a B cell presents the specific antigen recognized by the T_H2 cell, the adhesive interactions are strengthened and the B cell becomes trapped by the T cell. This interaction gives rise to a primary focus of activated B cells and helper T cells (see Figure 4.16, p.100).

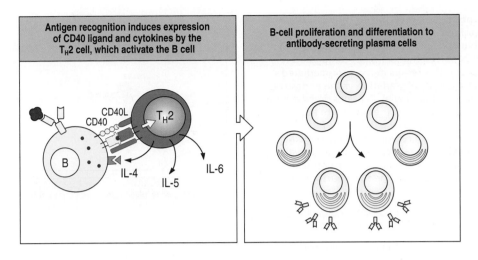

Figure 6.30 T$_H$2 cells stimulate the proliferation and differentiation of naive B cells. The specific interaction of an antigen-binding B cell with a helper T$_H$2 cell leads to the expression of CD40 ligand (CD40L) and the secretion of IL-4, IL-5, and IL-6. In concert, these T$_H$2 products drive the proliferation of B cells and their differentiation to form plasma cells dedicated to the secretion of antibody.

When the T-cell receptor of a helper T cell recognizes peptide:MHC class II complexes on the surface of a naive B cell, the T cell responds by synthesizing CD40 ligand. This molecule is involved in all T-cell interactions with B cells, which express the corresponding receptor molecule CD40. The interaction of CD40 ligand with CD40 drives the resting B cell into the cycle of cell-division. The characteristic cytokine secreted by T$_H$2 cells on stimulation by their target cell is IL-4, which works in concert with CD40 ligand to initiate the proliferation and clonal expansion of B cells that precede their differentiation into antibody-secreting plasma cells. T$_H$2 cells also produce IL-5 and IL-6, which drive further B-cell differentiation to plasma cells (Figure 6.30).

The principle governing T-cell help to B cells is that co-operation occurs only between B and T cells that are specific for the same antigen, although they usually recognize different epitopes. Such interactions are called **cognate interactions**. The peptide recognized by the T cell must be part of the same physical entity bound by the B cell's surface immunoglobulin. For example, the T cell might recognize a peptide derived from an internal protein of a virus, whereas the B cell recognizes an exposed external carbohydrate epitope of a viral capsid glycoprotein. The specialized antigen-presenting function of a B cell makes it supremely efficient at presenting peptides that derive from any protein, virus, or microorganism that specifically binds to its surface immunoglobulin. Only those B cells that selectively internalize a pathogen antigen by receptor-mediated endocytosis will present enough of the pathogen-derived peptide to engage and stimulate an antigen-specific T$_H$2 cell. It is estimated that a B cell that can use receptor-mediated endocytosis to capture a particular antigen is 10,000-fold more efficient at presenting peptides derived from that antigen than is a B cell that cannot.

Knowledge of the mechanism by which B and T cells cooperate helps in the design of vaccines. An example illustrated in Figure 6.31 is the vaccine against *Haemophilus influenzae* B, a bacterial pathogen that is life-threatening to young children. It infects the lining of the brain—the meninges—producing a meningitis that in severe cases causes lasting neurological damage or death. Protective immunity against *H. influenzae* is provided by antibodies specific for the capsular polysaccharides. However, a child's antibody response is weakened by the lack of associated peptide epitopes that could engage T$_H$2 cells and provide help to polysaccharide-specific B cells. To enable the immune system to make antibodies against *H. influenzae*, a vaccine was made in which the immunizing antigen was the bacterial polysaccharide covalently coupled to tetanus toxoid, a protein containing good peptide epitopes that are bound by MHC class II molecules and presented to T$_H$2 cells.

Figure 6.31 Molecular complexes recognized by both B and T cells make effective vaccines. The first panel shows a naive B cell's surface immunoglobulin binding a carbohydrate epitope on a vaccine composed of a *Haemophilus* polysaccharide (blue) conjugated to tetanus toxoid (red), a protein. This results in receptor-mediated endocytosis of the conjugate and its degradation in the endosomes and lyosomes, as shown in the second panel. Peptides derived from degradation of the tetanus toxoid part of the conjugate are bound by MHC class II molecules and presented on the B cell's surface. In the third panel, the receptor of a T_H2 cell recognizes the peptide:MHC complex. This induces the T cell to secrete cytokines that activate the B cell to differentiate into plasma cells, which produce protective antibody against the *Haemophilus* polysaccharide (fourth panel).

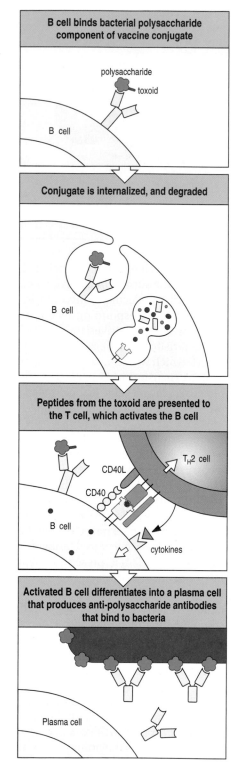

Summary

The three types of effector T cell—CD8 cytotoxic cells, CD4 T_H1 cells, and CD4 T_H2 cells—have complementary roles in the immune response. The common principle by which they function is to affect the behavior of other types of cell through intimate contact and the action of effector molecules. Naive T cells are activated to develop effector function in the secondary lymphoid tissues, whereupon cytotoxic CD8 T cells and CD4 T_H1 cells enter the blood and travel to sites of infection. In contrast, CD4 T_H2 cells remain in the secondary lymphoid tissue, where they activate naive B cells that have specificity for the same antigen as themselves. Linked recognition between T_H2 cells and B cells arises because a B cell efficiently internalizes and processes the antigen to which its surface immunoglobulin binds, and then presents peptide antigens to a T cell. When the T-cell receptor of the T_H2 cell binds to peptide:MHC complexes on the B-cell surface, the B cell becomes activated by interactions between CD40 ligand on the T cell and CD40 on the B cell, and by IL-4, the characteristic cytokine secreted by the T_H2 cell. Cytotoxic CD8 T cells induce cells overwhelmed by viral infection to die by apoptosis. This mode of death ensures that the infected cell's load of viruses is also destroyed rather than being released to infect healthy cells. Apoptosis is induced by the cytotoxic enzymes contained in secretory lytic granules that are stored by the cytotoxic T cell and are released onto the target cell membrane once contact has been established. After granule release, the T cell rapidly synthesizes new granules so it can kill several targets in succession. In contrast, the effector molecules of effector CD4 T cells are made to order once contact with an antigen-bearing target cell has been established. A major role of T_H1 CD4 cells is to activate macrophages, helping them to become more competent at destroying extracellular pathogens that they have taken up into their vesicular system, including those that exploit the phagocytic pathway for their own survival. Interferon-γ is the characteristic cytokine of T_H1 cells and is instrumental in activating macrophages. Both T_H1 and T_H2 cells influence isotype switching in B cells through the cytokines they produce.

Summary to Chapter 6

All aspects of the adaptive immune response are initiated and controlled by effector T cells—CD4 T_H1 cells, CD4 T_H2 cells and CD8 cytotoxic T cells. These cells differentiate from naive recirculating T cells that have been trapped and activated by antigen presented by professional antigen-presenting cells in the secondary lymphoid tissues. Dendritic cells, macrophages, and B cells are the three types of professional antigen-presenting cell and each is adapted to presenting particular categories of antigen. Activation leads to cell proliferation and differentiation, all of which proceeds while the T cell remains in contact with the antigen-presenting cell.

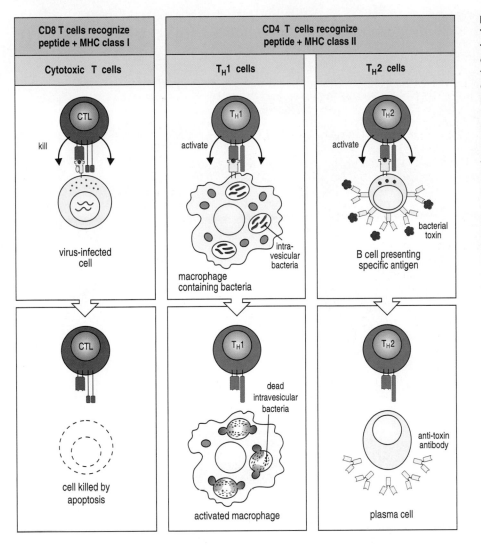

Figure 6.32 Three classes of effector T cell are specialized to deal with three classes of pathogens. CD8 cytotoxic T cells (CTL) (left panels) kill target cells that present peptides derived from viruses and other cytosolic pathogens bound to MHC class I molecules. T_H1 CD4 T cells (middle panels) recognize peptides derived from pathogens or their products that have been phagocytosed by macrophages. These T cells induce macrophages to become activated, which enhances their general capacity to eliminate extracellular infection and more specifically to eliminate organisms colonizing the vesicular system of the macrophage. T_H2 CD4 T cells (right panels) activate naive B cells and control many aspects of the development of the antibody response.

Each of the three types of effector T cell has a distinct role in the immune response, but all function through interactions with another type of cell (Figure 6.32). Effector CD4 T_H1 cells mostly migrate from the secondary lymphoid tissues to sites of infection. There they activate tissue macrophages. This both increases the macrophages' capacity to phagocytose and kill pathogenic organisms infecting the extracellular spaces, and increases their capacity to act as professional antigen-presenting cells. Activation also enhances the capacity of macrophages to eliminate microorganisms that deliberately parasitize macrophage vesicles. Within the secondary lymphoid tissues, effector CD4 T_H2 cells activate naive B cells that are specific for the same antigen as the T cell. Activated B cells divide and differentiate into antibody-secreting plasma cells, and also undergo isotype switching under the influence of effector CD4 T cells of both T_H2 and T_H1 types. The three types of effector T cell enable the human immune system to respond to different categories of infection and to different developmental stages in the course of the same infection.

Immunity Mediated by B Cells and Antibodies 7

The production of antibodies is the sole function of the B-cell arm of the immune system. Antibodies are useful in defense against any pathogen that is present in the extracellular spaces of the body's tissues. Some human pathogens, such as many species of bacteria, live and reproduce entirely within the extracellular spaces, whereas other pathogens, such as viruses, replicate inside cells but are carried through extracellular spaces as they spread from one cell to the next. Antibodies secreted by plasma cells in secondary lymphoid tissues and bone marrow find their way into the fluids filling the extracellular spaces.

Antibodies are not in themselves toxic or destructive to pathogens; their role is simply to bind tightly to them. This can have several consequences. One way in which antibodies reduce infection is by covering up sites on a pathogen's surface that are necessary for growth or replication, for example the viral glycoproteins that viruses use to bind to the surface of human cells and initiate infection. Such antibodies are said to **neutralize** the pathogen. In the development of a vaccine against an infectious agent or its toxic products, the gold standard at which a company aims is the induction of a neutralizing antibody. Antibodies also act as molecular adaptors that bind to pathogens with their antigen-binding arms and to receptors on phagocytic cells with their Fc regions. **Opsonization**, or coating of a pathogen with antibody, promotes its phagocytosis. The antibody-directed destruction of pathogens due to opsonization is enhanced by the actions of a set of proteins that do not discriminate between antigens and are present in blood and lymph. These proteins are collectively known as **complement**, because their functions complement the antigen-binding function of the antibody.

The structure, specificity and other properties of antibodies were discussed in Chapter 2, and the development of B cells from their origin in bone marrow to their differentiation into antibody-secreting plasma cells was the subject of Chapter 4. This chapter will focus on how antibodies clear infection by targeting destructive but non-specific components of the immune system onto an infecting pathogen. In the first part of the chapter we consider the antigens that provoke a B-cell response, how the response develops, and the generation of the different antibody isotypes. The structural differences between antibody isotypes provide a variety of adaptor functions that can target antibody-bound pathogen to different types of non-specific effector cell; these aspects of the antibody-mediated immune response will be discussed in the second part of the chapter. In the last part of the chapter, we shall look at the functions of the complement system, one of the principal mechanisms of targeting extracellular pathogens for destruction, and how it is activated by antibody.

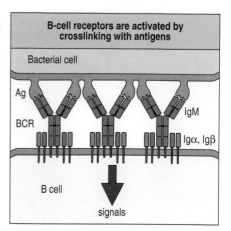

Figure 7.1 Crosslinking of antigen receptors is the first step in B-cell activation. The B-cell receptors (BCR) on B cells are physically crosslinked by the repetitive epitopes of antigens (Ag) on the surface of a bacterial cell. The B-cell receptor on a mature naive B cell is composed of surface IgM, which binds antigen, and associated Igα and Igβ chains, which provide the signaling capacity.

Antibody production by B lymphocytes

The antibodies most effective in combating infection are those that are made early in infection and bind strongly to the pathogen. On first exposure to an infectious agent, these two goals make competing demands on the immune system. As we saw in Chapters 4 and 6, B cells generally require help from activated T cells to mature into antibody-secreting plasma cells; this delays the onset of antibody production until around a week after infection. In addition, B cells take time to switch isotype and undergo affinity maturation, processes that are necessary for production of the high-affinity antibodies that are most effective at dealing with pathogens. Thus, during the course of an infection, the effectiveness of the antibodies produced improves steadily. This experience is retained in the form of memory B cells and high-affinity antibodies, which provide long-term immunity to re-infection.

A faster primary response is made to certain bacterial antigens that are able to activate B cells without the need for T-cell help. However, the antibodies produced in such a response are predominantly of the IgM isotype and of generally low affinity. They do, however, provide an early defense, helping to keep the infection at a relatively low level until a better antibody response can be developed.

7-1 B-cell activation requires crosslinking of surface immunoglobulin

On binding to protein or carbohydrate epitopes on the surface of a microorganism, the surface IgM molecules of a naive mature B cell become physically crosslinked to each other and are drawn into the localized area of contact with the microbe (Figure 7.1). This clustering and aggregation of B-cell receptors sends signals from the receptor complex to the inside of the cell. Signal transduction from the B-cell receptor complex resembles, in many ways, the signaling from the T-cell receptor complex discussed in Section 6-5, p. 138. Both types of

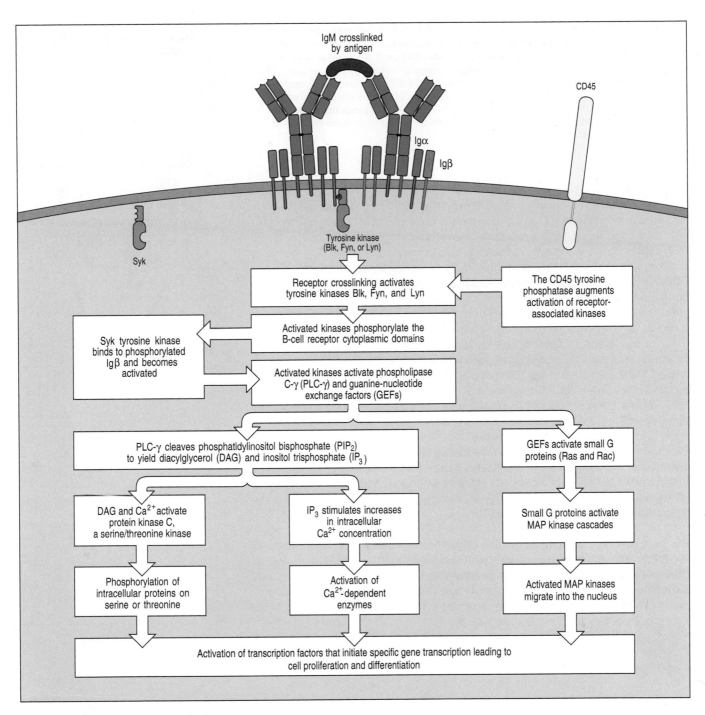

Figure 7.2 Signals from the B-cell receptor initiates a cascade of intracellular signals.

receptor are associated with cytoplasmic protein tyrosine kinases that are activated by receptor clustering, and both receptors activate similar intracellular signaling pathways.

Interaction of antigen with surface immunoglobulin is communicated to the interior of the B cell by the proteins Igα and Igβ, which are associated with IgM in the B-cell membrane to form the functional B-cell receptor. Like the CD3 polypeptides of the T-cell receptor complex, the cytoplasmic tails of Igα and Igβ contain immunoreceptor tyrosine-based activation motifs (ITAMs) with which the tyrosine kinases associate. The ITAMs become phosphorylated on tyrosine residues, and the active tyrosine kinases and the phosphorylated receptor tails then initiate intracellular signaling pathways that lead to changes in gene expression in the nucleus (Figure 7.2).

Figure 7.3 Signals generated from the B-cell receptor and co-receptor synergize in B-cell activation. Binding of CR2 to complement fragments (C3d) deposited on the surface of a pathogen crosslinks the B-cell co-receptor complex with the B-cell receptor. This causes them to cluster together on the B-cell surface. The cytoplasmic tail of CD19 is then phosphorylated by tyrosine kinases associated with the B-cell receptor. Phosphorylated CD19 binds intracellular signaling molecules whose signals synergize with those generated by the B-cell receptor.

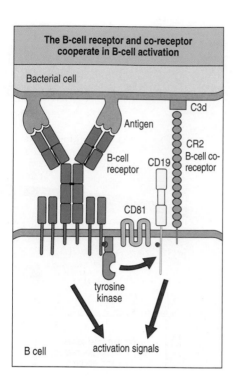

Crosslinking of the B-cell receptor by antigen generates a signal that is necessary but not sufficient to activate a naive B cell. The additional signals required are delivered in several ways. One set of signals is delivered when the B-cell receptor becomes crosslinked with a cell-surface protein complex known as the **B-cell co-receptor**, causing the two receptors to cluster together. The B-cell co-receptor is a complex of at least three proteins: one is the complement receptor 2 (CR2 or **CD21**), which binds to complement components deposited on a pathogen; one is the protein **CD19**, which acts as the signaling chain of the receptor, and the other is the protein **CD81** (TAPA-1), whose function is not yet known. It does act, however, as a cell-surface receptor for hepatitis C virus. Co-ligation of the B-cell receptor and the co-receptor brings the Igα-bound tyrosine kinase Lyn into close proximity with the CD19 cytoplasmic tail, which it phosphorylates. The phosphorylated CD19 can then bind intracellular signaling molecules which generate signals that synergize with those generated by the B-cell receptor complex (Figure 7.3). Simultaneous ligation of the B-cell receptor and co-receptor increases the signals by 1000- to 10,000-fold.

Even the combined effects of B-cell receptor and co-receptor signals are generally insufficient to activate a naive B cell. This requires additional signals provided by CD4 helper T cells, the effector cells produced upon antigen-activation of naive CD4 T cells. Whether a B cell needs T-cell help or not depends on the nature of the antigen, as we shall see in the next section.

The final outcome of B-cell activation is the proliferation and differentiation of the B cell into plasma cells and the secretion of antibodies. The morphological effects of activation are striking: the small resting B cell, which in appearance is all nucleus and no cytoplasm, gives rise to plasma cells committed to antibody secretion and whose large active cytoplasm packed with rough endoplasmic reticulum is testimony to this function (Figure 7.4).

7-2 The antibody response to certain antigens does not require T-cell help

In their cell walls and capsules, bacterial pathogens possess complex polysaccharides, lipopolysaccharides, and proteoglycans that are chemically and antigenically distinct from those of mammalian cells and are characterized by repetitive epitopes. These cell-surface molecules are a major target of the antibody response against extracellular bacterial pathogens, and some of these antigens can activate naive B cells without help from CD4 T cells. Such antigens are known as **thymus-independent antigens** (**TI antigens**), because immunodeficient patients born

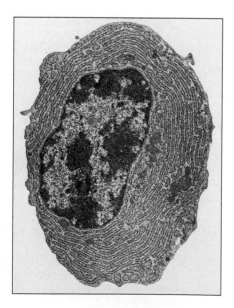

Figure 7.4 Plasma cell. Electron micrograph of a plasma cell. Note the characteristic 'clockface' pattern in the nucleus, which resembles the hands and face of a clock as well as the extensive endoplasmic reticulum. Photograph courtesy of C. Grossi.

Figure 7.5 T-cell-independent type 1 antigens (TI-1 antigens) are polyclonal B-cell activators at high concentrations. At high concentrations of a TI-1 antigen the mitogen (pink triangle) is sufficient to induce proliferation and antibody secretion by all B cells in the absence of antigen binding to surface immunoglobulin (top two panels). This gives a non-specific antibody response. At low concentrations, only those B cells whose B-cell receptors are specific for the TI-1 antigen can bind sufficient antigen to benefit from its B-cell activating properties. This gives an antibody response that is specific for epitopes (red circles) of the TI-1 antigen (bottom two panels).

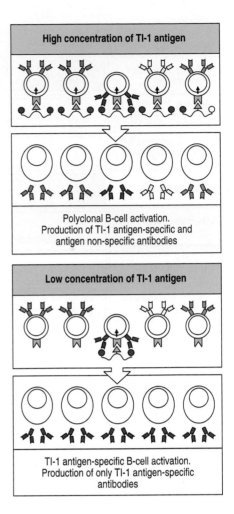

without a thymus are able to raise an antibody response against them. In contrast, the antibody response to other antigens requires help from an antigen-activated helper T cell (see Section 6-16, p. 155); these antigens are known as **thymus-dependent antigens (TD antigens)**. For thymus-independent antigens, the need for T-cell help can be overcome in two different ways.

Certain thymus-independent antigens (called **TI-1 antigens**) have an intrinsic capacity for inducing B cells to proliferate. As well as binding to the B-cell receptor, they bind to other receptors on B cells whose stimulation induces proliferation and differentiation. At high concentrations, these antigens cause the proliferation and differentiation of most B cells, regardless of the B cells' antigen specificity, and can induce the proliferation of both mature and immature B cells. Antigens that can cause such **polyclonal activation** of B cells are known as **B-cell mitogens**, because they drive B cells into DNA replication, mitosis, and cell division. An example of a B-cell mitogen is the lipopolysaccharide (LPS) of Gram-negative bacteria. The concentration of antigen required to induce polyclonal activation is, however, much higher than what is normally encountered. In a natural infection, the lower mitogenic stimulus will activate B cells only in combination with signaling through the B-cell receptor. Thus only B cells specific for the antigen are activated (Figure 7.5). Without the participation of cytokines produced by activated T cells, only IgM antibodies are produced.

The second type of thymus-independent antigen (**TI-2 antigen**) activates only mature B cells. TI-2 antigens are typically composed of repetitive carbohydrate or protein epitopes present at high density on the surface of a microorganism. They probably act by crosslinking the B-cell receptor and co-receptors to an extent where the need for additional signals is overridden. Responses to TI-2 antigens are usually seen around 48 hours after antigen was encountered. Typical antigens of this kind are bacterial cell-wall polysaccharides, and the responding B cells are often of the B-1 subpopulation (see Section 4-6, p. 95). Human B-1 cells develop their full function only when a person is about 5 years old, perhaps explaining why infants make relatively poor antibody responses to polysaccharide antigens. Both IgM and IgG antibodies are induced by TI-2 antigens and they are likely to be an important part of the early B-cell response to some common bacterial infections. Examples of TD, TI-1, and TI-2 antigens and the responses against them are presented in Figure 7.6.

Although the TI-2 antigens on some bacteria induce an early antibody response that helps to contain the infection, this response has considerable limitations. There is no significant isotype switching and so the antibodies are predominantly IgM. There is also no somatic hypermutation and so the antibodies produced do not increase their affinity for the antigen. Lastly, TI-2 antigens do not induce long-term immunological memory and so provide no long-lasting immunity against re-infection. The development of all these attributes requires T-cell help, as we shall see in the next section.

Property	TD antigen	TI-1 antigen	TI-2 antigen
Antibody response in infants	Yes	Yes	No
Antibody production in congenitally athymic individual	No	Yes	Yes
Antibody response in absence of cognate T cells	No	Yes	Yes
Activates T cells	Yes	No	No
Polyclonal B-cell activation	No	Yes	No
Requires repeated epitopes	No	No	Yes
Induces immunological memory	Yes	No	No
	Diphtheria toxin Viral hemagglutinin Purified protein derivative (PPD) of *Mycobacterium tuberculosis*	Bacterial lipopoly-saccharide *Brucella abortus*	Pneumococcal poly-saccharide Polymerized flagellin (*Salmonella*) Dextran Hapten-conjugated Ficoll (polysucrose)

Figure 7.6 Properties of different classes of antigen that elicit antibody responses. Responses to TI-2 antigens are generally enhanced by the presence of T cells but they do not require cognate interactions. The effect of the T cells might be due to the action of cytokines they produce, for example IL-5. Alternatively the effect could be due to γ:δ T cells, which recognize antigens other than conventional peptide:MHC complexes and do not need a thymus for their development.

7-3 B cells needing T-cell help are activated in secondary lymphoid tissues where they form germinal centers

Although the antibody response to a pathogen may be initiated by thymus-independent antigens, the bulk of the pathogen-specific antibody eventually comes from B cells stimulated by thymus-dependent antigens. Activation of these B cells occurs in the secondary lymphoid tissues where B cells, specific antigen and helper CD4 T cells are all brought together. We will describe these processes using the lymph node as an example.

Antigens arrive at a node in the lymph draining the infected tissue whereas antigen-specific lymphocytes enter the node from the blood (see Section 4-9, p. 100). The antigens can be transported by dendritic cells or be carried along in the fluid; either way they are trapped in the node and processed by professional antigen presenting cells. In the T-cell zone antigen-specific CD4 T lymphocytes are activated to become effector helper T cells by engagement with dendritic cells presenting the cells' specific antigen on MHC class II molecules. B cells pass through the T-cell zone and when their specific antigen is present they are activated through cognate interactions with effector helper T cells (Figure 7.7, see also Section 6-16, p. 155).

In B-cell activation the B-cell receptor has two distinct roles: binding antigen, which sends a signal to the B cell's nucleus, and internalizing antigen by receptor-mediated endocytosis, which facilitates the processing and presentation of antigen to helper T cells. When the antigen receptors of a CD4 T cell bind complexes of peptide antigen and MHC class II molecules on the B-cell surface, interaction also occurs between CD40 ligand on the T cell and CD40 on the B cell. The latter interaction signals the B cell to activate the transcription factor NFκB and increase surface expression of the adhesion molecule ICAM-1. This strengthens

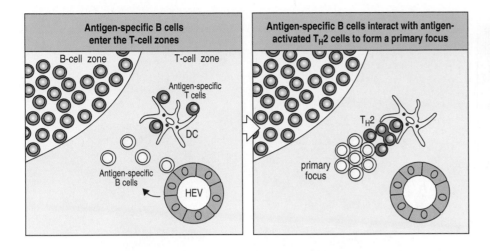

Figure 7.7 If they encounter cognate helper T cells, naive B cells become trapped in the T-cell zone of secondary lymphoid tissues. Recirculating naive B cells enter the T-cell zone of a lymph node from the blood through high endothelial venules (HEV). If they encounter helper T_H2 cells specific for the same antigen, they interact with them to form a primary focus of proliferating activated B cells and T_H2 cells.

the cognate interaction between the B cell and the helper CD4 T cell. A signal produced in the T cell causes cytokines to be secreted. Of these IL-4 is characteristic of T_H2 cells and essential for B-cell proliferation. By binding to receptors on the B-cell surface cytokines drive the B cell to proliferate and differentiate into plasma cells (Figure 7.8).

B cells activated by interactions with cognate helper T cells in the T-cell areas of a lymph node, form a primary focus of dividing B lymphoblasts (see Figure 7.7). Some of these B lymphoblasts move to the medullary cords and differentiate directly into plasma cells under the influence of the cytokines IL-5 and IL-6, which are secreted by T_H2 cells. They secrete predominantly IgM antibody but some isotype switching can also occur in the primary focus. For infections where the pathogen carries no thymus-dependent antigens, this will be the first antibody produced. It will start to appear several days after the onset of infection.

Other B lymphoblasts from the primary focus move into primary follicles as conjugates with their effector helper T cells. Their rate of division increases to about once every 6 hours and they become large metabolically active cells called centroblasts (see Section 4-9, p. 100). With expansion in centroblast cell number the morphology of a follicle changes and it becomes dominated by the **germinal center** which contains the newly formed B cells (Figure 7.9). Germinal centers appear in secondary lymphoid tissues about one week after the start of infection and they cause the characteristic swelling of lymph nodes draining an infection.

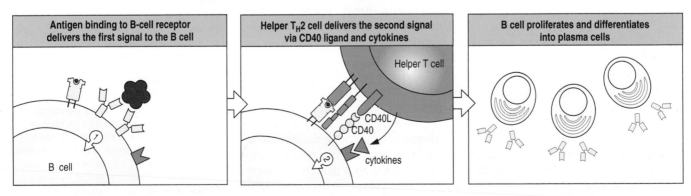

Figure 7.8 B-cell activation in response to thymus-dependent antigens requires cognate T-cell help. The first signal required for B-cell activation is delivered through the antigen receptor (left panel). With thymus-dependent antigens, the second signal is delivered by a cognate helper T cell that recognizes a peptide fragment of the antigen bound to MHC class II molecules on the B-cell surface (center panel). The two signals together drive B-cell proliferation and differentiation into plasma cells (right panel).

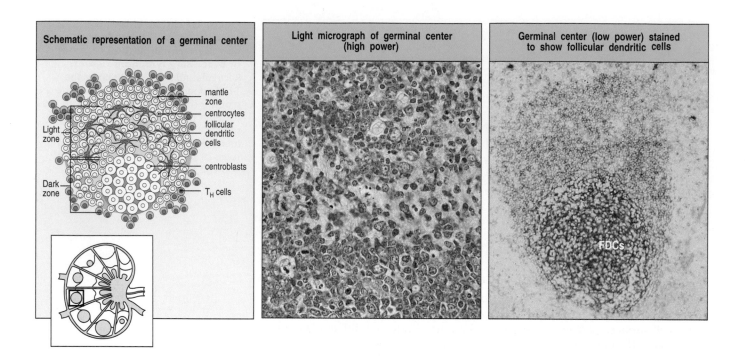

Schematic representation of a germinal center	Light micrograph of germinal center (high power)	Germinal center (low power) stained to show follicular dendritic cells

As they divide, the centroblasts become increasingly closely packed and form a region that is darkly staining in histological sections and is called the **dark zone** of the germinal center. The centroblasts give rise to non-dividing centrocytes, which leave the close-packed lymphocytes to interact with **follicular dendritic cells** (**FDCs**), in the **light zone** of the germinal center (see Figure 7.9). Follicular dendritic cells are the characteristic stromal cell of primary lymphoid follicles. They pick up antigen but do not internalize it, and it remains bound to their surface for long periods of time. They interact with B cells through a dense network of antigen-loaded dendrites. The follicular dendritic cells are quite distinct from the dendritic cells that present antigen to naive T cells and activate them (see Section 6-4, p. 135); they do not derive from a hematopoietic stem cell and do not express MHC class II molecules.

Those helper T cells that migrated to the primary follicle along with the activated B cells also proliferate in the light zone and are intermingled with the centrocytes. With time, the vast majority of lymphocytes present in a germinal center are clones derived from one or a few founder pairs of antigen-activated B and T cells. B cells that were present in the primary follicle before entry of the activated B cell–T cell conjugates, and are not specific for the antigen, are pushed to the outside of the germinal center, forming the **mantle zone** (see Figure 7.9).

7-4 Activated B cells undergo somatic hypermutation and affinity maturation in the specialized microenvironment of the germinal center

As we have seen in Chapters 4–6, a common theme in lymphocyte development is for a phase of activation and proliferation to be followed by one of selection. This is precisely what happens to the B cells maturing in a germinal center. Somatic hypermutation, initiated by T-cell cytokines, takes place in centroblasts dividing within the germinal center, and gives rise to non-dividing centrocytes with mutated surface immunoglobulin.   After hypermutation, the surface immunoglobulin expressed by an individual centrocyte can have an affinity for its specific antigen that is higher, lower, or the same as that of the unmutated

Figure 7.9 Germinal centers are formed when activated B cells enter lymphoid follicles. The germinal center is a specialized microenvironment in which B-cell proliferation, somatic hypermutation, and selection for antigen binding all occur. Rapidly proliferating B cells in germinal centers are called centroblasts. Closely packed centroblasts form the so-called 'dark zone' of the germinal center. This can be seen in the lower part of the center panel, which shows a light micrograph of a section through a germinal center, and in the accompanying diagram (left panel). As these cells mature, they stop dividing and become small centrocytes, moving out into an area of the germinal center called the 'light zone' (in the upper part of the center panel), where the centrocytes make contact with a dense network of follicular dendritic cell processes. The follicular dendritic cells are not stained in the center panel but can be seen clearly in the right panel, in which both follicular dendritic cells (stained blue with antibody against Bu10, a marker of follicular dendritic cells) in the germinal center as well as the mature B cells in the mantle zone (stained brown with an antibody against IgD) can be seen. The plane of this section reveals mostly the dense network of follicular dendritic cells in the light zone, although the less dense network in the dark zone can just be seen at the bottom of the figure. Photographs courtesy of I. MacLennan.

Figure 7.10 B cells recognize antigen as immune complexes bound to the surface of follicular dendritic cells. On the follicular dendritic cells (FDCs) the antigen is in the form of antigen:antibody complexes bound to Fc receptors (top panel) or in the form of antigen:antibody:complement complexes bound to complement receptors (bottom panel).

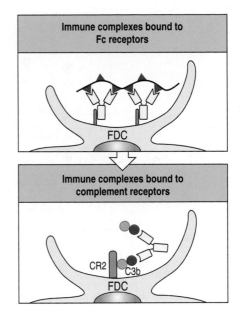

immunoglobulin. The population of centrocytes in a germinal center thus expresses immunoglobulins with a range of affinities for the specific antigen.

Centrocytes are programmed to die by apoptosis within a short period unless their surface immunoglobulin is bound by antigen and they are subsequently contacted by a helper T cell bearing CD40 ligand. To engage such a helper T cell, the centrocyte must first bind and process antigen, then present antigenic peptides at its surface in association with MHC class II molecules. The mutated centrocytes now compete with each other, first for access to antigen on follicular dendritic cells and then for antigen-specific helper T cells.

Follicular dendritic cells provide a source of intact antigen. They bind antigen in the form of complexes either with antibody or with antibody and complement. Such complexes are called **immune complexes**. The first source of these complexes is the IgM produced early in the primary immune response. Later in the immune response the immune complexes contain IgG. Follicular dendritic cells bear receptors for complement and for the Fc region of IgG (Figure 7.10). The immune complexes are not internalized and they persist for long periods at the surface of follicular dendritic cells, where they can be bound by antigen-specific B cells. Bundles of membrane coated with immune complexes also bud off from the surface of follicular dendritic cells. These bundles, called **iccosomes (immune-complex coated bodies)** (Figure 7.11), are bound and taken up by antigen-specific B cells, which then process and present the antigen.

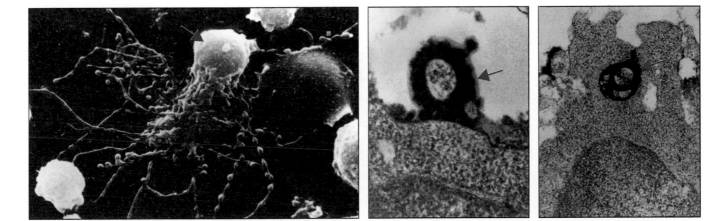

Figure 7.11 Immune complexes bound to follicular dendritic cells form iccosomes, which are released and can be taken up by B cells in the germinal center. Follicular dendritic cells have a prominent cell body and many dendritic processes. Immune complexes are bound to Fc receptors on the follicular dendritic cell surface and become clustered, forming prominent beads along the dendrites (left panel). In this scanning electron micrograph the cell body of the follicular dendritic cell is arrowed. The beads are shed from the cell as iccosomes, one of which is shown and arrowed in the center panel. Iccosomes are taken up by B cells within the germinal center, as shown by the arrow in the right panel. The immune complexes in the transmission electron micrographs in the center and right panels contain horseradish peroxidase, which generates the dense staining. Photographs courtesy of A.K. Szakal.

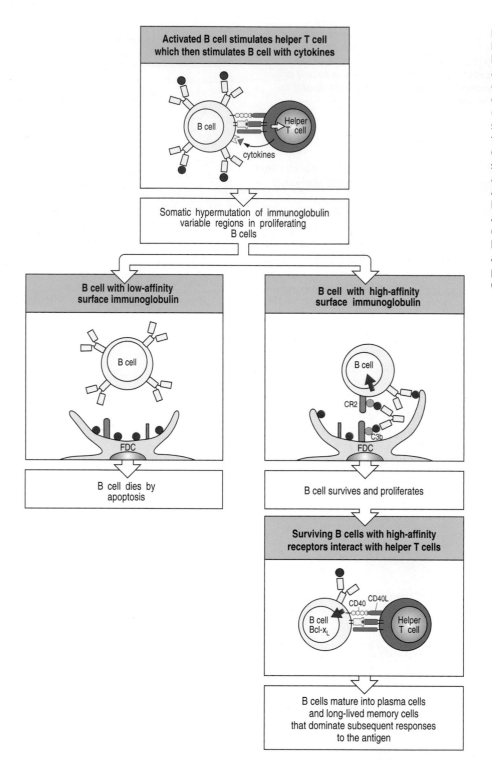

Figure 7.12 After somatic hypermutation, B cells with high-affinity receptors for antigen are rescued from apoptosis. In the germinal center, helper T cells induce B cells to undergo somatic hypermutation (top panel). B cells that have undergone somatic hypermutation interact with follicular dendritic cells (FDCs) that display immune complexes on their surface. B cells whose receptors bind antigen poorly, or do not bind antigen at all because they have mutated beyond recognition, cannot compete for access to the FDCs and die by apoptosis (left panel). B cells with receptors that bind well receive signals from the FDC and are induced to express Bcl-x$_L$, which prevents apoptosis; these cells survive (right panels).

Newly formed centrocytes move from the dark zone of the germinal center to contact follicular dendritic cells in the light zone. If a centrocyte captures sufficient antigen from the follicular dendritic cells or iccosomes, it then moves to the outer regions of the light zone, where helper T cells are concentrated. Engagement of peptide:MHC class II by the T-cell receptor complex, and of CD40 on the centrocyte by CD40 ligand on the T cell, induces the centrocyte to express the Bcl-x$_L$ protein, which prevents its death by apoptosis (Figure 7.12).

Property						
	Intrinsic			Inducible		
B-lineage cell	Surface Ig	Surface MHC class II	High-rate Ig secretion	Growth	Somatic hyper-mutation	Isotype switch
Resting B cell	Yes	Yes	No	Yes	Yes	Yes
Plasma cell	No	No	Yes	No	No	No

Figure 7.13 Comparison of resting B cells and plamsa cells. The resting B cell expresses an antigen receptor in the form of surface immunoglobulin, and can also take up protein antigen and present it as a peptide:MHC class II complex. It can thus activate helper T cells. Its immunoglobulin genes can also undergo somatic hypermutation, giving rise to progeny with altered immunoglobulin specificity. The plasma cell, in contrast, is a terminally differentiated B cell that is dedicated to the synthesis and secretion of soluble antibody. It no longer divides and its antibody specificity cannot be changed.

Centrocytes with the highest-affinity antigen receptors are thus selected for survival and further differentiation into antibody-producing plasma cells or into long-lived memory cells. In this way, the affinity of antibodies for the specific antigen increases during the course of an immune response and in subsequent exposures to the same antigen. This process is known as **affinity maturation**.

Under the influence of T cells, isotype switching also takes place in B cells within the germinal center. Thus not only does the affinity of the antibodies produced increase, but antibodies of different isotypes are made, principally IgG in the case of B cells that differentiate in lymph nodes and spleen.

For mutated centrocytes that survive selection, the interaction with an antigen-specific helper T cell serves several purposes. The mutual engagement of ligands and receptors on the two cells generates an exchange of signals that induces the further proliferation of both B and T cells. This serves to expand the population of selected high-affinity, isotype-switched B cells. Individual B cells are also directed along pathways of differentiation leading either to plasma cells or to memory B cells. At the height of the adaptive immune response, when the main need is for large quantities of antibodies to fight the infection, centrocytes that win in this selection leave the germinal center and differentiate into antibody-producing plasma cells. The differences between resting B cells and plasma cells are summarized in Figure 7.13. In the later stages of a successful immune response, as the infection subsides, centrocytes are thought to differentiate into long-lived memory B cells, which now possess isotype-switched, high-affinity antigen receptors. Plasma cells are the effector B cells that provide antibody for dealing with today's infection, whereas the memory B cells represent an investment in prevention of future infection with the same pathogen should the current infection be successfully resolved.

If a centrocyte fails to obtain, internalize, and present antigen, it dies by apoptosis and is phagocytosed by macrophages in the germinal center. Macrophages that have recently engulfed apoptotic centrocytes are a characteristic feature of germinal centers and, because of their contents, are called **tingible body macrophages**. Somatic hypermutation can produce centrocytes bearing immunoglobulin that reacts with a self antigen on the surface of cells in the germinal center. When this happens, contact with helper T cells or other cells in the germinal center will render such centrocytes inactive or anergic—a mechanism similar to the one whereby self-reactive immature B cells are inactivated in the bone marrow (see Section 4-7, p. 97).

7-5 Interactions with T cells are required for isotype switching in B cells

In Chapter 2 we saw how the first immunoglobulins made by B cells are of the IgM and IgD classes, but that after activation by antigen, B cells can switch the heavy-chain isotype to produce IgG, IgA, or IgE. Isotype switching takes place in activated B cells mainly within the germinal center, and the isotype to which an individual B cell switches is determined by cognate interactions with helper T cells. The particular isotype to which a switch is made depends on the cytokines secreted by the helper T cell. The roles of individual cytokines in switching the isotype of mouse immunoglobulin heavy chains are summarized in Figure 7.14. Cytokines secreted by T_H2 cells—IL-4, IL-5, and TGF-β—are the predominant players. They initiate the antibody response by activating naive B cells to differentiate into plasma cells secreting IgM, and also induce the production of other antibody isotypes including, in humans, the weakly opsonizing antibodies IgG2 and IgG4 as well as IgA and IgE. However, IFN-γ, the characteristic cytokine produced by T_H1 cells, switches B cells to making the IgG2a and IgG3 classes of immunoglobulin (in mice), and the strongly opsonizing antibody IgG1 in humans.

T-cell cytokines induce isotype switching by stimulating transcription from the switch regions that lie 5′ to each heavy-chain C gene. For example, when activated B cells are exposed to IL-4, transcription from a site upstream of the switch regions of $C_\gamma1$ and C_ε can be detected a day or two before switching occurs. As with the low-level transcription that occurs in immunoglobulin loci before rearrangement (see Section 4-2, p. 88), this transcription could be opening up the chromatin and making the switch regions accessible to the somatic recombination machinery that will place a new C gene in juxtaposition to the V-region sequence.

The induction of isotype switching by cognate helper T cells also requires the ligation of CD40 on the B-cell surface by CD40 ligand on the T cells. The importance of helper T cells and the CD40–CD40 ligand interaction for isotype switching is apparent from the immunodeficiency of patients who lack CD40 ligand. These patients have abnormally high levels of IgM in their blood serum, which gives the name **hyper-IgM syndrome** to their condition, but almost no IgG and IgA because of the inability of their B cells to switch isotype. They cannot make responses to thymus-dependent antigens, showing the general importance of CD40–CD40 ligand interactions in T-cell help. Aspects of cell-mediated immunity are also impaired in these patients, who are mostly male because the gene for CD40 ligand is on the X chromosome.

Influence of cytokines on antibody isotype switching in mice							
Cytokine	IgM	IgG3	IgG1	IgG2b	IgG2a	IgA	IgE
IL-4	Inhibits	Inhibits	Induces		Inhibits		Induces
IL-5						Augments production	
IFN-γ	Inhibits	Induces	Inhibits		Induces		Inhibits
TGF-β	Inhibits	Inhibits		Induces		Induces	

Figure 7.14 Different cytokines induce B cells to switch to different immunoglobulin isotypes. Individual cytokines can either induce (green), augment (bright yellow), or inhibit (red) the switching of immunoglobulin synthesis to a particular isotype. The inhibitory effects are largely due to the positive effect of the cytokine on switching to another isotype. This compilation is drawn from experiments on mouse B cells. There are differences in humans. For example switching to IgA involves TGF-β and IL-10, not IL-5.

Summary

B cells respond to specific antigen with activation, proliferation and differentiation. They then become plasma cells that synthesize and secrete massive amounts of antibody. Activation of a mature but naive B-cell requires signals delivered through its antigen receptor and most B cells also need additional signals that are delivered only on cognate interaction with an antigen-specific helper T cell. Activating signals are also delivered through the B-cell co-receptor when this is simultaneously ligated with the antigen receptor. The first antibodies made are always IgM; further contact with effector helper T cells is required for activated B cells to undergo isotype switching, somatic hypermutation, and affinity maturation within the germinal centers of secondary lymphoid organs. All this takes time, during which the pathogen can multiply, spread from the focus of infection and cause disease. However, if the host survives, there will remain in the circulation expanded populations of high-affinity antibodies and of memory B cells programmed to make them again should the need arise. Some antigens, notably certain components of bacterial cell walls and capsules, are capable of inducing a rapid antibody response that does not require T-cell help. These thymus-independent antigens are of two types. TI-1 antigens contain a moiety that stimulates mitosis and antibody production independently of stimulation through the B-cell antigen receptor. TI-2 antigens are generally microbial cell-surface macromolecules with repetitive epitopes that are present at high density on microbial surfaces and extensively crosslink the antigen receptors and co-receptors on the B-cell surface. Antibodies produced against TI antigens are predominantly IgM and the cells producing them are often of the B-1 lineage. Responses to TI antigens neither induce immunological memory nor long-lasting immunity.

Antibody effector functions

As the B-cell response to an infection gets under way, isotype switching diversifies the functional properties of the antibody Fc region, which contains binding sites for other proteins and cells of the immune system. Fc regions serve two distinct functions: they deliver antibody to anatomical sites that would otherwise be inaccessible, and they link bound antigen to molecules or cells that will effect its destruction. Such cells carry receptors called **Fc receptors**, which bind to the Fc regions of antibodies of a particular class or subclass irrespective of the antibody's antigen specificity. In this part of the chapter we shall consider how antibodies of different isotypes recruit non-specific effector cells such as macrophages and neutrophils into the immune response by interaction with their Fc receptors.

7-6 IgM, IgG, and IgA antibodies protect the blood and extracellular fluids

In any antibody response, IgM is the first antibody to be made. It is secreted as a pentamer by plasma cells in the bone marrow, the spleen, and the medullary cords of lymph nodes. IgM enters the blood and is carried to sites of tissue damage and infection throughout the body. The pentameric nature of IgM enables it to bind strongly to microorganisms and particulate antigens, but its large size decreases the extent to which this antibody isotype can passively leave the blood and penetrate infected tissues. There are no receptors for the IgM Fc region on phagocytic cells or other leukocytes, so IgM cannot directly recruit the destructive capabilities of these cells into the immune response. The Fc region of IgM can, however, bind complement and activate the complement system, with consequences that we shall consider in the last part of this chapter.

Later in an immune response, the dominant blood-borne antibody is the smaller IgG molecule. An important function of circulating IgM and IgG is to prevent blood-borne infection—**septicemia**—and the spread of microorganisms by neutralizing those that enter the blood. Because the blood circulation is so effective in distributing cells and molecules to all parts of the body, infections of the blood itself can have grave consequences.

IgA is synthesized by plasma cells in secondary lymphoid tissues. Monomeric IgA is made by plasma cells derived from B cells that switched their antibody isotype in lymph nodes or spleen. In contrast, dimeric IgA is made in the secondary lymphoid tissues underlying mucosal surfaces, as we shall see in the next section. Like IgG, monomeric IgA enters extracellular spaces and helps IgG to protect them against infection by bacteria and virus particles.

7-7 IgA and IgG are transported across epithelial barriers by specific receptor proteins

Whereas IgM, IgG, and monomeric IgA provide antigen-binding functions within the fluids and tissues of the body, dimeric IgA protects the surfaces of the epithelia that communicate with the external environment and are particularly vulnerable to infection. These epithelia include the linings of the gastrointestinal tract, the eyes, the nose, the throat, the respiratory, urinary and genital tracts, and the mammary glands. Dimeric IgA is made in patches of mucosal-associated lymphoid tissues in the lamina propria, the connective tissue that underlies the basement membrane of the mucosal epithelium. In these tissues, antigen-specific B-cell and T-cell responses to local infections are developed. However, the IgA-secreting plasma cells are on one side of the epithelium and their target pathogens are on the other. To reach their targets, dimeric IgA molecules are transported individually across the epithelium by means of a receptor on the basolateral surface of the epithelial cells.

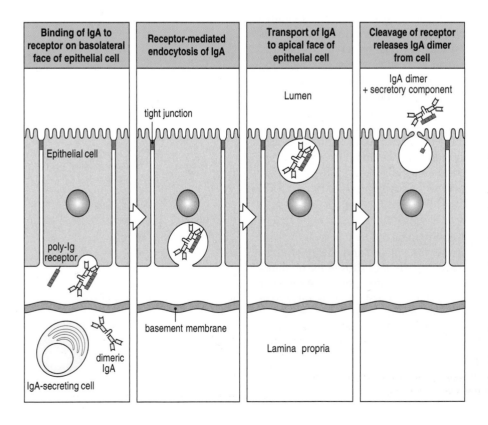

Figure 7.15 Transcytosis of dimeric IgA antibody across epithelia is mediated by the poly-Ig receptor. Dimeric IgA is made mostly by plasma cells lying just beneath epithelial basement membranes of the gut, respiratory tract, tear glands, and salivary glands. The IgA dimer bound to the J chain diffuses across the basement membrane and is bound by the poly-Ig receptor on the basolateral surface of an epithelial cell. Binding to the receptor is by C_H3 constant domains of the IgA heavy chains. The bound complex undergoes transcytosis across the cell in a membrane vesicle and is finally released onto the apical surface. There the poly-Ig receptor is cleaved, releasing the IgA from the epithelial cell membrane while still being bound to a fragment of the receptor called the secretory component or secretory piece. The residual membrane-bound fragment of the poly-Ig receptor is non-functional and is degraded.

The dimeric form of IgA, but not the monomer, binds to a cell-surface receptor on the basolateral surface of epithelial cells that is called the **poly-Ig receptor** because of its specificity for IgA polymers and, to a lesser extent, for pentameric IgM (Figure 7.15). The poly-Ig receptor itself is made up of a series of immunoglobulin-like domains. On being bound, the IgA dimer is taken into the cell by receptor-mediated endocytosis and the antibody:receptor complex is carried across the cell to the apical surface in endocytic vesicles. Receptor-mediated transport of a macromolecule from one side of a cell to the other is known as **transcytosis**. Once receptor-bound IgA appears on the apical surface, a protease cleaves the poly-Ig receptor at sites between the membrane-anchoring region and the IgA-binding site. Dimeric IgA is released from the membrane still bound to a small fragment of the poly-Ig receptor, which is called the **secretory component**, or **secretory piece**, of IgA.

IgG is actively transported from the blood into the extracellular spaces within tissues by an Fc receptor present on the endothelial cells (Figure 7.16). This receptor is sometimes called the **Brambell receptor** (**FcRB**) after the scientist who first described its function. FcRB is similar in structure to an MHC class I molecule, with the α_1 and α_2 domains forming a site that binds to the Fc region of the antibody. In the antibody:receptor complex, two molecules of FcRB bind to the Fc region of one IgG molecule. The delivery of IgG to the extracellular spaces in connective tissue helps to protect tissues against infection and also protects IgG from the degradation pathways to which serum proteins are subject. As a consequence, IgG molecules have a relatively long half-life in relation to most other plasma proteins.

During pregnancy, the fetus is physically protected by the mother from the microorganisms that inhabit the external environment. Upon birth the baby is suddenly exposed to numerous pathogens. Because of their lack of actively acquired immunity, newborn infants are particularly vulnerable to infection arising from the microbial colonization of epithelia. To help the newborn baby counter such attack, it receives IgA from its mother. The IgA is first secreted into breast milk and then transferred on breast-feeding into the baby's gut. Within this transferred IgA are antibodies against microorganisms to which the mother has previously mounted an IgA response. Within the child's gut, the IgA molecules bind to microorganisms and their products, preventing their attachment to the gut epithelium and facilitating their expulsion in feces. The transfer of preformed IgA from mother to child in breast milk is an example of the **passive transfer of immunity**.

IgA is not the only immunoglobulin isotype that mothers donate to their children. During pregnancy, IgG from the maternal circulation is transported across the placenta and is delivered directly into the fetal bloodstream. The efficiency of this mechanism is such that human babies at birth have as high a level of IgG in their plasma as their mothers, and as wide a range of antigen specificities. Transport of IgG across the placenta is performed by FcRB (the Brambell receptor) (see Figure 7.16).

Mice and rats express a homolog of FcRB called FcRn, but its function is somewhat different. In rodents, the receptor is expressed in the intestine for a short

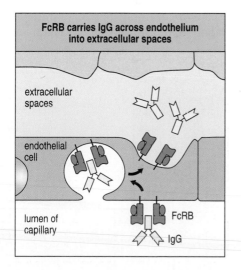

Figure 7.16 The Brambell receptor (FcRB) transports IgG from the bloodstream into extracellular spaces. An IgG molecule binds to two FcRB molecules at the apical (luminal) side of the endothelial cell. After receptor-mediated endocytosis, the IgG molecule is carried in a vesicle across the endothelial cell to the basal side of the cell. There it is released into the extracellular space.

FcRB carries IgG across endothelium into extracellular spaces

extracellular spaces

endothelial cell

lumen of capillary

FcRB

IgG

Figure 7.17 Immunoglobulin isotypes are selectively distributed in the body. IgG and IgM predominate in plasma, whereas IgG and monomeric IgA are the major isotypes in extracellular fluid within the body. Dimeric IgA predominates in secretions across epithelia, including breast milk. The fetus receives IgG from the mother by transplacental transport. IgE is associated mainly with mast cell surfaces and is therefore found beneath epithelial surfaces (especially the respiratory tract, gastrointestinal tract, and skin). The brain is normally devoid of immunoglobulin.

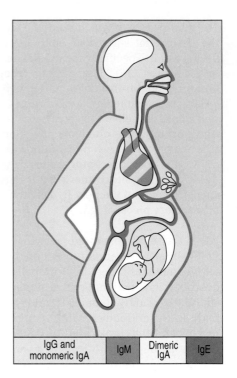

| IgG and monomeric IgA | IgM | Dimeric IgA | IgE |

period after birth. During this time, the newborn rodent ingests maternal IgG in colostrum, the protein-rich fluid in the postnatal mammary gland, which is then transported across the intestinal epithelium into the tissues by the FcRn.

By means of these specialized transport systems mammals are supplied from birth with antibodies against common pathogens in their environment. As the young mature and make their own antibodies of all the isotypes, these are each distributed to selected sites in the body (Figure 7.17). Thus, throughout life, the production of different isotypes provides protection against infection in the extracellular spaces throughout the body.

7-8 Antibody production is deficient in very young infants

During the first year of life there is a window of time when all infants are relatively deficient in antibodies and specially vulnerable to infection. During pregnancy, maternal IgG antibodies are transported across the placenta into the fetal circulation, providing newborn infants with antibody levels comparable to those of their mothers. As the maternally derived IgG is catabolized, the antibody level gradually decreases until the immune system of the infant begins to produce its own antibody at about 6 months of age (Figure 7.18).

Consequently, IgG levels are lowest in infants aged 3–12 months, and this is when they are most susceptible to infection. This problem is particularly acute in babies born prematurely, who begin life with lower levels of maternal IgG and take longer to attain immunocompetence after birth than babies born at term.

7-9 High-affinity IgG and IgA antibodies are used to neutralize microbial toxins and animal venoms

Many bacteria secrete protein toxins that cause disease by disrupting the normal function of human cells (Figure 7.19). To have this effect, a bacterial toxin must first bind to a specific receptor molecule on the surface of the human cell. In some toxins, for example those of diphtheria and tetanus, the receptor-binding activity is carried by one polypeptide chain and the toxic function by another. Antibodies

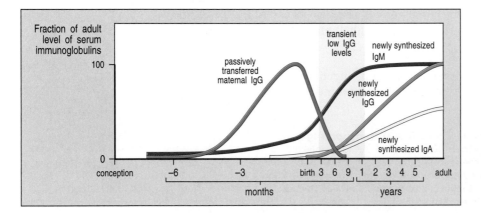

Figure 7.18 In the first year of life infants have a transient decrease in levels of IgG. Before birth, high levels of IgG are provided by the mother, but after birth maternally derived IgG declines. Although infants produce IgM soon after birth, secretion of IgG does not begin for about 6 months. The overall level of IgG reaches a minimum within the first year and then gradually increases until adulthood.

Disease	Organism	Toxin	Effects *in vivo*
Tetanus	*Clostridium tetani*	Tetanus toxin	Blocks inhibitory neuron action leading to chronic muscle contraction
Diphtheria	*Corynebacterium diphtheriae*	Diphtheria toxin	Inhibits protein synthesis leading to epithelial cell damage and myocarditis
Gas gangrene	*Clostridium perfringens*	Clostridial-α toxin	Phospholipase activation leading to cell death
Cholera	*Vibrio cholerae*	Cholera toxin	Activates adenylate cyclase, elevates cAMP in cells, leading to changes in intestinal epithelial cells that cause loss of water and electrolytes
Anthrax	*Bacillus anthracis*	Anthrax toxic complex	Increases vascular permeability leading to edema, hemorrhage and circulatory collapse
Botulism	*Clostridium botulinum*	Botulinum toxin	Blocks release of acetylcholine leading to paralysis
Whooping cough	*Bordetella pertussis*	Pertussis toxin	ADP-ribosylation of G proteins leading to lymphocytosis
		Tracheal cytotoxin	Inhibits ciliar movement and causes epithelial cell loss
Scarlet fever	*Streptococcus pyogenes*	Erythrogenic toxin	Causes vasodilation leading to scarlet fever rash
		Leukocidin Streptolysins	Kill phagocytes enabling bacteria to survive
Food poisoning	*Staphylococcus aureus*	Staphylococcal enterotoxin	Acts on intestinal neurons to induce vomiting. Also a potent T-cell mitogen (SE superantigen)
Toxic-shock syndrome	*Staphylococcus aureus*	Toxic-shock syndrome toxin	Causes hypotension and skin loss. Also a potent T-cell mitogen (TSST-1 superantigen)

Figure 7.19 Many common diseases are caused by bacterial toxins. Several examples of exotoxins, or secreted toxins, are shown here. Bacteria also make endotoxins, or non-secreted toxins, which are usually only released when the bacterium dies. Endotoxins, such as bacterial lipopolysaccharide (LPS), are important in the pathogenesis of disease, however, their interactions with the host are more complicated than those of the exotoxins and less clearly understood.

that bind to the receptor-binding polypeptide can be sufficient to neutralize a toxin (Figure 7.20). The vaccines for diphtheria and tetanus work on this principle. They are modified toxin molecules, called **toxoids**, in which the toxic chain has been denatured to remove its toxicity. On immunization protective neutralizing antibodies are made against the receptor-binding chain.

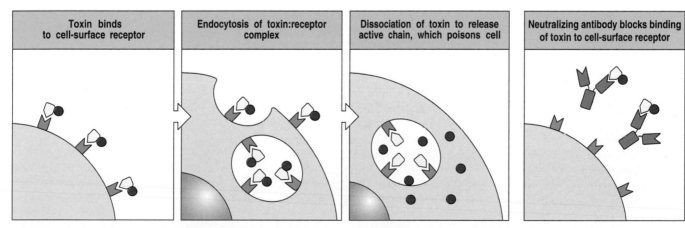

Toxin binds to cell-surface receptor	Endocytosis of toxin:receptor complex	Dissociation of toxin to release active chain, which poisons cell	Neutralizing antibody blocks binding of toxin to cell-surface receptor

Figure 7.20 Neutralization by IgG antibodies protects cells from toxin action. Many species of bacteria cause their harmful effects by producing protein toxins. These toxins are usually of modular construction. One part of the toxin binds to a cellular receptor, which allows the toxin to be internalized whereupon the second part poisons the cell. Antibodies that inhibit the toxin from poisoning the cell are called neutralizing antibodies.

Bacterial toxins are potent at low concentrations, a single molecule of diphtheria toxin being sufficient to kill a cell. To neutralize a bacterial toxin, an antibody must be of high affinity and essentially irreversible in its binding to the toxin; it must also be able to penetrate tissues and reach the sites where toxins are being released. High-affinity IgG is the main source of neutralizing antibodies for the tissues of the human body, whereas high-affinity IgA serves a similar purpose for the mucosal surfaces.

Poisonous snakes, scorpions, and other animals introduce venoms containing toxic polypeptides into humans through a bite or sting. For some venoms, a single exposure is sufficient to cause severe tissue damage or even death, and in such situations the primary response of the immune system is too slow to help. As exposure to such venoms is rare, protective vaccines against them have not been developed. For patients who have been bitten by poisonous snakes or other venomous creatures, the preferred therapy is to infuse them with antibodies specific for the venom. These antibodies are produced by immunizing large domestic animals—such as horses—with the venom. Transfer of protective antibodies in this manner is known as **passive immunization** and is analogous to the way in which newborn children acquire passive immunity from their mothers.

7-10 High-affinity neutralizing antibodies prevent viruses and bacteria from infecting cells

The first step in the infection of a human cell by a virus is its attachment to the cell by means of a cell-surface protein, which is used as the virus receptor. The influenza virus, for example, binds to oligosaccharides on cell-surface glycoproteins on epithelial cells of the respiratory tract. The virus binds through a protein in its outer envelope, which is known as the **influenza hemagglutinin** because the protein can **agglutinate**, or clump together, red blood cells by binding to oligosaccharides on the red cell surface. Neutralizing antibodies that have been developed during primary immune responses to influenza and other viruses are the most important aspect of subsequent immunity to these viruses. Such antibodies coat the virus, inhibit its attachment to human cells, and prevent infection (Figure 7.21, upper panels).

Some bacteria who exploit mucosal surfaces maintain their populations by binding to and colonizing the surface of the epithelial cells, as does the bacterium *Neisseria gonorrhoeae*, which causes gonorrhea. Others enter epithelial cells, as do the species of *Salmonella* that cause food-borne gastrointestinal infections. IgA antibodies against the adhesion proteins (adhesins) responsible for binding to epithelial cells limit bacterial populations within the gastrointestinal, respiratory, urinary, and reproductive tracts and prevent disease-causing infections in these tissues (see Figure 7.21, lower panels).

7-11 The Fc receptors of hematopoietic cells are signaling receptors that bind the Fc regions of antibodies

Although the binding of a neutralizing antibody to a pathogen or toxin prevents further infection, it does not in itself remove the antigen from the body. This is accomplished by phagocytic effector cells, principally neutrophils, blood monocytes and tissue macrophages. These cells express various receptors that bind to the Fc regions of antibodies of different isotypes and are known generally as Fc receptors.

The Fc receptors of phagocytes and other hematopoietic cells are functionally and structurally distinct from the FcRB of endothelial cells. They each consist of several polypeptide chains, the most important of which is an α chain made up of

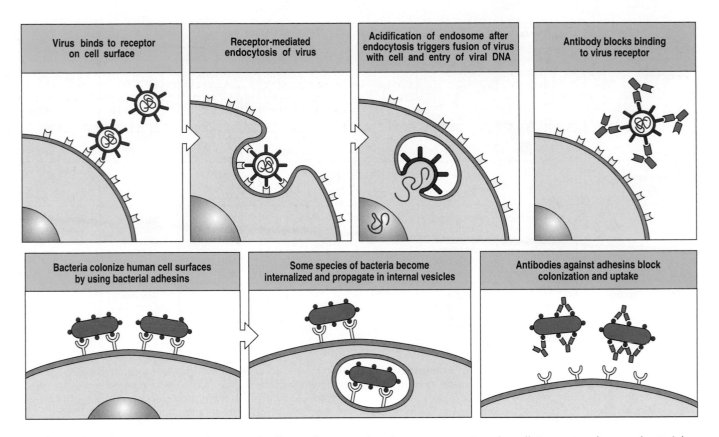

Figure 7.21 Viral and bacterial infection of cells can be blocked by neutralizing antibodies. Upper panels: for a virus to infect a cell, it must gain entry to the cytoplasm. This requires binding of the virus to the cell surface, internalization in an endosome and fusion of viral and cell membranes to release viral nucleic acid into the cytoplasm. Antibodies binding to viral surface proteins can inhibit either the initial binding of virus or its subsequent entry into the cell. Lower panels: many bacterial infections require an interaction between the bacterium and the surface of a human cell. This is particularly true for infections of mucosal surfaces. The attachment process involves very specific molecular interactions between bacterial adhesins and their ligands on human cells. Antibodies specific for epitopes of the bacterial adhesins can therefore block infection.

immunoglobulin-like domains (Figure 7.22). This chain binds to the Fc region of the antibody and determines the isotype specificity of the receptor. Associated with the α chain are other polypeptide chains that function either in the folding of the Fc receptor and its movement to the cell surface, or signal the cell once the receptor has bound its ligand. One of the signaling components, the γ chain, is closely related in amino-acid sequence to the ζ chain of the T-cell receptor complex.

FcγRII-B1 and -B2 are inhibitory receptors that help to control the activation of naive B cells, mast cells, macrophages, and neutrophils. These receptors bear **immunoreceptor tyrosine-based inhibition motifs** (**ITIMs**) in their cytoplasmic tails, which associate with intracellular proteins that develop inhibitory signals.

7-12 Phagocyte Fc receptors facilitate the recognition, uptake, and destruction of antibody-coated pathogens

As we saw in Chapter 6, phagocytic cells can recognize, ingest, and destroy bacteria in the absence of specific antibody. This capacity is of paramount importance in containing infection during the period before an antigen-specific immune response has been made and in enabling macrophages to take up, process, and present antigen to T cells in the early phases of an adaptive immune response. However, the speed with which pathogens can be bound and engulfed by

Receptor	Fcγ RI (CD64)	Fcγ RII-A (CD32)	Fcγ RII-B2 (CD32)	Fcγ RII-B1 (CD32)	Fcγ RIII (CD16)	FcεRI	FcαRI (CD89)
Structure	α 72 kDa γ	α 40 kDa γ-like domain	ITIM	ITIM	α 50–70 kDa or γ or ζ	α 45 kDa β 33 kDa γ 9 kDa	α 55–75 kDa γ 9 kDa
Relative binding strength	IgG1 200	IgG1 4	IgG1 4	IgG1 4	IgG1 1	IgE 20,000	IgA1, IgA2 20
Cell type	Macrophages Neutrophils* Eosinophils* Dendritic cells	Macrophages Neutrophils Eosinophils Platelets Langerhans' cells	Macrophages Neutrophils Eosinophils	B cells Mast cells	NK cells Eosinophils Macrophages Neutrophils Mast cells FDCs	Mast cells Eosinophils* Basophils FDCs	Macrophages Neutrophils Eosinophils†
Effect of ligation	Uptake Stimulation Activation of respiratory burst Induction of killing	Uptake Granule release (eosinophils)	Uptake Inhibition of stimulation	No uptake Inhibition of stimulation	Induction of killing (NK cells)	Secretion of granules	Uptake Induction of killing

Figure 7.22 Receptors for the Fc regions of immuno-globulins are present on a variety of immune system cells. The subunit structure, relative binding strength, and cellular distribution of the Fc receptors are shown. The complete multimolecular structure of most receptors is not yet known but they might all be multichain molecular complexes similar to the Fcε receptor I (FcεRI). Receptor structure can vary slightly from one cell type to another. For example, FcγRIII in neutrophils is expressed as a protein with a glycophosphoinositol membrane anchor and has no associated γ chains, whereas in natural killer (NK) cells it is a transmembrane protein associated with γ chains as shown. The information in the figure is based on the Fc receptors of mouse cells, with the exception of the relative binding strengths which pertain to the human receptors. The Fcγ receptors also bind the other subclasses of IgG. FcγRIII binds IgG1 and IgG3 with equal strength. For the other FcγRs, IgG1 binds most strongly, IgG2 least strongly, and IgG3 and IgG4 with intermediate affinity. FDCs, follicular dendritic cells. *In these cases Fc receptor expression is inducible rather than constitutive. † In eosinophils, the molecular weight of CD89α is 70–100 kDa.

phagocytes greatly increases when the pathogens are coated with antibodies, or opsonized. This is because the principal phagocytic cells of the body—the macrophages and neutrophils—express Fc receptors, called **Fcγ receptors**, which are specific for the Fc regions of IgG antibodies, particularly that of IgG1 (see Figure 7.22).

When IgG molecules specific for the surface components of a pathogen bind to the pathogen with their Fab arms, the Fc regions are left exposed on the outside of the antibody-coated particle. The pathogen becomes coated with many IgG molecules, presenting multiple Fc regions to the Fc receptors on a phagocyte. On contact with a phagocyte, multiple ligand–receptor interactions are made, producing a stable and strong binding from interactions that are individually of low affinity and short-lived. The low affinity of Fcγ receptors for individual IgG molecules means that they bind only transiently to free IgG molecules in the absence of antigen. This property enables high concentrations of IgG of diverse antigenic specificities to circulate in the body's fluids and not clog up the Fc receptors of phagocytes in the absence of antigen.

After the pathogen has been bound to the phagocyte, interactions between antibody Fc regions and their receptors facilitate the engulfment of the antibody-coated pathogen (Figure 7.23). The surface of the phagocyte gradually extends around the surface of the opsonized pathogen through cycles of binding and release between the Fc receptors of the phagocyte and the Fc regions projecting from the pathogen surface. Bearing some similarities to walking, the engulfment is an active process triggered by signals from the Fc receptors.

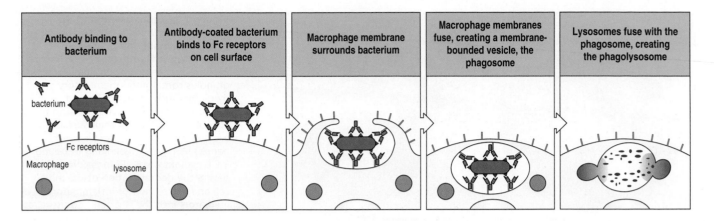

Figure 7.23 Fc receptors on phagocytes trigger the uptake and breakdown of antibody-coated bacteria. Specific IgG molecules coat the bacterial surface and then tether the bacterium to the phagocyte cell surface by binding to the Fc receptors. Signals from the Fc receptors enhance phagocytosis of the bacterium and the fusion of lysosomes containing degradative enzymes with the phagosome.

A coating of antibodies makes different sorts of microorganisms appear similar to the macrophage, and thus enables it to deal with them all by using a single effector mechanism. Encapsulated bacteria such as *Streptococcus pneumoniae* have evolved cell-surface structures that are resistant to direct phagocytosis; for these species a coating with antibody that masks their surface is essential if they are to be phagocytosed.

Once an opsonized bacterium has been endocytosed, it becomes enclosed in an acidified vesicle called a phagolysosome, formed from the fusion of the phagosome with lysosomes and neutrophil granules, which contain hydrolytic enzymes and antimicrobicidal peptides. Activated neutrophils and macrophages also produce oxygen radicals, nitric oxide, and other oxidizing agents with powerful microbicidal actions. The engulfed bacteria are killed by the combined effects of these substances.

As well as destroying microorganisms intracellularly, activated macrophages also attack larger antibody-coated parasites, such as worms, that they have bound via their Fc receptors but which are too big for them to engulf. In this case, the toxic contents of the lysosomes and diffusible metabolites such as nitric oxide are secreted by the macrophage and poured onto the parasite.

7-13 IgE binds to high-affinity Fc receptors on mast cells, basophils, and activated eosinophils

IgE antibodies against a wide variety of different antigens are normally present in small amounts in all human beings. They are produced in responses dominated by CD4 T_H2 cells, in which the cytokines produced favor switching to the IgE isotype. A consequence of the low affinity of Fc receptors for IgG is that free IgG molecules do not form stable interactions with cells expressing these receptors. The Fc receptor for IgE on **mast cells**, **basophils**, and activated **eosinophils** has quite the opposite properties. This receptor, called FcεRI, has such a high affinity ($\sim 10^{10}$ M^{-1}) for the Fc region of IgE that IgE molecules are tightly bound in the absence of antigen and the cells are almost always coated with antibody. In the absence of allergy or parasitic infection, a single mast cell carries IgE molecules specific for many different antigens.

Mast cells are sentinels posted throughout the body's tissues, particularly in the connective tissues lying in the mucosa of the gastrointestinal and respiratory

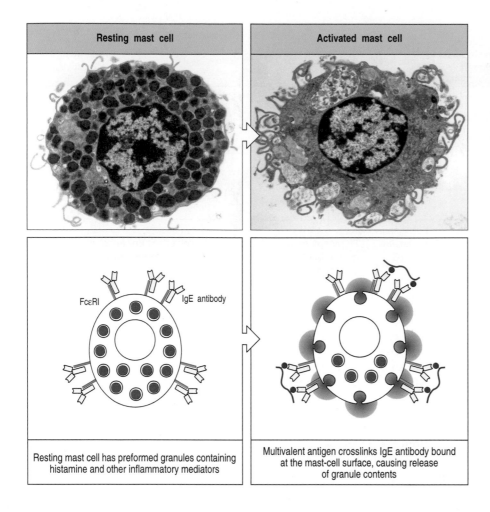

Resting mast cell	Activated mast cell

Resting mast cell has preformed granules containing histamine and other inflammatory mediators

Multivalent antigen crosslinks IgE antibody bound at the mast-cell surface, causing release of granule contents

Figure 7.24 IgE crosslinking on mast-cell surfaces leads to the rapid release of mast-cell granules containing inflammatory mediators. Resting mast cells contain numerous granules containing inflammatory mediators such as histamine and serotonin. The cells have high-affinity Fc receptors (FcεRI) on their surface that are occupied by IgE molecules (left panels). Antigen crosslinking of bound IgE crosslinks the FcεRI molecules, triggering the degranulation of the mast cell and the release of inflammatory mediators into the surrounding tissue, as shown in the right panels. Photographs courtesy of A.M. Dvorak.

tracts and in connective tissues along blood vessels—especially those in the dermis of the skin. The cytoplasm of the resting mast cell is filled with large granules containing **histamine** and other molecules that contribute to inflammation, which are known generally as **inflammatory mediators**. Mast cells become activated to release their granules when antigen binds to the IgE molecules bound to FcεRI on the mast cell surface (Figure 7.24). To activate the cell, the antigen must crosslink at least two IgE molecules and their associated receptors, which means that the antigen must have at least two topographically separate epitopes recognized by the cell-bound IgE. Crosslinking of FcεRI generates the signal that initiates the release of the mast cell granules. After degranulation the mast cell synthesizes and packages a new set of granules.

Inflammatory mediators secreted into the tissues by activated mast cells, basophils, and eosinophils increase the permeability of the local blood vessels, enabling other cells and molecules of the immune system to move out of the bloodstream and into tissues. This causes the local accumulation of fluid and the swelling, reddening, and pain that characterize inflammation. Inflammation in response to an infection is beneficial as it recruits cells and proteins required for host defense into sites of infection.

The prepackaged granules and the high-affinity FcεRI receptor already armed with IgE make the mast cell's response to antigen impressively fast. The infections that are the 'natural' targets of IgE-activated mast cells and eosinophils are thought to be those caused by parasites.

Parasites are a heterogeneous set of organisms that include the unicellular protozoa and multicellular invertebrates, notably the helminths—intestinal worms and the blood, liver, and lung flukes—and ectoparasitic arthropods such as ticks and mites. As a group, parasites establish long-lasting, persistent infections in human hosts and are well practised in the avoidance and subversion of the human immune system. Most parasites are much larger than any microbial pathogen. The largest human parasite is the tapeworm *Diphyllobothrium latum*, which can reach 9 meters in length and lives in the small intestine, causing vitamin B$_{12}$ deficiency and, in some patients, megaloblastic anemia. Multicellular parasites cannot be controlled by the cellular and molecular mechanisms of destruction that work for microorganisms, so a different strategy based on IgE has evolved.

Inflammatory mediators released by mast cells, basophils, and eosinophils cause the contraction of smooth muscle surrounding the airways and the gut. In addition to violent muscular contractions that can expel parasites from the airways or gut, the increased permeability of local blood vessels supplies an outflow of fluid across the epithelium which can help to flush parasites out. In summary, the combined actions of IgE, mast cells, basophils, and eosinophils serve to physically remove parasite pathogens and other material from the body.

Eosinophils can also use their Fcε receptors to act directly against multicellular parasites. Such organisms, even small ones such as the blood fluke *Schistosoma mansoni*, which causes schistosomiasis, cannot be ingested by phagocytes. However, if the parasite induces an antibody response and becomes coated with IgE, activated eosinophils will bind to it through FcεRI and then pour the toxic contents of their granules directly onto its surface (Figure 7.25).

For human populations in developed countries where parasite infections are rare, the mast cell's response is most frequently seen as a detriment, because its actions are the cause of allergy and asthma. People with these conditions make IgE in response to relatively innocuous substances, for example grass pollens or shellfish, which are often either airborne or eaten. Such substances are known as allergens. Having made specific IgE, any subsequent encounter with the allergen leads to massive mast-cell degranulation and a damaging response that is quite inappropriate to the threat posed by the antigen or its source. In extreme cases, the ingestion of an allergen can lead to a systemic life-threatening inflammatory response called anaphylaxis.

Parasite infections that invoke a protective IgE response are not major health problems in the developed world, whereas allergy and asthma seem not to be prevalent in the developing countries where infection with parasites is endemic. This is one of various pieces of circumstantial evidence suggesting that if elements of the immune system are left unstimulated by infection, they can respond in ways that are frankly unhelpful.

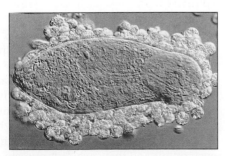

Figure 7.25 Eosinophils attacking a schistosome larva in the presence of serum from an infected patient. Large parasites, such as worms, cannot be ingested by phagocytes; however, when the worm is coated with antibody, especially IgE, eosinophils can attack it by using their high-affinity FcεRI. Similar attacks can be mounted by other Fc-receptor-bearing cells on various large targets. Photograph courtesy of A. Butterworth.

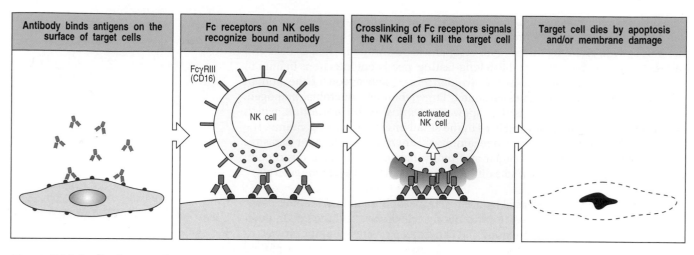

| Antibody binds antigens on the surface of target cells | Fc receptors on NK cells recognize bound antibody | Crosslinking of Fc receptors signals the NK cell to kill the target cell | Target cell dies by apoptosis and/or membrane damage |

Figure 7.26 Antibody-coated target cells can be killed by natural killer cells (NK cells) in antibody-dependent cell-mediated cytotoxicity (ADCC). NK cells are large granular lymphocytes that are distinct from B and T cells and have FcγRIII receptors (CD16) on their surface. When these cells encounter cells coated with IgG antibody, they rapidly kill the target cell. The importance of ADCC in host defense or tissue damage is uncertain.

7-14 Fc receptors activate natural killer cells to destroy antibody-coated human cells

Human natural killer cells (NK cells) are large lymphocytes (see Figure 1.6, p. 8) whose chief role is in innate immunity. However, they also express an Fc receptor called FcγRIII, or CD16, which is specific for IgG1 and IgG3. In experimental situations, NK cells have been shown to recognize and kill human cells coated with antibody against cell-surface components (Figure 7.26). This **antibody-dependent cell-mediated cytotoxicity** (**ADCC**) requires the presence of preformed antibody. This is not available during a primary immune response but could have a role in secondary responses. Another situation in which preformed antibody is present is in the newborn, who have passively acquired IgG against many pathogens to which they have yet to be exposed.

Summary

Secreted antibodies are the only effector molecules produced by B cells; their principal function is as adaptor molecules that neutralize the pathogen and bring together pathogens or their products with the effector cells that destroy them. Antibodies can become bound through their Fc regions to Fc receptors on various types of effector cell. These interactions with Fc receptors are specific for immunoglobulin isotype and are used by the immune system for two main purposes. The first is to deliver antibodies to sites where they would not be carried by the circulation of the blood and lymph. To this end the poly-Ig receptor of epithelium provides the lumen of the intestines and other mucosal surfaces with a continual supply of IgA that binds to the microorganisms that inhabit and infect those tissues. FcRB delivers IgG from plasma into the extracellular fluid in tissues and also, during pregnancy, delivers maternal IgG to the fetal circulation, which is useful after birth against many types of infection. The second purpose of Fc receptors is to attach pathogens or antigens that have bound to specific antibody to effector cells that will respond in ways that eliminate infection. Fc receptors for IgG can deliver antibody-bound bacteria to phagocytes. In these cases, antibody binds first to the antigen and then to the Fc receptor. In contrast, the IgE antibody binds first to the Fc receptor of mast cells, basophils, and activated eosinophils and then awaits its antigen.

Complement tags microorganisms for destruction

The binding of antibodies to antigens involves non-covalent bonds. Interactions between them are potentially reversible, depending on local concentrations of antibody and antigen, and are therefore susceptible to disruption, especially where low-affinity antibodies are concerned. Once a pathogen or antigen has been identified as foreign, it becomes advantageous to mark it for destruction in a more permanent manner. This is accomplished by a system of blood proteins known collectively as the **complement system**. Activation of the complement system initiates a series of enzymatic reactions in which the proteolytic cleavage and activation of successive complement proteins leads to the covalent bonding or 'fixing' of complement fragments to the pathogen surface. A major consequence of this is the uptake and destruction of complement-coated microbes by phagocytes that bear receptors for these complement fragments. Phagocytic cells and complement-like proteins are present in both vertebrates and invertebrates, whereas the antigen-binding molecules of the adaptive immune system are found only in vertebrates. This suggests that complement pre-dates antibody in the evolution of the immune system and that the antibody response evolved to enhance and be compatible with existing mechanisms of complement activation and phagocytosis. This part of the chapter will describe the activation of the complement system by antibody, and its consequences.

7-15 The complement system is activated by evidence of infection

The complement system uses three different strategies for recognizing microorganisms, each of which initiates a pathway of complement activation that leads to the covalent bonding of complement proteins to microbial surfaces (Figure 7.27). The **classical pathway** of complement activation is triggered by antibodies bound to antigens on a microbial surface. In this pathway, which was the first to be discovered, the complement proteins work together with antibodies to enhance the clearance of antigen:antibody complexes from the body. The two other pathways provide for complement activation in the absence of antibody and are considered part of the innate immune defenses. They are the **alternative pathway** of

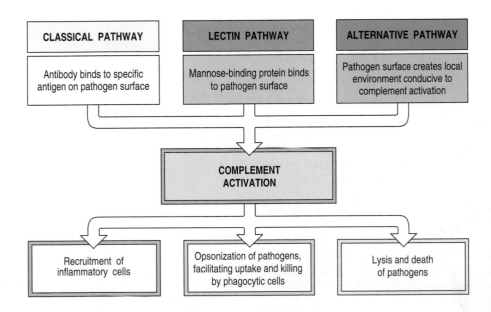

Figure 7.27 Schematic overview of the three pathways of complement activation. The classical pathway is initiated by the binding of either IgM or IgG antibodies to a microbial surface, whereas the lectin-mediated pathway is initiated by the mannose-binding protein of plasma, which binds to carbohydrates found on bacterial cells. The alternative pathway is triggered by the local physicochemical environment created by the constituents of some bacterial surfaces.

Figure 7.28 The components of the complement system and their functions. The soluble complement proteins are organized into five groups on the basis of their functions; the members of each functional group are often structurally similar. Proteins of all three pathways are listed here. As in many areas of immunology, the nomenclature for complement proteins has arisen rather haphazardly. The soluble complement proteins of the classical pathway are named with a capital letter C followed by a numeral. When a complement protein is cleaved into two fragments, the individual fragments are given the same name as the parent protein followed by a lowercase letter. In this book we call the larger of the fragments 'b' and the smaller 'a'. Thus, C3b is the larger of the protein fragments formed by the cleavage of C3, whereas C3a is the smaller. For historical reasons, the small cleavage product of C2 is sometimes called C2b, and the larger product C2a. We do not use that nomenclature here. Some components of the classical pathway, including all the components that form the membrane-attack complex, also participate in the lectin-mediated and alternative pathways. Only components unique to a particular pathway are given separate names. The groups are color-coded and the color applies to the flow diagram of the pathways of complement activation in Figure 7.29.

Functionally distinct classes of complement protein	
Function	**Protein**
Binding to antigen:antibody complexes	C1q
Activating enzymes	C1r C1s C2b Bb D
Membrane-binding proteins and opsonins	C4b C3b
Peptide mediators of inflammation	C5a C3a C4a
Membrane-attack proteins	C5b C6 C7 C8 C9

complement activation, which is triggered directly by constituents of bacterial cell surfaces, and the **lectin-mediated pathway**, which is activated by the binding of a mannose-binding protein present in blood plasma to mannose-containing proteoglycans on the surfaces of bacteria and yeast.

All three pathways lead to the covalent bonding of a particular fragment of a complement component (the C3b fragment of C3) to the pathogen surface. Macrophages and neutrophils bear receptors for C3b and so the opsonization of a pathogen with complement facilitates its phagocytosis and destruction in the same way as does opsonization with antibody. Other inflammatory cells such as eosinophils also bear receptors for complement components; complement-coated pathogens can therefore also activate and recruit these cells into the inflammatory response to infection. Inflammation is also induced by the actions of other small complement fragments produced in the course of the response. Complement deposited on the surface of certain pathogens can also lead directly to the lysis of the coated cells through the assembly of a complex of so-called terminal complement components that makes a hole in the cell membrane and destroys its integrity.

The soluble complement components present in plasma are synthesized by the liver; the enzymatically active components are secreted in inactive forms called **zymogens**. In the course of complement activation, individual components are activated by highly selective proteolytic cleavage, usually at a single site. In the first part of the classical pathway, a series of proteases is activated in this way, each enzyme cleaving and activating the next in the pathway. Cleavage usually produces two fragments, a larger, enzymatically active, fragment and a smaller fragment that often has an inflammatory effect. The soluble complement proteins present in plasma are listed in Figure 7.28 and can be divided into groups on the basis of their functions.

The activation of complement takes place in the blood and in the tissues, into which complement proteins leak from the blood. Of all the complement components, C3 is supremely important and the most abundant. All three pathways of complement activation result in the formation of a **C3 convertase** enzyme that cleaves C3 into C3b and C3a fragments (Figure 7.29). The C3b fragment becomes covalently bound to the pathogen surface. The importance of C3 is also

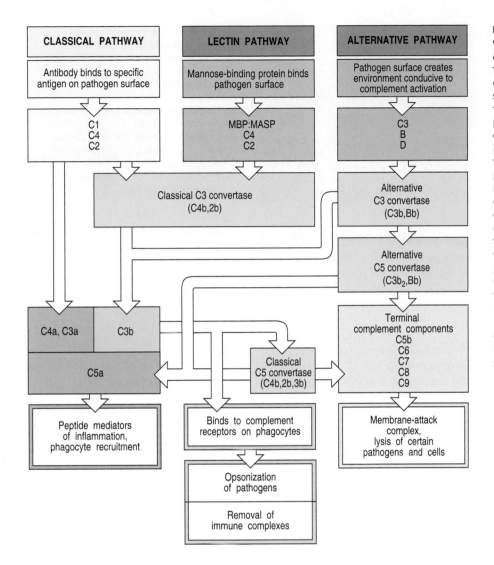

Figure 7.29 Overview of the pathways of complement activation and action. The main difference between the three pathways is the recognition event that initiates activation and the steps that immediately succeed it. The function of the early part of all three pathways is to produce an enzyme called C3 convertase, which cleaves C3 into C3a and C3b fragments. The C3b fragment binds covalently to the pathogen's surface. C3b bound to microbial surfaces can either bind to complement receptors on phagocytic cells, which facilitates the phagocytosis of the pathogen, or can activate the terminal components of complement, which attack the integrity of the pathogen's cell membrane. The smaller C3a fragment, together with similar fragments cleaved from C4 and C5, induces inflammation by recruiting inflammatory cells into the area of complement activation. MBP, mannose-binding protein; MASP, MBP-associated serum protease.

understood from a comparison of the immunodeficiencies suffered by patients who lack particular complement components. Deficiencies of all the complement components have now been described: they vary considerably in their effects on resistance to infection. Most severe is the increased susceptibility to bacterial infections in patients with C3 deficiency.

The following sections deal mainly with the classical pathway of complement, which is activated when the complement component C1 binds to the Fc region of an antibody bound to an antigen. Whenever IgM or IgG binds to antigen, the classical pathway of complement is activated.

7-16 The classical pathway of complement activation tags antigens that have been bound by antibodies

The classical pathway of complement activation is triggered when the complement component C1 binds to the Fc region of an antibody that is part of an antibody:antigen complex. C1 is a complex of three proteins, one of which—C1q—specifically recognizes and binds to the Fc region of the antibody, whereas the other two—C1r and C1s—are inactive proteases. The structure of C1 is dominated by C1q, a large molecule of 18 polypeptides that resembles a bunch of six

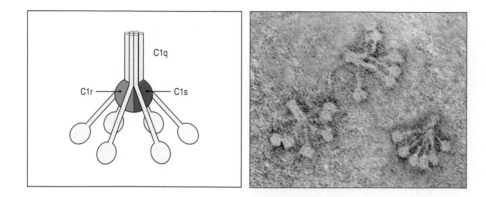
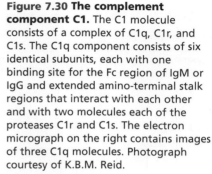

Figure 7.30 The complement component C1. The C1 molecule consists of a complex of C1q, C1r, and C1s. The C1q component consists of six identical subunits, each with one binding site for the Fc region of IgM or IgG and extended amino-terminal stalk regions that interact with each other and with two molecules each of the proteases C1r and C1s. The electron micrograph on the right contains images of three C1q molecules. Photograph courtesy of K.B.M. Reid.

tulips when viewed in the electron microscope (Figure 7.30). Each tulip is composed of three similar polypeptides. The amino-terminal two-thirds of the polypeptides form the stalk, while the carboxy-terminal third of the polypeptides form the globular flower, which contains the binding site for antibody. The stalks can flex so that the antibody-binding sites can move with respect to each other to allow multipoint attachment.

Free antibodies cannot activate complement in solution because of the structural requirements of binding to C1q. The complement cascade is initiated when antibody is bound to multiple sites on a cell surface, normally that of a pathogen. IgM is the isotype that is most efficient at activating complement. The other human isotypes which activate the complement system are IgG1 and IgG3 and, to a lesser extent, IgG2. Pentameric IgM has five Fc regions, each of which can provide a binding site for one of the six binding sites of C1q. Multipoint attachment of C1q

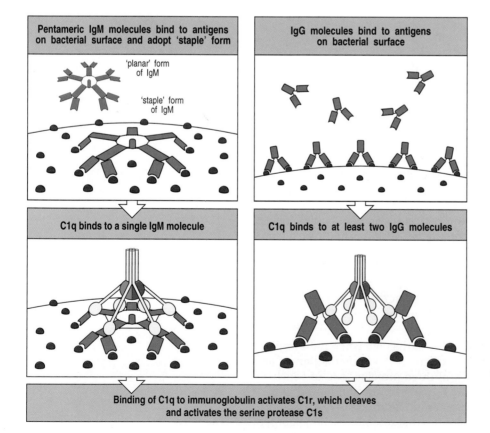

Figure 7.31 The classical pathway of complement activation is initiated by binding of C1q to antibody on a bacterial surface. The binding of C1q to a molecule of pentameric IgM is shown in the left panels. On establishing multipoint binding to bacterial cell-surface antigens, the IgM molecule adopts a less planar conformation, the so-called staple conformation (upper panel). This distortion allows the C1q molecule to establish multipoint attachment to the Fc regions of a single IgM molecule, using the hinges in the C1q stalks to position the globular Fc-binding sites (lower panel). It also exposes binding sites for the C1q heads. The binding of C1q to IgG is shown in the right panels. The C1q molecule needs to find pathogen-bound IgG molecules that are close enough to each other for the C1q molecule to span between them. As a consequence, the activation of complement by IgG depends more on the amount and density of antibodies bound to a pathogen surface than does complement activation by IgM.

to IgM is required for a stable interaction; this can readily be satisfied by a single molecule of each type. IgG can also bind C1q, but it has only one Fc region; consequently, at least two molecules of IgG bound to a microbial surface within 30–40 nm of each other are required to bind one molecule of C1q (Figure 7.31). This difference explains why IgM activates complement much more effectively than does IgG.

C1r and C1s are serine proteases which are activated when C1q binds to an antibody Fc region. Serine proteases are proteolytic enzymes that have a serine residue at the active site. They are typically synthesized in an inactive form and become enzymatically active only after proteolytic cleavage by another protease. On binding to antibody, one molecule of C1r is induced to cleave itself, thereby becoming enzymatically active. It then cleaves and activates the second C1r molecule and both C1s molecules. Activated C1s is the protease that binds, cleaves, and activates the next two components of the classical pathway, the serine proteases C4 and C2.

7-17 The C2 and C4 components of the classical pathway generate a C3 convertase covalently bound to the pathogen surface

C4 is a large globular protein that is encoded and synthesized as a single polypeptide. As part of the intracellular folding and post-translational modification of C4, this single polypeptide is cleaved into three chains called α, β, and γ. A more important event in the biosynthesis is the introduction of a highly reactive thioester bond into the C4 α chain between cysteine and glutamine residues separated by glycine and glutamate residues. Thioesters are high-energy bonds used transiently in enzyme catalysis and are readily hydrolysed by water and other nucleophiles. The thioester bond of C4 is sequestered within the hydrophobic interior of the molecule, where it is protected from the watery milieu of the blood in which the C4 molecule circulates. For a protein to contain a stable thioester bond is unusual, a property described only for the homologous complement components C4 and C3 and the related protease inhibitor of plasma, α_2-macroglobulin.

When a C4 molecule interacts with the activated C1s protease, it is cleaved at a specific site in the α chain, releasing a small fragment of the α chain called C4a. The cleavage also alters the conformation of the cleaved C4 molecule, now called C4b, and exposes the thioester bond to the environment (Figure 7.32). As a result, C4b is released from C1 and then the amino and hydroxyl groups of proteins and carbohydrates compete with water to hydrolyze the thioester by nucleophilic attack. The reaction is so energetically favorable that hydrolysis takes place before C4b has diffused away from C1 bound to antibody on the pathogen surface. The thioester bonds of the vast majority of C4b molecules are hydrolysed by water, but a minority become covalently bonded to the activating C1, to the antibody bound to C1, or to surface molecules of the pathogen bound by the antibody (see Figure 7.32).

As well as cleaving C4, the activated C1s protease also binds and cleaves C2, another inactive serine protease, into an enzymatically active C2b fragment and a smaller C2a fragment. On being released from C1, the C2b fragment forms a complex with a C4b fragment covalently bonded to the pathogen. This C4b,2b complex, called the **C3 convertase of the classical pathway**, is a surface-associated protease whose function is solely to cleave and activate the complement component C3, a reaction central to complement function in the immune response.

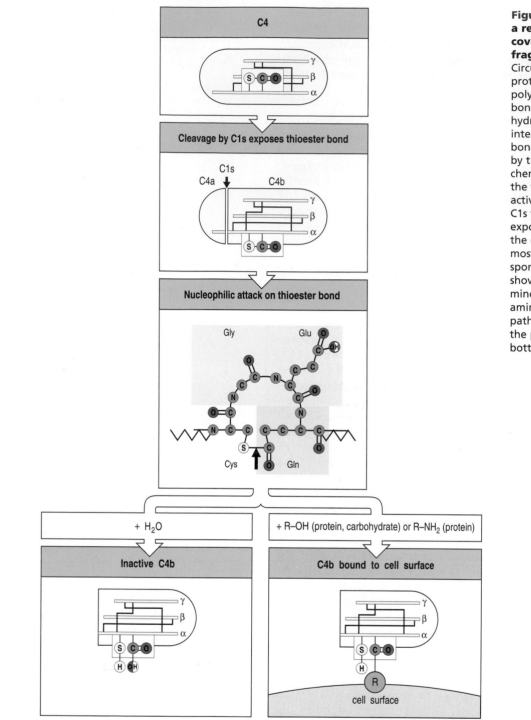

Figure 7.32 Cleavage of C4 exposes a reactive thioester bond that covalently attaches the C4b fragment to the pathogen surface. Circulating C4 is an inactive serine protease consisting of α, β, and γ polypeptide chains in which a thioester bond in the α chain is protected from hydrolysis within the hydrophobic interior of the protein. The thioester bond is denoted in the top two panels by the encircled letters SCO and its chemistry is shown in greater detail in the third panel. The C4 molecule is activated by cleavage of the α chain by C1s to give fragments C4a and C4b. This exposes the thioester bond of C4b to the environment. The thioester bonds of most of the C4b fragments will be spontaneously hydrolyzed by water as shown in the bottom left panel, but a minority will react with hydroxyl and amino groups on molecules on the pathogen's surface, bonding the C4b to the pathogen surface as shown in the bottom right panel.

C3 is structurally very similar to C4 and, like C4, it contains a sequestered thioester bond that becomes activated on cleavage of the molecule. So the reaction that occurs when C3 binds to and is cleaved by C4b,2b is essentially the same as the C1s-mediated cleavage of C4. The α chain of C3 is cleaved to release a smaller C3a fragment and a larger C3b fragment in which the thioester bond is activated, leading to C3b molecules being bonded onto the pathogen's surface around each site where a C4b,2b complex is active (Figure 7.33).

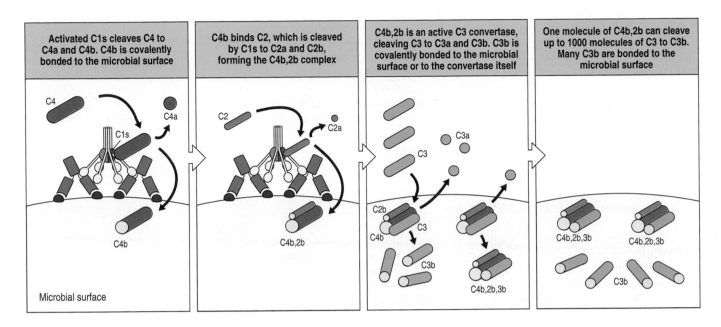

| Activated C1s cleaves C4 to C4a and C4b. C4b is covalently bonded to the microbial surface | C4b binds C2, which is cleaved by C1s to C2a and C2b, forming the C4b,2b complex | C4b,2b is an active C3 convertase, cleaving C3 to C3a and C3b. C3b is covalently bonded to the microbial surface or to the convertase itself | One molecule of C4b,2b can cleave up to 1000 molecules of C3 to C3b. Many C3b are bonded to the microbial surface |

7-18 C3b formation by the classical pathway is amplified by the C3 convertase of the alternative pathway

C3, the most abundant complement protein, is present in plasma at a concentration of 1.2 mg ml^{-1}, far in excess of the concentration of C4. This concentration of substrate ensures that large numbers of pathogen-bonded C3b molecules will be produced by each molecule of pathogen-bonded C4b. The combined actions of C4 and C3 amplify the covalent bonding of complement fragments to the pathogen surface.

Once some C3b is covalently bonded to the pathogen, a second type of C3 convertase can assemble that further amplifies the cleavage and deposition of C3b. Formation of this second C3 convertase involves factors B and D of the alternative pathway of complement activation and thus the second convertase is called the **C3 convertase of the alternative pathway** (Figure 7.34). Factor B binds to C3 bound to the pathogen surface and becomes susceptible to cleavage by complement factor D, another serine protease, which cleaves off a small Ba fragment

Figure 7.33 The classical pathway of complement activation generates a C3 convertase that deposits C3b molecules covalently bonded to the surface of the pathogen. The steps in the reaction are outlined here and detailed in the text.

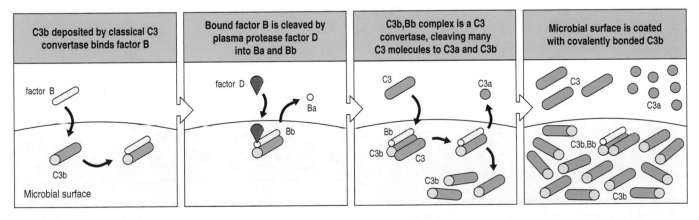

| C3b deposited by classical C3 convertase binds factor B | Bound factor B is cleaved by plasma protease factor D into Ba and Bb | C3b,Bb complex is a C3 convertase, cleaving many C3 molecules to C3a and C3b | Microbial surface is coated with covalently bonded C3b |

Figure 7.34 The alternative pathway of complement amplifies the deposition of C3b on the pathogen achieved by the classical pathway. The C3 convertase of the alternative pathway is assembled from C3b and the active Bb fragment of

factor B. Because C3b, the product of the alternative C3 convertase, actually makes more convertase, it means that the alternative convertase is inherently more active than the classical C3 convertase in depositing C3b on pathogen surfaces.

leaving the larger, and now proteolytically active, Bb fragment associated with C3b. The C3b,Bb complex is homologous in both structure and function to C4b,2b. It cleaves C3 molecules, exposing their thioester bonds to attack by water, plasma proteins and more importantly components of the microbial surface.

The cleavage and activation of C3 by C3b,Bb is an accelerating reaction with positive feedback, because the product of the reaction itself forms more enzyme. This provides dramatic amplification of the reaction started by C4b,2b and would rapidly consume the supply of C3 were there not control mechanisms that dampen the response, as we shall see later. By the end of the reaction, most of the C3b fragments that now saturate the surface of the pathogen close to the initiating antigen:antibody complex are due to the action of the C3 convertase of the alternative pathway (see Figure 7.34).

Complement factor B is related to C2 and these two serine proteases are encoded by closely linked genes in the same region of the major histocompatibility complex that contains the C4 genes. Here we have seen how activation of the classical pathway by antibody and antigen leads to formation of the C3 convertase of the alternative pathway. This enzyme is also formed as a result of activation of the alternative pathway, which is triggered by the surface components of certain bacteria and other pathogens and is not dependent on the presence of antibodies.

7-19 Complement components binding to specific receptors on phagocytes facilitate the uptake and destruction of pathogens

Macrophages and neutrophils express a cell-surface receptor that binds to C3b and is called **complement receptor 1** (**CR1**). The covalent attachment of C3b molecules to the outside of a pathogen facilitates its uptake by phagocytes through this receptor; this is the most important function of the complement system. The interaction of C3b with CR1 cannot by itself stimulate phagocytosis and the respiratory burst, but it enhances these functions once they have been initiated by the binding of IgG to an Fcγ receptor or by the T-cell-derived cytokine IFN-γ. In this context, C3b is acting as an opsonin (Figure 7.35). CR1 also binds C4b, but this interaction is of less functional significance because so much more C3b than C4b is deposited on pathogen surfaces.

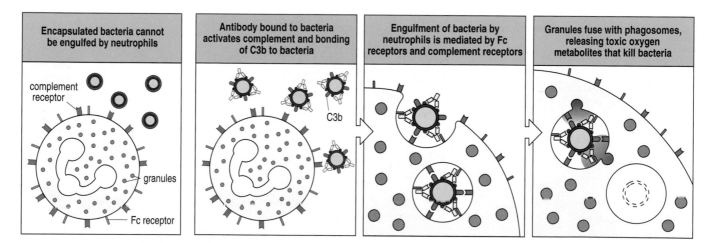

Figure 7.35 Encapsulated bacteria are more efficiently engulfed by phagocytes when the bacteria are coated with antibody and C3b. Encapsulated bacteria are naturally resistant to uptake by phagocytes, here represented by a neutrophil. When such bacteria are coated with antibody and C3b they become susceptible to phagocytosis mediated by Fc receptors and C3b receptors. Fc receptors bind IgG, whereas complement receptors bind C3b, inducing efficient phagocytosis and also the activation of the neutrophil. When the neutrophil is activated, its granules fuse with the phagosome containing the bacteria, releasing bactericidal metabolites. Macrophages also phagocytose and kill encapsulated bacteria in the same manner.

Receptor	Ligand	Functions	Cell types
CR1	C3b, C4b	Promotes C3b and C4b decay Stimulates phagocytosis Erythrocyte transport of immune complexes	Erythrocytes, macrophages, monocytes, polymorphonuclear leukocytes, B cells, FDCs
CR2 (CD21)	C3d, C3dg, iC3b Epstein– Barr virus	Part of B-cell co-receptor Epstein–Barr virus receptor	B cells, FDCs
CR3 (CD11b/ CD18)	iC3b	Stimulates phagocytosis	Macrophages, monocytes, polymorphonuclear leukocytes, FDCs
CR4 (gp150, 95) (CD11c/ CD18)	iC3b	Stimulates phagocytosis	Macrophages, monocytes, polymorphonuclear leukocytes
C1q receptor	C1q stalk	Binding of immune complexes to phagocytes	B cells, macrophages, monocytes, platelets, endothelial cells

Figure 7.36 Distribution and function of receptors for complement proteins. There are several different complement receptors, which are specific for different complement components or their fragments. CR1 and CR3 are especially important for causing the phagocytosis of complement-coated bacteria. CR1 on erythrocytes clears immune complexes from the circulation, and CR2 is mainly present on B cells, where it is also part of the B-cell co-receptor complex. C3d, C3dg and iC3b are cleavage products of C3b. FDC, follicular dendritic cell.

While covalently bound to the pathogen surface, C3b can undergo further cleavage to produce the fragments iC3b, C3d, or C3dg. The CR1 receptor cannot recognize them, but three other types of complement receptor can (Figure 7.36). **Complement receptor 2** (**CR2**, also known as CD21) is a component of the B-cell co-receptor and binds iC3b, C3d, and C3dg. As we saw in Section 7-1, whereas the B-cell receptor interacts with an antigen on the pathogen's surface, the CR2 component of the co-receptor can simultaneously engage a C3b product covalently bound nearby. In this way the B-cell co-receptor and B-cell receptor are simultaneously crosslinked so that their signals synergize. CR2 is also expressed on the follicular dendritic cells in lymphoid follicles. This enables them to bind antigens that have been tagged with C3b or its products (see Figure 7.12) and to retain them for long-term stimulation of B cells. The Epstein–Barr virus (EBV), which causes infectious mononucleosis and certain lymphomas, also binds to CR2 and exploits this interaction to infect B cells. In this context, CR2 is known as the EBV receptor of human B cells.

Complement receptors 3 and **4** (**CR3** and **CR4**) bind iC3b and are expressed on phagocytes (macrophages and neutrophils), where they augment the activities of Fc receptors and CR1 in activating phagocytosis. Unlike the binding of C3b to CR1, the binding of iC3b to CR3 is sufficient in itself to stimulate phagocytosis.

7-20 Complement receptors remove immune complexes from the circulation

In the previous sections we have been principally concerned with the binding of antibody and complement to pathogenic microorganisms, which are large particles. High-affinity antibodies also bind to soluble protein antigens such as bacterial toxins, forming complexes that cannot be engulfed by phagocytes because they contain too few molecules of IgG to form a stable interaction with Fcγ receptors. Such soluble immune complexes are present in the circulation after the immune response to most infections and they are removed through the action of complement. The number of IgG molecules in an immune complex is sufficient to bind C1 and activate the enzymes that cleave first C4 and then C3, so

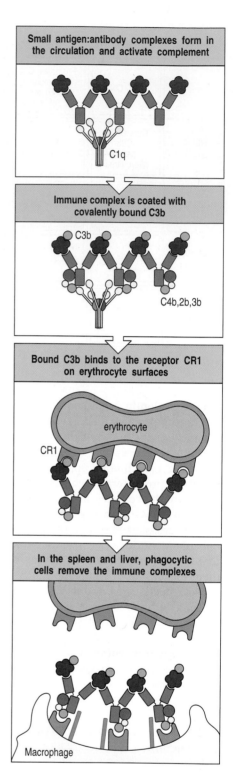

Figure 7.37 Erythrocyte CR1 helps to clear immune complexes from the circulation. Immune complexes bind to CR1 on erythrocytes, which transport them to the liver and spleen. Here they are removed by macrophages expressing receptors for Fc regions and for bound complement components.

that the antigen and antibody molecules within the complex become covalently tagged with C4b and C3b. Having been tagged in this way, the complex can now be bound by circulating cells that express CR1. Of these, the most numerous is the erythrocyte, and the vast majority of immune complexes become bound to the surface of red blood cells. During their circulation in the blood, erythrocytes pass through areas of the liver and the spleen where tissue macrophages remove and degrade the complexes of complement, antibody, and antigen from the erythrocyte surface while leaving the erythrocyte unscathed (Figure 7.37).

If immune complexes are not removed, they have a tendency to enlarge by aggregation and to precipitate at the basement membrane of small blood vessels, most notably those of the kidney glomeruli, where blood is filtered to form urine and is under particularly high pressure. Immune complexes that pass through the basement membrane bind to CR1 receptors expressed by podocytes, specialized epithelial cells that cover the capillaries. Deposition of immune complexes within the kidney probably occurs at some level all the time, and mesangial cells within the glomerulus are specialized in the elimination of immune complexes and in stimulating the repair of the tissue damage they cause.

A feature of the autoimmune disease **systemic lupus erythematosus (SLE)** is a level of immune complexes in the blood sufficient to cause massive deposition of antigen, antibody, and complement on the renal podocytes. These deposits damage the glomeruli, and kidney failure is the principal danger for patients with this disease. A similar deposition of immune complexes can also be the major problem for patients who have inherited deficiencies in the early components of the complement pathway and cannot tag their immune complexes with C4b or C3b. Such patients cannot clear immune complexes; these accumulate with successive antibody responses to infection and inflict increasing damage on the kidneys.

7-21 The terminal complement proteins lyse pathogens by forming a membrane pore

As we have seen, the most important product of complement activation is C3b bonded to pathogen surfaces and soluble antigens. However, the cascade of complement reactions does extend beyond this stage. C3b can bind to either of the C3 convertases to produce enzymes that act on the C5 component of complement and are called **C5 convertases**. The C5 convertase of the classical pathway consists of C4b, C2b, and C3b and is designated C4b,2b,3b, whereas the C5 convertase of the alternative pathway consists of Bb plus two C3b fragments and is designated C3b$_2$,Bb (Figure 7.38).

Complement component C5 is structurally similar to C3 and C4 but lacks the thioester bond and has a different function. It is cleaved by one or other convertase into a smaller C5a fragment and a larger C5b fragment. The function of C5b is to initiate the formation of a **membrane-attack complex**, which can make holes in the membranes of bacterial pathogens and eukaryotic cells. In succession, C6 and C7 bind to C5b—interactions that expose a hydrophobic site in C7

Figure 7.38 Complement component C5 is activated by a C5 convertase. C5 convertases are formed when C3b binds to either of the two C3 convertases, as shown in the top panel. The C5 convertase of the classical pathway (left) consists of a complex of C3b, C4b, and C2b, whereas the C5 convertase of the alternative pathway (right) consists of two molecules of C3b and one of Bb. In the center panel, C5 is shown binding to the C3b component of the convertase enzymes. As shown in the bottom panel, the bound C5 is cleaved into fragments C5a and C5b, of which C5b initiates the assembly of the terminal complement components to form the membrane-attack complex.

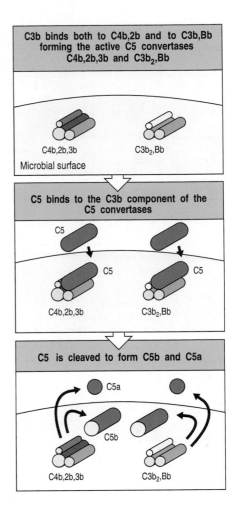

which inserts into the lipid bilayer. Complement factor C8 consists of two associated proteins—C8β, which binds to C5b, and C8α–γ, which has a hydrophobic site that becomes exposed and inserted into the membrane. The inserted C8α–γ component then initiates polymerization of the C9 complement component, which forms the transmembrane pores. C9 is structurally similar to perforin, the protein in the lytic granules of cytotoxic T cells, but forms pores of about 100 Å diameter compared to the 160 Å pores formed by perforin (Figure 7.39).

Although the effect of the membrane-attack complex on membranes is dramatic, the contribution of the C5–C9 components to host defense seems at present to be quite limited. The only clear effect of deficiency in any of these components is to increase susceptibility to bacteria of the genus *Neisseria*, different species of which cause the sexually transmitted disease gonorrhea and a common form of bacterial meningitis.

7-22 Small peptides released during complement activation induce local inflammation

During complement activation, C3, C4, and C5 are each cleaved into two fragments; we have seen how the larger fragments (C3b, C4b, C5b) continue the pathway of complement activation. The smaller C3a, C4a, and C5a fragments produce inflammation at the site of complement activation by binding to receptors on a number of cell types. In some circumstances, these small polypeptides can induce anaphylaxis, an acute systemic inflammatory response, and they are therefore referred to as **anaphylatoxins**. Of the anaphylatoxins, C5a is the most stable and the most potent, followed by C3a and then C4a. All three induce smooth muscle contraction and the degranulation of mast cells and basophils, with the consequent release of histamine and other vasoactive substances that increase capillary permeability. They also have vasoactive effects directly on local blood vessels, increasing blood flow and vascular permeability. These changes make it easier for antibody, complement proteins, phagocytic cells, and lymphocytes to pass out of the blood into the site of an infection (Figure 7.40). Meanwhile, the increased fluid in the tissues hastens the passage of pathogen-containing antigen-presenting cells to the draining lymph nodes and the initiation of B and T lymphocyte responses.

C5a also acts directly on neutrophils and monocytes to increase their adherence to vessel walls and, as a chemoattractant, to direct their migration towards sites of antigen deposition. It also increases their capacity for phagocytosis, as well as raising the expression of CR1 and CR3 on their surfaces. In these ways, the anaphylatoxins act in concert with other complement components to speed the destruction of pathogens by phagocytes.

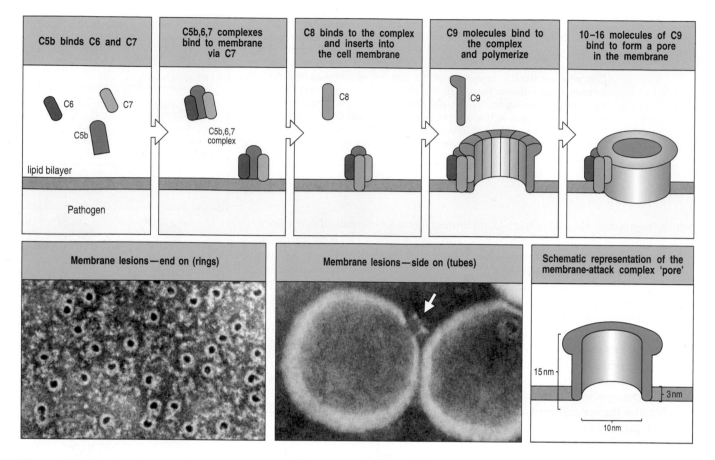

Figure 7.39 The membrane-attack complex assembles to generate a pore in the lipid bilayer membrane. The sequence of steps and their approximate appearance is shown here in schematic form. C5b is generated by the cleavage of C5 by both the classical C5 convertase C4b,2b,3b and the alternative C5 convertase C3b₂,Bb. C5b then forms a complex by the successive binding of one molecule each of C6, C7, and C8. In forming the complex, C7 and C8 undergo a conformational change that exposes hydrophobic sites, which insert into the membrane. This complex causes some membrane damage and also induces the polymerization of C9. As each molecule of C9 is added to the polymer, it exposes a hydrophobic site and inserts into the membrane. Up to 16 molecules of C9 can be added to generate a transmembrane channel of 100 Å diameter. The channel disrupts the bacterial outer membrane, killing the bacterium. In the laboratory, the erythrocyte is a convenient cell with which to measure complement-mediated lysis. The electron micrographs show erythrocyte membranes with membrane-attack complexes in two orientations, end on and side on. Photographs courtesy of S. Bhakdi and J. Tranum-Jensen.

7-23 Regulatory proteins limit the extent of complement activation and its effect on cells of the body

The positive feedback loop by which existing C3 convertases produce more C3 convertase gives the system the tendency to run out of control. To prevent this from happening, there are regulatory proteins that dampen the response by disrupting enzymes that act at key stages in the activation pathways. The supply of C4b and C2b components of the classical C3 convertase is shut down by the plasma protein **C1 inhibitor** (**C1INH**), which binds to activated C1r:C1s, causing them to dissociate from C1q (Figure 7.41). In this way, C1INH limits the time that any C1 molecule is active. The inhibitor also controls the spontaneous activation of C1 in the plasma, which occurs constantly at a low rate independently of any immune response. Patients who lack C1INH suffer from a disease called **hereditary angioneurotic edema** (**HANE**) in which chronic spontaneous complement activation leads to excessive amounts of cleaved fragments of C4 and C2. The C2a fragment is further cleaved into a peptide called the C2 kinin, which causes

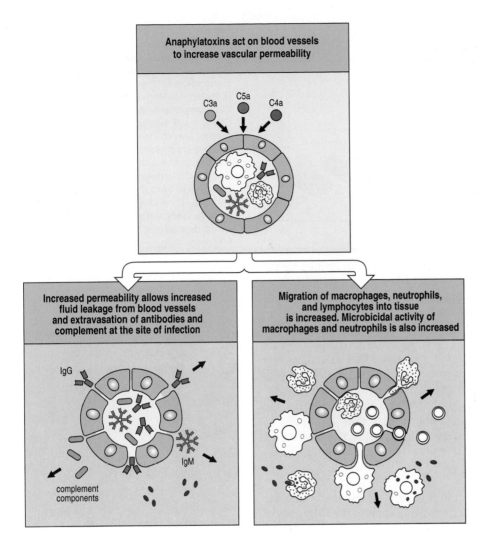

Figure 7.40 Local inflammatory responses can be induced by the small complement fragments C3a, C4a, and especially C5a. These small anaphylatoxic peptides are produced by complement cleavage at the site of infection and cause local inflammatory responses by acting on local blood vessels. They cause increased blood flow, increased binding of phagocytes to endothelial cells, and increased vascular permeability, leading to the accumulation of fluid, plasma proteins, and cells in the local tissues. The antibodies, complement, and cells recruited by this inflammatory stimulus remove pathogen by enhancing the activity of phagocytes, which are themselves also directly stimulated by the anaphylatoxins.

fluid-induced swelling or **edema**. If this occurs in the vicinity of the trachea it can lead to suffocation. The peptide bradykinin, which has similar effects to those of C2 kinin, is also present in unusually high amounts in this disease, because C1INH normally inhibits another plasma protease involved in bradykinin production. The symptoms of hereditary angioneurotic edema can be cured simply by replacement therapy with C1INH.

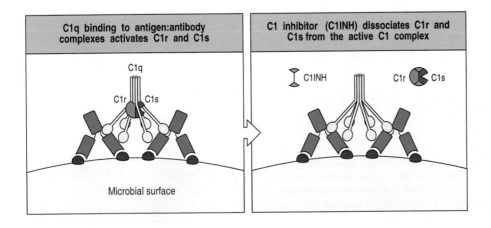

Figure 7.41 C1INH inhibits the first stages of the classical pathway of complement activation.

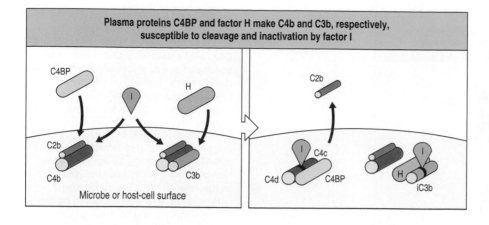

Figure 7.42 C4-binding protein and factor H inactivate classical C3 and C5 convertases respectively. The C4-binding protein (C4BP) binds the C4b component of the classical C3 convertase (C4b,2b), causing displacement of the C2b component. When bound to C4BP, C4b is made susceptible to attack by factor I, which cleaves it into the inactive fragments C4c and C4d. Factor H binds to the C3b component of the classical C5 convertase (C4b,2b,3b) and displaces it from the complex with C4b and C2b. When bound to factor H, C3b is made susceptible to attack by factor I, which degrades it to the inactive fragment iC3b.

The production and duration of classical C3 convertase are also limited by the plasma protein **C4-binding protein** (**C4BP**), which on binding to C4b displaces the C2b component of the convertase and renders the C4b component susceptible to degradation by the plasma protease **factor I**. The functions of C3b are similarly regulated by the plasma protein **factor H**, which on binding to C3b makes it susceptible to inactivation by factor I, with the formation of iC3b (Figure 7.42). Factor I deficiency in humans leads to the depletion of C3, as the convertase activity of the C3b,Bb complex runs away unchecked. This deficiency also prevents the formation of iC3b, which is the ligand for the complement receptor CR3, through which neutrophils and macrophages can be activated in the absence of antibody (see Figure 7.36). Fragment iC3b is thus important in opsonizing bacteria during innate immunity, before antibodies are made. Patients with factor I deficiency are more susceptible than usual to ear infections and abscesses caused by pyogenic bacteria.

An important aspect of complement regulation is to prevent the deadly effector function recruited by activated complement from being aimed at healthy cells of the body. As well as the protection afforded by regulatory proteins in the plasma, human cells express cell-surface molecules that inactivate any C4b and C3b fragments that become covalently bonded to their surfaces. These proteins have two types of activity: one is to dissociate any C3 convertases that become assembled on these fragments; the other is to facilitate their inactivation by proteolytic degradation.

A regulatory protein of the first kind is **decay-accelerating factor** (**DAF**), which binds to C4b and C3b components of C3 convertases, causing their dissociation and inactivation. A regulatory protein of the second kind is **membrane co-factor protein** (**MCP**), which by binding to C3b and C4b makes them susceptible to cleavage and inactivation by factor I. This function is similar to that of the soluble complement regulator, factor H. Indeed, factor H can itself become membrane-associated on any type of human cell because it has a binding site for sialic acid,

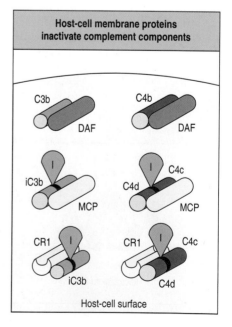

Host-cell membrane proteins inactivate complement components

Host-cell surface

Figure 7.43 Regulatory proteins on human cells protect them from complement-mediated attack. Decay-accelerating factor (DAF) binds to C3b and C4b, causing the dissociation of existing C3 convertases and preventing the assembly of new ones. CR1 and MCP bind to C3b and C4b, making them susceptible to proteolytic cleavage by factor I. DAF, CR1, MCP, factor H, C4-binding protein (C4BP), and other regulatory proteins that bind to C3b and C4b contain one or more copies of a structural element called the complement control protein (CCP) element. This element is thought to provide the structural basis for the regulatory interactions.

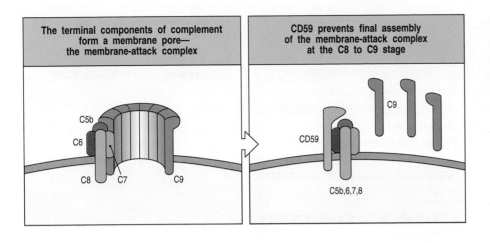

Figure 7.44 CD59 prevents assembly of terminal complement components into a membrane pore. By binding to the C5b,6,7,8 complex, CD59 prevents the polymerization of C9 in the membrane to form a pore.

a component of the oligosaccharides of human glycoproteins and glycolipids that is absent from most bacteria. In human cells that express CR1, this receptor can disrupt C3b fragments deposited on the cells' own surfaces and also engage the C3b deposited on a pathogen. In addition, CR1 can also engage C4b and disrupt its function. CR1 acts like factor H and MCP, making C3b and C4b susceptible to cleavage by factor I (Figure 7.43).

The activity of the terminal complement components on human cells is also regulated by cell-surface proteins: the best known is CD59 (also called protectin) (Figure 7.44). Both CD59 and DAF are linked to the plasma membrane by a phosphoinositol glycolipid tail. The disease **paroxysmal nocturnal hemoglobinuria**, which is characterized by episodes of intravascular red blood cell lysis by complement, is most often caused by deficiencies of both DAF and CD59 in patients who fail to generate the phosphoinositol glycolipid linkage.

Summary

The complement system is a major defense mechanism against microbial infection, particularly infection by extracellular bacteria. The complement system responds to the presence of infection by depositing covalently bonded C3b fragments on bacterial surfaces. These C3b fragments are recognized by specific receptors of phagocytic cells that facilitate the engulfment and breakdown of bacteria and other pathogens that have been so tagged by complement activation. The complement system can be activated at all stages of an immune response and by different mechanisms for recognizing antigen. At the beginning of an infection, complement activation relies on the mannose-binding protein of plasma, which binds to mannose-containing carbohydrates on bacteria, and the spontaneous activation of C3. Later in the immune response, after lymphocytes have been activated, antibodies specific for the infecting organism provide a more directed method of targeting complement to the source of infection. In this role, complement provides the major effector function for the IgM and IgG antibody isotypes. Although simple in principle, the complement system has been refined by the evolution of many different complement proteins that increase the effectiveness of the basic reaction of tagging antigens with C3b (Figure 7.45).

Complement activation is a kinetic process in which stable circulating plasma proteins are turned into enzymes by highly specific proteolytic cleavages. In these cleavage reactions, almost all of the fragments produced have activities that either contribute directly to complement activation or to the recruitment of phagocytes and other inflammatory cells to the local site of infection. The activated product from one step of complement activation serves as the activator for

Proteins of the classical pathway of complement activation		
Component in serum	Active form	Function
C1 (C1q: C1r$_2$:C1s$_2$)	C1q	Binds to antibody that has bound antigen: activates C1r
	C1r	Cleaves C1s to activate its protease
	C1s	Cleaves C4 and C2
C4	C4b	Covalently binds to pathogen and acts as opsonin. Binds C2, which is then cleaved by C1s
	C4a	Peptide mediator of inflammation (weak)
C2	C2b	Active enzyme of classical pathway C3 and C5 convertases: cleaves C3 and C5
	C2a	Precursor of vasoactive C2 kinin
C3	C3b	Binds covalently to pathogen surface and acts as opsonin. Initiates amplification via the alternative pathway. Binds C5, which is then cleaved by C2b
	C3a	Peptide mediator of inflammation (moderate)
Proteins of the alternative pathway of complement activation		
Component in serum	Active fragments	Function
C3	C3b	Binds covalently to pathogen surface and acts as opsonin. Binds B, which is then cleaved by D. C3b,Bb is C3 convertase and C3b$_2$,Bb is C5 convertase
Factor B (B)	Ba	Small fragment of B, no known function
	Bb	Active enzyme of the C3 convertase C3b,Bb and C5 convertase C3b$_2$,Bb
Factor D (D)	D	B cleaves to Ba and Bb when it is bound to C3b

Figure 7.45 Summary of the names and functions of the complement proteins of the classical pathway and the alternative pathway. The color-coding is as in Figure 7.28.

the next step, a strategy that provides for amplification of the response so that a single antigen:antibody interaction can trigger the deposition of thousands of C3b fragments. Although accelerated amplification of complement activation is valuable in developing a rapid and forceful attack on a pathogen, its inherent tendency to run out of control has driven the evolution of regulatory molecules that can dampen the response and divert complement deposition away from healthy human cells.

Summary to Chapter 7

The response of B lymphocytes to infection is the secretion of antibodies. These molecular adaptors bind and link the pathogen to effector molecules or cells that will destroy it. In developing an antibody response, the population of responding B cells can combine a quick but less than optimal response in the short term with a more effective response that takes time to develop. B-1 cells, which do not require cognate helper T cells, and IgM molecules, which are not dependent on high-affinity binding sites, both represent short-term strategies. In contrast, B-cell activation driven by cognate T-cell help, with resultant isotype switching and somatic hypermutation, is the long-term strategy that provides effective protection from subsequent infection by the pathogen. One function of the isotypic diversification of immunoglobulins is to provide antibody responses in different compartments of the human body. IgM, IgG, and monomeric IgA work in the blood, lymph, and connective tissues, providing antibody responses to infections within the body's tissues. In contrast, dimeric IgA is transported to the luminal side of the gut wall and other mucosal surfaces, where it provides antibody responses against the microorganisms that colonize these surfaces. A second function of antibody isotype is to recruit different effector functions into the immune response: receptors for IgG on neutrophils and macrophages deliver pathogens for phagocytosis, whereas the high-affinity receptor for IgE on mast cells, neutrophils, and eosinophils ensures that antigens binding IgE provoke an inflammatory response. Proteins of the complement system work synergistically with IgM and IgG to modify pathogens and make them more susceptible to phagocytosis. Activation of the complement system covers the pathogen surface with covalently bonded complement components—C3b and C4b—which are bound by specific receptors on phagocytic cells, facilitating the uptake and destruction of the pathogen. Other complement fragments produced in the course of activation help induce an inflammatory response at the site of infection. The activation of phagocytes and inflammatory reactions by complement provides protection both before and after the antibody response develops. The defense mechanism of complement-mediated phagocytosis of pathogens evolved long before the existence of antibodies; in molecular history it was the antibodies that actually provided the complementary function, rather than the complement.

The Body's Defenses Against Infection

8

Most infectious diseases suffered by the human species are caused by pathogens smaller than a single human cell. For these agents, the human body constitutes a vast resource-rich environment in which to live and reproduce. In the face of such threats, the body deploys multifarious defense mechanisms that have accumulated over hundreds of millions of years of invertebrate and vertebrate evolution. In Chapter 1 we took a broad look at the nature of infections and of the body's defenses against them. We saw how these defenses fall into three categories: physical barriers that prevent pathogens from entering the body's tissues; the fixed or 'hard-wired' mechanisms of **innate immunity**, which attack an infection from its very beginning; and the mechanisms of the **adaptive immune response**, which respond to the infection in a very specific fashion. Much of medical practice is concerned with the diseases that result from the small proportion of infections that innate immunity fails to terminate, and in which the spread of the pathogen to lymphoid tissues stimulates an adaptive immune response. In such situations the attending physicians and the adaptive immune response work together to effect a cure, a partnership that has historically favored the scientific investigation of adaptive immunity over innate immunity.

The adaptive immune response is the component of the immune defenses that is enhanced by vaccination and provides long-term protection against many infectious diseases. Chapters 2–7 of this book focus on the cellular and molecular basis of the adaptive immune response. They describe how lymphocytes recognize pathogens through antigen-specific receptors, how B and T cells develop, and how they fulfill their effector functions. In this chapter we see how, in practice, the adaptive immune response cooperates with innate immunity to fight infections. In the first part of the chapter we examine the mechanisms of innate immunity and why they are an essential prerequisite for any adaptive immune response. The second part shows how the various arms of the immune system cooperate in producing an adaptive immune response. In the last part of the chapter we see how a primary adaptive immune response produces immunological memory—**immunity**—which lessens the impact of subsequent encounters with the same pathogen.

Innate immunity

Although we become conscious of infectious agents only when suffering from the diseases they cause, microorganisms are always with us. Fortunately, the vast majority of microorganisms that we come into contact with are prevented from ever causing an infection by barriers at the body's surface. Furthermore, almost all infectious agents that penetrate those barriers and start an infection are eliminated quickly by the innate immune response before causing any obvious symptoms of disease. The physical barriers and the mechanisms of innate immunity can be considered as fixed defenses, completely determined by genes inherited from our parents and ready for action at all times. This part of the chapter introduces the agents that cause disease and describes how the fixed defenses cope with infections by pathogens to which the immune system has had no previous exposure.

The proportion of infections that are successfully eliminated by innate immunity is difficult to assess, mainly because such infections are overcome before they have caused symptoms severe enough to command the attention of those infected or their physicians. Intuitively, it seems likely to be a high proportion, especially in light of the human body's capacity to sustain vast populations of resident microorganisms without them causing symptoms of disease. The importance of innate immunity is also implied by the rarity of inherited deficiencies in non-adaptive immune mechanisms, and the considerable impairment of protection when they do occur (Figure 8.1).

8-1 Infectious diseases are caused by pathogens of diverse types that live and replicate in the human body

Human pathogens are of four kinds: viruses, bacteria, fungi, and parasites. The first three categories each correspond to a single taxonomic group of microorganisms, whereas the term 'parasites' refers to various disease-causing eukaryotic organisms that are mostly unicellular protozoa and multicellular worms (Figure 8.2). Pathogens are diverse in their structure, in the manner in which they exploit

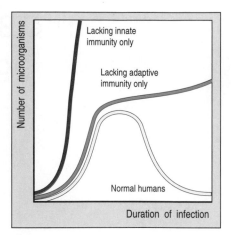

Figure 8.1 The time course of infections in immunocompetent and immunodeficient individuals. An infection is initially contained by innate immunity and then cleared from the body by the adaptive immune response. In a person who lacks innate immunity, uncontrolled infection occurs, as the adaptive immune response cannot be deployed without the preceding innate response. In a person who lacks adaptive immune responses, the infection is initially contained by innate immunity but cannot be cleared from the body.

Some common causes of disease in humans			
Viruses	DNA viruses	Adenoviruses	Human adenoviruses (eg types 3, 4, and 7)
		Herpesviruses	Herpes simplex, varicella-zoster, Epstein–Barr virus, cytomegalovirus
		Poxviruses	Vaccinia virus
		Parvoviruses	Human parvovirus
		Papovaviruses	Papilloma virus
		Hepadnaviruses	Hepatitis B virus
	RNA viruses	Orthomyxoviruses	Influenza virus
		Paramyxoviruses	Mumps, measles, respiratory syncytial virus
		Coronaviruses	Some common cold viruses
		Picornaviruses	Polio, coxsackie, hepatitis A, rhinovirus
		Reoviruses	Rotavirus, reovirus
		Togaviruses	Rubella, arthropod-borne encephalitis
		Flaviviruses	Arthropod-borne viruses, (yellow fever, dengue fever)
		Arenaviruses	Lymphocytic choriomeningitis, Lassa fever
		Rhabdoviruses	Rabies
		Retroviruses	Human T-cell leukemia virus, HIV
Bacteria	Gram +ve cocci	Staphylococci	*Staphylococcus aureus*
		Streptococci	*Streptococcus pneumoniae, S. pyogenes*
	Gram –ve cocci	Neisseriae	*Neisseria gonorrhoeae, N. meningitidis*
	Gram +ve bacilli		*Corynebacterium, Bacillus anthracis, Listeria monocytogenes*
	Gram –ve bacilli		*Salmonella, Shigella, Campylobacter, Vibrio, Yersinia, Pasteurella, Pseudomonas, Brucella, Haemophilus, Legionella, Bordetella*
	Anaerobic bacteria	Clostridia	*Clostridium tetani, C. botulinum, C. perfringens*
	Spirochetes		*Treponema pallidum, Borrelia burgdorferi, Leptospira interrogans*
	Mycobacteria		*Mycobacterium tuberculosis, M. leprae, M. avium*
	Rickettsias		*Rickettsia prowazekii*
	Chlamydias		*Chlamydia trachomatis*
	Mycoplasmas		*Mycoplasma pneumoniae*
Fungi			*Candida albicans, Cryptococcus neoformans, Aspergillus, Histoplasma capsulatum, Coccidioides immitis, Pneumocystis carinii*
Parasites	Protozoa		*Entamoeba histolytica, Giardia, Leishmania, Plasmodium, Trypanosoma, Toxoplasma gondii, Cryptosporidium*
	Worms	Intestinal	*Trichuris trichura, Trichinella spiralis, Enterobius vermicularis, Ascaris lumbricoides, Ancylostoma, Strongyloides*
		Tissues	*Filaria, Onchocerca volvulus, Loa loa, Dracuncula medinensis*
		Blood, liver	*Schistosoma, Clonorchis sinensis*

Figure 8.2 Diverse microorganisms cause human disease. Pathogenic organisms are of four main types— viruses, bacteria, fungi, and parasites, which are mostly protozoa or worms. Some pathogens in each category are listed in the column on the right.

	Direct mechanisms of tissue damage by pathogens			Indirect mechanisms of tissue damage by pathogens		
	Exotoxin release	Endotoxin release	Direct cytopathic effect	Immune complexes	Anti-host antibody	Cell-mediated immunity
Pathogenic mechanism						
Infectious agent	*Streptococcus pyogenes* *Staphylococcus aureus* *Corynebacterium diphtheriae* *Clostridium tetani* *Vibrio cholerae*	*Escherichia coli* *Haemophilus influenzae* *Salmonella typhi* *Shigella* *Pseudomonas aeruginosa* *Yersinia pestis*	Variola Varicella-zoster Hepatitis B virus Polio virus Measles virus Influenza virus Herpes simplex virus	Hepatitis B virus Malaria *Streptococcus pyogenes* *Treponema pallidum* Most acute infections	*Streptococcus pyogenes* *Mycoplasma pneumoniae*	*Mycobacterium tuberculosis* *Mycobacterium leprae* Lymphocytic choriomeningitis virus *Borrelia burgdorferi* *Schistosoma mansoni* Herpes simplex virus
Disease	Tonsilitis, scarlet fever Boils, toxic shock syndrome, food poisoning Diphtheria Tetanus Cholera	Gram-negative sepsis Meningitis, pneumonia Typhoid Bacillary dysentery Wound infection Plague	Smallpox Chickenpox, shingles Hepatitis Poliomyelitis Measles, subacute sclerosing panencephalitis Influenza Cold sores	Kidney disease Vascular deposits Glomerulonephritis Kidney damage in secondary syphilis Transient renal deposits	Rheumatic fever Hemolytic anemia	Tuberculosis Tuberculoid leprosy Aseptic meningitis Lyme arthritis Schistosomiasis Herpes stromal keratitis

Figure 8.3 Pathogens damage tissues in different ways. Representative pathogens and the diseases they cause are grouped according to the type of damage they cause. Pathogens can directly kill cells and damage tissues in three ways. Exotoxins released by microorganisms act at the surfaces of host cells, usually via a cell-surface receptor (panel 1). When phagocytes degrade certain microorganisms, endotoxins are released that induce the phagocytes to secrete cytokines, causing local or systemic symptoms (panel 2). Cells infected by pathogens are usually killed or damaged in the process (panel 3). Tissue damage can also be caused in an indirect fashion. Antibodies produced by B cells in response to pathogens can form large antigen:antibody complexes that damage blood vessels and kidneys (panel 4). Antibodies against a pathogen can react with a human cell, causing it to be treated by the immune system as though it were a foreign invader (panel 5). In clearing pathogens from an infected site, inflammatory cells of the immune system damage tissue and leave debris (panel 6).

the human body, and in the type of damage that they cause (Figure 8.3). For purposes of host defense, a fundamental distinction is between pathogens that replicate in the spaces between human cells to produce extracellular infections and pathogens that replicate inside human cells to produce intracellular infections. The latter category can be further subdivided. Viruses and some bacteria (certain chlamydias and rickettsias as well as *Listeria*) replicate within the cytosol and/or nucleus, whereas mycobacteria and the protozoan parasite *Leishmania*, for example, replicate in membrane-bounded vesicles derived from endosomes (Figure 8.4).

These different sites of replication have implications for the types of immune mechanism that can be used to eliminate the pathogen. Extracellular forms of pathogens are accessible to soluble molecules of the immune system, whereas intracellular forms are not. Those intracellular pathogens that live in the nucleus or cytosol are attacked by killing the infected cell, which interferes with the pathogen's life cycle and exposes pathogens that are released from the killed cells to the soluble molecules of the immune system. Those that live in vesicles, in contrast, can be attacked by activating the infected cell to intensify its antimicrobial activity.

	Extracellular		Intracellular	
	Interstitial spaces, blood, lymph	Epithelial surfaces	Cytoplasmic	Vesicular
Site of infection				
Organisms	Viruses Bacteria Protozoa Fungi Worms	*Neisseria gonorrhoeae* Worms *Mycoplasma* *Streptococcus pneumoniae* *Vibrio cholerae* *Escherichia coli* *Candida albicans* *Helicobacter pylori*	Viruses *Chlamydia* spp. *Rickettsia* spp. *Listeria monocytogenes* Protozoa	Mycobacteria *Salmonella typhimurium* *Leishmania* spp. *Listeria* spp. *Trypanosoma* spp. *Legionella pneumophila* *Cryptococcus neoformans* *Histoplasma* *Yersinia pestis*
Protective immunity	Antibodies Complement Phagocytosis Neutralization	Antibodies, especially IgA Anti-microbial peptides	Cytotoxic T cells NK cells	T-cell and NK-cell dependent macrophage activation

Figure 8.4 Pathogens exploit different compartments of the body which are defended in different ways. Virtually all pathogens have an extracellular stage in their life cycle. For some pathogens, all stages of their life cycle are extracellular, whereas others exploit intracellular sites as places to grow and replicate. Different components of the immune system contribute to protective immunity against different types of microorganism in different locations. NK, natural killer.

Every day the human body is exposed to vast numbers of infectious microorganisms. Most of these contacts do not result in infection and even then only a small fraction progress to causing disease. The outer epithelial surfaces of the body provide an effective physical barrier against most organisms. They are also colonized by non-pathogenic resident microorganisms that compete with the invading pathogen for nutrients and living space.

The pathogens that cause severe disease are not normally resident in the human body; neither are they ubiquitously present in the environment. Most of these agents infect only a few related host species and for this reason human disease infrequently arises from transmission from other vertebrate species, such as the domesticated animals with which humans are often in contact. The vast majority of human infections result from transmission of the pathogen from another person who is already infected. Transmission can be directly from one person to another, or it can require intermediate passage through a distantly related organism, for example an insect or mollusc, that is necessary for completing the pathogen's life cycle.

The stability of pathogens to environmental conditions varies, and determines the ease with which diseases are spread. For example, the bacterial disease anthrax is spread by spores that are resistant to heat and desiccation and can therefore be passed over long distances from one person to another. It is these properties that make anthrax a 'hot topic' in discussions of germ warfare. In contrast, the human immunodeficiency virus is very sensitive to changes in its environment and can be passed between individuals only by intimate contact and the exchange of infected body fluids and cells.

Experiments involving deliberate infection show that a large initial dose of pathogenic microorganisms is usually necessary to cause disease. To establish an infection, a microorganism must colonize a tissue in sufficient numbers to overwhelm the cells and molecules of innate immunity that are rapidly recruited to the site of invasion. Even in these circumstances the effects will be minor unless

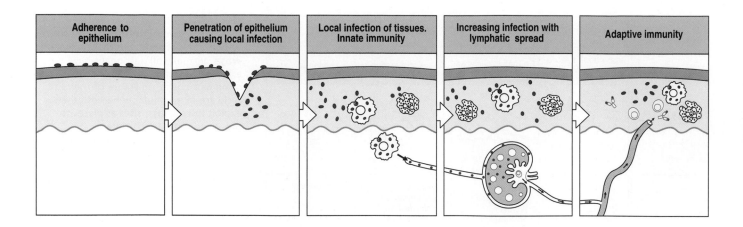

| Adherence to epithelium | Penetration of epithelium causing local infection | Local infection of tissues. Innate immunity | Increasing infection with lymphatic spread | Adaptive immunity |

the infection can spread within the body. Extracellular pathogens usually spread to other tissues by carriage in lymph or blood, which leads to the initiation of an adaptive immune response in secondary lymphoid tissue. Intracellular infections spread from one cell to its neighbors within a tissue, or by the release of infectious microorganisms that can be spread over greater distances in the extracellular fluids, again leading to the activation of adaptive immunity (Figure 8.5).

Recovery from an infectious disease involves the clearance of infectious organisms from the body and the repair of the damage caused both by the infection and by the immune response. A cure is not always possible. Infection can overwhelm the immune system with death as the consequence, as was once common in smallpox infections. In intermediate situations the infection persists but its pathological effects are controlled by the adaptive immune response, as usually occurs on infection with herpes viruses.

8-2 Surface epithelia present a formidable barrier to infection

The surface of the human body is protected by a continuous covering of epithelial cells, which forms a barrier between the internal milieu of the body's organs and tissues and the external world containing pathogens. These epithelia comprise the skin and the epithelial linings of the respiratory, gastrointestinal, and urogenital tracts (see Figure 1.4, p. 6). Infections occur only when these barriers are breached or when pathogens colonize the surfaces or interiors of the epithelial cells themselves. Figure 8.6 summarizes some barrier properties of epithelia. The skin principally provides a physical barrier, whereas the mucosal tissues permit

Figure 8.5 Stages in infection and host defense. To enter the body through the skin, a pathogen must first adhere to it, as shown in the first panel. In the second panel, a wound allows the pathogen to penetrate the connective tissue underlying the protective epithelium. In the third panel, the pathogen replicates while the defenses of innate immunity develop a state of inflammation at the infected site. In the fourth panel, pathogens are carried to the draining lymph node, where they are trapped and stimulate antigen-specific T and B cells to mount an adaptive immune response. After lymphocyte division and differentiation, antibodies and effector T cells arrive at the infected site and work with the cells and molecules of innate immunity to clear the pathogen, as shown in the fifth panel.

Epithelial barriers to infection	
Mechanical	Epithelial cells joined by tight junctions Flow of air or fluid over epithelial surface Movement of mucus by cilia
Chemical	Fatty acids (skin) Enzymes: lysozyme (saliva, sweat, tears), pepsin (gut) Low pH (stomach) Antibacterial peptides; cryptidins (intestine)
Microbiological	Normal flora compete for nutrients and attachment to epithelium and can produce antibacterial substances

Figure 8.6 Surface epithelia have physical, chemical, and microbiological barriers to infection.

various forms of communication between the body and its environment, including interactions with other members of the human species. Because the skin provides almost all of the outer covering of the body, it is much more frequently breached by wounds or burns than are the mucosa, and it is thus more susceptible to the passive entry of pathogens. In contrast, the relatively protected mucosa are more susceptible to pathogens that actively exploit the communication functions of these tissues to gain entry to the body.

Surface epithelia are more than just a physical barrier to infection: they also secrete chemical weapons that inhibit the attachment and growth of microorganisms. For example, the acid and hydrolytic enzymes secreted by the lining of the stomach create an environment unfriendly to bacterial growth, and antibacterial peptides called cryptidins are made by Paneth cells in the epithelium of the small intestine. Lysozyme in tears and saliva degrades bacterial cell walls. Most epithelia also have a normal flora of non-pathogenic microorganisms that compete with pathogenic organisms for nutrients and for attachment sites on epithelial cells. For *Escherichia coli*, a major bacterial component of the gut flora, this competition involves the secretion of antibacterial proteins called colicins that prevent the colonization of the gut by other bacteria. When a patient takes a course of antibiotic drugs, the non-pathogenic flora is killed along with the pathogens that caused the disease. After such treatment the body is recolonized by microorganisms; in this situation other pathogenic organisms can establish themselves, causing further disease.

8-3 Complement activation by the alternative pathway provides a first line of molecular defense against many microorganisms

As soon as a pathogen penetrates an epithelial barrier and starts living in the tissues, the defense mechanisms of innate immunity are brought into play. One of the first components of innate immunity to be activated is complement. Complement proteins are ubiquitous in blood and lymph, making complement activation a molecular defense mechanism that can be used immediately after an infection begins. At this point, long before any pathogen-specific antibody is made, complement activation occurs mainly by the **alternative pathway**.

The first step in the alternative pathway involves the spontaneous hydrolysis and activation of complement component C3, a process that occurs continually at a low rate in blood, lymph, and extracellular fluids. The rate increases in the vicinity of some pathogens. The product of spontaneous C3 hydrolysis, $C3(H_2O)$, can bind to **factor B** in the blood or extracellular fluid, making factor B susceptible to cleavage by **factor D**. This reaction forms a fluid-phase C3 convertase $(C3(H_2O),Bb)$, which cleaves C3 into C3a and C3b. Although most of the C3b fragments produced are either hydrolyzed by water or become attached to serum proteins, some bond to the invading pathogen as described in Section 7-16, p. 185. Bound C3b then binds other complement components to form C3 convertases, which cause more C3b to be bonded to the pathogen surface. The complex of C3b and Bb, designated C3b,Bb, forms the C3 convertase of the alternative pathway (Figure 8.7). C3a, the other product of C3 cleavage, is an anaphylatoxin (see Section 7-22, p. 193).

Once a few C3 convertase molecules have been assembled, they activate more C3 and fix more C3b at the pathogen's surface, leading to the assembly of yet more convertase. This process is one of progressive amplification, which from the initial deposition of a few molecules can rapidly coat the pathogen with C3b. To facilitate the reaction, the plasma protein **properdin** (**factor P**) binds to the convertase and stabilizes it against inhibition by factor H, a plasma protein that

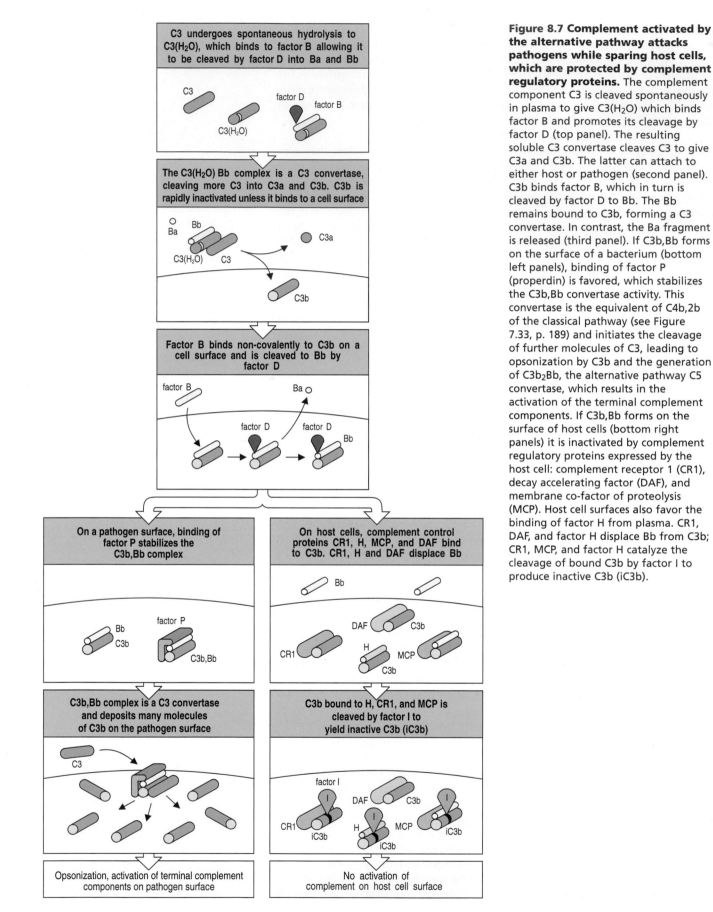

Figure 8.7 Complement activated by the alternative pathway attacks pathogens while sparing host cells, which are protected by complement regulatory proteins. The complement component C3 is cleaved spontaneously in plasma to give C3(H$_2$O) which binds factor B and promotes its cleavage by factor D (top panel). The resulting soluble C3 convertase cleaves C3 to give C3a and C3b. The latter can attach to either host or pathogen (second panel). C3b binds factor B, which in turn is cleaved by factor D to Bb. The Bb remains bound to C3b, forming a C3 convertase. In contrast, the Ba fragment is released (third panel). If C3b,Bb forms on the surface of a bacterium (bottom left panels), binding of factor P (properdin) is favored, which stabilizes the C3b,Bb convertase activity. This convertase is the equivalent of C4b,2b of the classical pathway (see Figure 7.33, p. 189) and initiates the cleavage of further molecules of C3, leading to opsonization by C3b and the generation of C3b$_2$Bb, the alternative pathway C5 convertase, which results in the activation of the terminal complement components. If C3b,Bb forms on the surface of host cells (bottom right panels) it is inactivated by complement regulatory proteins expressed by the host cell: complement receptor 1 (CR1), decay accelerating factor (DAF), and membrane co-factor of proteolysis (MCP). Host cell surfaces also favor the binding of factor H from plasma. CR1, DAF, and factor H displace Bb from C3b; CR1, MCP, and factor H catalyze the cleavage of bound C3b by factor I to produce inactive C3b (iC3b).

dampens down complement reactions. When C3b fragments are bound at high density to a pathogen's surface they form effective ligands for the complement receptors (for example CR1) on neutrophils and macrophages. The covalently coupled C3b fragments thereby target the pathogen for phagocytosis.

The binding of C3b to existing C3b,Bb complexes at the pathogen's surface forms the complex called C3b$_2$,Bb, the **C5 convertase of the alternative pathway**. This enzyme activates C5 and the terminal components of complement (C6–C9), which form a pore in the membrane and can induce the lysis of certain pathogens. The activation of C5 also releases C5a, a small soluble peptide that facilitates pathogen destruction by the recruitment of phagocytes into the site of infection. C5a is the most potent of the anaphylatoxins (see Section 7-22, p. 193).

Initiation of the alternative pathway on the surface of human cells is stopped at an early stage by the human cell-surface proteins CR1, DAF (decay-accelerating factor), and MCP (membrane co-factor of proteolysis), as well as by the plasma protein factor H. As described in Section 7-23, p. 194, these control proteins all act to destroy C3 convertase activity by binding to C3b and displacing Bb, and/or by rendering C3b susceptible to cleavage by factor I (see Figure 8.7).

Pathogens differ in the extent to which complement activated by the alternative pathway can lead to their elimination. In general, bacteria with an abundance of sialic acids on their surface, such as *Streptococcus pyogenes* and *Staphylococcus aureus*, tend to evade the effects of complement when it acts in the absence of specific antibodies. This is because C3b bound to their surfaces is more readily attacked by factor H. It is also possible that the physicochemical properties of the pathogen's surface influence the rate of spontaneous C3 activation.

8-4 Macrophages and neutrophils provide innate cell-mediated immunity and initiate the inflammatory response

By engulfing and killing microorganisms, phagocytic cells provide the principal means by which the immune system destroys invading pathogens. Two kinds of phagocyte—the macrophage and the neutrophil—serve this function throughout an immune response. These two cell types have distinct properties and complementary functions. **Macrophages** are the mature forms of circulating monocytes that have left the blood and taken up residence in the tissues. They are prevalent in the connective tissues, the linings of the gastrointestinal and respiratory tracts, the alveoli of the lungs, and in the liver, where they are known as **Kupffer cells**. Macrophages are long-lived cells. In addition to their role as phagocytic microbicidal cells, activated macrophages release a variety of cytokines that have important functions in innate immunity and in setting the scene for the adaptive immune response to develop. Macrophages also act as professional antigen-presenting cells in the development of adaptive immunity. They are also instrumental in the removal of damaged or unwanted human cells, thus facilitating the repair of tissues damaged by wounds, infections, inflammation, and the immune response.

Neutrophils, the second kind of phagocyte, were historically called microphages because of their smaller size compared with macrophages. Neutrophils are the most abundant of the white blood cells, with large numbers being formed and lost each day. They are a type of granulocyte, having numerous granules in the cytoplasm, and are also known as **polymorphonuclear leukocytes**, because of the variable and irregular shapes of the nuclei (see Figure 1.6, p. 8). Neutrophils are the most abundant white cells in the blood and are normally excluded from healthy tissues. At sites of infection, however, actively phagocytosing macrophages release cytokines that attract neutrophils to leave the blood and

Figure 8.8 Bacteria binding to neutrophil receptors induce phagocytosis and microbial killing. The figure illustrates this type of mechanism for two such receptors, CD14 and CR4, which are specific for bacterial lipopolysaccharide (LPS). Bacteria binding to these receptors stimulate the phagocytosis and degradation of the bacterium.

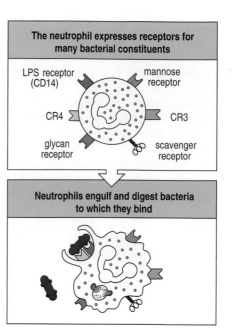

enter the infected area in large numbers, where they soon become the dominant phagocytic cell. For example, each day some three billion (3×10^9) neutrophils enter tissues of the oral cavity, the most contaminated site in the body. The arrival of neutrophils is the first of a series of reactions, called the **inflammatory response**, by which cells and molecules of innate immunity are recruited into sites of wounding or infection. Neutrophils are specialized for working under the anaerobic conditions that prevail in damaged tissues.

Cytokines produced in infected and inflamed tissues by macrophages and natural killer cells (NK cells) also induce the bone marrow to release its reserves of neutrophils into the blood. In addition, the population of neutrophils that is loosely attached to blood vessel walls, and said to be **marginating**, is also recruited into the circulating pool of leukocytes. Neutrophils are short-lived cells, dying within a few hours of entering a tissue. In doing so, they form the creamy **pus** that characteristically develops at infected wounds and other sites of infection. This is why the extracellular bacteria responsible for the superficial infections and abscesses that neutrophils tackle in large numbers are known as pus-forming or **pyogenic** bacteria.

Phagocytosis of bacteria by neutrophils and macrophages at times early in infection depends upon the recognition molecules of innate immunity. Tissue macrophages and neutrophils have surface receptors that recognize molecular components common to many bacterial pathogens. These receptors include the mannose receptor, the scavenger receptor, which is specific for molecules rich in sialic acid, and the glycan receptor, which recognizes polysaccharides. Bacterial lipopolysaccharide (LPS) is also recognized by the serum protein **LPS-binding protein**, which delivers LPS to the cell-surface receptor CD14 on macrophages and neutrophils (Figure 8.8). Phagocytes also possess receptors for the complement component C3b, or its degradation product iC3b, and thus can readily phagocytose pathogens that have been coated with complement produced by the alternative pathway.

Some of the complement receptors also recognize microbial constituents directly. For example, the receptors CR3 and CR4 recognize several microbial cell-surface molecules, including LPS, the lipophosphoglycan of *Leishmania*, the filamentous hemagglutinin of *Bordetella*, and cell-surface structures on yeasts such as *Candida* and *Histoplasma*.

Almost immediately after a pathogen is engulfed by a neutrophil, a battery of degradative enzymes and other toxic substances are brought to bear upon it. Phagosomes containing recently captured microorganisms are fused with the preformed neutrophil granules, which contain hydrolytic degradative enzymes, NADPH-dependent oxidases, and antimicrobial peptides such as the defensins (Figure 8.9). The NADPH-dependent oxidases generate toxic oxygen radicals and hydrogen peroxide, the latter being quickly converted into hypochlorous acid. Activated phagocytes also produce nitric oxide (NO), generated from arginine and oxygen by the inducible enzyme nitric oxide synthase (iNOS). Nitric oxide reacts with oxygen radicals within the phagolysosome to produce highly toxic peroxynitrite.

Antimicrobial mediators	Specific products
Acid	pH=~3.5–4.0, bacteriostatic or bactericidal
Toxic oxygen derivatives	Superoxide (O_2^-), hydrogen peroxide (H_2O_2), singlet oxygen ($^1O_2^-$), hydroxyl radical ($OH^·$), hypohalite (OCl^-), nitric oxide (NO)
Peptides and proteins	Defensins, cationic proteins
Enzymes	Myeloperoxidase: generation of toxic oxygen derivatives Lysozyme: dissolves cell walls of some Gram-positive bacteria Acid hydrolases: further digest bacteria
Competitors	Lactoferrin (binds iron), Vitamin B_{12}-binding protein

Figure 8.9 Bactericidal agents produced or released by phagocytic cells on the ingestion of microorganisms. Most of these agents are present in both macrophages and neutrophils. Some of them are toxic; others, such as lactoferrin, work by binding essential nutrients and preventing their uptake by bacteria.

This general oxidative attack on ingested microorganisms is accompanied by a transient increase in oxygen consumption called the **respiratory burst**. The combination of toxic oxygen metabolites, proteases, phospholipases, lysozyme, and antimicrobial peptides is able to kill both Gram-positive and Gram-negative bacteria, fungi, and even some of the enveloped viruses. A neutrophil is unable to synthesize more granules after it has been activated; once all the granules have been used up, the neutrophil's usefulness is at an end and it dies. The dependence of the body's defenses upon neutrophils is illustrated by the miserable effects of their absence. Patients with neutrophil deficiencies suffer recurrent infections, often by bacteria and fungi that form part of the normal flora of healthy people.

In macrophages, the phagosomes containing ingested microorganisms fuse with lysosomes, which contain very similar degradative enzymes to those of neutrophil granules (see Figure 8.9). The resultant vesicle is a phagolysosome. A respiratory burst accompanied by the generation of toxic oxygen metabolites and nitric oxide occurs, and ingested microorganisms are killed. Unlike neutrophils, macrophages are long-lived cells and continue to generate more lysosomes as needed.

The toxic oxygen species produced by activated neutrophils and macrophages can diffuse out of the cell and damage host cells. To limit the damage, the respiratory burst is also accompanied by the synthesis of enzymes that inactivate these potent small molecules: superoxide dismutase converts superoxide to hydrogen peroxide, and catalase degrades hydrogen peroxide to water and oxygen.

In some circumstances the contents of phagocyte lysosomes and granules are released to the outside of the cell. This enables neutrophils and macrophages to attack antibody-coated pathogens that are too big for them to ingest, such as parasitic worms. The phagocyte adapts to this situation by secreting the contents of its granules onto the surface of the worm at the region of contact. This secretory reaction can also be provoked by contact with host cells, and phagocyte activation can thus cause extensive tissue damage during the course of an infection.

8-5 The innate immune response induces inflammation at sites of infection

Complement activation and phagocytosis of complement-coated bacteria occur at sites of infection within a few hours of microbial invasion. Many infections are terminated by the combination of these mechanisms, but when they are insufficient to prevent microbial growth, additional defenses are brought into play.

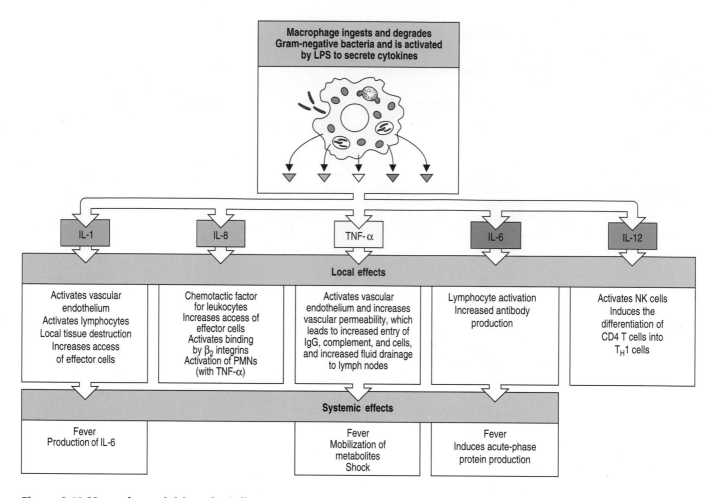

Figure 8.10 Macrophages initiate the inflammatory response. The cytokines secreted by macrophages in response to bacterial lipopolysaccharide (LPS) include IL-1, IL-6, IL-8, IL-12, and TNF-α. The local and systemic effects of each cytokine are shown. PMN, polymorphonuclear leukocyte.

In an infection with extracellular bacteria, macrophages within the infected tissue are the first cells to respond. In addition to their phagocytic function, stimulated macrophages secrete a battery of cytokines and other molecules that recruit effector cells, especially neutrophils, into the infected area. There they develop a state of **inflammation** within the tissue. Inflammation describes the local accumulation of fluid accompanied by swelling, reddening, and pain. These effects stem from changes induced in the local blood capillaries that lead to an increase in their diameter (a process called dilation), reduction in the rate of blood flow, and increased permeability of the blood vessel wall. The increased supply of blood to the region causes the local redness and heat associated with inflammation. The increased permeability of blood vessels allows the movement of fluid, plasma proteins, and white blood cells from the blood capillaries into the adjoining connective tissues, causing the swelling and pain. Prominent cytokines produced by activated macrophages are **IL-1**, **IL-6**, **IL-8**, **IL-12**, and **tumor necrosis factor-α (TNF-α)**. These cytokines have potent effects that can be localized to the infected tissue or manifested systemically throughout the body (Figure 8.10).

Other molecules released by macrophages are plasminogen activator, phospholipase and other enzymes, prostaglandins, oxygen radicals, peroxides, nitric oxide, leukotrienes, and platelet-activating factor (PAF), which all contribute to inflammation and tissue damage. In the course of complement activation, the soluble complement fragments C3a and C5a recruit neutrophils from the blood into infected tissues and stimulate mast cells to degranulate, releasing the inflammatory molecules histamine and TNF-α, among others. Molecules involved in the induction of inflammation are known generally as **inflammatory mediators**. The combined effect of all this activity is to produce a local state of inflammation with its characteristic symptoms.

8-6 The migration of leukocytes into tissues is induced by inflammatory mediators

Within the blood capillaries, white blood cells move more slowly than in the larger vessels, particularly in the region close to the endothelial wall. Inflammatory mediators stimulate the endothelium of the capillaries in infected tissue to express adhesion molecules and MHC class II molecules (see Figure 6.2, p. 132). This is known as the activation of the vascular endothelium. Inflammatory mediators also induce the expression of complementary adhesion molecules on circulating monocytes and neutrophils. These changes enable monocytes and neutrophils in the blood to bind to activated vascular endothelium within an infected site and then to squeeze through the spaces between endothelial cells and enter the infected tissue. Infiltrating monocytes then differentiate and add to the local population of tissue macrophages. The release of inflammatory mediators by the increasing numbers of macrophages and neutrophils steadily increases the inflammation within the infected tissue.

At various stages in their maturation, activation, or function, all types of white blood cell leave the blood and home to particular target tissues. In all cases, the white blood cell is directed to its destination by the combinations of adhesion molecules expressed by the vascular endothelium in the tissue and by the white blood cell (see Figure 6.2, p. 132). The process by which leukocytes migrate out of blood capillaries and into tissues is called **extravasation** and it occurs in four steps. The first step is an interaction between circulating leukocytes and blood vessel walls that slows the leukocytes down. This interaction is mediated by selectins, adhesion molecules that bind to particular types of carbohydrate (see Figure 6.2, p. 132). In healthy tissue, vascular endothelial cells contain granules, known as **Weibel–Palade bodies**, which contain **P-selectin**. On exposure to inflammatory mediators, including leukotriene LTB_4, complement fragment C5a, and histamine, the P-selectin in the Weibel–Palade bodies is mobilized and within a few minutes is delivered to the cell surface. A second selectin, **E-selectin**, is also expressed on the endothelial cell surface a few hours after exposure to LPS or TNF-α. The two selectins bind to the sialyl-Lewis^x carbohydrates of cell-surface glycoproteins on the leukocyte. These reversible interactions allow the leukocytes to adhere to the blood vessel walls and to 'roll' slowly along them by forming new adhesive interactions at the front of the cell while breaking them at the back (Figure 8.11, top panel).

The second step in extravasation depends on interactions between the integrins LFA-1 and CR3 on the phagocyte and adhesion molecules on the endothelium, for example ICAM-1, whose expression is also induced by TNF-α. Under normal conditions, LFA-1 and CR3 interact only weakly with endothelial adhesion molecules, but exposure to chemoattractant cytokines such as IL-8, produced by cells in the inflamed tissues, induces conformational changes in the LFA-1 and CR3 on a rolling leukocyte that strengthen their adhesion. As a result, the leukocyte holds tightly to the endothelium and stops rolling (see Figure 8.11, bottom panels).

In the third step, the leukocyte crosses the blood vessel wall. LFA-1 and CR3 contribute to this movement, as does adhesion involving the immunoglobulin super-family protein CD31, which is expressed by both leukocytes and endothelial cells at their junctions with one another. The leukocyte squeezes between neighboring

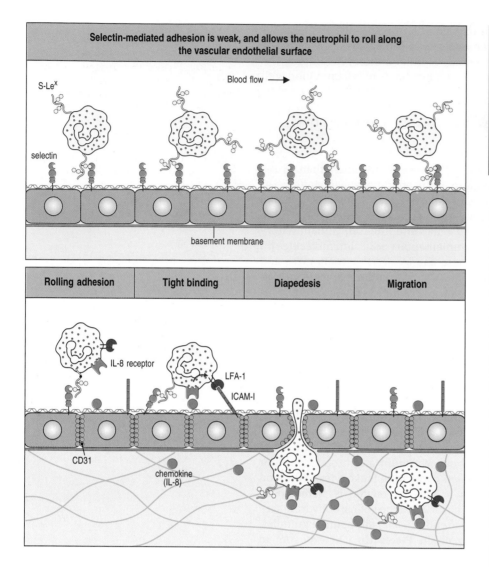

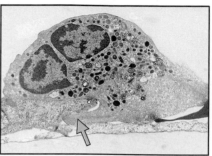

Figure 8.11 Neutrophils are directed to sites of infection through interactions between adhesion molecules. Inflammatory mediators and cytokines produced as the result of infection induce the expression of selectin on vascular endothelium, which enables it to bind leukocytes. The top panel shows the rolling interaction of a neutrophil with vascular endothelium due to transient interactions between selectin on the endothelium and sialyl-Lewis^x (s-Le^x) on the leukocyte. The bottom panel shows the conversion of rolling adhesion into tight binding and subsequent migration of the leukocyte into the infected tissue. The four stages of extravasation are shown. Rolling adhesion is converted into tight binding by interactions between integrins on the leukocyte (LFA-1 is shown here) and adhesion molecules on the endothelium (ICAM-1). Expression of these adhesion molecules is also induced by cytokines. A strong interaction is induced by the presence of chemoattractant cytokines (the chemokine IL-8 is illustrated here) that have their source at the site of infection. They are held on proteo-glycans of the extracellular matrix and cell surface to form a gradient along which the leukocyte can travel. Under the guidance of these chemoattractant cytokines, the neutrophil squeezes between the endothelial cells and penetrates the connective tissue (diapedesis). It then migrates to the center of infection along the IL-8 gradient. The electron micrograph shows a neutrophil that has just started to migrate between adjacent endothelial cells but has yet to break through the basement membrane, which is at the bottom of the photograph. The blue arrow points to the pseudopod that the neutrophil is inserting between the endothelial cells. The dark mass in the bottom right-hand corner is an erythrocyte that has become trapped under the neutrophil. Photograph ($\times$ 5500) courtesy of I. Bird and J. Spragg.

endothelial cells, a maneuver known as **diapedesis**, and reaches the basement membrane, a part of the extracellular matrix. It then crosses the basement membrane by secreting proteases that break down the membrane. The fourth and final step in extravasation is movement of the leukocyte toward the center of infection in the tissue. This migration is accomplished along gradients of chemoattractants such as IL-8, which have their source within the infected site (see Figure 8.11).

8-7 TNF-α is a powerful inflammatory mediator with both beneficial and detrimental effects

In addition to inducing adhesion molecules on vascular endothelium within infected areas, TNF-α also induces the endothelial cells to produce molecules such as platelet-activating factor that trigger blood clotting and blockage of the local blood vessels. This restricts the leakage of plasma from the blood and prevents pathogens from entering the blood and disseminating infection throughout the body. However, in those instances in which infection does spread to the blood, the very actions of TNF-α that so effectively contain local infection can become catastrophic (Figure 8.12).

Blood-borne pathogens induce the release of TNF-α by macrophages in the liver, spleen, and other sites. Such systemic release of TNF-α produces **sepsis**, a condition

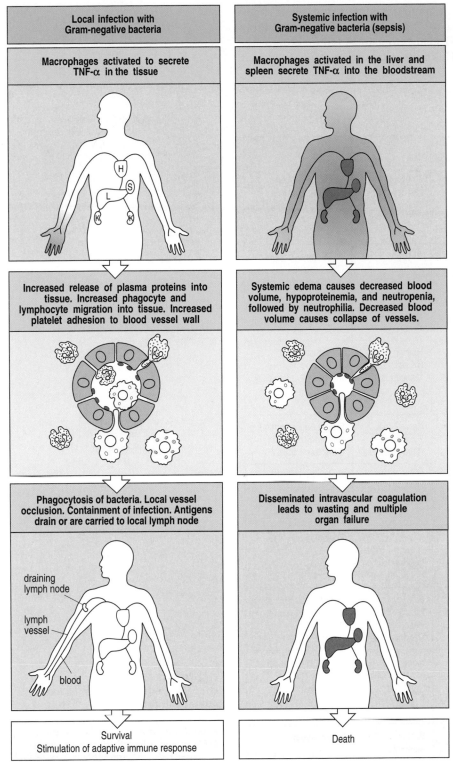

Figure 8.12 TNF-α released by macrophages induces protection at the local level but can lead to catastrophe when released systemically. The panels on the left describe the causes and consequences of the release of TNF-α within a local area of infection. In contrast, the panels on the right describe the causes and consequences of the release of TNF-α throughout the body. The initial effects of TNF-α are on the endothelium of blood vessels, especially venules. It causes increased blood flow, vascular permeability, and endothelial adhesiveness for white blood cells and platelets. These events cause the blood in the venules to clot, preventing the spread of infection and directing extracellular fluid to the lymphatics and lymph nodes, where the adaptive immune response is activated. When an infection develops in the blood, the systemic release of TNF-α and the effect it has on the venules in all tissues simultaneously induce a state of shock that can lead to organ failure and death. H, heart; L, liver; S, spleen; k, kidney.

in which the systemic vasodilation of blood vessels causes massive leakage of fluid into tissues. This leads to a state of **septic shock** in which widespread blood clotting in capillaries—disseminated vascular coagulation—exhausts the supply of clotting proteins. More critically, this condition frequently leads to the failure of vital organs such as the kidneys, liver, heart, and lungs, which are soon compromised by the lack of a normal blood supply. Consequently, patients suffering from septic shock have a high mortality rate. The importance of TNF-α in septic shock was sharply demonstrated by the observation that mice lacking a functional TNF-α receptor are resistant to the induction of septic shock. This benefit, however, was offset by their inability to control local infections.

Class	Chemokine	Source	Receptors	Target cells	Major effects
CXC	IL-8	Monocytes Macrophages Fibroblasts Keratinocytes Endothelial cells	CXCR1 CXCR2	Neutrophils Naive T cells	Mobilizes, activates and degranulates neutrophils Angiogenesis
	PBP β-TG NAP-2	Platelets	CXCR2	Neutrophils	Activates neutrophils Clot resorption Angiogenesis
	GROα, β, γ	Monocytes Fibroblasts Endothelium	CXCR2	Neutrophils Naive T cells Fibroblasts	Activates neutrophils Fibroplasia Angiogenesis
	IP-10	Keratinocytes Monocytes T cells Fibroblasts Endothelium	CXCR3	Resting T cells NK cells Monocytes	Immunostimulant Anti-angiogenic Promotes T$_H$1 immunity
	SDF-1	Stromal cells	CXCR4	Naive T cells Progenitor (CD34$^+$) B cells	B-cell development Lymphocyte homing Competes with HIV-1
CC	MIP-1α	Monocytes T cells Mast cells Fibroblasts	CCR1, 3, 5	Monocytes NK and T cells Basophils Dendritic cells	Competes with HIV-1 Anti-viral defense Promotes T$_H$1 immunity
	MIP-1β	Monocytes Macrophages Neutrophils Endothelium	CCR1, 3, 5	Monocytes NK and T cells Dendritic cells	Competes with HIV-1
	MCP-1	Monocytes Macrophages Fibroblasts Keratinocytes	CCR2B	Monocytes NK and T cells Basophils Dendritic cells	Activates macrophages Basophil histamine release Promotes T$_H$2 immunity
	RANTES	T cells Endothelium Platelets	CCR1, 3, 5	Monocytes NK and T cells Basophils Eosinophils Dendritic cells	Degranulates basophils Activates T cells Chronic inflammation
	Eotaxin	Endothelium Monocytes Epithelium T cells	CCR3	Eosinophils Monocytes T cells	Role in allergy
C	Lymphotactin	CD8>CD4 T cells	?	Thymocytes Dendritic cells NK cells	Lymphocyte trafficking and development
CXXXC (CX$_3$C)	Fractalkine	Monocytes Endothelium Microglial cells	CX$_3$CR1	Monocytes T cells	Leukocyte–endothelium adhesion Brain inflammation

Figure 8.13 Properties of some chemokines. Chemokines fall into two related but distinct groups. The CC chemokines, which in humans are all encoded in one region of chromosome 4, have two adjacent cysteine residues, whereas the CXC chemokines, which are found in a cluster on chromosome 17, have an amino acid residue between the equivalent two cysteines. A C chemokine with only one cysteine at this location, and fractalkine, a CX$_3$C chemokine, are encoded elsewhere in the genome. Each chemokine interacts with one or more receptors, and affects one or more types of cell.

8-8 Chemoattractant cytokines recruit phagocytes to infected tissues

The chemoattractant cytokines, or **chemokines**, comprise a family of proteins that recruit effector cells into sites of infection. IL-8, which attracts neutrophils from the blood into sites of infection and inflammation, is one example. The chemokines are a family of small soluble proteins with similar structures, and are secreted by a variety of cell types in response to tissue damage or infection. Two major subfamilies of chemokines are defined according to the position of certain cysteine residues in the amino-acid sequence and the type of cells that they attract (Figure 8.13). Sequence motifs with adjacent cysteines distinguish the CC chemokines, which attract monocytes, effector T cells, and memory T cells. The CXC chemokines have motifs with two cysteines separated by another amino acid; they attract neutrophils and naive T cells.

In recruiting neutrophils from the blood, IL-8 serves two purposes. First, it converts the rolling adhesion of leukocytes to vascular endothelium into a stable binding. Second, it directs neutrophil migration along a gradient of the chemokine that increases in concentration towards the infection. The gradient is set up by the binding of chemokines emanating from the infected area to proteoglycan molecules of the extracellular matrix and the endothelial cell surfaces. This means that the chemokines are immobilized on a solid substrate along which the white blood cells can migrate. White blood cells sense the chemokine gradient through chemokine receptors on their surface.

8-9 Cytokines released by macrophages cause a rise in body temperature and also activate the acute-phase response

The cytokines IL-1, IL-6, and TNF-α produced by macrophages have various effects (Figure 8.14), some of which have already been mentioned. Another systemic effect of these inflammatory cytokines is to cause the rise in body temperature that is called **fever**. The cytokines and other mediators act on temperature control sites in the hypothalamus, and on muscle and fat cells, altering energy mobilization to generate heat. In this context, such molecules are called 'endogenous pyrogens' to distinguish them from the 'exogenous' products of pathogens that also raise the body's temperature. On balance, a raised body temperature helps the immune system in the fight against infection, because most bacterial and viral pathogens grow better at temperatures lower than those of the human body, and adaptive immunity becomes more potent at higher temperatures. At elevated temperatures,

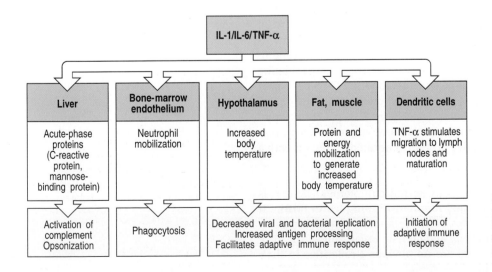

Figure 8.14 The macrophage-produced cytokines TNF-α, IL-1, and IL-6 have a spectrum of biological activity that helps to coordinate the body's response to infection.

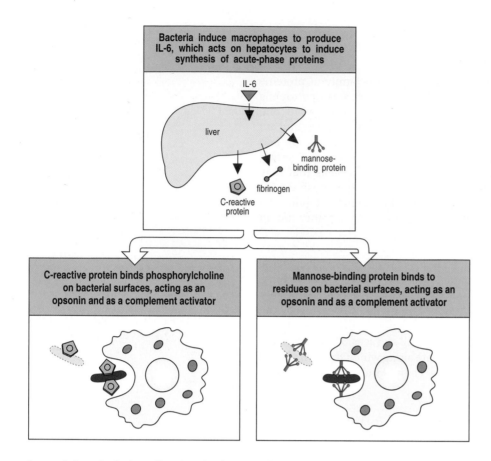

Bacteria induce macrophages to produce IL-6, which acts on hepatocytes to induce synthesis of acute-phase proteins

IL-6

liver

mannose-binding protein

fibrinogen

C-reactive protein

C-reactive protein binds phosphorylcholine on bacterial surfaces, acting as an opsonin and as a complement activator

Mannose-binding protein binds to residues on bacterial surfaces, acting as an opsonin and as a complement activator

Figure 8.15 The acute-phase response increases the supply of the recognition molecules of innate immunity. Acute-phase proteins are produced by liver cells in response to the cytokines released by phagocytes in the presence of bacteria. In humans they include C-reactive protein, fibrinogen, and mannose-binding protein. Both C-reactive protein and mannose-binding protein bind to structural features of bacterial cell surfaces that are not found on human cells. On binding to bacteria, they act as opsonins and also activate complement, facilitating phagocytosis and also direct lysis (dashed lines) of the bacteria.

bacterial and viral replication is decreased, whereas processing of antigen is enhanced. In addition, human cells become more resistant to the deleterious effects of TNF-α at raised temperatures.

A second systemic effect of TNF-α, IL-1, and IL-6 is to change the spectrum of soluble plasma proteins secreted by hepatocytes in the liver. The changes induced by acute infection are collectively called the **acute-phase response**. Those proteins whose synthesis and secretion is increased during the acute-phase response are called **acute-phase proteins** (Figure 8.15).

Two of the acute-phase proteins—C-reactive protein and mannose-binding protein—enhance the fixation of complement at pathogen surfaces. **C-reactive protein** is a pentamer of identical subunits and a member of the **pentraxin** family of proteins. C-reactive protein binds to the phosphorylcholine component of lipopolysaccharides in bacterial and fungal cell walls, but not to the phosphorylcholine present in the phospholipids of human cell membranes. In binding to bacteria, C-reactive protein acts as an opsonin, which can then bind C1q, initiating the classical pathway of complement fixation. The interaction of C-reactive protein with C1q involves the stalks of the C1q molecule, whereas the interaction of antibodies with C1q involves the globular heads (see Figure 7.31, p. 186). Despite these differences in the initial binding mechanism, the same sequence of complement reactions occurs when either C-reactive protein or antibody interacts with a pathogen.

The second acute-phase protein that can activate the complement system is **mannose-binding protein** (**MBP**), also called **mannan-binding lectin** (**MBL**). It binds to mannose-containing carbohydrates of bacteria and yeast and is a calcium-dependent lectin. In its overall structure and domain organization, the MBP molecule is similar to C1q, although there are no obvious similarities in their

Step in pathway	Protein serving function in pathway			Relationship
	Alternative	Lectin	Classical	
Initiation	D	MBP-associated serine protease	C1s	None known
Covalent binding to cell surface	C3b	C4b		Homologous
C3/C5 convertase	Bb	C2b		Homologous
Control of activation	CR1 H	CR1 C4bp		Identical Homologous
Opsonization	C3b			Identical
Initiation of membrane-attack complex formation	C5b			Identical
Local inflammation	C5a, C3a			Identical

Figure 8.16 There are close structural relationships between components of the alternative, lectin-mediated, and classical pathways of complement activation. Although there are many different complement components they are built from a small number of structural modules. C3, C4, and C5 form one set of related components; C6–C9 form another. A third family comprises the serine proteases, which include factor D of the alternative pathway, C1r and C1s of the classical pathway, and the serine protease associated with the mannose-binding protein. Factor B of the alternative pathway and C2 of the classical pathway are highly homologous. The major divergence between the three pathways is in their initiation. The antibodies that initiate the classical pathway, the mannose-binding protein that initiates the lectin pathway, and the spontaneous hydrolysis of C3 that initiates the alternative pathway, are distinct molecular mechanisms.

amino-acid sequences. MBP resembles a bunch of flowers in which the carbohydrate-recognition domains form the flowers. Each flower contains three binding sites and each MBP molecule has five or six flowers, giving MBP either 15 or 18 potential sites for attachment to a pathogen's surface. Even relatively weak individual interactions with a carbohydrate structure can be developed into high-avidity interactions through the use of multipoint attachments. Although some carbohydrates of human cells contain mannose, they do not bind MBP because their geometry does not permit multipoint attachment to MBP.

Like C1q, MBP activates a proteolytic enzyme complex (MBP-associated serine protease or MASP) that cleaves C4 and C2 and thus initiates complement activation. This pathway is known as the **lectin-mediated pathway** of complement activation. MBP also serves as an opsonin that facilitates the uptake of bacteria by monocytes in the blood. These cells do not express the macrophage mannose receptor but have receptors that can bind to MBP molecules coating a bacterial surface. MBP is a member of a protein family called the **collectins**; the family also includes the pulmonary surfactant proteins A and D (SP-A and SP-D), which work in the lungs to opsonize pathogens such as *Pneumocystis carinii*.

In the absence of infection, C-reactive protein and MBP are present at low levels in plasma; their levels increase during the acute-phase response. Both proteins bind to structures that are commonly found on pathogens but not on human cells, and initiate complement activation and fixation by a pathway that is almost identical to the classical pathway used by antibodies. Such comparisons reveal how the antibodies of the adaptive immune response have a role analogous to that of C-reactive protein and MBP in innate immunity and serve to expand the range of pathogen-recognition molecules. Furthermore, nearly all of the complement components used in the classical and lectin pathways are structurally and functionally related to components of the alternative pathway, which was probably the first sequence of complement reactions to evolve (Figure 8.16).

8-10 Type 1 interferons inhibit viral replication and activate host defenses

When human cells become infected with viruses they respond with the production and secretion of proteins called **interferon-α (IFN-α)** and **interferon-β (IFN-β)**, which block the spread of virus to uninfected cells. The α and β interferons are

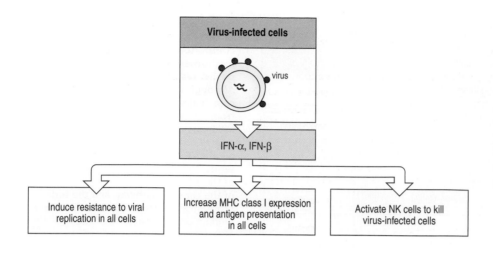

Figure 8.17 Major functions of the type I interferons. The interferons-α and -β have three major functions. First, they induce resistance to viral replication by activating cellular genes that destroy viral mRNA and inhibit the translation of viral proteins. Second, they induce MHC class I expression in most cells of the body, thus increasing the level of antigen presentation to CD8 cytotoxic T cells by infected cells, and the resistance of uninfected cells to NK cells. Third, they activate NK cells to kill virus-infected cells.

sometimes called type 1 interferons to distinguish them from interferon-γ (IFN-γ), an unrelated protein with cytokine function, which is typically secreted by activated NK cells, CD8 T cells, and CD4 T_H1 cells. Expression of the type 1 interferon genes is induced by double-stranded RNA, a type of nucleic acid not found in healthy human cells but which is a component of some viral genomes, and might be an intermediary nucleic acid in viral life cycles.

Interferons inhibit the spread of viral infection in several ways. They prevent the replication of virus within infected cells and its spread to uninfected cells, and they also activate innate and adaptive immune responses that target virus-infected cells (Figure 8.17). Interferons act through a single cell-surface receptor that is common to many human cell types, and through this receptor they induce changes in the expression of a variety of human genes.

One interferon-induced protein is the enzyme oligoadenylate synthetase, which polymerizes ATP by 2′-5′ linkages rather than the 3′-5′ linkages normally present in human nucleic acids. These unusual oligomers activate an endoribonuclease that degrades viral RNA. Also activated by IFN-α and IFN-β is a serine/threonine protein kinase called P1 kinase. By phosphorylating the protein-synthesis-initiation-factor eIF-2, P1 kinase prevents the synthesis of viral proteins and consequently the assembly of new infectious virions. The importance of the interferon-induced inhibition of viral replication can be appreciated from the phenotype of mice lacking an interferon-inducible protein called Mx, which is required for cellular resistance to influenza virus replication. Such animals are much more susceptible to influenza virus infection than mice that have Mx.

The second effect of IFN-α and IFN-β is to stimulate innate and adaptive immune responses that are specifically targeted to virus infections. NK cells provide innate immunity to viral infections, as we shall see in the next section, and can be activated by IFN-α and IFN-β. The presentation of viral antigens to CD8 T cells is enhanced by interferon-mediated increases in the expression of MHC class I molecules, TAP peptide transporter proteins, and the LMP2 and LMP7 proteasome subunits, which skew intracellular protein degradation toward the production of peptides that bind to MHC class I molecules (see Section 3-9, p. 66). All these interferon-regulated genes are located in the MHC. In response to activation by interferon, NK cells secrete IFN-γ, a cytokine that is unrelated in both structure and function to IFN-α and IFN-β.

8-11 NK cells serve as an early defense against intracellular infections

In Chapter 6, we saw how cytotoxic CD8 T cells kill virus-infected cells on recognition of virus-derived peptides presented by MHC class I molecules. The population of circulating lymphocytes contains a second type of cytotoxic lymphocyte known as the **natural killer cell** (**NK cell**). This cell has many similarities in effector mechanisms to the cytotoxic T lymphocyte. What most clearly distinguishes NK cells from T cells is that the former do not rearrange their T-cell receptor genes; neither do NK cells rearrange or express immunoglobulin genes. NK cells therefore represent a third type of lymphocyte distinct from B and T cells, and active in innate immunity.

NK cells are large lymphocytes that circulate in the blood, but unlike circulating B and T cells they have a well developed cytoplasm containing cytotoxic granules. When first discovered, NK cells were called 'large granular lymphocytes'. NK cells provide innate immunity against intracellular infections and migrate from the blood into infected tissues in response to inflammatory cytokines. Patients who lack NK cells suffer from persistent virus infections, particularly of herpes viruses, which they cannot clear despite making a normal adaptive immune response. These rare individuals demonstrate the importance of NK cells in managing virus infections, and show how the response of NK cells complements that of cytotoxic T cells. NK cells have two types of effector function—cell killing and the secretion of cytokines—which are used in different ways depending on the pathogen.

When NK cells are freshly isolated from blood they can kill certain types of target cell. This base level of cytotoxicity is increased 20–100-fold on exposure to the IFN-α and IFN-β produced in response to viral infection. Type 1 interferons also induce the proliferation of NK cells. NK cells are also activated by IL-12, which particularly targets NK cells, and by TNF-α; both of these are produced by macrophages early in many infections. The actions of these four cytokines produce a wave of activated NK cells during the early part of a virus infection that can either terminate the infection or contain it during the time required to develop effector CD8 cytotoxic T cells (Figure 8.18).

Stimulation of NK cells with IFN-α and IFN-β favors the development of cytotoxic effector function, whereas stimulation with IL-12 favors the production of cytokines. The principal cytokine released by NK cells is IFN-γ, which activates macrophages (see Section 6-14, p. 151). Macrophage secretion of IL-12 and NK-cell secretion of IFN-γ create a system of positive feedback that increases the activation of both types of cell within an infected tissue. Early in infection, NK cells are the major producers of IFN-γ and by activating macrophages to secrete cytokines that help to activate T cells they promote the initiation of T-cell responses. When effector T cells are formed and enter the infected site, they become the major source of IFN-γ and of cell-mediated cytotoxicity. With the arrival of effector T cells, NK cell functions are turned off by IL-10, an inhibitory cytokine made by cytotoxic T cells.

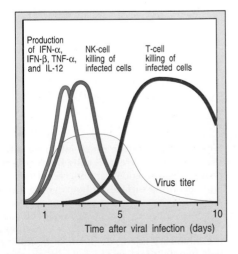

Figure 8.18 Natural killer cells (NK cells) provide an early response to virus infection. The kinetics of the immune response to an experimental virus infection of mice are shown. As a result of infection, a burst of cytokines is secreted, including IFN-α, IFN-β, TNF-α, and IL-12 (green curve). These induce the proliferation and activation of NK cells (blue curve), which are seen as a wave emerging after cytokine production. NK cells control virus replication and the spread of infection while effector CD8 T cells (red curve) are developing. The level of virus (the virus titer) is given by the curve described by the yellow shading.

8-12 NK cells express inhibitory cell-surface receptors for MHC class I molecules

A general feature of NK cells is their sensitivity to the expression of MHC class I molecules. Cells that have lost the expression of one or more MHC class I isoforms can become susceptible to attack by NK cells; loss of MHC class I expression is a common property of virus-infected cells. Many viruses have evolved mechanisms that interfere with MHC class I expression because this enables them to evade the cytotoxic T-cell response.

The receptor mechanisms by which NK cells recognize cells infected with viruses or other types of intracellular pathogen are poorly understood. Better defined are the recognition systems that prevent NK cells from killing healthy human cells. NK cells express cell-surface receptors that bind to MHC class I molecules. Binding delivers inhibitory signals that turn off NK cell cytotoxicity and cytokine secretion. Every NK cell has at least one inhibitory receptor that recognizes self MHC class I molecules and so NK cells are prevented from attacking uninfected cells. In contrast, cells that have lost MHC class I expression become susceptible to attack by NK cells (Figure 8.19). Viral infections lead to the synthesis of IFN-α and IFN-β, which induce increased MHC class I expression on healthy cells, making them more resistant to NK cells, which are themselves also activated by interferons.

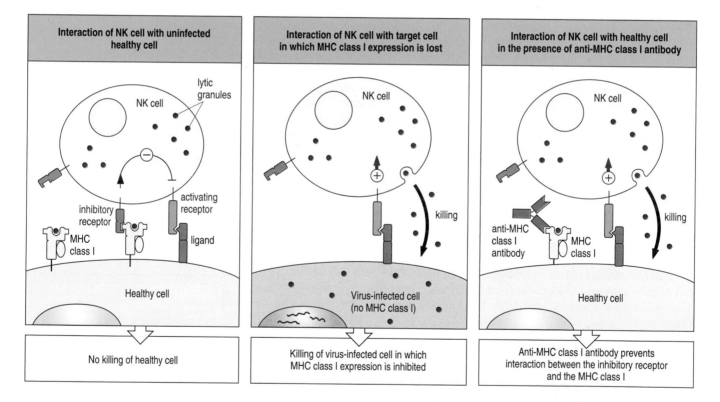

Interaction of NK cell with uninfected healthy cell	Interaction of NK cell with target cell in which MHC class I expression is lost	Interaction of NK cell with healthy cell in the presence of anti-MHC class I antibody
No killing of healthy cell	Killing of virus-infected cell in which MHC class I expression is inhibited	Anti-MHC class I antibody prevents interaction between the inhibitory receptor and the MHC class I

Figure 8.19 Possible mechanism by which NK cells distinguish uninfected cells from infected cells. NK cells have activating and inhibitory receptors that interact with ligands on the surfaces of human cells. The nature of the activating receptors and their ligands is uncertain. The inhibitory receptors interact with MHC class I molecules, which are present on virtually all healthy cells. On interaction with a healthy cell the inhibitory signals dominate and inhibit NK-cell cytotoxicity (left panel). Infection of a cell with an intracellular parasite might change the expression or conformation of MHC class I molecules such that the interaction of the cell with the receptors on the NK cell generates a stimulatory signal that causes the NK cell to kill the infected cell (center panel). In the laboratory this situation can be mimicked by using a specific antibody to cover up the MHC class I molecules on a healthy cell's surface so that they no longer bind to inhibitory receptors. Consequently, the NK cell kills the healthy cell (right panel).

Figure 8.20 The inhibitory receptors of human NK cells. One type of receptor is a heterodimer of the CD94 and NKG2A polypeptides, which both contain extracellular domains with sequence similarities to the carbohydrate-recognition domains of C-type lectins, for example the mannose-binding protein. The ligands for the receptor CD94:NKG2A are complexes of HLA-E, a non-classical MHC class I molecule, and peptides derived from the leader peptides of most, but not all, HLA-A, B, and C allotypes. The second type of receptor has extracellular domains that resemble immunoglobulin domains. These receptors are collectively called KIR, for killer cell immunoglobulin-like receptor. KIR3DL has specificity for certain HLA-B allotypes. The KIR2DL receptors are specific for HLA-C heavy chains.

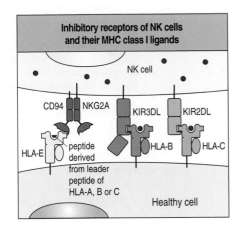

In human NK cells, the inhibitory receptors are of two different types (Figure 8.20). The first kind are heterodimers of CD94 and NKG2, and have characteristics of C-type lectins. The second type are members of the immunoglobulin superfamily and are known as **killer cell immunoglobulin-like receptors** (**KIRs**). The lectin-like receptors are encoded by a family of genes on human chromosome 12, whereas the KIRs are encoded by a family of genes on human chromosome 19. For both types of receptor, binding of an MHC class I ligand to the extracellular domains produces an inhibitory signal.

Both kinds of inhibitory receptor recognize ligands determined by the polymorphisms of HLA class I molecules. The CD94:NKG2A molecule recognizes composite ligands in which a peptide derived from the leader sequence of an HLA-A, B, or C allotype is bound by the monomorphic class I molecule HLA-E. The KIR receptors consist of polypeptides with either two or three extracellular domains that are immunoglobulin-like. KIRs with two domains bind to HLA-C allotypes, whereas those with three domains bind to HLA-B allotypes. Determining the interactions with KIR are polymorphic sequence motifs in the α helix of the α_1 domain of the HLA class I molecule (see Figure 8.20).

Individual NK cells express different types of inhibitory receptor on their surfaces. However, every mature NK cell in a person's circulation expresses at least one inhibitory receptor that can bind to one or more of the HLA class I molecules expressed by the person. This property implies that the HLA class I type of an individual either selects such cells from a repertoire of immature NK cells or actually directs receptor expression to conform to this pattern. As a consequence, mature NK cells do not lyse healthy cells from the same individual. However, because of differences in HLA type, NK cells from one person are often alloreactive toward the cells of another person. In mouse models of transplantation, alloreactions mediated by NK cells can cause the rejection of bone marrow transplants.

8-13 γ:δ T cells and B-1 cells have limited receptor diversity and could be specialized cells of innate immunity

The γ:δ T cells are the first T cells to appear in embryonic development, with the early populations being characterized by limited diversity in their antigen receptors. In particular, the γ:δ cells found in a given epithelial tissue almost all bear identical antigen receptors (see Section 5-5, p. 115). In this respect γ:δ receptors resemble the recognition molecules of innate immunity rather than those of adaptive immunity. This suggests that γ:δ receptors recognize common features associated with infection, perhaps surface changes on epithelial cells caused by stress.

The B-1 lineage of B cells similarly arises early in embryonic development and uses a limited set of V regions to produce surface IgM that is characterized by a rather general affinity for polysaccharide antigens (Figure 8.21). B-1 cells respond rapidly to antigen, producing detectable levels of antibody within 48 hours of

Figure 8.21 B-1 cells provide a quick response to carbohydrate antigens such as bacterial polysaccharides. Almost all of the antibody made by B-1 cells is of the IgM isotype and its principal effector function is the activation of complement. Although the B-1 cell response is stimulated by IL-5 made by T cells, it does not depend on cognate interactions with T cells.

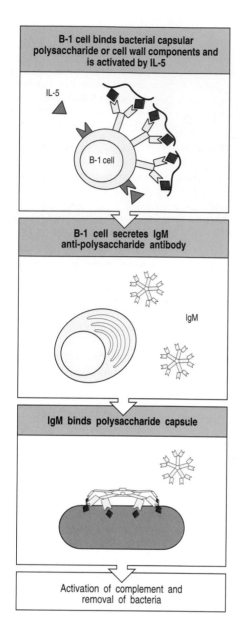

infection. This speed is possible because large numbers of B-1 cells with identical specificity are present, and so little clonal expansion is required to produce amounts of antibodies that are effective in clearing the infection. These responses are independent of cognate T-cell help, do not involve isotype switching or somatic hypermutation, and do not produce immunological memory. B-1 cells are characteristically found in the peritoneal cavity and do not depend on the bone marrow for renewal. As with γ:δ T cells, the receptors of B-1 cells function more like recognition molecules of innate immunity than of adaptive immunity.

Mice deficient in B-1 cells are more susceptible to infection with *Streptococcus pneumoniae* because they fail to produce antibodies specific for phosphoryl-choline, which protect against this bacterium. Many B-1 cells make antibodies of this specificity and because helper T cells are not required for the response it can be made early in infection. Whether human B-1 cells have the same role in infection as mouse B-1 cells is not known.

Although the precise functions of B-1 cells and γ:δ T cells remain uncertain, γ:δ T cells seem to defend the body's surfaces, whereas B-1 cells defend the body cavity. Cells of these types might therefore represent early phases in the evolution of the immune system, when lymphocyte antigen receptors were less diverse but more focused on particular types of pathogen and their products.

Summary

Innate immunity provides a defense against infection that works immediately a pathogen is first confronted or very soon after. These fixed defenses are always available and include physical barriers, provided by skin and other epithelia, microbicidal proteins, complement and the acute-phase proteins of blood plasma, the phagocytic macrophages and neutrophils and the cytokines that they produce, the anti-viral interferons, and natural killer (NK) cells. The limited antigen receptor diversity of γ:δ T cells and B-1 cells suggests that they too are on the front line of defense and relatively fixed in their purpose. Innate immunity has no specificity for particular antigens in the way that adaptive immune responses do, and does not produce a specific immunological memory of an infection. The components of innate immunity act in concert to clear an infection or to contain it until the adaptive immune respose develops. Extracellular pathogens, bacteria in particular, activate the complement system by the alternative pathway, whereas acute-phase proteins bind to bacteria and initiate the classical pathway. Opsonization by complement flags the pathogen as something to be destroyed by macrophages and neutrophils. Macrophages resident within an infected tissue are the first cells to battle with an extracellular pathogen; in doing so they release cytokines, chemokines, and inflammatory mediators that help activate other cells, create a state of inflammation in the infected tissue, and recruit neutrophils and monocytes into the site of infection. Inflammation is central to the innate immune response whether it be directed against extracellular or intracellular pathogens. NK cells that can recognize and kill virus-infected cells are activated by cytokines and inflammatory mediators produced by macrophages and neutrophils. NK cells recognize virus-infected cells that have lost expression of MHC molecules. α- and β-interferons secreted by infected cells inhibit viral replication within cells and also activate NK cells.

Adaptive immune responses to infection

Through the activation of phagocytes, the production of cytokines, and the creation of a state of inflammation, innate immunity sets the scene for the development of the adaptive immune response. In this part of the chapter we shall consider the adaptive immune response to pathogens to which an individual has never been previously exposed. Such infections stimulate a **primary immune response**, involving the coordinated activation of naive T cells and B cells and the production of effector cells and molecules, including antibodies specific against the pathogen. These effectors help to clear the pathogen from the body and prevent re-infection in the short term, while the pathogen is still infecting others in the population. Usually, an adaptive immune response also leads to long-lasting immunity against the pathogen, which we shall consider in the last part of this chapter. Adaptive responses build on a foundation of effector mechanisms that are all part of innate immunity. Indeed, the inflammation produced by the innate immune response provides the essential environment for starting the adaptive immune response. The adaptive response refines and amplifies these effector mechanisms, providing the extra dimension of antigen specificity. In this part of the chapter we shall see how the component parts of adaptive immunity, which have been discussed in Chapters 2–7 work together in clearing infections.

8-14 Adaptive immune responses start with T-cell activation in draining lymphoid tissues

Extracellular pathogens and other antigens are carried by the flow of lymph and blood from the infected site to the secondary lymphoid tissue, where macrophages process and present them to T cells. Macrophages have been activated to become professional antigen-presenting cells as one of the consequences of the innate response. Pathogens also reach secondary lymphoid tissue by carriage within cells specialized for this purpose. Dendritic cells transport both viral infections and endocytosed extracellular antigens from epithelial tissues into lymph nodes and tonsils (see Section 6-4, p. 133) whereas, in the gut, M cells deliver antigens from the gut lumen to Peyer's patches and other lymphoid tissues lining the gut wall. Infections of the skin drain through the lymphatics to nearby lymph nodes, blood-borne pathogens are trapped in the spleen, and infections of the gastrointestinal and respiratory tracts are dealt with by the lymphoid tissues that line their mucosal surfaces. In this section we shall use the lymph node to illustrate the events that proceed in all secondary lymphoid tissues.

Secondary lymphoid tissues provide places where naive T cells can encounter specific antigen. T cells arriving from the blood through the high endothelial venules scan the surface of dendritic cells and macrophages for specific peptide:MHC complexes. The dendritic cells express a chemokine, DC-CK, that specifically attracts naive T cells. The small subset of T cells that are specific for the antigens in the node are activated on engagement of their T-cell receptors; proliferation and differentiation begins and eventually leads to large clones of antigen-specific effector T cells.

Lymph nodes are remarkably efficient at removing pathogens from afferent lymph and bringing them into contact with their specific lymphocytes. In an experiment in which 3 billion streptococci were perfused through a single lymph node in a small volume of fluid (5 ml), over 99% of the organisms were removed. A complementary experiment showed that within 2 days of antigen's being presented in a lymph node, all the antigen-specific T cells in the circulation were trapped there (Figure 8.22).

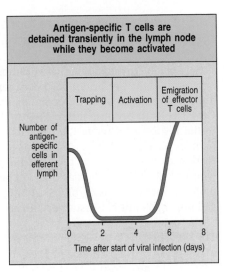

Figure 8.22 Antigen-specific T cells become trapped in lymph nodes. This graph charts the presence of virus-specific T cells in efferent lymph during the course of an experimental virus infection. Before infection, virus-specific T cells recirculate between blood, lymph nodes, and lymphatics. On infection, antigen-specific T cells are trapped, and collect in the lymph nodes draining the infected tissue. After 2 days they can no longer be detected in the efferent lymph. For 2–4 days from the start of the infection, the T cells remain in the lymph node, where they divide and differentiate into effector T cells. At that time, effector T cells leave in the efferent lymph to travel to the infected tissue. Because of clonal expansion, the number of antigen-specific T cells in the efferent lymph after day 4 becomes much larger than the numbers detected before infection.

8-15 Primary CD4 T-cell responses are influenced by the cytokines made by cells of innate immunity

Virus-specific cytotoxic T cells are almost always produced during viral infection, as a result of the presentation of viral peptides to naive CD8 T cells by professional antigen-presenting cells in secondary lymphoid tissues. However, the decision for CD4 T cells to become either T_H1 or T_H2 effector cells is made only during the initial response of naive CD4 T cells to antigen (see Section 6-8, p. 142). Such choices have profound implications for the overall development of the adaptive response, because they tend to be self-reinforcing and thereby determine whether CD4 T-cell populations will help principally macrophages or B cells. A major influence is the cytokine environment experienced by the activated CD4 T cell during its initial burst of cell division.

The presence of IL-12, which at this early stage of infection is mainly the product of dendritic cells and macrophages, drives the differentiation of T_H1 cells (Figure 8.23). IL-12 also stimulates NK cells to produce IFN-γ, which in turn activates macrophages and stimulates T_H1 differentiation. IL-12 and IFN-γ also function to prevent the development of T_H2 cells. Conversely, differentiation of T_H2 cells is promoted by IL-6 and IL-4 and these cells in turn activate B cells and their production of antibodies. Further favoring a T_H2 response is the property of IL-4 to inhibit the generation of T_H1 cells.

The source of the cytokines that drive T_H2 cell differentiation is not established. For mice, one possibility is a subset of CD4 T cells that express the cell-surface molecule NK1.1, a marker usually associated with mouse NK cells. On activation, NK1.1 T cells secrete large amounts of IL-4 and in experimental situations they can bias the differentiation of naive CD4 T cells toward the T_H2 phenotype. In mice infected with parasitic worms, NK1.1 T cells produce IL-4, which drives CD4 T cells to become T_H2 cells, and induces isotype switching to IgE in B cells. IgE is considered, but not proven, to be a useful defense against parasites (see Section 7-13, p. 179). Cells with similar properties as NK1.1 cells have been found in humans. Mast cells activated during innate immune responses might also be a source of the IL-4 that drives T_H2 development.

The commitment of the CD4 T-cell response toward a T_H1 or T_H2 phenotype probably depends on the way that pathogens interact with immature dendritic cells, macrophages, and NK cells during the early phases of an infection, and the mix of cytokines that is synthesized at that time. In this manner, the course of the innate immune response early in infection determines the strategy taken by the adaptive immune response.

The direction of CD4 T-cell differentiation in the primary immune response is also influenced by the abundance and density of the peptide:MHC complexes at the surface of professional antigen-presenting cells and their affinity for T-cell receptors. Peptides that are abundantly presented, or have strong affinity for T-cell receptors, tend to stimulate T_H1 responses, whereas peptides that are at low abundance, or have weak affinity for T-cell receptors, tend to stimulate T_H2 responses. Allergic responses, in which minute amounts of antigen preferentially stimulate T_H2 cells and IgE production, are of the latter type.

8-16 T_H1 cells, T_H2 cells, and cytotoxic T cells regulate each other's growth and effector functions

CD4 T cells differ in the cytokines that they secrete and in the type of immune response that those cytokines produce. Animal models of infectious diseases have shown that two very different types of CD4 T-cell response can be produced,

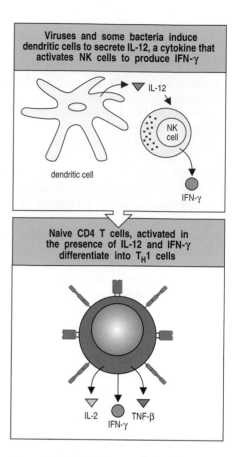

Figure 8.23 Cytokines elicited by the pathogen affect the differentiation of CD4 T cells. The innate immune response to different types of pathogen involves the secretion of cytokines that influence the course of the adaptive immune response. In response to viruses and intracellular bacteria, macrophages and dendritic cells produce IL-12. This cytokine stimulates NK cells to produce IFN-γ, which drives proliferating CD4 T cells to become of the T_H1 type.

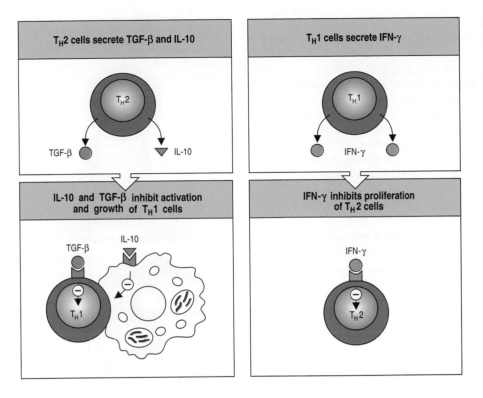

Figure 8.24 The two types of CD4 T cell each produce cytokines that inhibit the other type of cell. T$_H$2 cells make IL-10 and TGF-β (top left panel). IL-10 prevents macrophages from activating T$_H$1 cells, perhaps by preventing the synthesis of IL-12 by the macrophage. TGF-β acts directly on T$_H$1 cells to inhibit their growth (bottom left panel). T$_H$1 cells make IFN-γ (top right panel), which blocks the growth of T$_H$2 cells (bottom right panel).

depending on the experimental conditions. For example, when the C57/BL6 strain of mouse is infected with the intracellular protozoan parasite *Leishmania major*, the CD4 T-cell response consists entirely of T$_H$1 cells secreting IFN-γ. This response activates macrophages. In contrast, infection of the BALB/c strain of mice with exactly the same pathogen produces a CD4 T-cell response that consists entirely of T$_H$2 cells. These cells enhance the production of antibodies. The critical difference between the two responses to *Leishmania* is that the T$_H$1 response clears the infection but the T$_H$2 response does not.

Highly polarized responses of this type can develop because the cytokines produced by one type of effector CD4 T cell inhibit the differentiation and/or the function of the other type. The IL-10 produced by T$_H$2 cells prevents the development of a T$_H$1 response through a general inhibition of antigen processing and presentation by dendritic cells and macrophages. Thus, IL-12 is never produced, which in turn prevents activation of NK cells, differentiation of T$_H$1 cells, and IFN-γ production. The lack of IFN-γ further inhibits macrophages and their ability to act as professional antigen-presenting cells. Conversely, the IFN-γ produced by T$_H$1 cells prevents the activation of T$_H$2 cells (Figure 8.24). From comparison of mouse and human immune responses to a range of pathogens and antigens it seems likely that the highly polarized T$_H$1 and T$_H$2 responses triggered by *Leishmania* in mice define the extremes of the spectrum of CD4 T-cell responses that can actually occur. In general, T$_H$1 cells promote a state of inflammation in which macrophages are activated and opsonizing antibodies are produced, whereas T$_H$2 cells promote a broad-based antibody response.

Certain human diseases seem to result from a CD4 T-cell response that is polarized in ways that do not serve to clear the infection. For example, the chronic infection that causes lepromatous leprosy correlates with the development of a T$_H$2 response against *Mycobacterium leprae* (see Section 6-8, p. 143). A possible therapy against this form of leprosy would be to infuse patients with cytokines that could inhibit the existing T$_H$2 response and promote a T$_H$1 response. The aim being to encourage macrophages to clear the infection. IL-2 and IFN-γ adminis-tered at sites of infection have indeed been shown to resolve lesions and to induce a change in the CD4 T-cell response.

In addition to their cytotoxic function, CD8 T cells also secrete cytokines, which can influence the immune response. The polarization of the CD4 T-cell response to T_H2 in lepromatous leprosy is thought to arise from the presence of CD8 T cells secreting IL-10 and TGF-β, which suppress the T_H1 response. In virus infections, both cytotoxic T cells and antibodies help to clear the infection, and antibodies are of particular importance for long-lived protective immunity. In this situation, suppression of the T_H1 response by cytotoxic T cells is a helpful response.

8-17 Effector T cells are guided to sites of infection by newly expressed cell adhesion molecules

The activation of naive T cells and their differentiation into effector T cells takes 4–5 days. In the process their complement of cell-surface adhesion molecules changes, enabling them to leave the secondary lymphoid tissue and travel to the site of infection. The expression of L-selectin, which enables naive T cells to enter lymph nodes, is lost and the expression of other adhesion molecules is increased. A crucial change is the increased expression on effector T cells of the integrin VLA-4 (see Figure 6.2, p. 132), which binds to VCAM-1 adhesion molecules induced on vascular endothelium by inflammatory cytokines (Figure 8.25).

The differential expression of adhesion molecules is also thought to direct subsets of effector T cells to distinct anatomical sites. Homing to Peyer's patches and to the lamina propria of the gut involves simultaneous binding of L-selectin and LPAM-1 (the integrin $\alpha_4\beta_7$) on the T cell to the MAdCAM-1 adhesion molecule on the vascular endothelium. T cells that home to gut epithelium express the integrin $\alpha_E\beta_7$ and bind to E-cadherin on gut epithelial cells. In contrast, T cells that home to the skin express the cutaneous lymphocyte antigen (CLA) and are thought to bind E-selectin on vascular endothelium.

Not all infections activate local endothelial cells; in such cases it is not known how effector T cells are guided to the infected areas. Effector T cells enter all tissues in small numbers, whether the tissue is infected or not. Such uninvited visits to an infected site could provide the opportunity for effector T cells to meet specific antigen and produce cytokines such as TNF-α, which activate local endothelial cells and facilitate the entry of additional effector T cells. In this manner, the entry of one or a few effector T cells into an infected site could initiate the inflammatory response required to make the site a destination for the rest of the effector T-cell population. Most of the tissues that effector T cells enter in a non-specific manner will be uninfected, and so specific antigen will not be encountered. In these cir-cumstances, effector T cells either die in the tissues by apoptosis or leave in the draining lymphatics to return to the blood.

8-18 Antibody responses develop in lymphoid tissues under the direction of T_H2 cells

Whereas effector CD8 T cells and T_H1 cells work mainly at sites of infection, effector T_H2 cells activate B cells within the secondary lymphoid tissues, from whence secreted antibodies then travel in blood or lymph or across epithelia to the sites of infection. Early in an infection, pathogen-derived antigens are presented by the dendritic cells and macrophages in the draining lymph node. Clones of anti-gen-specific T cells are expanded within the T-cell areas of the node and some differentiate into effector T_H2 cells. Naive B cells also recirculate through the blood and lymph, entering lymph nodes by the same mechanisms used by naive

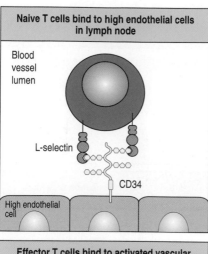

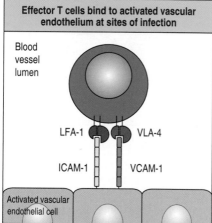

Figure 8.25 Changes in the adhesion molecules expressed by effector T cells enable them to leave the blood and enter infected tissues. As shown in the upper panel, naive T cells express L-selectin, which binds to the sulfated carbohydrates of vascular addressins (such as CD34) on high endothelial vessels in lymph nodes. As shown in the lower panel, effector T cells no longer express L-selectin but express VLA-4 and increased amounts of LFA-1. These adhesion molecules bind to VCAM-1 and ICAM-1 respectively, whose expression is induced on vascular endothelium at sites of inflammation. This change enables effector T cells to home to infected tissues.

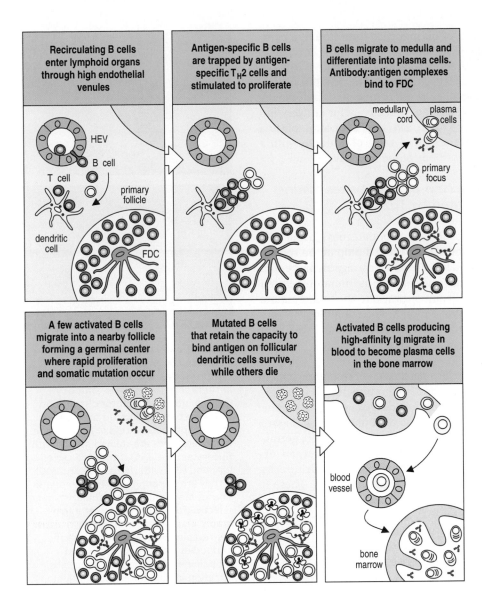

Figure 8.26 Secondary lymphoid tissues provide an environment in which antigen-specific B cells can interact with helper T_H2 cells specific for the same antigen. Top row: Naive B cells leave the blood at high endothelial venules (HEV) and meet effector T_H2 cells specific for the same antigen in the T-cell areas of the lymphoid tissues (a lymph node is illustrated here). B cells that engage with their cognate T_H2 cells become activated and proliferate. Some of the activated B cells move to the medullary chords of the lymph node, where they divide further and differentiate into plasma cells that over a few days secrete antibodies of the first wave. Bottom row: Other activated B cells migrate with their effector T_H2 cells to primary lymphoid follicles, where they are stimulated both by the cytokines released by the T_H2 cells and by the antigen:antibody complexes bound to the surface of follicular dendritic cells (FDC). The resultant B-cell proliferation forms a germinal center, the site of somatic hypermutation and selection of B cells expressing mutant immunoglobulins with higher affinity for antigen. The surviving B cells move to the medullary cords of the lymph node or migrate to the bone marrow, where they become plasma cells secreting a second wave of antibodies with higher overall affinity than those in the first wave.

T cells. B cells enter lymph nodes in the T-cell areas, where they bind and endocytose specific antigen by capturing it with their cell-surface immunoglobulin. The endocytosed antigen is then processed by the B cells and presented on MHC class II molecules to the effector T_H2 cells, which in response secrete the cytokines necessary for B-cell proliferation and differentiation.

After encounter with antigen, some 5 days are required for effector T_H2 cells to be produced from naive T cells. At about this time after the start of an infection, antigen-specific B cells in contact with T_H2 cells begin to proliferate in the T-cell area to form a primary focus. Some of the B cells activated in these primary foci migrate to the medullary cords of the node, where they become short-lived plasma cells that produce a first wave of pathogen-specific antibody (Figure 8.26, top row). The corresponding B cells in the spleen move to those parts of the red pulp directly adjoining the T-cell zones of the white pulp. In these sites, B-cell numbers increase exponentially for 2–3 days; the cells undergo six or seven cell divisions before their progeny differentiate into antibody-producing plasma cells. Most of these plasma cells die by apoptosis after 2–3 days. Other activated B cells migrate to the follicles, where they proliferate further, forming a germinal center in which the B cells undergo somatic hypermutation (see Figure 8.26, bottom row).

The antibodies produced by the first wave of plasma cells form immune complexes with antigen and activate complement. Some of the immune complexes bind to follicular dendritic cells in the lymphoid follicles (see Figure 8.26). The variant antibodies produced by the mutated B cells are tested against the antigen in the immune complexes collected on the follicular dendritic cells. B cells having the highest-affinity antigen receptors compete most effectively for antigen and will be selected to survive, to be helped by CD4 T cells, and to become plasma cells (see Figure 7.12, p. 168). They secrete a second wave of antibody with a higher overall affinity for antigen than the antibody of the first wave.

8-19 Antibody secretion by plasma cells occurs at sites distinct from those where B cells are activated by T_H2 cells

As we saw in the previous section, the first B cells to differentiate into plasma cells migrate to the medullary cords in lymph nodes or to the red pulp of the spleen. The other activated B cells enter the lymphoid follicles and form a germinal center. About 10% of the activated B cells in germinal centers leave as immature plasma cells called **plasmablasts**, and migrate to distant sites where they differentiate into plasma cells that live on average for about a month. The remaining 90% die in the germinal center, having failed to capture antigen and cognate T-cell help (see Figure 7.12, p. 168).

Plasmablasts originating in the follicles of Peyer's patches and mesenteric lymph nodes migrate to the lamina propria of the gut and other epithelia. Those originating in peripheral lymph nodes migrate to the medullary cords or the bone marrow, and those originating in the spleen migrate to the bone marrow (Figure 8.27). In immune responses that successfully terminate an acute infection, germinal centers are present for only 3–4 weeks after the supply of antigen is turned off. After this time, small amounts of antigen are retained on follicular dendritic cells and these continue to stimulate the proliferation of B cells in follicles. These B cells are likely to be the precursors of pathogen-specific plasma cells that can be found in the mucosa and bone marrow over periods of months to years. They are thus the source of the antibodies that provide protective immunity.

8-20 The effector mechanisms used to clear an infection depend on the pathogen

The primary adaptive immune response terminates most infections that elude the innate immune response. However, within the population suffering from an infectious disease, some individuals succumb to its effects before the immune system can gain the upper hand. Those at greatest risk are the young, the old, the malnourished, and those already suffering from other diseases. Some infections can never be cleared, because certain pathogens successfully evade or subvert the immune response so that they can persist in the body for life. In this latter category are the parasites *Leishmania* and *Toxoplasma*, and the herpes viruses, including Epstein–Barr virus, herpes simplex virus, and cytomegalovirus. Depending on the way in which a pathogen exploits the human body, the type of immune response that eliminates or contains it is different. The different types of infection and the way in which they can be eliminated by a primary immune response are summarized in Figure 8.28.

Summary

Most infections are highly localized and can be cleared by the innate immune response. Only if infection spreads to secondary lymphoid tissues is an adaptive immune response made, and its initiation is dependent on the inflammatory cytokines produced by the cells of innate immunity. In the secondary lymphoid

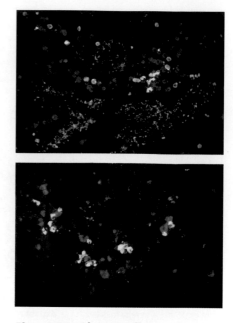

Figure 8.27 Plasma cells are dispersed in medullary cords of lymph nodes and in bone marrow. In these sites they secrete antibody directly into the blood for distribution to the rest of the body. In the upper micrograph, plasma cells in lymph node medullary cords are stained green (with fluorescein anti-IgA) if they are secreting IgA, and red (with rhodamine anti-IgG) if they are secreting IgG. The lymphatic sinuses are outlined by granular staining selective for IgA. In the lower micrograph, plasma cells in the bone marrow are revealed with light-chain specific antibodies (fluorescein anti-λ and rhodamine anti-κ stain). Plasma cells secreting immunoglobulins containing λ light chains are stained yellow-green; those secreting immunoglobulins containing κ light chains stain red. Photographs courtesy of P. Brandtzaeg.

Infectious agent	Disease	Humoral immunity				Cell-mediated immunity	
		IgM	IgG	IgE	IgA	CD4 T cells (+macrophages)	CD8 cytotoxic T cells
Viruses							
Variola	Smallpox						
Varicella-zoster	Chickenpox						
Epstein–Barr virus	Mononucleosis						
Influenza virus	Influenza						
Mumps virus	Mumps						
Measles virus	Measles						
Polio virus	Poliomyelitis						
Human immunodeficiency virus	AIDS						
Bacteria							
Staphylococcus aureus	Boils						
Streptococcus pyogenes	Tonsilitis						
Streptococcus pneumoniae	Pneumonia						
Neisseria gonorrhoeae	Gonorrhea						
Neisseria meningitidis	Meningitis						
Corynebacterium diphtheriae	Diphtheria						
Clostridium tetani	Tetanus						
Treponema pallidum	Syphilis			Transient			
Borrelia burgdorferi	Lyme disease			Transient			
Salmonella typhi	Typhoid						
Vibrio cholerae	Cholera						
Legionella pneumophila	Legionnaire's disease						
Rickettsia prowazekii	Typhus						
Chlamydia trachomatis	Trachoma						
Mycobacteria	Tuberculosis, leprosy						
Fungi							
Candida albicans	Candidiasis						
Protozoa							
Plasmodium spp.	Malaria						
Toxoplasma gondii	Toxoplasmosis						
Trypanosoma spp.	Trypanosomiasis						
Leishmania spp.	Leishmaniasis						
Worms							
Schistosome	Schistosomiasis						

Figure 8.28 The effector mechanisms used to clear primary infections depend on the pathogen. The classes of pathogens are listed in order of their increasing anatomical complexity. The defense mechanisms used to clear a primary infection, where they are known, are identified by the red shading of the boxes. Yellow shading indicates some role in protective immunity; paler shades indicate less well-established mechanisms. Much has still to be learnt about the interactions between pathogens and the human immune system. It is apparent that pathogens which exploit the cells and tissues of the body in similar ways elicit similar types of immune response.

	Cytotoxic CD8 T cell	T_H1 cell	T_H1/T_H2 cell
Typical pathogens	Vaccinia virus Influenza virus Rabies virus *Listeria*	*Mycobacterium tuberculosis* *Mycobacterium leprae* *Leishmania donovani* *Pneumocystis carinii*	*Clostridium tetani* *Staphylococcus aureus* *Streptococcus pneumoniae* Polio virus *Pneumocystis carinii*
Location	Cytosol	Macrophage vesicles	Extracellular fluid
Antigen recognition	Peptide:MHC class I on infected cell	Peptide:MHC class II on infected macrophage	Peptide:MHC class II on antigen-specific B cell
Effector action	Killing of infected cell	Activation of infected macrophages	Activation of specific B cell to make antibody

Figure 8.29 The role of the three kinds of effector T cell in the response to typical pathogens.

tissues the presentation of pathogen-derived antigens to naive circulating T cells causes pathogen-specific T cells to divide and differentiate into effector T cells within the lymphoid tissue. The relative production of T_H1 and T_H2 effector CD4 T cells and of CD8 T cells depends on the nature of the infection and the mix of inflammatory cytokines produced at the site of infection. Effector T_H1 cells and CD8 T cells leave the lymphoid tissues and travel in the lymph and blood to the infected tissues. Once there, T_H1 cells activate the destruction of extracellular pathogens by macrophages, while CD8 T cells kill infected human cells. Effector T_H2 cells remain mainly in the lymphoid tissues, where they help to activate pathogen-specific B cells and drive isotype switching and affinity maturation of their immunoglobulins. The roles of effector T cells in immune responses to various pathogens are summarized in Figure 8.29.

Immunological memory and the secondary immune response

As well as clearance of the infection, a successful primary adaptive immune response also establishes a state of long-term **protective immunity**. A second or subsequent encounter with the same pathogen will provoke a faster, stronger, **secondary immune response**. This is produced by circulating antibody and by clones of long-lived B and T cells that have been formed during the primary response. In a person possessing protective immunity, the infection is usually cleared before it produces any symptoms. The purpose of vaccination is to produce a state of protective immunity against a particular pathogen.

8-21 Immunological memory after infection is long-lived

In the period immediately after recovery from an infection, a person will usually encounter the pathogen again as it is passed around family, friends, colleagues, and other members of the community. Such re-infections are terminated quickly by the effector cells and antibodies built up during the first infection. Specific antibodies present in blood and extracellular fluids as a result of a primary immune response are one of the first immune defenses encountered by a pathogen on a subsequent infection. Their rapid action prevents the infection from becoming established. As a result of the re-infection, new supplies of antibody and effector cells will be made. In the absence of further infection the levels of antibody and effector T cells gradually decline, eventually reaching very low levels that are no longer effective. However, the B- and T-cell populations retain a 'memory' of the infection that enables a rapid and powerful secondary adaptive immune response to be made should re-infection occur many years after the

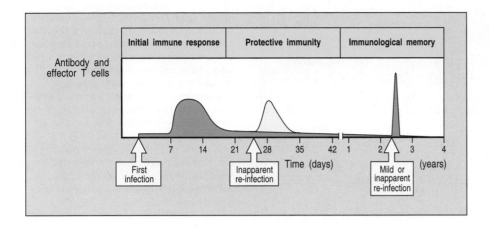

Figure 8.30 Protective immunity consists of preformed antibodies and existing effector T cells, and of immunological memory. The abundance of pathogen-specific antibodies and effector T cells is tracked here through the course of a person's successive exposures to a pathogen. During first infection the level rises reaching a plateau as the immune system gains control. After clearance the levels decline but for a time are sufficient to quickly terminate any re-infection. After an extended period without exposure to the pathogen, levels decline to the point where protective immunity is mediated by memory B and T cells, which respond rapidly to infection by producing new supplies of antibody and effector T cells.

primary infection. Memory B and T cells can be activated more quickly and in greater numbers than are naive lymphocytes activated in the primary response. As a consequence, the infection is cleared quickly by the secondary response, with few or no symptoms of disease (Figure 8.30).

This phenomenon of **immunological memory** is classically illustrated by epidemiological studies of the inhabitants of one of the Faroe Islands in the North Atlantic Ocean, who were severely affected by a measles epidemic that infected the entire population. Many years later, when measles virus was again introduced to the island, only those members of the population born after the original epidemic developed disease. All those who had survived the first measles epidemic retained a protective immunity that prevented their second measles infection from becoming established and causing disease.

The molecular and cellular mechanisms that maintain immunological memory remain incompletely understood and much debated. In one type of model, memory is sustained by long-lived lymphocytes formed during the primary response, which persist in a resting state and can be reactivated when confronted with the pathogen. A second type of model envisages that lymphocytes activated by the original exposure to the pathogen are periodically restimulated, even in the absence of further encounters with the pathogen. This prevents their death by apoptosis and maintains their numbers. Such restimulation could be caused in three ways: by small amounts of pathogen antigen that persist in the body after recovery from infection; by cross-reactive recognition of other pathogens or antigens that enter the body; or by the cytokines generated in responses to other pathogens and antigens.

Knowledge of the mechanisms of immunological memory comes largely from the study of lymphocytes which have been taken from mice immunized with a protein antigen and transferred to unimmunized mice. In this situation immunological memory is transferred with the lymphocytes. Thus when the recipient mice are subsequently challenged with the antigen they produce a response which has the characteristic kinetics, strength, and quality of a secondary immune response. Such experiments have shown that memory B cells reach their maximum number one month after a primary encounter with an antigen and that this level is maintained with little alteration for the lifetime of the animals—about 2 years.

8-22 Memory B cells are resting B cells with isotype-switched, high-affinity immunoglobulins

The presence of memory cells in the B-cell population is relatively easy to demonstrate, because the antibodies they produce at the beginning of a secondary response differ in several ways from those made by activated naive B cells at the

	Source of B cells	
	Unimmunized mouse	**Immunized mouse**
Frequency of specific B cells	$1:10^4 - 1:10^5$	$1:10^3$
Isotype of antibody produced	IgM > IgG	IgG, IgA
Affinity of antibody	Low	High
Somatic hypermutation	Low	High

Figure 8.31 The antibodies produced on the activation of memory B cells are distinct from those produced on the activation of naive B cells. B cells taken from unimmunized and immunized mice are stimulated with antigen *in vitro* to model a primary and a secondary response, respectively. In the secondary response the first antibodies made differ in their isotype, affinity and extent of somatic hypermutation from those produced in the primary response. In both these experiments antigen-specific helper T cells were provided *in vitro*.

start of a primary response. This can be shown by taking B cells from an unimmunized mouse and from a mouse several weeks after immunization and stimulating them both with antigen *in vitro* in the presence of effector helper T cells specific for the same antigen (Figure 8.31).

The antigen-specific B-cells that respond in the latter case are the memory B cells. There are 10–100 times more of them than in the naive B-cell population as a result of the proliferation of a relatively small number of clones of antigen-specific B cells during the primary response. Furthermore, the average antibody affinity for antigen is greater in the secondary response. Memory B cells are also distinguished by their ability to make antibodies of isotypes other than IgM. The population of memory B cells generated by the primary response includes cells that have switched to IgG, IgA, and IgE, so these isotypes emerge at the very beginning of a secondary response.

Both the numbers of antigen-specific B cells (as judged by the amount of antibody produced) and the average affinity of the antibodies continue to increase through subsequent encounters with antigen (Figure 8.32). As can be seen from the figure, increases in affinity are much greater for IgG than for IgM. This is because IgG derives from isotype-switched B cells that have had time to undergo affinity maturation in germinal centers. IgM, in contrast, is produced chiefly from B cells that differentiate early in the immune response and have not undergone affinity maturation. The progressive increase in affinity is due to the processes of somatic hypermutation and affinity maturation (see Section 7-4, p. 166), which continue in reactivated memory B cells when they enter germinal centers. B cells making higher-affinity immunoglobulins are selected for further clonal expansion and differentiation.

The higher affinity of their antigen receptors makes memory B cells more efficient than naive B cells in binding and internalizing antigen for processing and presentation to helper T cells. Memory B cells also express high levels of MHC class II molecules on their surface compared with naive B cells, which also makes their presentation of antigen more efficient. These factors enable the secondary

Figure 8.32 The amount and affinity of antibody increase after successive immunizations with the same antigen. The upper panel shows the increase in the amount of antibody present in blood serum over a course of three immunizations with the same antigen. The relative amounts of the IgM (green) and IgG (blue) isotypes are shown. The lower panel shows the changes in average antibody affinity that occur. Note that the vertical axis of each graph has a logarithmic scale.

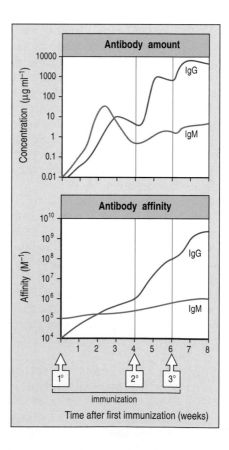

response to be more sensitive to the presence of antigen. Smaller pathogen populations are sufficient to trigger a B-cell response, which will therefore occur at a time earlier during an infection than was the case for the primary response.

By the generation and maintenance of populations of antigen-specific memory B cells, the immune system makes a heavy investment in protection against reexposure to infections already experienced and survived. For these, a strong memory B-cell response can be recalled years after infection. There is an obvious advantage to this strategy. Infections are passed from person to person within human populations, often over extended periods, and the likelihood of repeated exposures to the same pathogen remains high. Antibodies produced in the primary response remain in the blood and lymph for some time after the infection has cleared. They are important contributors to a secondary response because they can bind to pathogens immediately the pathogen enters the body. The strength of B-cell memory is such that even with organisms that produce high mortality on first infection, for example the smallpox virus, the mortality on secondary exposure is very low.

8-23 T-cell memory is maintained by T cells that have different cell-surface markers from naive T cells

As the T-cell receptor undergoes neither isotype switching nor somatic hypermutation, these properties cannot be used to distinguish between T cells of the primary and secondary responses, as is possible for B cells. Furthermore, because effector T cells are known to survive for fairly long periods, it is uncertain whether the 10–100-fold increase in antigen-reactive T cells that follows a primary immune response, and persists throughout life, arises only from long-lived effector cells or also involves a distinct type of memory T cell. Some cell-surface differences between naive T cells and cells that provide T-cell memory are shown in Figure 8.33.

Cell-surface molecule	Other names	Relative expression on cells of indicated subset		Comments
		Naive	Memory	
LFA-3	CD58	1	>8	Ligand for CD2, involved in adhesion and signaling
CD2	T11	1	3	Mediates T-cell adhesion and activation
LFA-1	CD11a/CD18	1	3	Mediates leukocyte adhesion and signaling
α_4 integrin	VLA4	1	4	Involved in T-cell homing to tissues
CD44	Ly24 Pgp-1	1	2	Lymphocyte homing to tissues
CD45RO		1	30	Lowest molecular weight isoform of CD45
CD45RA		10	1	High molecular weight isoform of CD45
L-selectin		High	Most high, some low	Lymph node homing receptor
CD3		1.0	1.0	Part of T-cell receptor complex

Figure 8.33 Naive and memory T cells differ in their surface phenotype. Many of the cell-surface molecules that distinguish memory T cells from naive T cells are also shared with effector T cells. The changes serve to increase the adhesion of the T cell to antigen-presenting cells and to endothelial cells. They also increase the sensitivity of the memory T cell to antigen stimulation. This occurs, for example, with the cell-surface phosphatase CD45, for which there is a change in the isoforms expressed. In naive T cells, high molecular weight isoforms are used that do not associate with either the T-cell receptor or co-receptor. In memory T cells, alternative mRNA splicing generates a low molecular weight isoform that associates with the T-cell receptor and co-receptor complex and enables it to transduce signals more effectively than the receptor on naive T cells.

Direct evidence for a memory T cell with distinctive properties comes from the study of CD8 cytotoxic T cells. Memory CD8 T cells differ from effector cytotoxic CD8 T cells in that they need an induction period of 1–2 days before they have the capacity to lyse target cells. Effector CD8 T cells on the other hand can program a target cell for lysis in a matter of minutes. Memory CD8 T cells are also unlike naive CD8 T cells because their induction period is shorter than that for naive T cells and no cell division is involved.

The evidence for memory CD4 T cells is less direct, principally because helper function (for example, the stimulation of antibody production by B cells) requires several days after antigen stimulation to become measurable, a period comparable to that needed for the presumed induction. By contrast, the cytolytic function of CD8 T cells can be determined within a few hours after completion of the induction period. What has been defined is a subset of CD4 T cells in immunized animals that have the characteristic cell-surface marker molecules of effector cells but different activation requirements. Unlike effector CD4 T cells, these presumed memory T cells can be stimulated to secrete effector cytokines only by antigen presented by a professional antigen-presenting cell.

8-24 Antigens retained long after infection can maintain immunological memory and protective immunity

Immediately after a successful adaptive immune response, the pathogen population is largely cleared from the body and the activation of naive lymphocytes no longer occurs. Antibody levels gradually decline and effector T-cell activities are no longer detectable. In the absence of reliable methods for detecting single particles or molecules, it is impossible to decide whether all pathogens or pathogen-derived antigens have been eliminated or not. There is evidence that small amounts of antigen are retained over long periods, attached to follicular dendritic cells. Some immunologists propose that such residual antigen is essential for the maintenance of memory B and T cells.

In addition to targeting antigens to phagocytes for destruction, antibodies also deliver pathogen antigens to the secondary lymphoid tissues in the form of immune complexes, where they become associated with the surfaces of follicular dendritic cells (Figure 8.34). The sequestering of antigen into secondary lymphoid tissues by antigen:antibody complexes in this way means that small amounts of antigen can be used efficiently to stimulate immune responses. Follicular dendritic cells retain antigen on their surface for very long periods of time and, in the form of iccosomes, provide a long-term depot of antigen for continued B-cell stimulation and affinity maturation (see Section 7-4, p. 166). This in turn leads to the restimulation of antigen-specific T cells, resulting in the formation of memory B and T cells. If antigen were not retained by follicular dendritic cells, all immune complexes formed in the course of the immune response would be degraded by phagocytes and there would be no long-term depot of antigen available to restimulate B and T cells.

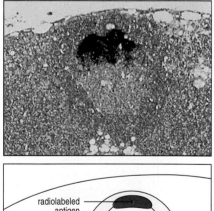

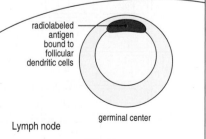

Figure 8.34 Antigen injected into immune animals localizes to follicular dendritic cells in lymphoid follicles. Antigen localization within immune animals has been studied by using radioactive antigens that are detected by autoradiography. Radiolabeled antigen localizes to, and persists in, lymphoid follicles of draining lymph nodes, as shown in the light micrograph (top panel) and the accompanying schematic representation (bottom panel). Radiolabeled antigen was injected 3 days previously and its localization in part of the germinal center rich in follicular dendritic cells is shown by the intense dark staining. Photograph courtesy of J. Tew.

Figure 8.35 Encounter with antigen generates effector T cells and long-lived memory T cells. Most of the effector T cells generated from antigen-stimulated naive T cells are relatively short lived, dying either from antigen overload or from lack of antigenic stimulus or necessary cytokines. Some become long-lived memory T cells; repeated antigenic stimulation might be required for these cells to persist. APC, antigen-presenting cell.

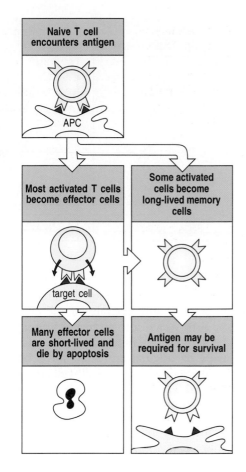

In this model, the derivation of memory T cells from either activated naive T cells directly or from effector T cells is irrelevant. The main proposal is that antigen is critical in determining the fate of activated T cells. What it suggests is that high doses of antigen trigger the apoptosis of effector T cells, as does the complete absence of antigen. Intermediate levels of antigen permit the long-term persistence of clones of antigen-specific T cells, either because such doses of antigen program individual cells for an extended lifespan or because the clone is preserved by continual generation of new cells which replace the old (Figure 8.35).

8-25 The second and subsequent responses to a pathogen are mediated solely by memory lymphocytes and not by naive lymphocytes

The secondary adaptive immune response to a pathogen involves the reactivation of clones of B and T cells that were first activated and expanded during the primary response. Activation of naive lymphocytes with specificity for the pathogen does not occur in a secondary immune response. This is because the cells and molecules made during the primary response prevent the activation of naive lymphocytes. Such suppression can be demonstrated by the transfer to unimmunized mice of antigen-specific antibody or effector T cells produced by a primary immune response in immunized mice. When the recipient mice are subsequently immunized with the antigen, the naive antigen-specific lymphocyte populations known to be present in these mice are not activated.

This phenomenon is put to practical use in prevention of the **hemolytic anemia of the newborn** caused by antibodies specific for the rhesus (Rh) antigen on erythrocytes. If a Rh⁻ mother carrying a Rh⁺ child is infused with anti-Rh IgG antibody before she begins to make an immune response to her child's red cells, the response will be inhibited. Fetal red cells that enter the maternal circulation bind the anti-Rh IgG antibodies. When these red cells subsequently bind to the surface immunoglobulin of naive maternal B cells that are Rh specific, a crosslinking of B-cell receptors and Fc receptors occurs. This crosslinking delivers an inhibitory signal to the naive B cell. In contrast, when the same antigen:antibody complexes interact with the antigen receptors and the Fc receptors on the surface of memory B cells, an activating signal is delivered (Figure 8.36).

Such a mechanism, whereby the primary response to a pathogen or antigen prevents the further activation of naive cells during a secondary response, is believed to underlie the phenomenon of **original antigenic sin**. This implicitly puritanical term was coined on the basis of observations that the first influenza strain encountered by a person constrained the types of antibody made in response to future infections with different strains. During a second infection only those epitopes that are common to the infecting strain and the original strain stimulate antibody production. This dependence on the memory cells that developed during the first infection is seen even when subsequent infections involve influenza strains with highly immunogenic epitopes that are different from those on the original strain. The strategy of the immune system is to use only B cells that are

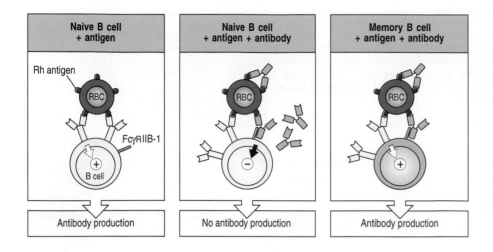

Figure 8.36 IgG antibody suppresses the activation of naive B cells by crosslinking the B-cell receptor to the receptor FcγRIIB-1 on the B-cell surface. Antigen binding to the B-cell antigen receptor delivers an activating signal (left panel) via the B-cell receptor; simultaneous signaling via the antigen receptor and FcγRIIB-1 delivers a negative signal to naive B cells (middle panel). Such crosslinking does not affect memory B cells (right panel). In this example the antigen is the Rh+ alloantigen of erythrocytes. In pregnancy a fetus that is Rh+ can stimulate an Rh− mother to make Rh−-specific antibody. This response can be inhibited by prior infusion of some anti-Rh IgG antibody.

rapidly and easily mobilized. This allows the highly mutable influenza virus to increasingly escape from the protective immunity of a host while the host's immune system is paralyzed and prevented from responding to the changes in the virus. As a result, protective immunity to influenza is not retained for life, and the virus successfully infects millions of human beings every year. The imprint made by original antigenic sin is broken only on infection with a strain of influenza that lacks all the B-cell epitopes of the strain to which a person was first exposed (Figure 8.37).

Summary

In the course of the primary immune response, effector T cells and antibodies accumulate. Once an infection has been cleared, these cells and molecules will prevent re-infection by the same pathogen in the short term. In the long term, persistent clones of memory B and T cells expanded during the primary response afford protective immunity if the same antigen is encountered. When that happens they mount a secondary response that is both stronger and more rapid than the primary response. Antibodies produced in the secondary resonse are of higher affinity and of isotypes other than IgM. Activation of naive lymphocytes is suppressed and only memory lymphocytes are reactivated. In a secondary immune response, the immune system therefore devotes all its resources to producing high-affinity antibodies and specific T cells that rapidly clear the invading pathogen before the infection can become established. Long-term immunity might require periodic restimulation by small amounts of the original antigen retained on follicular dendritic cells.

Summary to Chapter 8

The human body has three types of defense, all of which must be overcome if a pathogen is to establish an infection and then exploit its human host for the remainder of that person's life. The first defense is made up of the protective epithelial surfaces of the body, which successfully prevent most pathogens from ever gaining entry to the rich resources of the body's interior. Any pathogen that succeeds in penetrating an epithelial surface is immediately faced by the recognition molecules and effector cells of innate immunity. Some of these elements of immunity have been refined over hundreds of millions of years to respond to the distinguishing features of bacteria and other groups of pathogens. Viruses pose a different challenge, because they are made by human cells and, unlike bacteria,

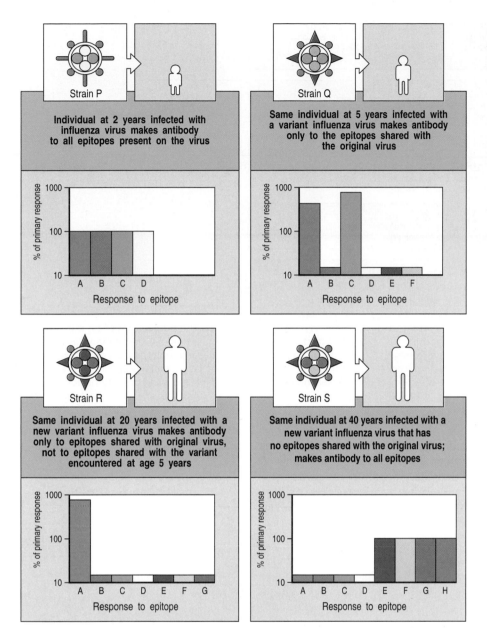

Figure 8.37 Highly mutable viruses such as influenza can escape from the protective immunity of the human host without eliciting new immunity to the changes in the virus. A child 2 years old is infected with influenza virus strain P and responds by making antibodies against all the viral epitopes: A, B, C, and D (top left panels). At 5 years old the child is infected with the Q strain of the virus, which shares epitopes A and C with strain P but has new epitopes E and F. The child's immune system uses only memory B cells resulting from the earlier infection with strain P in defense against strain Q and so the response is restricted to antibodies against epitopes A and C (top right panels). At 20 years old, the same individual is infected with a third strain of influenza, R, which has epitope A in common with strain P, epitopes E and F in common with strain Q, and a new epitope, G. Even 18 years after the first influenza infection, the antibody response concentrates solely on epitope A, which is shared with strain P (bottom left panels). Only when the person is infected with influenza strain S, which lacks epitopes A, B, C, and D, can a primary response be made to epitopes E, F, G, and H of strain S (bottom right panels). Strain S is seen as a novel virus to the person's immune system and as a consequence the person will likely suffer a full blown influenza as a result of this infection.

do not have chemically distinctive surface macromolecules that non-specifically distinguish them from human cells. However, NK cells have receptor systems that can sense subtle changes in the surface of infected cells and, by killing the cells, eliminate or contain viral infections.

Most infections are efficiently cleared by the innate immune response and lead to neither disease nor incapacitation. In the minority of infections that escape innate immunity and spread from their point of entry, the pathogen then faces the combined forces of innate and adaptive immunity. The week or so required for the primary adaptive immune response to develop is the time during which the body is most vulnerable and during which infections can progress to the point of causing disease, damage, and even death. This delay, during which the pathogens have the upper hand and can be transmitted to other people, is largely responsible for the devastation caused by epidemic infections. Most infections, however, never reach that stage, but are successfully terminated by the adaptable

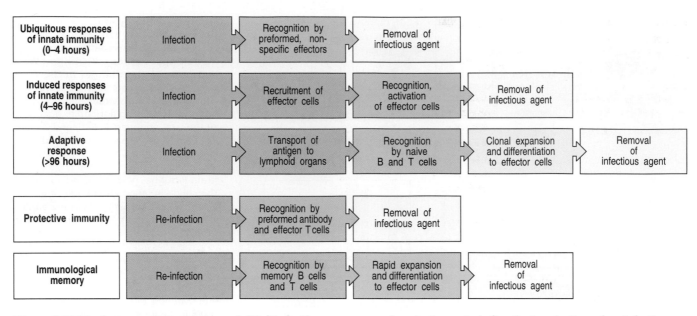

Figure 8.38 The immune response to an initial infection. The innate immune response can be divided into mechanisms that start immediately on infection, for example complement activation and macrophage phagocytosis, and those induced within a few hours by inflammatory cytokines. The effector functions of adaptive immunity begin to work only after 4 or more days. In the period after the termination of an infection, the effector cells and molecules made in the primary response continue to provide protective immunity but gradually decline. In the long term, immunological memory enables the strength of the secondary response to far surpass the primary response.

and specific recognition systems of adaptive immunity. A successful recovery from infectious disease and the consequent development of immunological memory provide the body with enhanced defenses that will repel future attack from the causal agent (Figure 8.38). Once a state of immunity has been established, a subsequent encounter with the pathogen provokes a stronger and more rapid attack that clears the pathogen from the body before the infection can take hold.

Failures of the Body's Defenses

9

In Chapter 8 we saw how most pathogens are prevented from establishing infection and how most infections are terminated through the combined actions of innate and adaptive immunity. In this situation there is strong pressure on pathogens to evolve ways of escaping or subverting the immune response. Microorganisms with such advantages can compete more successfully with other potential pathogens to exploit the resources of the human body. The first part of this chapter describes examples of the different types of mechanisms they use.

The body's defenses against infection can also fail because of inherited deficiencies of the immune system. Some of these are described in the second part of the chapter. Within the human population are mutant alleles for many of the genes encoding components of the immune system. These mutant genes cause immunodeficiency diseases, which vary in severity depending on which gene is defective.

Correlation of the molecular defects in immunodeficiency diseases with the types of infection to which patients become vulnerable reveals the effectiveness of the various arms of the immune response against different kinds of pathogen.

In the third part of the chapter we explore one particular host–pathogen relationship that combines themes from the first two parts of the chapter. This concerns the human immunodeficiency virus (HIV), which is extraordinarily effective at both escaping and subverting the immune response. During the course of an infection, which can last for decades, HIV gradually but inexorably wears down the immune system to the point where it no longer works. The long-term consequence of HIV infection is that patients become severely immunodeficient and develop the fatal disease known as acquired immune deficiency syndrome (AIDS).

Evasion and subversion of the immune system by pathogens

The immune response to any pathogen involves complex molecular and cellular interactions between the pathogen and its host, and any stage in this interaction could be targeted by a pathogen and used for its own benefit. The systematic study of pathogen genomes reveals that most, if not all, pathogens have means of escaping or subverting immune defenses, and that some of them have many genes devoted to this purpose.

9-1 Genetic variation within some species of pathogen prevents effective long-term immunity

Antibodies directed against macromolecules on the surfaces of pathogens are the most important source of long-term protective immunity to many infectious diseases. Some species of pathogen evade such protection by existing as numerous different strains, which differ in the antigenic macromolecules on their outer surfaces. One such pathogen is the bacterium *Streptococcus pneumoniae*, which causes pneumonia. Different genetic strains of *S. pneumoniae* differ in the structure of the capsular polysaccharides and compete with each other to infect humans. These strains, of which at least 90 are known, are called **serotypes** because antibody-based serological assays are used to define the differences between them. After resolution of infection with a particular serotype, or type, of *S. pneumoniae*, a person will have made antibodies that prevent re-infection with that type but which will not prevent primary infection with another type (Figure 9.1). *S. pneumoniae* is a common cause of bacterial pneumonia because its genetic variation prevents individuals from developing an effective immunological memory against all strains. Genetic variation in *S. pneumoniae* has evolved as a result of selection by the immune response of its human hosts.

9-2 Mutation and recombination allows influenza virus to escape from immunity

Some viruses also display genetic variation, influenza virus being a well-studied example. This virus infects epithelia of the respiratory tract and passes easily from one person to another in the aerosols generated by coughs and sneezes. Protective immunity to influenza is provided principally by antibodies that bind to the hemagglutinin and neuraminidase glycoproteins of the viral envelope. These antibodies are made during the primary immune response to the virus. The course of a primary infection is short (1–2 weeks) and the virus is cleared from the system by a combination of cell-mediated immunity and antibody. The pattern of infection of influenza virus characteristically causes **epidemics**, in which the virus

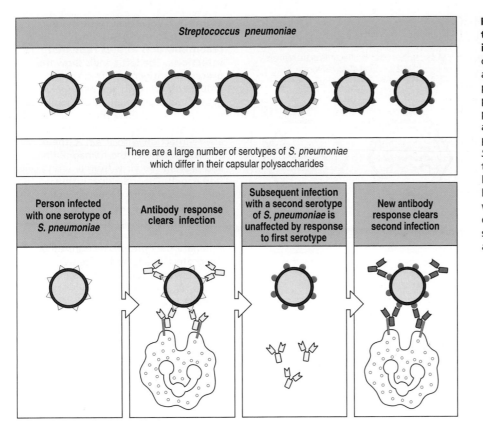

Streptococcus pneumoniae

There are a large number of serotypes of *S. pneumoniae* which differ in their capsular polysaccharides

| Person infected with one serotype of *S. pneumoniae* | Antibody response clears infection | Subsequent infection with a second serotype of *S. pneumoniae* is unaffected by response to first serotype | New antibody response clears second infection |

Figure 9.1 Protective immunity towards *Streptococcus pneumoniae* is serotype-specific. Different strains, or serotypes, of *S. pneumoniae* have antigenically different capsular polysaccharides, as shown in the upper panel. Antibodies against capsular polysaccharides opsonize the pathogen and enable it to be phagocytosed. A person infected with one serotype of *S. pneumoniae* clears the infection with type-specific antibody, as shown in the lower panels. These antibodies, however, have no protective effect when the same person is infected with a different type of *S. pneumoniae*. The second infection can be cleared only by a new primary response.

spreads rapidly through the population and then quickly subsides. Long-term survival of the influenza virus is ensured by the generation of new viral strains that evade the protective immunity generated during past epidemics.

Influenza is an RNA virus with a genome consisting of eight RNA molecules. RNA replication is relatively error-prone and generates many point mutations on which selection can act. New viral strains which lack the hemagglutinin or neuraminidase epitopes that induced protective immunity in the previous epidemic regularly emerge and cause an influenza epidemic every other winter or so. An individual's protective immunity against influenza is determined by the strain of virus to which they were first exposed—the phenomenon of 'original antigenic sin' (see Section 8-25, p. 237). The history of exposure to particular strains of the virus differs within the population, largely according to age, and so there are subpopulations of people with differing degrees of immunity to the current strain of influenza. The people to suffer most at any particular time will be those whose protective immunity has been lost because of the new mutations present in the current strain. This type of evolution of influenza, which causes relatively mild and limited disease epidemics, is called **antigenic drift** (Figure 9.2, left panels).

In contrast, every 10–50 years an influenza virus emerges that is structurally quite different from its predecessors and is able to infect almost everyone. Besides spreading more widely to cause a **pandemic** (a worldwide epidemic), such viruses cause more severe disease and a greater number of deaths than do the viruses that result from antigenic drift. The influenza strains that cause pandemics are recombinant viruses that have some of their RNA genome derived from an avian influenza virus and the remainder from a human influenza virus. In these recombinant strains, the hemagglutinin and/or the neuraminidase are encoded by RNA

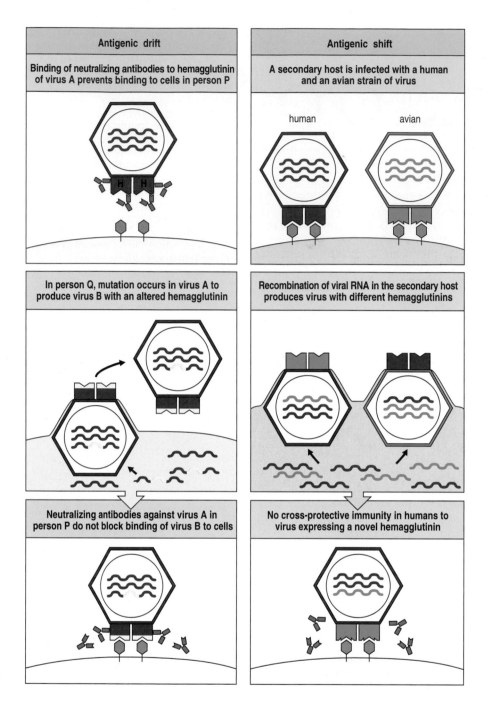

Figure 9.2 Variation in influenza virus due to mutation and recombination allows repeated infections. The left panels show the process of antigenic drift in the influenza virus. Only the hemagglutinin (H) is shown for simplicity. An individual infected with strain A produces antibodies against the viral hemagglutinin. In any re-infection with strain A, these antibodies bind to the hemagglutinin and prevent the virus from entering cells (top panel). The individual is thus immune to strain A. However, point mutations (yellow) within the hemagglutinin gene of strain A that originate in infections elsewhere eliminate the epitopes recognized by the protective antibodies (center panel). This allows the variant virus, strain B, to successfully re-infect the individual immune to strain A (bottom panel). The right panels show the process of antigenic shift. A human influenza virus (red) and an avian influenza virus (blue) simultaneously infect a cell within the secondary host, a pig (top panel), and their RNA segments become reassorted to produce a recombinant virus (center panel). The recombinant virus expresses hemagglutinin of avian origin that is antigenically very different from that in the original human virus. The recombinant virus readily infects humans because antibodies protective against the original hemagglutinin cannot bind to the new hemagglutinin (bottom panel). For simplicity, only three of the eight RNA molecules are drawn.

molecules of avian origin and are antigenically very different from those against which people have protective immunity. New pandemic strains often arise in parts of south-east Asia where farmers live in close proximity with their livestock such as pigs, chickens, and ducks. One theory is that the recombinant viruses arise in pigs that have become simultaneously infected with both avian and human viruses. If such a recombinant jumps back into the human population it has a tremendous competitive advantage, and in sweeping through the human population will rapidly replace other viral strains. Recombinant influenza viruses can similarly sweep through bird populations and are greatly feared by poultry farmers. This mode of evolution is called **antigenic shift** (see Figure 9.2, right panels).

9-3 Trypanosomes use gene rearrangements to change their surface antigens

Mutation and recombination are not the only means by which pathogens can change the face that they present to the immune system. Certain protozoans regularly change their surface antigens by a process of gene rearrangement, the most striking example being the African trypanosomes (eg *Trypanosoma brucei*), which cause sleeping sickness. The trypanosome life cycle involves both mammalian and insect hosts. Insect bites transmit trypanosomes to humans, in whom the trypanosomes replicate in the extracellular spaces. The trypanosome's surface is formed by a glycoprotein, of which there are numerous variants, each encoded by a different gene. The trypanosome genome contains more than 1000 genes encoding these **variable surface glycoproteins** (**VSGs**). At any one time, an individual trypanosome expresses only one form of VSG. This is because rearrangement of the VSG gene into a unique site in the genome—the expression site—is required for its expression. Rearrangement occurs by a process of gene conversion (Figure 9.3). The vast majority of the rapidly replicating trypanosomes that emerge after initial infection will express the same dominant form of VSG. A very small minority will, however, have changed the expressed VSG gene, and now express other forms. The host makes an antibody response to the dominant form of VSG, but not to the minority forms. Antibody-mediated clearance of trypanosomes expressing the dominant VSG facilitates the growth of those expressing the minority forms, of which one will come to dominate the trypanosome population. In time, the numbers of trypanosomes expressing the new dominant form are sufficient to stimulate antibody production, whereupon they too are cleared and yet other variants selected.

This mechanism of immune evasion causes trypanosome infections to produce a dramatic cycling in the number of parasites within an infected person (see Figure 9.3, bottom panel). The chronic cycle of antibody production and antigen clearance leads to a heavy deposition of immune complexes, and inflammation. Neurological damage occurs and eventually leads to coma, the so-called sleeping sickness. Trypanosome infections are a major health problem for humans and cattle in large parts of Africa. Indeed it is largely because of trypanosomes that wild populations of big game animals still survive in Africa and have not been replaced by domestic cattle. Malaria, another disease caused by a protozoan parasite that escapes immunity by varying its surface antigens, is also a major cause of human mortality in equatorial Africa.

Trypanosomes change their coats by gene rearrangements similar to the gene conversions used by rabbit and chicken B lymphocytes to produce their primary repertoire of immunoglobulins (see Section 2-11, p. 45). Similar strategies of antigenic variation are also used by several species of bacteria whose ability to escape from the human immune response makes them successful pathogens and major public health problems. *Salmonella typhimurium*, a common cause of food

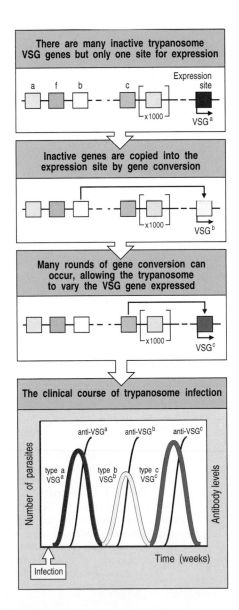

Figure 9.3 Antigenic variation by African trypanosomes allows them to escape from adaptive immunity. In the top panel, the VSGa gene (shown in red) is in the expression site and the VSGb gene (yellow) and VSGc gene (blue) are inactive. In the second panel, gene conversion has replaced VSGa with VSGb at the expression site; in the third panel, VSGc has replaced VSGb at the expression site. The bottom panel shows how the patient's antibody response (thin black lines) to the VSG proteins selects for low-frequency variants and causes a cycle in which parasite populations (red, yellow, and blue lines) alternate between boom and bust.

poisoning, can alternate expression of two antigenically distinct flagellins, proteins of the bacterial flagella. This occurs by reversible inversion of part of the promoter of one of the flagellin genes, which inactivates that gene and allows the expression of the second gene. *Neisseria gonorrhoeae*, the cause of the widespread sexually transmitted disease gonorrhea, has several variable antigens, the most impressive being the pilin protein, a component of the adhesive pili on the bacterial surface. Like the VSG of African trypanosomes, pilin is encoded by a family of variant genes, only one of which is expressed at a time. Different versions of the pilin gene introduced into the expression site provide a minority population of variant bacteria. When the host's immune response places pressure on the dominant type, another is ready to take its place.

9-4 Herpes viruses persist in human hosts by hiding from the immune response

To terminate an established viral infection, infected cells must be killed by cytotoxic CD8 T cells. For this to occur, some of the peptides presented by MHC class I molecules at the surface of infected cells must be of viral origin, a condition easily fulfilled by rapidly replicating viruses such as influenza. Consequently, influenza infections are efficiently cleared by the immune system by a combination of cytotoxic T cells and antibodies, the latter neutralizing extracellular virus particles. In contrast, some other viruses are difficult to clear because they enter a quiescent state within human cells, one in which they neither replicate nor generate enough virus-derived peptides to signal their presence to cytotoxic T cells. Development of this dormant state, which is called **latency** and does not cause disease, is a favored strategy of the herpes viruses. Later on, when the initial immune response has subsided, the virus will reactivate, causing an episode of disease.

Herpes simplex virus, the cause of cold sores, first infects epithelial cells and then spreads to sensory neurons serving the area of infection. The immune response clears virus from the epithelium, but the virus persists in a latent state in the sensory neurons. Various stresses can reactivate the virus, including sunlight, bacterial infection, or hormonal changes. After reactivation, the virus travels down the axons of the sensory neurons and re-infects the epithelial tissue (Figure 9.4). Viral replication in epithelial cells and production of viral peptides restimulates CD8 T cells, which kill the infected cells, creating a new sore. This cycle can be repeated many times throughout life. Neurons are a favored site for latent viruses to lurk because they express very small numbers of MHC class I molecules, further reducing the potential for presentation of viral peptides to CD8 T cells.

The herpes virus varicella-zoster remains latent in one or a few ganglia, chiefly dorsal root ganglia, after the acute infection of epithelium—chickenpox—is over. Stress or immunosuppression can reactivate the virus, which moves down the nerve and infects the skin. Re-infection causes the reappearance of the classic varicella rash of blisters, which cover the area of skin served by the infected ganglia.

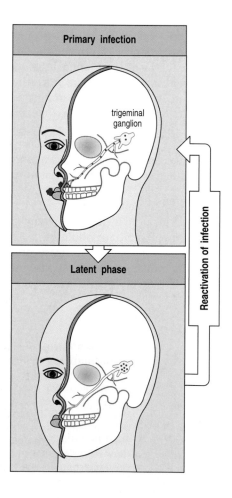

Figure 9.4 Persistence and reactivation of herpes simplex virus infection. The initial infection around the lips is cleared by the immune response and the resulting tissue damage is manifest as cold sores (upper panel). The virus (small red dots) has meanwhile entered sensory neurons, for example those in the trigeminal ganglion whose axons innervate the lips, where it persists in a latent state (lower panel). Various types of stress can cause the virus to leave the neurons and re-infect the epithelium, once again reactivating the immune response and causing cold sores. People infected with herpes simplex viruses periodically get cold sores as a result of this process. During its active phase the virus can be passed from one person to another.

The disease caused by reactivation of varicella-zoster (also called herpes zoster) is commonly known as shingles. In contrast to herpes simplex virus, reactivation of varicella-zoster usually occurs only once in a lifetime.

A third herpes virus that causes a persistent infection is the Epstein–Barr virus (EBV), to which most humans are exposed. First exposure in childhood produces a mild cold-like disease, whereas adolescents or adults encountering EBV for the first time develop infectious mononucleosis (also known as glandular fever), an acute infection of B lymphocytes. EBV infects B cells by binding to CR2 (also called CD21), a component of the B-cell co-receptor complex (see Figure 7.3, p. 162). Most of the infected cells proliferate and produce virus, leading in turn to the proliferation of EBV-specific T cells. The result is an unusually large number of mononuclear white blood cells (lymphocytes), which gives the disease its name. After some time, the acute infection is brought under control by CD8 cytotoxic T cells, which kill the infected B cells. The virus persists in the body, however, because a minority of B cells become latently infected. To maintain the EBV genome in these cells a virally encoded protein called EBNA-1 is expressed. One feature of EBNA-1 is that it interacts with the proteasome to prevent its own degradation into peptides that could be presented to T cells.

After recovery from the initial exposure to EBV it is unusual for reactivation of the virus to lead to disease. It seems likely that CD8 T cells quickly control episodes of viral reactivation. In immunosuppressed patients, however, reactivation of the virus can cause a disseminated EBV infection, and infected B cells can also undergo malignant transformation, causing B-cell lymphoproliferative disease.

9-5 Certain pathogens sabotage or subvert immune defense mechanisms

Pathogens also exploit the immune system cells that are ranged against them. For example, *Mycobacterium tuberculosis* commandeers the macrophage's pathway of phagocytosis for its own purposes. On being phagocytosed, *M. tuberculosis* prevents fusion of the phagosome with the lysosome, thus protecting itself from the bactericidal actions of the lysosomal contents. It then survives and flourishes within the cell's vesicular system. *Listeria monocytogenes*, in contrast, escapes from the phagosome into the macrophage's cytosol, where it grows and replicates. However, the intracytosolic way of life elicits cytotoxic CD8 T-cell responses against *L. monocytogenes*, which eventually terminate the infection.

The protozoan parasite *Toxoplasma gondii*, the cause of **toxoplasmosis**, creates its own specialized environment within the cells that it infects. It surrounds itself with a membrane-bounded vesicle that does not fuse with other cellular vesicles or cell membranes. Such isolation prevents the binding of *T. gondii*-derived peptides to MHC molecules and their presentation to T cells. The spirochete *Treponema pallidum*, the cause of syphilis, evades specific antibody by coating itself with human proteins. This is also a strategy pursued by the schistosome, a parasitic helminth.

Of the four groups of pathogens (see Figure 1.2, p. 3), viruses have evolved the greatest variety of mechanisms for subverting or escaping immune defenses. This is because their replication and life cycle depend completely on the metabolic and biosynthetic processes of human cells. Viral self-defense strategies include: capture of cellular genes encoding cytokines or cytokine receptors, which when expressed by the virus can divert the immune response; synthesis of proteins that inhibit complement fixation; and synthesis of proteins that inhibit antigen processing and presentation by MHC class I molecules. Examples of defensive mechanisms used by herpes viruses and pox viruses are shown in Figure 9.5.

Strategy	Mechanism	Result	Examples
Inhibition of humoral immunity	Virally encoded Fc receptor	Blocks effector functions of antibodies bound to infected cells	Herpes simplex Cytomegalovirus
	Virally encoded complement receptor	Blocks complement-mediated effector pathways	Herpes simplex
	Virally encoded complement control protein	Inhibits complement activation on infected cell	Vaccinia
Inhibition of inflammatory response	Virally encoded cytokine homolog, eg β-chemokine receptor	Sensitizes infected cells to effects of β-chemokine; advantage to virus unknown	Cytomegalovirus
	Virally encoded soluble cytokine receptor, eg IL-1 receptor homolog, TNF receptor homolog, γ-interferon receptor homolog	Blocks effects of cytokines by inhibiting their interaction with host receptors	Vaccinia Rabbit myxoma virus
	Viral inhibition of adhesion molecule expression, eg LFA-3, ICAM-1	Blocks adhesion of lymphocytes to infected cells	Epstein–Barr virus
Blocking of antigen processing and presentation	Inhibition of MHC class I expression	Impairs recognition of infected cells by cytotoxic T cells	Herpes simplex Cytomegalovirus
	Inhibition of peptide transport by TAP	Blocks peptide association with MHC class I	Herpes simplex
Immunosuppression of host	Virally encoded cytokine homolog of IL-10	Inhibits T_H1 lymphocytes Reduces γ-interferon production	Epstein–Barr virus

Figure 9.5 Mechanisms used by herpes and pox viruses to subvert the immune response.

9-6 Too much or too little immune activity can contribute to persistent disease

Some pathogens induce a general suppression of a person's immune response. For example, staphylococci produce toxins like the staphylococcal enterotoxins and toxic shock syndrome toxin-1 that act as superantigens (see Section 3-13, p. 72). Through their capacity to bind to so many different T-cell receptors, superantigens stimulate excessive non-specific proliferation of T cells and an overproduction of cytokines. The massive cytokine release, particularly of IL-1, IL-2, and TNF-α, causes systemic shock (see Figure 8.12, p. 215). At the same time, a useful adaptive immune response is suppressed. After proliferation, T cells that have bound superantigens undergo apoptosis in the absence of any further specific stimulation, removing many T-cell clones from the peripheral circulation.

Other pathogens cause mild or transient immunosuppression in the course of acute infection, making the patient more susceptible to further infection. Such secondary infections are often due to organisms with which humans are continually in contact. Trauma from burns or major surgery, for example, can also produce a state in which there is a diminished capacity to combat infection, and in these patients generalized infection with opportunistic pathogens such as *Pseudomonas aeruginosa* can be fatal.

Mycobacterium leprae, the leprosy bacillus, causes immunosuppression that is alternatively targeted to different arms of the immune response, depending on the conditions that develop at the very beginning of infection. In this disease, either cell-mediated immunity due to T_H1 CD4 cells or antibody responses due to CD4 T_H2 cells tends to be suppressed. This bias stems from the particular mix of cytokines produced at the start of infection, which depends on factors such as the amount of antigen present and the cells that are activated in the innate immune response (see Sections 6-8, p. 142 and 8-16, p. 226). As a consequence, leprosy consists of two quite different forms of disease. Suppression of cell-mediated immunity causes the more severe disease, lepromatous leprosy, whereas suppression of the antibody response leads to a more localized and generally less damaging tuberculoid leprosy (Figure 9.6). In neither instance, however, is the patient able to get rid of the infection.

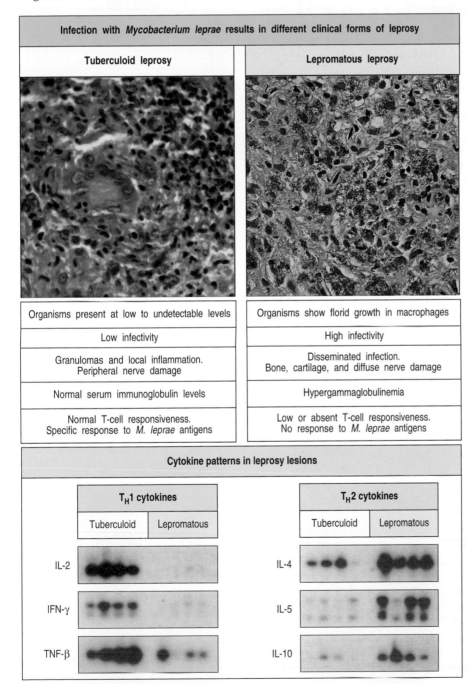

Figure 9.6 Infection with *Mycobacterium leprae* causes two diseases associated with very different host immune responses. The top panels show light micrographs of sections of biopsied leprosy lesions in which the *M. leprae* organisms are stained red. In tuberculoid leprosy (upper left panel), growth of the organism is controlled by T_H1 cells and the cytokines they produce (lower left panels), which activate infected macrophages. Inflammation and granulomas within the tuberculoid lesion cause only local damage within the tissues. In contrast, in lepromatous leprosy, infection is widely disseminated (upper right panel) and the bacilli (small red dots) grow in host macrophages in an uncontrolled fashion. The disease is slowly progressive and in the late stages there is severe damage to connective tissue and the peripheral nervous system. In lepromatous leprosy, the host immune response is dominated by T_H2 cells and the cytokines they produce (lower right panel), leading to an antibody response that is ineffective in controlling disease. Forms of disease that are intermediate between these two extremes are also observed in patients infected with *M. leprae*. In the lower panels the presence of cytokines in leprosy lesions is assayed by Northern blotting of extracted mRNA. Micrographs courtesy of G. Kaplan; Northern blots courtesy of R.L. Modlin.

9-7 Immune responses can contribute to disease

In tuberculoid leprosy the disease symptoms, or **pathology**, are due largely to the immune response. This is true to some degree of all infections. A good example of a viral infection whose symptoms are due to the immune response is **respiratory syncytial virus (RSV)**. This causes a wheezy bronchiolitis that accounts for a large proportion of hospital admissions for infants in the developed countries—some 90,000 admissions and 4500 deaths each year in the United States alone. A failed attempt at developing a vaccine against RSV revealed that infants vaccinated with a precipitated killed virus preparation suffered worse disease on subsequent infection with RSV than did unvaccinated infants. This was because the vaccine had failed to induce neutralizing antibodies but had successfully induced virus-specific activated T_H2 cells. On subsequent infection with RSV, the release of IL-3, IL-4, and IL-5 during the secondary response by the T_H2 cells induced bronchospasm, increased secretion of mucus and the recruitment of tissue-damaging eosinophils into the tissues of the respiratory tract.

Tissue damage and disease symptoms can also result from the immune response to parasites, as illustrated by *Schistosoma mansoni*. These blood flukes lay their eggs in the hepatic portal vein. Some of the eggs reach the intestine and are shed in the feces, thereby enabling the infection to spread to other people. Other eggs lodge in the portal circulation of the liver, where they elicit a powerful immune response that leads to chronic inflammation, hepatic fibrosis, and eventual liver failure. Underlying this progression is an excessive activation of T_H2 cells. Thus, with experimental infections of mice, the progression to severe disease can be moderated by increasing the actions of T_H1 cells, IFN-γ, or CD8 T cells.

Summary

From the human perspective the ideal immune system would be one that terminates infection before the pathogen damages tissues or saps the body's resources. In contrast, an ideal situation for a pathogen is one in which the immune system does not interfere with growth and replication, while other parts of the body provide food and shelter. To further their cause, pathogens have evolved ways of reducing the effectiveness of the human immune response. Antigenic variation prevents the maturation of the adaptive response and the development of useful immunological memory. Latency, a means of avoiding the immune response, allows viruses to lie low within cells until immunity has waned. More active strategies are for pathogens to interfere with key elements of the immune response, either to inhibit normal immune function or to recruit the response to the pathogen's advantage. Immune responses to pathogens can themselves be a significant cause of pathology.

Inherited immunodeficiency diseases

Inherited defects in genes for components of the immune system cause **immunodeficiency diseases**, which reveal themselves by enhanced susceptibility to infection. Before the advent of antibiotic therapy during the 1940s, most individuals with inherited immune defects died from infection during infancy or early childhood. Because many normal infants also succumbed to infection in that earlier era, death from immunodeficiency did not stand out until the 1950s when the first such disease was described. Since then, many inherited immunodeficiency diseases have been identified and correlated with susceptibility to particular classes of pathogen. Each disease is due to a defect in a particular protein or glycoprotein, and the precise symptoms depend on the role of that component in the immune response (Figure 9.7).

Deficiency syndrome	Abnormality	Immune defect	Susceptibility
Severe combined immune deficiency	ADA deficiency	No T or B cells	General
	PNP deficiency	No T or B cells	General
	γ_c chain deficiency	No T cells	General
	DNA repair defect	No T or B cells	General
DiGeorge syndrome	Thymic aplasia	Variable numbers of T and B cells	General
MHC class I deficiency	Mutant TAP	No CD8 T cells	Viruses
MHC class II deficiency	Lack of expression of MHC class II	No CD4 T cells	General
Wiskott–Aldrich syndrome	X-linked; defective WASP gene	Defective polysaccharide antibody responses	Encapsulated extracellular bacteria
Common variable immunodeficiency	Unknown; MHC-linked	Defective antibody production	Extracellular bacteria
X-linked agamma-globulinemia	Loss of Btk tyrosine kinase	No B cells	Extracellular bacteria, viruses
X-linked hyper-IgM syndrome	Defective CD40 ligand	No isotype switching	Extracellular bacteria
Selective IgA and/or IgG deficiency	Unknown; MHC-linked	No IgA synthesis	Respiratory infections
Phagocyte deficiencies	Many different	Loss of phagocyte function	Extracellular bacteria and fungi
Complement deficiencies	Many different	Loss of specific complement components	Extracellular bacteria especially *Neisseria* spp.
Natural killer (NK) cell defect	Unknown	Loss of NK function	Herpes viruses
X-linked lympho-proliferative syndrome	Unknown; X-linked	EBV-triggered immunodeficiency	EBV
Ataxia telangiectasia	Gene with PI-3 kinase homology	T-cell numbers reduced	Respiratory infections
Bloom's syndrome	Defective DNA helicase	T-cell numbers reduced Reduced antibody levels	Respiratory infections

Figure 9.7 Inherited immuno-deficiency syndromes of humans, their gene defects, and their pathogen susceptibilities. ADA, adenosine deaminase; PNP, purine nucleotide phosphorylase; TAP, transporter associated with antigen processing; WASP, Wiskott–Aldrich syndrome protein; EBV, Epstein–Barr virus; NK cell, natural killer cell; PI-3 kinase, phosphoinositol-3 kinase.

9-8 Most inherited immunodeficiency diseases are caused by recessive gene defects

Before the 1950s, any dominant trait causing a severe immunodeficiency would probably have been eliminated from the population with the death of the child in whom it first occurred. Thus, most of the inherited immunodeficiency syndromes that have been identified are due to recessive mutations in single genes. One of the few dominant immune defects known is a defect in the receptor for interferon-γ (IFN-γ), in which the defective receptor still binds IFN-γ but cannot produce an intracellular signal. Infants with this defect are susceptible to disseminated infection with mycobacteria, including the normally innocuous strain used for vaccination against tuberculosis.

For recessive defects in autosomal genes, that is genes on chromosomes other than the sex chromosomes, only children who inherit a defective allele from both parents are immunodeficient. The parents are heterozygous for the defect and are usually healthy. Such people are referred to as **carriers**. Recessive defects in genes on the X chromosome are a special case, because women have two copies of the X chromosome whereas men have only one. As there is no other copy of the gene to compensate, inheritance of one defective copy of an X-linked gene by a male child is sufficient to cause disease. A woman who inherits a single defective gene will also have the normal gene and usually will be healthy. Because of this mode of inheritance, X-linked immunodeficiencies are far more common in men than in women and are also more common than immunodeficiencies linked to autosomes.

Modern genetic techniques now enable immunologists to engineer strains of mice in which a chosen gene is deliberately rendered inactive. These **gene knock-outs** are analogous to the experiments of nature that cause human immunodeficiency diseases. The effect of inactivating different genes in the mouse immune system varies greatly. At one end of the spectrum are knock-outs that are lethal; at the other end are ones that seem to have no effect. A similarly wide range of defects in immune system genes exists in the human population. Whereas those listed in Figure 9.7 were discovered in patients with severe disease, other immune system gene defects were found first in healthy blood donors. Examples of the latter are the lack of certain complement components or of one of the MHC class I molecules. Those immune system genes that can be lost with little effect are often members of multigene families in which one family member can compensate for another.

9-9 Antibody deficiency leads to an inability to clear extracellular bacteria

The first immunodeficiency disease to be described is characterized by antibody deficiency and X-linked inheritance and is named **X-linked agammaglobulinemia (XLA)**. The defect in XLA is in a protein tyrosine kinase, which is called Bruton's tyrosine kinase (Btk) in honor of the discoverer of the syndrome. Btk contributes to intracellular signaling from the B-cell receptor and is necessary for the growth and differentiation of pre-B cells (see Section 4-4, p. 93). Btk is normally also expressed in monocytes and T cells, but these cells are not functionally compromised by its absence from XLA patients. Other immunodeficiency diseases affecting antibody production have been described, and almost all are caused by failure of either the development or the activation of B lymphocytes.

In XLA, women with one functional and one non-functional copy of the *Btk* gene are themselves healthy, but are carriers who pass the disease on to half their male children. In females, one X chromosome is randomly inactivated in each cell. In carriers, those B-cell precursors in which the inactivated X carries the defective *Btk* gene develop normally, whereas B-cell precursors in which the inactivated X carries the functional *Btk* gene fail to develop. Thus, all circulating B cells in heterozygous females have as their active X chromosome the one carrying the functional *Btk* gene; consequently, carrier females make entirely normal B-cell responses.

Patients who lack antibodies are vulnerable to infection with extracellular pyogenic bacteria that have polysaccharide capsules resistant to phagocytosis. These include *Haemophilus influenzae*, *Streptococcus pneumoniae*, *Streptococcus pyogenes*, and *Staphylococcus aureus*. In people who make normal antibody responses, such organisms are cleared by phagocytosis after opsonization by antibody and complement. Antibody-deficient patients are also more susceptible to viral infections, particularly those caused by enteroviruses, which enter the body through the gut and which in normal individuals are neutralized by antibodies.

Genetic defects can also cause deficiency in a particular isotype or class of antibody. The most common kind of inherited deficiency of this type is IgA deficiency, which is experienced by about 1 person in 800. No obvious increased susceptibility to infectious disease is associated with this condition, which might reflect the fact that modern humans tend to eat cooked and processed foods, which greatly reduces the threat posed by pathogens entering the gastrointestinal tract. In contrast, chronic lung disease is more common in people with IgA deficiency, suggesting that the trend towards poorer air quality in the cities, where most people live, makes the role of IgA in the respiratory tract of increasing importance.

Patients who have immunodeficiencies confined to B-cell functions are able to resist many pathogens successfully, and those to which they are susceptible can be treated with antibiotics. However, although pyogenic infections can be cured in this way, the successive rounds of infection and treatment sometimes lead to permanent tissue damage caused by the excessive release of proteases from both the infecting bacteria and the defending phagocytes. These effects are particularly pronounced in the airways, where the bronchi lose their elasticity and become sites of chronic inflamation. This condition, called **bronchiectasis**, can lead to chronic lung disease and eventual death. To prevent such developments, patients with XLA are given monthly injections of gamma globulin, an antibody-containing preparation made from the plasma of healthy blood donors.

9-10 Diminished antibody production also results from inherited defects in T-cell help

Diminished antibody production will also result from defects in the gene for the membrane-associated cytokine CD40 ligand. As discussed in Chapter 7, interaction of CD40 ligand on activated T cells with CD40 on B cells is a crucial part of the T-cell help given to B cells. This stimulates their activation, the development of germinal centers, and isotype switching. CD40 ligand is encoded on the X-chromosome and so most patients with a hereditary deficiency in the CD40 ligand are males. In the absence of CD40 ligand, virtually no specific antibody is made against T-cell dependent antigens. IgG, IgA, and IgE levels are extremely low, and IgM levels are abnormally high. Recognition of this latter characteristic led to the condition being named X-linked hyper-IgM syndrome. Patients with this immunodeficiency are inherently susceptible to infection with pyogenic bacteria, but these infections can usually be prevented by regular injections of gamma globulin and cleared with antibiotics when they do occur. A further consequence of the disease is the absence of germinal centers in the lymph nodes and other secondary lymphoid tissues (Figure 9.8).

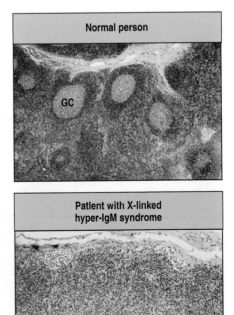

Figure 9.8 Defective B-cell activation in patients with X-linked hyper-IgM syndrome prevents the formation of germinal centers. These light micrographs show sections through a lymph node from a normal individual (top panel) and from a patient with X-linked hyper-IgM syndrome (lower panel). A typical germinal center is designated GC in the top panel. Top panel courtesy of A. Perez-Atayde; bottom panel courtesy of R. Geha.

Macrophage activation by T cells also depends on the interaction of CD40 ligand on the T cell with CD40 on the macrophage. The lack of this interaction in patients with X-linked hyper-IgM syndrome impairs the inflammatory response and the mobilization of leukocytes. Whereas an infection usually induces an increase in the number of white cells in the blood (**leukocytosis**), this cannot occur in patients lacking CD40 ligand. On the contrary, their blood can become profoundly deficient in neutrophils. This state, called **neutropenia**, leads to severe sores and blisters in the mouth and throat. These anatomical sites are infested with bacteria and their health depends on continual surveillance by phagocytes. These symptoms of neutropenia can be cured by the intravenous administration of granulocyte–macrophage colony-stimulating factor (GM-CSF), a cytokine that stimulates the production and release of phagocytes by the bone marrow. In normal individuals, GM-CSF is secreted by macrophages in response to activation by T cells through interactions with CD40 and CD40 ligand.

9-11 Defects in complement components impair antibody responses and cause the accumulation of immune complexes

The effector functions recruited by antibodies to clear pathogens and antigens are all facilitated by complement activation. Consequently, the spectrum of infections associated with complement deficiencies overlaps substantially with that associated with defective antibody production. Defects in the activation of C3, and in C3 itself, are associated with susceptibility to a wide range of pyogenic infections, emphasizing the important role of C3 as an opsonin that promotes the phagocytosis of bacteria by macrophages and neutrophils. In contrast, defects in C5–C9, the terminal complement components that form the membrane-attack complex, have more limited effects, of which susceptibility to *Neisseria* is the best example. The most effective defense against *Neisseria* is complement-mediated lysis of extracellular bacteria, and this requires all the components of the complement pathway (Figure 9.9).

The early components of the classical pathway are necessary for the elimination of immune complexes. As discussed in Section 7-20, p. 191, attachment of complement components to soluble immune complexes allows them to be transported, or ingested and degraded, by cells bearing complement receptors. Immune complexes are transported by erythrocytes; these capture them with the CR1 complement receptor, which binds to C4b or C3b. Deficiencies in complement components C1–C4 impair the formation of C4b and C3b and lead to an accumulation of immune complexes in the blood, lymph, and extracellular fluid

Complement protein	Effects of deficiency
C1, C2, C4	Immune-complex disease
C3	Susceptibility to capsulated bacteria
C5–C9	Only effect is susceptibility to *Neisseria*
Factor D, factor P	Susceptibility to capsulated bacteria and *Neisseria* but no immune-complex disease
Factor I	Similar effects to deficiency of C3
DAF, CD59	Autoimmune-like conditions including paroxysmal nocturnal hemoglobinuria

Figure 9.9 Diseases caused by deficiencies in the pathways of complement activation.

and their deposition within tissues. In addition to directly damaging the tissues in which they deposit, immune complexes activate phagocytes, causing inflammation and further tissue damage.

Deficiencies in the proteins that control complement activation can also cause immunodeficiency. People who lack the plasma protein factor I in effect lack C3. Because factor I is absent, the conversion of C3 to C3b is unchecked, and supplies of C3 are rapidly depleted (see Section 7-23, p. 196). Patients who lack properdin (factor P), a plasma protein that enhances the activity of the alternative pathway, have a heightened susceptibility to *Neisseria*, because reduced deposition of C3 prevents formation of the membrane-attack complex and bacterial lysis. In contrast, a deficiency in decay-accelerating factor (DAF) or CD59 causes an autoimmune-like condition. The cells of patients lacking DAF or CD59 are not protected from activating the alternative pathway of complement activation. The resultant complement-mediated lysis of erythrocytes causes paroxysmal nocturnal hemoglobinuria (see Section 7-23, p. 197).

Another disease caused by a deficiency of a complement regulator is hereditary angioneurotic edema. In this condition the absence of the C1 inhibitor C1INH leads to uncontrolled activation of the classical pathway (see Section 7-23, p. 194). One consequence is overproduction of the vasoactive C2a fragment, causing the accumulation of fluid in the tissues and epiglottal swelling, which can lead to death by suffocation.

9-12 Defects in phagocytes result in enhanced susceptibility to bacterial infection

Phagocytosis by macrophages and neutrophils is the principal method by which the immune system destroys bacteria and other microorganisms. Any defect that compromises phagocyte activity will therefore have a profound effect on the capacity to clear infections (Figure 9.10). One kind of deficiency arises from

Syndrome	Cellular abnormality	Immune defect	Associated infections and other diseases
Leukocyte adhesion deficiency	Defective CD18 (cell adhesion molecule)	Defective migration of phagocytes into infected tissues	Widespread infections with capsulated bacteria
Chronic granulomatous disease (CGD)	Defective NADPH oxidase. Phagocytes cannot produce O_2^-	Impaired killing of phagocytosed bacteria	Chronic bacterial and fungal infections. Granulomas
Glucose-6-phosphate dehydrogenase (G6PD) deficiency	Deficiency of glucose-6-phosphate dehydrogenase. Defective respiratory burst	Impaired killing of phagocytosed bacteria	Chronic bacterial and fungal infections. Anemia is induced by certain agents
Myeloperoxidase deficiency	Deficiency of myeloperoxidase in neutrophil granules and macrophage lysosomes and impaired production of toxic oxygen species	Impaired killing of phagocytosed bacteria	Chronic bacterial and fungal infections
Chédiak–Higashi syndrome	Defect in vesicle fusion	Impaired phagocytosis due to inability of endosomes to fuse with lysosomes	Recurrent and persistent bacterial infections. Granulomas. Effects on many organs

Figure 9.10 Defects in phagocytic cells cause persistent bacterial infections.

mutations in the gene encoding CD18, the common β subunit of the leukocyte integrins CR2, CR4, and LFA-1 (see Figure 6.2, p. 132). These cell adhesion molecules are needed for phagocytes to leave the blood and enter sites of infection (see Figure 8.11, p. 214). Phagocytes lacking functional integrins are therefore unable to migrate to where they are needed. This defect is known as **leukocyte adhesion deficiency**.

Leukocyte adhesion deficiency is associated with persistent infection with extracellular bacteria, which cannot be cleared because of the defective phagocyte function. Children with this defect have recurrent pyogenic infections and problems with wound healing; if they survive long enough, they develop severe inflammation of the gums. Their neutrophils and macrophages cannot migrate into tissues and also, because CR2 and CR4 are complement receptors as well as adhesion molecules, they cannot take up and destroy bacteria opsonized with complement (see Section 7-19, p. 190). Patients deficient in CD18 suffer from infections that respond poorly to antibiotic treatment and which persist despite the generation of normal B-cell and T-cell responses. Neutropenia caused by chemotherapy, malignancy, or aplastic anemia produces a similar susceptibility to severe pyogenic bacterial infections.

Other kinds of defect in phagocytes affect their ability to kill ingested bacteria. In **chronic granulomatous disease**, the antibacterial activity of phagocytes is compromised by their inability to produce the superoxide radical O_2^- (see Section 8-4, p. 209). Mutations affecting any of the four proteins of the NADPH oxidase system can produce this phenotype. Patients with this disease suffer from chronic bacterial infections, often leading to granuloma formation. Deficiencies in the enzymes glucose-6-phosphate dehydrogenase and myeloperoxidase also impair intracellular bacterial killing, leading to a similar but less severe phenotype. A different phenotype characterizes **Chédiak–Higashi syndrome**, in which phagocytosed materials are not delivered to lysosomes because of a defect in the vesicle fusion mechanism. The lack of phagocyte function has effects in many different organs as well as leading to persistent and recurrent bacterial infections. The gene defects that cause this disease are still unknown.

9-13 Defects in T-cell function result in severe combined immune deficiencies

Whereas B cells contribute only to the antibody response, T cells function in all aspects of adaptive immunity. This means that inherited defects in the mechanisms of T-cell development and T-cell function have a general depressive effect on the immune system's capacity to respond to infection. Patients with T-cell deficiencies tend to be susceptible to persistent or recurrent infections with a broader range of pathogens than patients with B-cell deficiences. Those patients who make neither T-cell dependent antibody responses nor cell-mediated immune responses are said to have **severe combined immune deficiency (SCID)**.

T-cell development and function depend on the action of many proteins and so the SCID phenotype can arise from defects in any one of a number of genes. Because of the inheritance pattern of the X-chromosome, X-linked diseases are more easily discovered, and at least two forms of SCID are of this type. The first is due to mutation in a gene on the X-chromosome that encodes the common γ chain of several cytokine receptors, including the receptors for IL-2, IL-4, IL-7, IL-9, and IL-15. In each of these receptors the γ chain transduces ligand binding into signals that alter gene expression. This is accomplished through interaction of the γ chain with the JAK3 kinase (see Section 6-11, p. 145). Indeed, as would be predicted, patients defective in the JAK3 kinase have an autosomally inherited immunodeficiency similar in phenotype to that of patients with X-linked SCID. The phenotype of SCID is so severe that affected children survive only if they are isolated in a pathogen-free environment, or if their immune system is replaced by bone marrow transplantation and the passive administration of antibodies.

Another X-linked deficiency of T-cell function is **Wiskott–Aldrich syndrome (WAS)**, which involves the impairment of platelets as well as lymphocytes. The relevant gene on the X-chromosome encodes a protein which has been called Wiskott–Aldrich syndrome protein (WASP). This protein is involved in the cytoskeletal reorganization needed before T cells can deliver cytokines and signals to the B cells, macrophages, and other target cells with which they routinely interact during the immune response.

SCID due to an absence of T-cell function is also caused by defects in **adenosine deaminase (ADA)** or **purine nucleotide phosphorylase (PNP)**, which are enzymes involved in purine degradation. The absence of these enzymes causes an accumulation of nucleotide metabolites that are highly toxic to developing T cells and, to a smaller extent, to developing B cells. These conditions, along with SCID caused by defects in DNA repair enzymes, are autosomally inherited.

Another kind of SCID is caused by the lack of expression of HLA class II molecules. The disease was named **bare lymphocyte syndrome** because the defect was first discovered on B lymphocytes, the major population of peripheral blood cells that expresses HLA class II. In these patients, CD4 T cells fail to develop (see Section 5-7, p. 119), which compromises all aspects of adaptive immunity. Bare lymphocyte syndrome arises from defects in the transcriptional regulators essential for the expression of all the HLA class II loci. Defects in one of two proteins, the class II transactivator (CIITA) or a promoter-binding protein called RFX, have been shown to underlie the disease in two subsets of patients with bare lymphocyte syndrome. Defects in at least three other genes are thought to contribute to causing this disease in other groups of patients.

A recently described immunodeficiency involving T cells is due to the expression of unusually small amounts of HLA class I molecules on cell surfaces. Here the defect lies in one of the two genes encoding the TAP peptide transporter. The defect prevents the delivery of peptides generated in the cytoplasm to the endoplasmic reticulum, where they normally bind to HLA class I molecules. This defect does not lead to SCID but to selective loss of CD8 T cells (see Section 5-7, p. 119) and of cytotoxic T-cell responses to intracellular infections.

9-14 Bone marrow transplantation is used to correct genetic defects of the immune system

Many immunodeficiencies are due to gene defects that principally affect hematopoietic cells. These conditions, in principle, can be corrected by bone marrow transplantation from a healthy donor. Such therapy is not undertaken lightly and the potential benefits from correcting the immunodeficiency must be weighed against the risks associated with the immunosuppressive treatments required for successful transplantation. The benefits of bone marrow transplantation are directly correlated with the degree of HLA matching between patient and donor. Matching serves two purposes: first, it reduces alloreactions that cause graft rejection and graft-versus-host disease (GVHD); second, it ensures effective presentation of foreign antigens by antigen-presenting cells derived from the transplant (Figure 9.11). After transplantation, all bone marrow derived cells in the recipient are of donor HLA type, whereas all other cells are of the recipient's HLA type. Positive selection of T cells is exclusively on thymic epithelium of the recipient's HLA type. Thus, the extent to which the mature T cells respond to antigens presented by professional antigen-presenting cells of donor HLA type is directly correlated with the number of shared HLA class I and II allotypes (see Section 5-6, p. 117).

The major complication of bone marrow transplantation is the **graft-versus-host disease** (**GVHD**) caused when mature donor T cells in the transplant attack the recipient's tissues. The occurrence of GVHD can be greatly reduced by depleting the transplanted marrow of mature T cells. However, this procedure increases the occurrence of graft rejection by the recipient's lymphocytes. This is less of a consideration for patients with SCID because they have no lymphocytes capable of mounting an alloreactive T-cell response (Figure 9.12).

Now that the genes responsible for immunodeficiency diseases are being identified, another type of therapeutic strategy is being explored. In **somatic gene therapy**, a functional copy of the defective gene is introduced into stem cells that have been isolated from the patient's bone marrow. The stem cells in which the defect has been corrected are then re-infused into the patient, where they provide

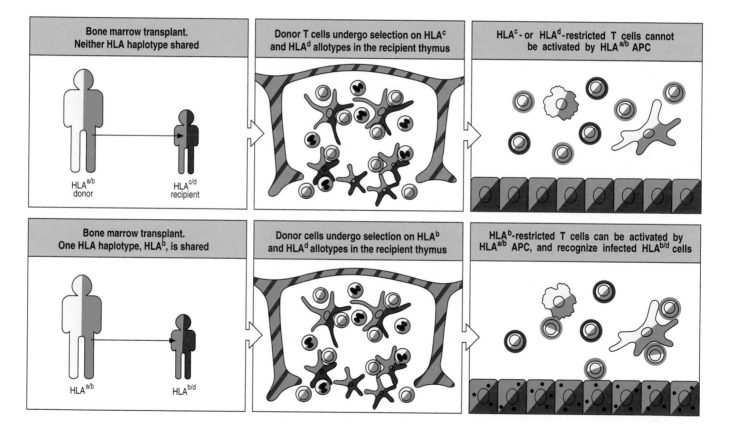

Figure 9.11 For bone marrow transplantation to restore immune function, the donor and recipient must share HLA allotypes. The top row portrays a hypothetical transplant in which the adult donor (yellow and green) shares neither HLA haplotype with the recipient child (blue and red), who has severe combined immune deficiency disease (left panel). After transplantation, stem cells from the donor reconstitute the recipient's immune system with macrophages, dendritic cells, B cells, T cells, and other hematopoietic cells. The immature T cells are selected on the HLA class I and II allotypes of the recipient's thymic epithelial cells (center panel). The mature T cells so selected cannot respond to antigens presented by the professional antigen-presenting cells, which express only HLA allotypes of donor type (right panel). The bottom row portrays a transplant in which the donor (green and yellow) shares one HLA haplotype (green) with the recipient (green and red), for example when a parent donates bone marrow to a child (left panel). After transplantation, immature T cells will be selected to recognize antigens presented by the child's HLA class I and II allotypes (green and red) (center panel). Those T cells that have been positively selected by allotypes of the green HLA haplotype will be able to recognize antigens presented by the donor professional antigen-presenting cells (right panel). See Sections 3-16–3-17, pp. 75–77 and Section 5-6, p. 117.

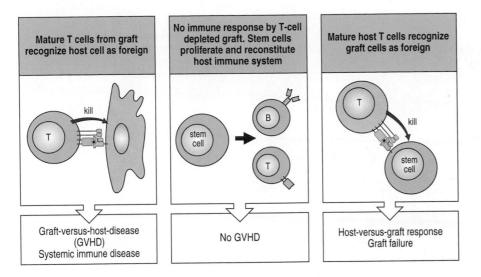

Figure 9.12 Alloreactions by either donor's or recipient's lymphocytes can prevent successful bone marrow transplantation. The left panel shows how mature T cells from the grafted bone marrow can attack the recipient's cells to produce graft-versus-host disease (GVHD). The center panel shows how the depletion of T cells from the bone marrow before transfusion can prevent GVHD. The right panel shows how T cells in the recipient can cause graft rejection.

a self-renewing source of immunocompetent lymphocytes and other hematopoietic cell types. Although highly attractive in principle, the practical development of gene therapy is at an early and experimental stage.

Fetal blood extracted from the placenta after birth is also a source of pluripotent stem cells. These samples, called cord blood, can be used to reconstitute a patient's hematopoietic system in a manner akin to that achieved by a bone marrow transplant. An advantage of cord blood transplant is that no invasive procedure is performed on the donor. A disadvantage is that fewer stem cells are obtained from each sample compared with bone marrow.

Summary

The best characterized gene defects affecting the immune system are those that show up in early childhood and confer exceptional vulnerability to common infections. The characterization of immunodeficiency diseases and the gene defects that cause them is almost the only way in humans of determining the relative importance of different cells and molecules in immune defenses, and of testing current models of how the human immune system works. The most severe immunodeficiencies are due to gene defects that cause an absence of all T-cell function and thus, directly or indirectly, impair B-cell function as well. Such deficiencies are known as severe combined immune deficiencies (SCID). The absence of antibodies due to genetic defects in B-cell development or function leads to particular susceptibility to pyogenic bacteria. Deficiencies in the early components of complement pathways, cause a failure to opsonize. This results in increased susceptibility to bacterial infection, as do defects in phagocytes.

Acquired immune deficiency syndrome

Acquired immune deficiency syndrome (AIDS) was first described by physicians early in the 1980s. The disease is characterized by a massive reduction in the number of CD4 T cells, which is accompanied by severe infections by pathogens that rarely trouble healthy people, or by aggressive forms of Kaposi's sarcoma or B-cell lymphoma. All patients diagnosed as having AIDS eventually die from the effects of the disease. In 1983 the virus now known to cause AIDS, the **human immunodeficiency virus (HIV)**, was first isolated. Two HIV viruses are now distinguished— HIV-1 and HIV-2. In most countries HIV-1 is the principal cause of AIDS. HIV-2 is less virulent, causing a slower progression to AIDS. It is endemic in West Africa and has spread widely through Asia.

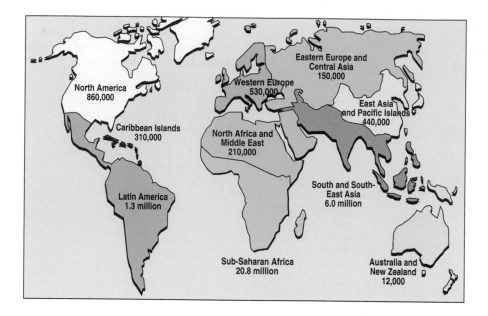

Figure 9.13 Incidence of HIV infection. Numbers of adults and children living with HIV/AIDS at the end of 1997, as estimated by the World Health Organization.

AIDS is a disease new to the medical profession and also to the human species. The earliest evidence for HIV comes from samples from African patients obtained in the late 1950s. It is believed that the viruses first infected humans in Africa by jumping from other primate species. HIV-1, for example, is believed to have originated from a simian immunodeficiency virus (SIV) carried by chimpanzees, which does not cause disease in the chimpanzee and has apparently co-existed with this species for many thousands of years.

As commonly occurs when a naive host population is hit with a new infectious agent, the effects on the human population have been immense and HIV infection is now at pandemic proportions. Although advances are being made in understanding the nature of the disease and its origins, the number of people infected with HIV continues to grow and many millions of people are likely to die from AIDS in the years to come (Figure 9.13).

9-15 HIV is a retrovirus that causes slowly progressing disease

HIV is an RNA virus with an RNA nucleoprotein core (the nucleocapsid) surrounded by a lipid envelope derived from the host-cell membrane and containing virally encoded envelope proteins (Figure 9.14). It is a retrovirus and the virion

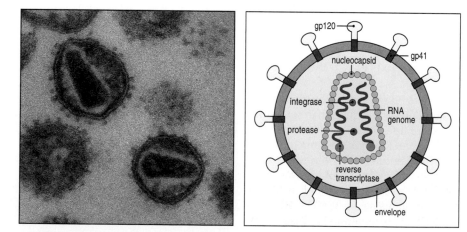

Figure 9.14 The virion of human immunodeficiency virus (HIV). The left panel is an electron micrograph showing three virions. The right panel is a diagram of a single virion. gp120 and gp41 are virally encoded envelope glycoproteins. Photograph courtesy of Hans Gelderblom.

Figure 9.15 The genes and proteins of HIV-1. HIV-1 has an RNA genome consisting of nine genes flanked by long terminal repeats (LTRs). The products of the nine genes and their known functions are tabulated. Several of the viral genes are overlapping and are read in different frames. Others encode large polyproteins that after translation are cleaved to produce several proteins having different activities. The *gag*, *pol*, and *env* genes are common to all retroviruses and their protein products are all present in the virion.

Gene		Gene product/function
gag	Group-specific antigen	Core protein
pol	Polymerase	Reverse transcriptase, protease, and integrase enzymes
env	Envelope	Transmembrane glycoproteins. gp120 binds CD4; gp41 is required for virus internalization
tat	Transactivator	Positive regulator of transcription
rev	Regulator of viral expression	Allows export of unspliced transcripts from nucleus
vif	Viral infectivity	Affects particle infectivity; helps assemble virions?
vpr	Viral protein R	Positive regulator of transcription. Augments virion production
vpu	Viral protein U	Unique to HIV-1. Downregulates CD4
nef	Negative-regulation factor	Augments viral replication *in vivo* and *in vitro*. Downregulates CD4

also contains a protease and the enzymes reverse transcriptase and integrase, which are required for viral replication. When a retrovirus infects a cell, the RNA genome is first copied into a complementary DNA (cDNA) by reverse transcriptase. The viral integrase then integrates the cDNA into the genome of the host cell to form a **provirus**, a process facilitated by repetitive DNA sequences called long terminal repeats (LTRs) that flank all retroviral genomes. Proviruses use the transcriptional and translational machinery of the host cell to make viral proteins and RNA genomes, which assemble into new infectious virions. The genes and proteins of HIV are illustrated in Figure 9.15. HIV belongs to a group of retroviruses that cause slowly progressing diseases. They are collectively called the **lentiviruses**, a name derived from the latin word *lentus*, meaning slow.

In almost all people, HIV produces an infection that cannot be successfully terminated by the immune system and continues for many years. Although the initial acute infection is controlled to the point at which disease is not apparent, the virus persists and replicates in a manner that gradually exhausts the immune system, leading to immunodeficiency and death.

9-16 HIV infects CD4 T cells, macrophages, and dendritic cells

Macrophages, dendritic cells, and CD4 T cells are vulnerable to HIV infection because they express CD4, which the virus uses as a receptor. Chimpanzees, our closest living relative, are resistant to HIV infection because of a small difference in the structure of their CD4 glycoprotein in comparison to ours. The gp120 envelope glycoprotein of HIV binds tightly to human CD4, enabling virions to attach to CD4-expressing human cells. Before entry of the virus into the cell, gp120 must also bind to a co-receptor in the host-cell membrane. Co-receptors for HIV have been identified as chemokine receptors. The principal ones used by the virus are

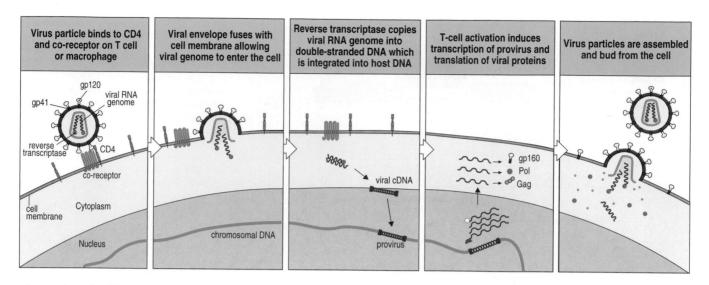

| Virus particle binds to CD4 and co-receptor on T cell or macrophage | Viral envelope fuses with cell membrane allowing viral genome to enter the cell | Reverse transcriptase copies viral RNA genome into double-stranded DNA which is integrated into host DNA | T-cell activation induces transcription of provirus and translation of viral proteins | Virus particles are assembled and bud from the cell |

Figure 9.16 The life cycle of HIV in human cells. The virus binds to CD4 on a cell surface by using the envelope protein gp120, which is altered by CD4 binding so that it now also binds a specific chemokine co-receptor on the cell. This binding releases gp41, which then causes fusion of the viral envelope with the plasma membrane and release of the viral core into the cytoplasm. The RNA genome is released and is reverse transcribed into double-stranded cDNA. This migrates to the nucleus in association with the viral integrase, and is integrated into the cell genome, becoming a provirus. New viral genomic RNAs and mRNAs are produced from the provirus and exported to the cytoplasm. Viral mRNA is translated into viral proteins. Envelope proteins travel to the plasma membrane, whereas other viral proteins and genomic RNA assemble into nucleocapsids. New virus particles bud from the cell, acquiring their lipid envelope and envelope glycoproteins in the process.

CCR5, which is expressed on dendritic cells, macrophages, and T cells, and CXCR4, which is expressed on T cells only. After co-receptor binding, the gp41 envelope glycoprotein mediates fusion of the viral envelope with the plasma membrane of the host cell, allowing the viral genome and associated proteins to enter the cytoplasm.

There are different variants of HIV, and the cell types they infect are determined to a large extent by which co-receptor they bind. Variants of HIV that infect macrophages, dendritic cells, and CD4 T cells via the CCR5 co-receptor are called 'macrophage-tropic'; those that infect CD4 T cells through the CXCR4 co-receptor are known as 'lymphocyte-tropic'. Macrophage-tropic HIV variants seem to be associated most commonly with the primary infection and do not require high levels of CD4 on the cells they infect. Macrophages and dendritic cells at the site of virus entry are the first cells to be infected. Subsequently, the virus produced by the macrophages starts to infect the CD4 T-cell population. In about 50% of cases, the viral phenotype switches to the lymphocyte-tropic type late in infection. This is followed by a rapid decline in CD4 T-cell count and a progression to AIDS.

The production of infectious virions from HIV provirus requires the activation of an infected CD4 T cell. Activation induces the synthesis of the transcription factor NFκB, which binds to promoters in the provirus. This directs the infected cell's RNA polymerase to transcribe viral RNAs. At least two of the proteins encoded by the virus serve to promote replication of the viral genome. Among other activities, the **Tat** protein binds to a sequence in the LTR of the viral mRNA, known as the transcriptional activation region (TAR), where it prevents transcription from shutting off and thus increases the transcription of viral RNA. The **Rev** protein controls the supply of viral RNA to the cytoplasm and the extent to which that RNA is spliced. At early times in infection Rev delivers RNA that encodes the proteins necessary for making virions. Later, complete viral genomes are supplied, which assemble with the proteins forming complexes that bud through the plasma membrane to give infectious virions (Figure 9.16).

9-17 Most people who become infected with HIV progress in time to develop AIDS

Infection with HIV usually occurs after the transfer of bodily fluids from an infected person to an uninfected recipient. Provirus can be carried in infected CD4 T cells, dendritic cells, and macrophages, whereas virions can be transmitted via blood, semen, vaginal fluid, or milk. Infection is commonly spread by sexual intercourse, intravenous administration of drugs with contaminated needles, or transfusion of human blood or blood components from HIV-infected donors. Evidence is accumulating that macrophage-tropic variants of HIV are preferentially transmitted by sexual contact, as they are the dominant viral phenotype found in newly infected individuals.

At the beginning of the AIDS epidemic, transmission through infected blood products caused infections of hemophiliacs and other patients dependent on blood products. Since then this route of infection has largely been eliminated, at least in the developed countries, by routine screening of individual units of blood for HIV. An important route of virus transmission is from an infected mother to her baby *in utero*, at birth, or through breast milk after birth. In Africa, this perinatal transmission rate is approximately 25%. Mothers who are newly infected and breastfeeding their infants transmit HIV 40% of the time.

Immediately after HIV infection a person can either be asymptomatic or experience a transient 'flu-like' illness. In either case, virus becomes abundant in the peripheral blood, while the number of circulating CD4 T cells markedly declines (Figure 9.17). This acute viremia is almost always accompanied by activation of an HIV-specific immune response, in which anti-HIV antibodies are produced and cytotoxic T cells are activated and kill virus-infected cells. This response reduces the load of virus carried by the infected person and causes a corresponding increase in the number of circulating CD4 T cells. When an infected person first exhibits detectable levels of anti-HIV antibodies in their blood serum, they are said to have undergone **seroconversion**. The amount of virus persisting in the blood after the symptoms of acute viremia have passed is directly correlated with the subsequent course of disease.

The initial phase of infection is followed by an asymptomatic period, also called 'clinical latency'. During this phase of the infection, which can last for 2–15 years,

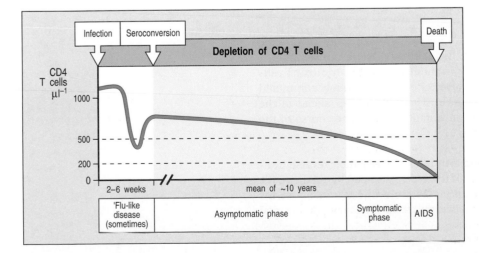

Figure 9.17 After infection with HIV there is a gradual extinction of CD4 T cells. The numbers of CD4 T cells (green line) refer to those present in peripheral blood. Opportunistic infections and other symptoms become more frequent as the CD4 T-cell count falls, starting at around 500 cells μl^{-1}. The disease then enters the symptomatic phase. When CD4 T-cell counts fall below 200 cells μl^{-1} the patient is said to have AIDS.

there is persistent infection and replication of HIV in CD4 T cells, causing a gradual decrease in T-cell numbers. Eventually, the number of CD4 T cells drops below that required to mount effective immune responses against other infectious agents. That transition marks the end of clinical latency, the beginning of the period of increasing immunodeficiency, and the onset of AIDS. Patients with AIDS become susceptible to a range of opportunistic infections and some cancers, and it is from the effects of these that they die.

Although the vast majority of infected people gradually progress to AIDS, a small minority do not follow this course and they are called long-term non-progressors. A small percentage seroconvert but their CD4 T-cell counts and other measures of immune competence are maintained. They have exceptionally low levels of circulating virus, and are being studied intensively to determine how they are able to control the infection. Another small group of people remain seronegative and disease-free despite extensive exposure to the virus. Some of this group have specific cytotoxic lymphocytes and T_H1 cells directed against infected cells, which suggests that at some time they have either been infected with the virus or have been exposed to non-infectious HIV antigens. The mechanisms that allow these people to resist AIDS are just beginning to be understood. As described in the next section, one way to resist infection is to lack a chemokine receptor that the virus uses as its co-receptor.

9-18 Genetic deficiency of the CCR5 chemokine co-receptor for HIV confers resistance to infection

Some people are resistant to HIV infection because they have an inherited defect in the chemokine receptor CCR5. These individuals are homozygous for a mutant allele of the CCR5 gene in which a 32-nucleotide deletion in the coding region leads to an altered reading frame, premature termination of translation, and a non-functional protein. So far, the mutant allele has been found only in Caucasoid populations, in which approximately 10% of the population are heterozygous for the mutant allele and 1% are homozygous.

CCR5 is the major macrophage and CD4 T-cell co-receptor used by HIV to establish primary infection. CCR5 deficiency therefore protects against primary infection with the macrophage-tropic variants of HIV. A few individuals who are homozygous for the non-functional variant of CCR5 do become infected with HIV. These people seem to have suffered from primary infection by lymphocyte-tropic strains of the virus.

9-19 HIV escapes the immune response and develops resistance to anti-viral drugs by rapid mutation

People infected with HIV make adaptive immune responses that can prevent the overt symptoms of disease for many years. Included in these responses are T_H1 and T_H2 cells, B cells that make neutralizing antibodies, and CD8 cytotoxic T cells that kill virus-infected cells (Figure 9.18). However, the virus is rarely eliminated by the immune response. One reason the virus is able to keep ahead of the human immune response is its high rate of mutation during the course of the infection.

HIV and other retroviruses have high mutation rates because their reverse transcriptases lack proof-reading mechanisms of the type possessed by cellular DNA polymerases. Consequently, reverse transcriptases are prone to making errors and these nucleotide substitutions soon accumulate to give new variant viral

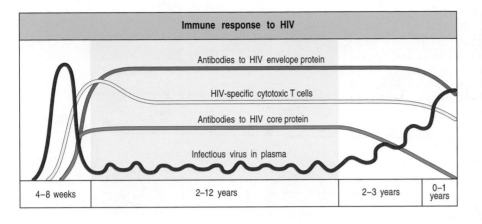

Figure 9.18 The adaptive immune response to HIV limits the effects of infection. During the course of infection, the levels of HIV virions and of components of adaptive immunity in the blood plasma change. Early in infection, while the adaptive immune response is being activated, the virus reaches high levels (red line). With the production of HIV-specific antibodies (blue lines) and HIV-specific cytotoxic T cells (yellow line), the virus is kept at a low level but is not eliminated. When the destruction of CD4 T cells outstrips their rate of renewal, adaptive immunity gradually collapses and virus levels increase again.

genomes. Even though the infection of a person might start from a single viral species, mutation throughout the infection produces many viral variants, called quasi-species, that co-exist within the infected person.

The presence of variant viruses increases the difficulty of terminating the infection by immune mechanisms. Immune attack by neutralizing antibody will select for the survival of viral variants that have lost the epitope recognized by the antibody. Similarly, pressure from virus-specific cytotoxic T cells selects for viruses in which the peptide epitope recognized by the cytotoxic T cell has changed. In some instances the homologous peptide derived from the variant virus interferes with the presentation of the antigenic peptide from the original virus, thereby allowing both viral species to escape the cytotoxic T cell.

The mutation rate of HIV is so high that different variants are isolated from every patient. Such diversity greatly complicates the task of developing a vaccine against HIV. Finding animal models that mimic the human disease is another problem for vaccine development. A potential model is that of the cynomolgus monkey, which on infection with a simian immunodeficiency virus (SIV) succumbs to an AIDS-like condition.

HIV mutability similarly limits the effectiveness of anti-viral drugs. Potential targets for drugs are the viral reverse transcriptase, which is essential for the synthesis of provirus, and the viral protease, which cleaves viral polyproteins after translation to give viral enzymes and proteins. Inhibitors of reverse transcriptase and protease have been found, and these drugs prevent further infection of healthy cells. Unfortunately, mutation inevitably produces HIV variants with proteins resistant to the action of the drugs. For some of the protease inhibitors, resistance can appear after only a few days. Resistance to the reverse transcriptase inhibitor zidovudine (AZT), in contrast, takes months to develop. This is because resistance to zidovudine requires the accumulated effect of three or four independent mutations in the reverse transcriptase, whereas the protease can acquire resistance with just a single mutation. Because HIV can so easily escape from the effects of any single drug, several anti-viral drugs are now used in combination. The hope is to destroy the entire population of viruses before any of them has accumulated enough mutations to resist all the drugs. One situation in which a relatively short period of drug treatment has long-term benefit is in pregnant women. A course of treatment with zidovudine can prevent HIV transmission to their children *in utero* and at birth.

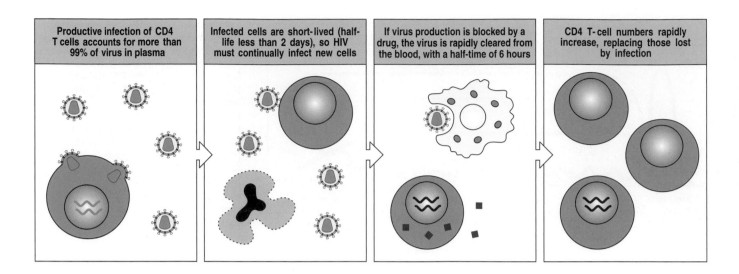

| Productive infection of CD4 T cells accounts for more than 99% of virus in plasma | Infected cells are short-lived (half-life less than 2 days), so HIV must continually infect new cells | If virus production is blocked by a drug, the virus is rapidly cleared from the blood, with a half-time of 6 hours | CD4 T-cell numbers rapidly increase, replacing those lost by infection |

9-20 Clinical latency is a period of active infection and renewal of CD4 T cells

The administration of anti-viral drugs to HIV-infected individuals has revealed the active nature of the infection during the period of clinical latency. Within two days of starting a course of drugs the amount of virus in the blood decreases dramatically. At the same time, the number of CD4 T cells increases substantially (Figure 9.19). This shows, first, that virus is being produced and cleared continuously within infected individuals, and second, that in the face of HIV infection the body continues to produce new CD4 T cells, which quickly become infected with HIV. Thus the period of clinical latency, when the overall numbers of CD4 T cells in the blood are gradually declining, is actually a time of immense immune activity. Vast numbers of T cells are produced and die and virions are neutralized, the latter most probably through antibody-mediated opsonization and phagocytosis.

Although measurement of lymphocyte and virion numbers in the blood is used clinically to monitor the progress of HIV infection, the secondary lymphoid tissues are the sites where most lymphocytes are found and where CD4 T cells are activated and produce virus. In HIV-infected people these tissues are loaded with virions, many of which are trapped on the surface of follicular dendritic cells.

9-21 HIV infection leads to immunodeficiency and death from opportunistic infections

Within a few days of being infected with HIV, a CD4 T cell dies. Three kinds of mechanism are thought to contribute to the death toll. One is direct killing as a result of the viral infection or virions binding to cell-surface receptors, the second is increased susceptibility of infected cells to apoptosis, and the third is killing by cytotoxic CD8 T cells specific for viral peptides presented by HLA class I molecules on the infected CD4 T cell.

Throughout clinical latency the daily loss of CD4 T cells is for the most part compensated for by the supply of new CD4 T cells. However, with time a steady decline in the CD4 T-cell number is evident, showing how the virus gradually wins this war of attrition. Eventually, the numbers of CD4 T cells become so low that immune responses to all foreign antigens are compromised and infected people become exceedingly susceptible to infection. Because CD4 T cells are required for all aspects of adaptive immunity, the vulnerability of patients with

Figure 9.19 Antiviral drugs rapidly clear virus from the blood and increase the number of circulating CD4 T cells. The first and second panels show that maintenance of HIV levels in the blood depends on the continual infection of newly produced CD4 T cells. This is because cells live for only a few days once infected. The third and fourth panels show the effects of administering a drug (red squares) that blocks the viral life cycle. The existing virions in the blood are rapidly cleared by the actions of neutralizing antibody, complement, and phagocytes. Newly produced CD4 T cells are not infected, whereupon they live longer and accumulate in the circulation.

Figure 9.20 A variety of opportunistic infections kill patients with AIDS. The most common of the opportunistic infections that kill AIDS patients in developed countries are listed. The malignancies are listed separately but they are also the result of impaired responses to infectious agents.

Infections	
Parasites	*Toxoplasma* species *Cryptosporidium* species *Leishmania* species *Microsporidium* species
Bacteria	*Mycobacterium tuberculosis* *Mycobacterium avium* 　*intracellulare* *Salmonella* species
Fungi	*Pneumocystis carinii* *Cryptococcus neoformans* *Candida* species *Histoplasma capsulatum* *Coccidioides immitis*
Viruses	Herpes simplex Cytomegalovirus Varicella–zoster

Malignancies
Kaposi's sarcoma (associated with 　herpes virus HHV8) Non-Hodgkin's lymphoma, including 　EBV-positive Burkitt's lymphoma Primary lymphoma of the brain

AIDS due to late-stage HIV infection resembles that of children with inherited severe combined immune deficiency.

The infections that most frequently affect AIDS patients are due to infectious agents that are already present in or on the body and that are actively kept under control in healthy people. Such agents are called opportunistic pathogens and the infections they cause are known as opportunistic infections (Figure 9.20). There is a rough hierarchy in the times at which particular opportunistic infections occur in AIDS, which is correlated with the order in which the different types of immunity collapse. Cellular immunity due to CD4 T_H1 cells tends to be lost before either antibody responses or cytotoxic CD8 T cells.

The oral and respiratory tracts are soft tissues loaded with organisms and in many AIDS patients they are the site of the first opportunistic infections. For example, *Candida* causes oral thrush and *Mycobacterium tuberculosis* causes tuberculosis. Later, patients can suffer from diseases caused by the reactivation of latent herpes viruses which are no longer under control by CD8 T cells. Such diseases include shingles due to varicella-zoster, B-cell lymphoma due to Epstein–Barr virus and an endothelial tumor called Kaposi's sarcoma, which is caused by the herpes virus HHV8. *Pneumocystis carinii*, a common environmental fungus that is hardly ever a problem to healthy people, frequently causes pneumonia and death in AIDS patients. In the later stages of AIDS, reactivation of cytomegalovirus, a herpes virus with similar effects to EBV, can cause B-cell lymphoproliferative disease. Infection with the opportunistic pathogen *Mycobacterium avium* also becomes prominent. The opportunistic infections suffered by individual AIDS patients vary greatly. With the collapse of the immune system, only drugs and other interventions can be used to treat the opportunistic infections. These are rarely completely effective or without their own deleterious effects. Eventually, the accumulated tissue damage resulting from the direct effects of HIV infection, opportunistic infections, and medical intervention causes death.

Summary

A major cause of acquired immunodeficiency in the human population today is the human immunodeficiency virus (HIV). This is a slow-acting retrovirus that remains latent for years after the initial infection. HIV infects CD4 T cells, macrophages, and dendritic cells, which all have the CD4 glycoprotein, the receptor for the virus. The effects of HIV are due chiefly to a gradual and sustained destruction of CD4 T cells, which leads eventually to a profound T-cell immunodeficiency known as acquired immune deficiency syndrome (AIDS). CD4 T cells proliferate as part of their normal function and are continually being renewed; these properties allow HIV to maintain a long-lasting infection in which individuals remain relatively healthy for years. Patients with AIDS succumb to a range of opportunistic infections and to some otherwise rare, virus-associated, cancers. HIV has only begun to infect humans in the past 50 years and the lack of any accommodation between host and pathogen is reflected in the strong correlation between infection and death. Genetic polymorphisms in the human population can also influence susceptibility to HIV infection. Because infection with common variants of the HIV virus is dependent on the presence of the chemokine receptor CCR5, a co-receptor for the virus, people who lack this cell-surface protein are resistant to infection by these variants.

A feature of HIV infection that encourages its dissemination in the population is that infected individuals can live normal lives for many years without knowing they are infected. This poses problems for social approaches to reducing the frequency of sexual transmission. A second approach is to develop vaccines that provide protective immunity, terminating the initial acute HIV infection. Determination of the most effective aspects of the initial immune response will be crucial to this approach, which is also bedeviled by the mutability of HIV. The third approach is that of designing drugs that interfere with the growth and replication of HIV while not affecting normal functions of human cells. This pharmacological approach is the one that has seen most investment and the results have been mixed. The mutability of the virus, which enables it to escape from individual drugs, is currently perceived as the major problem and its solution is the use of combination drug therapy.

Summary to Chapter 9

In the course of their long relationship with humans, successful pathogens have developed mechanisms that allow them to exploit the human body to the full. Indeed, a pathogen is, by definition, an organism that is habitually able to overcome the body's immune defenses to such an extent that it causes disease. One class of adaptations is those whereby the pathogen changes itself or its behavior to evade the ongoing immune response. This prevents the immune system from adapting to the pathogen and improving the response. In a second type of adaptation, the pathogen is able to impair or prevent the immune response. Pathogens can have more than one such adaptation, and for some pathogens a considerable fraction of their genome is devoted to foiling the immune system. Highly successful pathogens are not necessarily the most virulent. Non-lethal ubiquitous host–pathogen relationships, such as that of humans and the Epstein–Barr virus, have generally evolved over a long period of association. However, even these pathogens can cause life-threatening disease when the immune system is compromised.

Inherited immunodeficiencies are caused by a defect in one of the genes necessary for the development or function of the immune system. Depending on the gene involved, immunodeficiences range from manageable susceptibilities to particular pathogens to a general vulnerability created by the complete absence of adaptive immunity. The most severe inherited immunodeficiences are very rare—evidence of the importance of the immune system to human survival.

Immunodeficiency can also be acquired as the result of infection. The human immunodeficiency virus (HIV) infects macrophages, dendritic cells, and CD4 T cells and eventually reduces the number of CD4 T cells to a level at which severe immunodeficiency results—the acquired immune deficiency syndrome (AIDS). This retroviral pathogen has only just begun to exploit the human species, but it already has a variety of effective adaptions acquired during its history in other primate hosts. Most people infected with HIV have no overt symptoms of disease, which facilitates the spread of HIV by sexual transmission, and HIV infection has now reached epidemic proportions within the human population. The discovery of natural genetic traits that endow individual humans with resistance to HIV indicates possibilities for accommodation in the longer term.

Over-reactions of the Immune System 10

The protective functions of the immune system depend on recognition events that distinguish molecular components of infectious agents from those of the human body. In this context the immune system is said to recognize a pathogen's molecules as foreign. Besides infectious agents, humans come into daily contact with numerous other molecules that are equally foreign but do not threaten health. Many of these molecules derive from the plants and animals that we eat or are present in the environments where we live, work, and play. For most people for most of the time, contact with these molecules stimulates neither inflammation nor adaptive immunity.

However, in some circumstances certain kinds of innocuous molecule stimulate an adaptive immune response and the development of immunological memory in predisposed members of the population. On subsequent exposures to the antigen the immune memory produces inflammation and tissue damage that is at best an irritation and at worst a threat to life. The person feels ill, as though fighting off an infection, when no infection exists. The over-reactions of the immune system to harmless environmental antigens are called either **hypersensitivity**

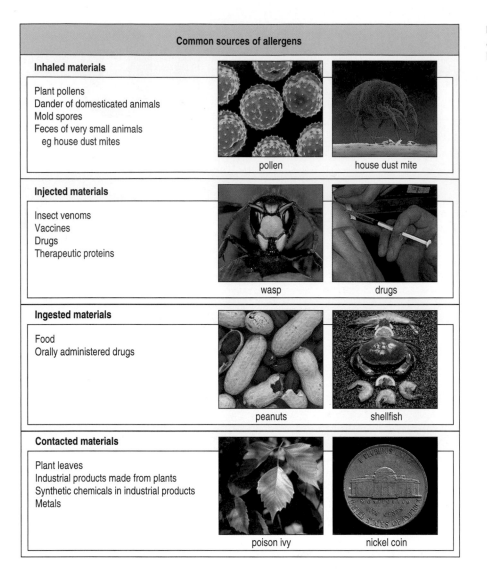

Figure 10.1 Some substances that are a common cause of hypersensitivity reactions.

reactions or **allergic reactions**. The environmental antigens that cause these reactions are termed **allergens** and they induce a state of hypersensitivity or **allergy**, the latter word being of Greek derivation and meaning 'altered reactivity'. Unfortunately, the antigens that provoke these over-reactions are often common in the human environment: in developed countries some 10–30% of the inhabitants are allergic to one or more environmental antigens. Some of the antigens responsible for common hypersensitivities are listed in Figure 10.1.

10-1 Four types of hypersensitivity reaction are caused by different effector mechanisms of adaptive immunity

Hypersensitivity reactions are conventionally grouped into four types according to the effector mechanisms that produce the reaction (Figure 10.2). It should be noted that a particular antigen (penicillin, for example) might cause several different types of hypersensitivity reaction, depending on the circumstances in which it is encountered. **Type I hypersensitivity reactions** result from the binding of antigen to antigen-specific IgE bound to its Fc receptor, principally on mast cells. This interaction causes the degranulation of mast cells and the release of inflammatory mediators. Type I reactions are commonly caused by inhaled particulate antigens, of which plant pollens are good examples. Type I reactions have effects of varying severity, ranging from a runny nose to breathing difficulties and even death by asphyxiation. Because of their prevalence and diversity, a large part of this chapter will be devoted to the IgE-mediated hypersensitivity reactions.

Type II hypersensitivity reactions are caused by small molecules that covalently bond to cell-surface components of human cells, producing modified structures that are perceived as foreign by the immune system. The B-cell response to these new epitopes produces IgG, which on binding to the modified cells causes their destruction through complement activation and phagocytosis. The antibiotic penicillin is an example of a small reactive molecule that can induce a type II hypersensitivity reaction.

Type III hypersensitivity reactions are due to small soluble immune complexes formed by soluble protein antigens binding to the IgG made against them. Some of these immune complexes become deposited in the walls of small blood vessels or the alveoli of the lungs. The immune complexes activate complement and initiate an inflammatory response that damages the tissue, impairing its physiological function. When antibodies or other proteins derived from non-human animal species are given therapeutically to patients, type III hypersensitivity reactions are a potential side-effect.

The effector molecules initiating type I, II, and III hypersensitivity reactions are all antibodies. In contrast, **type IV hypersensitivity reactions** are caused by the products of antigen-specific effector T cells. Most reactions are caused by CD4 T_H1 cells. For example, the inflammatory reaction around the site of an insect bite or sting is caused by CD4 T_H1 cells that respond to peptide epitopes derived from venom and other insect proteins introduced by the bite. A minority of type IV hypersensitivity reactions are due to cytotoxic CD8 T cells. They arise when small reactive lipid-soluble molecules pass through cell membranes and bond covalently to intracellular human proteins. Degradation of these chemically modified proteins yields abnormal peptides that bind to HLA class I molecules and stimulate a cytotoxic T-cell response. For example, the allergic response to poison ivy involves cytotoxic T cells. They recognize peptides derived from intracellular proteins that are modified by chemical reaction with pentadecacatechol, a chemical acquired by touching the plant's leaves.

	Type I	Type II		Type III	Type IV		
Immune reactant	IgE	IgG		IgG	T_H1 cells	T_H2 cells	CTL
Antigen	Soluble antigen	Cell- or matrix-associated antigen	Cell-surface receptor	Soluble antigen	Soluble antigen	Soluble antigen	Cell-associated antigen
Effector mechanism	Mast-cell activation	Complement, FcR+ cells (phagocytes, NK cells)	Antibody alters signaling	Complement Phagocytes	Macrophage activation	Eosinophil activation	Cytotoxicity
Example of hypersensitivity reaction	Allergic rhinitis, asthma, systemic anaphylaxis	Some drug allergies (eg penicillin)	Chronic urticaria (antibody to Fcε RIα)	Serum sickness, Arthus reaction	Contact dermatitis, tuberculin reaction	Chronic asthma, chronic allergic rhinitis	Contact dermatitis

Figure 10.2 Hypersensitivity reactions fall into four classes on the basis of their mechanism. The types of pre-existing immunity causing the reaction are listed along with the types of antigen that provoke the response, the underlying effector mechanism, and the nature of the ailment produced.

Type I hypersensitivity reactions

A prerequisite for a type I hypersensitivity reaction is that the person made IgE antibody when he or she first encountered the antigen. This is how a person becomes **sensitized** to the antigen. IgE differs from other antibody isotypes in being located predominantly in tissues, where it is bound to mast cells by high-affinity surface receptors known as FcεRI. We first look at the effector cells and molecules that produce a type I response, and then consider the characteristic properties of allergens and how sensitization occurs.

10-2 IgE binds irreversibly to Fc receptors on mast cells, basophils, and activated eosinophils

IgE antibody causes allergic reactions because of its exceptionally high affinity for its Fc receptor and the cellular distribution of the receptor. Binding of the IgE constant region to its high-affinity receptor—**FcεRI**—is the tightest of the antibody–Fc receptor interactions ($K_d = \sim 10^{10}$ M^{-1}, see Figure 7.22, p. 178) and can be considered for all practical purposes as irreversible. Also, unlike other isotypes, IgE binds to its receptor in the absence of antigen. FcεRI is expressed constitutively by mast cells and basophils, and by eosinophils after they have been activated by cytokines. After a primary IgE response has subsided and the antigen has been cleared by the usual means, all antigen-specific IgE molecules that have not encountered antigen will be bound by their Fc regions to FcεRI on these cells. The stable complexes of IgE and FcεRI effectively provide mast cells, basophils and eosinophils with antigen-specific receptors. All these cells have granules containing preformed inflammatory mediators. When the antigen is next encountered it will bind to the receptors and activate the cells to release their inflammatory mediators (Figure 10.3). Initial degranulation is followed by induced synthesis of a wider range of mediators.

Two important features distinguish the 'antigen receptors' on mast cells, basophils, and eosinophils from those of B cells and T cells. The first is that the cell's effector function becomes operational immediately after antigen binds to the 'receptor' and does not require a phase of cell proliferation and differentiation. The second is that each cell is not restricted to carrying receptors of a single antigen specificity but can carry a range of IgE representing that present in the circulation. These features combine to quicken and strengthen the response to any antigen to which a person has been sensitized. When any one of those antigens enters a tissue, all the mast cells in the vicinity bearing IgE specific for that antigen will be triggered to degranulate immediately.

IgE-mediated activation of mast cells, basophils, and activated eosinophils is thought to have evolved to provide protection against parasites (see Section 7-13, p. 179), which are prevalent in the tropical regions of the world. This type of environment is where the human species originated and where it has lived for most of

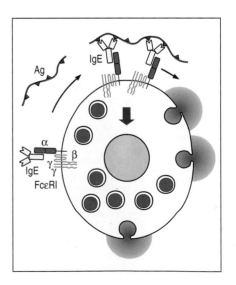

Figure 10.3 Crosslinking of FcεRI on the surface of mast cells by antigen and IgE causes mast-cell activation and degranulation. The high-affinity IgE receptor (FcεRI) on mast cells, basophils, and activated eosinophils is formed from one α chain, one β chain, and two γ chains. The binding site for IgE is formed by the two extracellular immunoglobulin-like domains of the α chain. The β chain and the two disulfide-bonded γ chains are largely intracellular and contribute to signaling. When receptors are crosslinked by antigen binding to IgE on the surface of a mast cell, a signal is transmitted that leads to the release of preformed granules containing histamine and other inflammatory mediators.

its existence. In the developing countries of the tropics, parasites are abundant and a third of the world's population is infected by one or more parasites.

In contrast, in the developed countries where parasite infections are rare, IgE responses tend to be stimulated by contact with non-threatening substances in the environment. In North America and Europe, the impact of allergic disease is increasingly being felt in all aspects of human life. Some physicians have called it an epidemic of allergy. Because allergy has directed medical research in this area we know far more about the allergic effects of IgE than we do about its benefits in controlling infection. It might be that IgE-mediated allergy is in fact a relatively new disease. People in industrialized countries might now be exposed to larger amounts of a small subset of environmental antigens, thus focusing the IgE response onto particular allergens. In contrast, a limited exposure to diverse environmental antigens during early life might tend to prevent sensitization to any particular allergen.

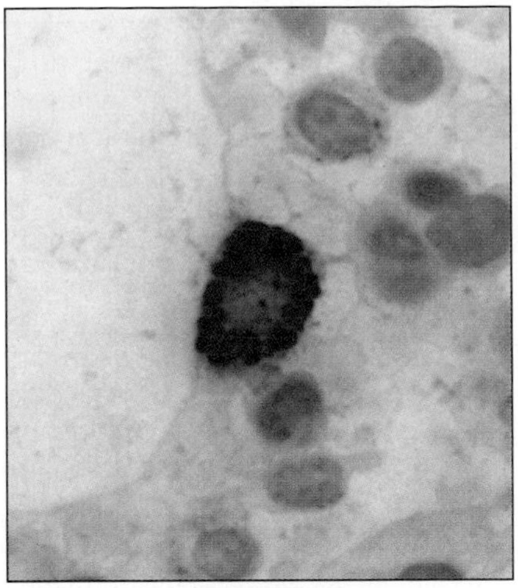

10-3 Tissue mast cells orchestrate IgE-mediated allergic reactions through the release of inflammatory mediators

Mast cells are resident in mucosal and epithelial tissues lining the body surfaces and serve to alert the immune system to local trauma and infection. Present in all vascularized tissues, except the central nervous system and the retina, their initial response to activation is to release inflammatory molecules that are stored in 50–200 large granules which fill the cytoplasm (Figure 10.4). The name **mast cell** derives from the German word *Mastzellen*, meaning 'fattened' or 'well-fed cells', which is how they appear under the microscope. Mast-cell granules contain principally histamine, heparin, tumor necrosis factor-α (TNF-α), chondroitin sulfate, neutral proteases, and other degradative enzymes and inflammatory mediators (Figure 10.5). Heparin, an acidic proteoglycan, is largely responsible for the characteristic staining of the mast-cell granules with basic dyes.

Figure 10.4 Light micrograph of mast cell stained for the granule protease chymase to show the numerous granules that fill its cytoplasm. The mast cell is in the center of the picture: its nucleus is stained pink and the cytoplasmic granules are stained red. Magnification × 1000. Photograph courtesy of D. Friend.

Class of product	Examples	Biological effects
Enzyme	Tryptase, chymase, cathepsin G, carboxypeptidase	Remodeling of connective tissue matrix
Toxic mediator	Histamine, heparin	Toxic to parasites Increase vascular permeability Cause smooth muscle contraction
Cytokine	IL-4, IL-13	Stimulate and amplify T_H2-cell response
	IL-3, IL-5, GM-CSF	Promote eosinophil production and activation
	TNF-α (some stored preformed in granules)	Promotes inflammation, stimulates cytokine production by many cell types, activates endothelium
Chemokine	MIP-1α	Chemotactic for monocytes, macrophages, and neutrophils
Lipid mediator	Leukotrienes C_4 and D_4	Smooth muscle contraction Increased vascular permeability Mucus secretion
	Platelet-activating factor	Chemotactic for leukocytes Amplifies production of lipid mediators Neutrophil, eosinophil, and platelet activation

Figure 10.5 Molecules released by mast cells on stimulation by antigen binding to IgE. Proteases, histamine, heparin, and TNF-α are prepackaged in granules and are released immediately the mast cell is stimulated by antigen binding. The other molecules are synthesized and released only as a result of mast-cell activation.

On leaving the bone marrow, immature mast cells travel in the blood as agranular cells that enter tissues and settle near small blood vessels. There they mature and form their characteristic granules. This differentiation is mediated by stem-cell factor (SCF), which interacts with CD117 (also called Kit) on the mast-cell surface. The importance of mast cells for IgE-mediated inflammatory reactions is well illustrated by the phenotype of mice lacking functional CD117. These mice have no differentiated mast cells and none of the inflammatory responses normally caused by IgE.

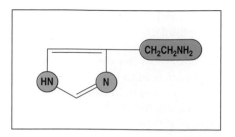

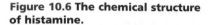

Figure 10.6 The chemical structure of histamine.

Two kinds of human mast cell are distinguished: a mucosal mast cell, which expresses the protease tryptase, and a connective tissue mast cell, which expresses chymotryptase. In patients with T-cell immunodeficiencies only connective tissue mast cells are present, suggesting that the development of mucosal mast cells is dependent on T cells. Mast-cell activation occurs in the presence of any antigen that can crosslink the IgE molecules bound to FcεRI at the cell surface. This can be accomplished by antigens with repetitive epitopes that crosslink IgE molecules of the same specificity, or by antigens possessing two or more different epitopes that crosslink IgE molecules of different specificities. Once a mast cell's receptors have been crosslinked, degranulation occurs within a few seconds, releasing the stored mediators into the immediate extracellular environment (see Figure 10.5).

Prominent among these chemicals is histamine, an amine derivative of the amino acid histidine (Figure 10.6). Histamine exerts a variety of physiological effects through three kinds of histamine receptor—H1, H2, and H3—which have been defined on different cell types. Acute allergic reactions involve histamine binding to the H1 receptor on nearby smooth muscle cells and on endothelial cells of blood vessels. Stimulation of the H1 receptor on endothelial cells induces vessel permeability and the entry of other cells and molecules into the allergen-containing tissue, causing inflammation. Smooth muscle cells are induced to contract on binding histamine, which constricts airways, for example, and histamine also acts on the epithelial linings of mucosa to induce the increased secretion of mucus. All these actions produce different effects depending upon the tissue exposed to allergen. Sneezing, coughing, wheezing, vomiting, and diarrhea can all be induced in the course of an allergic reaction.

Other molecules released from mast-cell granules include mast-cell chymotryptase, tryptase, and other neutral proteases that activate metalloproteases in the extracellular matrix. Collectively, these enzymes break down extracellular matrix proteins. The action of histamine is complemented by that of TNF-α, which is also released from mast-cell granules. TNF-α activates endothelial cells, causing an increased expression of adhesion molecules, thus promoting leukocyte traffic from the blood into the increasingly inflamed tissue (see Figure 8.11, p. 214). Mast cells are unique in their capacity to store TNF-α and release it on demand. At the start of an inflammatory response, mast cells are usually the main source of this cytokine.

Besides the preformed inflammatory mediators in granules, mast cells synthesize and secrete other mediators in response to activation (see Figure 10.5). These include chemokines, cytokines—IL-4 and more TNF-α—prostaglandins and leukotrienes. The latter are synthesized from fatty acids (Figure 10.7, top panel). All these mediators act locally within the area of tissue surrounding the activated mast cells. The leukotrienes have activities similar to those of histamine, but are more than a hundred times more potent on a molecule-for-molecule basis. The two kinds of mediator are therefore complementary. Histamine provides a rapid response while the more potent leukotrienes are being made. In the later stages of allergic reactions, leukotrienes are principally responsible for inflammation,

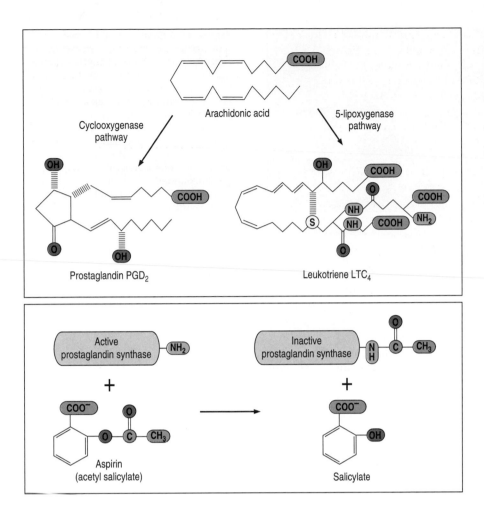

Figure 10.7 Mast cells synthesize prostaglandins and leukotrienes from arachidonic acid by different enzyme pathways. Arachidonic acid, an unsaturated fatty acid, is used as the substrate for the cyclooxygenase pathway to make prostaglandins, and as the substrate for the 5-lipoxygenase pathway to make leukotrienes, as shown in the top panel. Arachidonate is itself made from linolenic acid, one of the essential unsaturated fatty acids that must be provided in the diet because it cannot be made by human cells. Aspirin (acetyl salicylate) prevents the synthesis of prostaglandins by the cyclooxygenase pathway by irreversibly inhibiting the enzyme prostaglandin synthase, as shown in the bottom panel.

smooth muscle contraction, constriction of airways, and mucus secretion from mucosal epithelium. Mast cells also secrete prostaglandin D_2 (PGD$_2$), which promotes dilation and increased permeability of blood vessels and also acts as a chemoattractant for neutrophils. Aspirin reduces inflammation by inactivating prostaglandin synthase, the first enzyme in the cyclooxygenase pathway. The inactivation is irreversible because aspirin binds covalently to the active site of the enzyme (see Figure 10.7, bottom panel).

The combined effect of the chemical mediators released by mast cells is to attract circulating leukocytes to the site of mast-cell activation, where they amplify the reaction initiated by antigen and IgE on the mast-cell surface. These effector leukocytes include eosinophils, basophils, neutrophils, and T$_H$2 lymphocytes. In parasite infections these cells work together either to kill the parasite or to expel it from the body (see Section 7-13, p. 179). In allergic reactions the same strategy has a detrimental effect, damaging the body's tissues and impairing their function.

10-4 Eosinophils and basophils are specialized granulocytes that release toxic mediators in IgE-mediated responses

Eosinophils are granulocytes whose granules contain arginine-rich basic proteins. In histological sections these proteins stain heavily with eosin, hence the name eosinophil, meaning 'liking eosin' (Figure 10.8). Normally, only a minority of eosinophils are circulating in the blood. Most are resident in tissues, especially in the connective tissue immediately underlying epithelia of the respiratory,

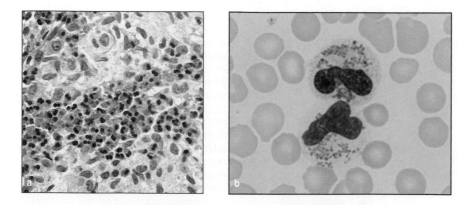

Figure 10.8 Eosinophils have a characteristic staining pattern in histological sections. Panel a shows a light micrograph of a section through a Langerhans' cell histiocytosis in skin that is heavily infiltrated with eosinophils. The eosinophils have bilobed nuclei and stain pink with the stain eosin. Panel b shows a higher-power light micrograph of a blood smear in which two partly degranulated eosinophils in the center are surrounded by erythrocytes. Panel a courtesy of T. Krausz; panel b courtesy of D. Swirsky and T. Krausz.

gastrointestinal, and urogenital tracts. As with mast cells, the activation of eosinophils by external stimuli leads to the staged release of toxic molecules and inflammatory mediators. Preformed chemical mediators and proteins present in the granules are released first (Figure 10.9, top panel). The normal function of these highly toxic molecules is to kill invading microorganisms and parasites directly. This is followed by a slower induced synthesis and secretion of prostaglandins, leukotrienes, and cytokines, which amplify the inflammatory response by activation of epithelial cells and leukocytes, including more eosinophils (see Figure 10.9, bottom panel).

The eosinophil response is highly toxic and potentially damaging to the host as well as to parasites. Limiting the damage are various control mechanisms. When the body is healthy, the number of eosinophils is kept low by limiting their production in the bone marrow. When infection or antigenic stimulation activates

Class of product	Examples	Biological effects
Enzyme	Eosinophil peroxidase	Toxic to targets by catalyzing halogenation Triggers histamine release from mast cells
	Eosinophil collagenase	Remodeling of connective tissue matrix
Toxic protein	Major basic protein	Toxic to parasites and mammalian cells Triggers histamine release from mast cells
	Eosinophil cationic protein	Toxic to parasites Neurotoxin
	Eosinophil-derived neurotoxin	Neurotoxin
Cytokine	IL-3, IL-5, GM-CSF	Amplify eosinophil production by bone marrow Cause eosinophil activation
Chemokine	IL-8	Promotes influx of leukocytes
Lipid mediator	Leukotrienes C_4 and D_4	Smooth muscle contraction Increased vascular permeability Mucus secretion
	Platelet-activating factor	Chemotactic to leukocytes Amplifies production of lipid mediators Neutrophil, eosinophil, and platelet activation

Figure 10.9 Activated eosinophils secrete toxic proteins contained in their granules and also produce cytokines and inflammatory mediators. The enzymes and toxic proteins are contained preformed in the granules and are released immediately on eosinophil activation. The cytokines, chemokines, and lipid mediators are synthesized only after activation.

Figure 10.10 The presence of an abnormally large number of circulating eosinophils causes damage to the heart. The photograph shows a section of endocardium from a patient with hypereosinophilic syndrome. The damaged tissue is characterized by an organized fibrous exudate and an endocardium thickened by fibrous tissue. Although the patient has large numbers of circulating eosinophils, these cells are not seen in the injured endocardium, which is thought to be damaged by the contents of granules released by circulating eosinophils. Photograph courtesy of D. Swirsky and T. Krausz.

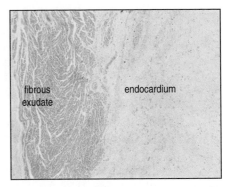

T$_H$2 cells, the IL-5 and other cytokines released by the T$_H$2 cells stimulate the bone marrow to increase both the production of eosinophils and their release into the circulation. The migration of eosinophils into tissues is controlled by a chemokine called **eotaxin**, which is produced by activated endothelial cells, T cells, and monocytes.

The activity of eosinophils is further controlled by modulation of their sensitivity to external stimuli. This is achieved through regulated expression of FcεRI. In the resting state, eosinophils do not express FcεRI, do not bind IgE, and cannot therefore be induced to degranulate by antigen. Once an inflammatory response is set in motion, cytokines and chemokines in the inflammatory site induce eosinophils to express FcεRI. The expression of Fcγ receptors and complement receptors on the eosinophil surface also increases, facilitating the binding of eosinophils to pathogen surfaces coated with IgG and complement.

The potential of eosinophils to cause tissue damage is clearly seen in patients who have abnormally high numbers of eosinophils. Certain T-cell lymphomas, for example, constitutively secrete IL-5, which continually expands the eosinophil population. This expansion is detected by a greatly increased number of eosinophils in the blood (**hypereosinophilia**). With so many circulating eosinophils, only a small proportion need to be activated for their effects to be seriously disruptive. Patients with hypereosinophilia can suffer damage to the endocardium of the heart (Figure 10.10) and to nerves, leading to heart failure and neuropathy. Both of these effects are thought to be due to cytotoxic and neurotoxic proteins released from eosinophil granules.

In localized allergic reactions, mast-cell degranulation and the activation of T$_H$2 cells cause the accumulation of activated eosinophils at the site. Their presence is characteristic of chronic allergic inflammation; for example, eosinophils are considered to be the principal cause of the damage to the airways that occurs in chronic asthma.

Basophils are granulocytes whose granules stain with basic dyes such as hematoxylin (Figure 10.11). The granules of basophils package a similar, but not identical, set of mediators to those of mast cells. Although basophils resemble mast cells in some ways, they are developmentally more closely related to eosinophils. They share a common stem-cell precursor, as well as requiring similar growth factors, including IL-3, IL-5, and GM-CSF. Production of eosinophils and basophils seems to be reciprocally regulated, so that a combination of TGF-β and IL-3 promotes the maturation of basophils while suppressing that of eosinophils. Basophils are normally present in very low numbers in the circulation and are therefore more difficult to study than other leukocytes. Their functions in defense against parasites seem similar to those of eosinophils. Like eosinophils, basophils are recruited into sites of allergic reactions, where they are activated to degranulate by antigen crosslinking the IgE bound to the FcεRI on the basophil cell surface.

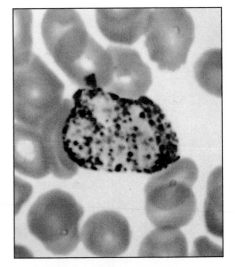

Figure 10.11 Light micrograph of basophil stained with Wright's Giemsa. The basophil in the center is surrounded by erythrocytes in this blood smear. Magnification × 1000. Photograph courtesy D. Friend.

Mast cells, eosinophils, and basophils often act in concert. Mast-cell degranulation initiates the inflammatory response, which then recruits eosinophils and basophils. Eosinophil degranulation releases **major basic protein**, which in turn causes the degranulation of mast cells and basophils. This latter effect is augmented by any of the cytokines—IL-3, IL-5, and GM-CSF—that affect the growth, differentiation, or activation of eosinophils and basophils.

10-5 Common allergens are small proteins inhaled in particulate form that stimulate an IgE response

In the previous sections we have considered the effector cells responsible for the symptoms of allergies. Here we turn to the types of antigen to which people become sensitized. IgE is responsible for most allergic responses (Figure 10.12). The production of IgE is favored when the immune system is challenged with small quantities of antigen and when the cytokine IL-4 is present at the time that naive CD4 T cells are presented with antigen. Under these circumstances CD4 T cells tend to make a T_H2 response (see Section 8-15, p. 226) which then produces more IL-4 and additional cytokines that stimulate B cells to switch their immunoglobulin isotype to IgE (see Section 7-5, p. 170). Initial sensitization to an allergen is thus favored by circumstances that promote the production of antigen-specific T_H2 cells and the production of IgE, and disfavored by those producing T_H1 cells. The principal function of IL-4 seems to be to facilitate the IgE response; in mice unable to make this cytokine the main defect is diminished IgE synthesis.

Because T cells are stimulated by antigen-derived peptides, the allergens that provoke type I responses are invariably proteins. Much human allergy is caused by a limited number of airborne proteins that are inhaled. Small amounts of the protein enter the body by crossing the mucosa of the respiratory tract and then stimulate

IgE-mediated allergic reactions			
Syndrome	**Common allergens**	**Route of entry**	**Response**
Systemic anaphylaxis	Drugs Serum Venoms Peanuts	Intravenous (either directly or following rapid absorption)	Edema Increased vascular permeability Tracheal occlusion Circulatory collapse Death
Wheal-and-flare	Insect bites Allergy testing	Subcutaneous	Local increase in blood flow and vascular permeability
Allergic rhinitis (hay fever)	Pollens (ragweed, timothy, birch) Dust-mite feces	Inhaled	Edema of nasal mucosa Irritation of nasal mucosa
Bronchial asthma	Pollens Dust-mite feces	Inhaled	Bronchial constriction Increased mucus production Airway inflammation
Food allergy	Shellfish Milk Eggs Fish Wheat	Oral	Vomiting Diarrhea Pruritis (itching) Urticaria (hives) Anaphylaxis

Figure 10.12 Allergic reactions mediated by IgE.

T_H2 responses in the local lymphoid tissues. Allergens are only a small proportion of all the proteins that humans inhale, and a central goal of research on allergy is to define what distinguishes proteins that become allergens from those that do not. A definitive answer has yet to be found. Figure 10.13 shows some of the properties that characterize inhaled allergens.

Most allergens are small, soluble proteins that are present in dried-up particles of material derived from plants and animals. Examples are pollen grains, the mixture of dried cat skin and saliva that forms dander, and the dried feces of the house dust mite *Dermatophagoides pteronyssimus*. The light dry particles become airborne and are inhaled by humans when they breathe. Once inhaled, the particles become caught in the mucus bathing the epithelia of the airways and lungs. They then rehydrate, releasing the antigenic proteins. These antigens are carried to professional antigen-presenting cells within the mucosa. The antigens are then processed and presented by these cells to CD4 T cells, stimulating a T_H2 response that leads to the production of IgE and its binding to mast cells (Figure 10.14). Small soluble protein antigens are more efficiently leached out of particles and penetrate the mucosa. A substantial proportion of allergens are proteases and it is likely that their enzymatic activities facilitate the breakdown of the particle, the release of allergen, and the generation of peptides that stimulate T_H2 cells.

The major allergen responsible for more than 20% of the allergies in the human population of North America is a cysteine protease derived from *D. pteronyssimus*. Advances in the heating and cooling of homes, offices, and other buildings are believed to be responsible for the prevalence of this allergy because they provide an environment which encourages both the growth of *D. pteronyssimus* and the desiccation of its feces. The air currents created by forced-air heating, air conditioners, and vacuum cleaners all help to move the particles into the air, where they will be breathed in by the buildings' human inhabitants.

The cysteine protease of *D. pteronyssimus* is related to the protease papain, which comes from the papaya fruit and is used in cooking as a meat tenderizer. Workers involved in the commercial production of papain become allergic to the enzyme, an example of an occupational immunological disease. Similarly, the protease subtilisin, the 'biological' component of some laundry detergents, causes allergy in laundry workers. Chymopapain, a protease related to papain, is used in medicine to degrade intervertebral disks in patients with sciatica. A rare complication

Features of inhaled allergens that may promote the priming of T_H2 cells that drive IgE responses	
Molecular type	Proteins, because only they induce T-cell responses
Function	Allergens are often proteases
Low dose	Favors activation of IL-4-producing CD4 T cells
Low molecular weight	Allergen can diffuse out of particle into mucus
High solubility	Allergen is readily eluted from particle
High stability	Allergen can survive in desiccated particle
Contains peptides that bind host MHC class II	Required for T-cell priming

Figure 10.13 Properties of inhaled allergens.

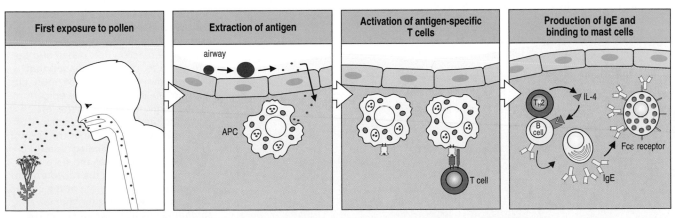

Figure 10.14 Sensitization to an inhaled allergen. Antigens that leach from inhaled pollen are taken up by antigen-presenting cells (APC) in the mucosa of the airways. These activate naive T cells to T_H2 effector cells, which secrete IL-4. Isotype switching by B cells in the presence of T_H2 help and IL-4 leads to plasma cells that secrete IgE. The IgE binds to FcεRI on mast cells.

of this procedure affects patients who are sensitized to chymopapain; they experience an acute systemic allergic response to the enzyme, an example of systemic anaphylaxis.

10-6 Predisposition to allergy has a genetic basis

In the caucasoid populations of Europe and North America, up to 40% of people are more likely to make IgE responses to common environmental antigens than the rest of the population. Allergists call this predisposed state **atopy**. As a group, atopic people have higher levels of soluble IgE and circulating eosinophils than non-atopic people. Family studies indicate a genetic basis for atopy, with an involvement of genes on chromosomes 5 and 11.

The candidate gene on chromosome 11 encodes the β subunit of FcεRI, and the region implicated on chromosome 5 contains a cluster of genes including those for IL-3, IL-4, IL-5, IL-9, IL-13, and GM-CSF. These cytokines are directly involved in isotype switching, eosinophil survival, and mast-cell proliferation. For example, an inherited sequence difference in the promoter region of the IL-4 gene is correlated with the raised IgE levels seen in atopic people. HLA class II polymorphism also affects the IgE response to certain allergens. An IgE response to several pollen antigens from ragweed is correlated with the expression of the HLA class II allotype DRB1*1501. Such associations imply that certain HLA class II:peptide combinations predispose to stimulation of a T_H2 response.

10-7 IgE-mediated allergic reactions consist of an immediate response followed by a late-phase response

One way in which clinical allergists detect a person's sensitivity to allergens is to observe the reaction when small quantities of common allergens are injected into the skin. Substances to which a person is sensitive produce a characteristic inflammatory reaction called a **wheal and flare** at the site of injection within a few minutes (Figure 10.15, left panel). Substances to which the person is not allergic produce no such reaction. Because of its rapid appearance, the wheal and flare is called an **immediate reaction**. Such reactions are the direct consequence of IgE-mediated mast-cell degranulation in the skin. Released histamine and other mediators cause increased permeability of local blood vessels, whereupon fluid leaves the blood and produces local swelling (edema). The swelling produces the wheal at the injection site, and the increased blood flow into the surrounding area produces the redness that is the flare. Immediate reactions can last for up to 30 minutes and the relative intensity of the wheal and flare varies.

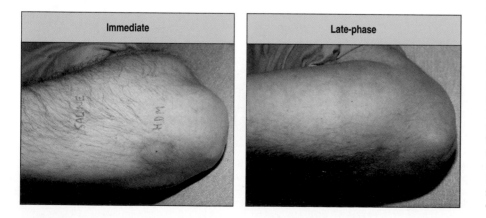

Figure 10.15 Allergic reactions consist of an immediate reaction followed later by a late-phase reaction. The photographs show the immediate reaction (left panel) and the late-phase reaction (right panel) that develop from the injection of house dust mite (HDM) antigen into the epidermis of an allergic individual. In the left panel the site of allergen injection is labeled HDM and the site of the control injection of saline is labeled 'saline'. HDM produces a wheal-and-flare reaction. The wheal is the raised area of skin with the injection site at the center, the flare is the redness (erythema) spreading out from the wheal. In the right panel, the swelling from the HDM injection has spread to involve surrounding tissue. Photographs courtesy of A.B. Kay.

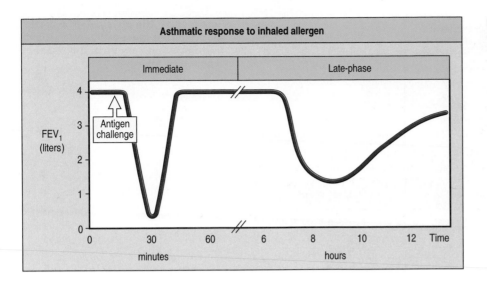

Figure 10.16 The course of an asthmatic response to inhaled allergen. The response is measured in terms of breathing capacity: the forced expiratory volume of air in one second (FEV_1). This is the maximum volume of air that a person can forcibly breathe out of the lungs in one second after taking a deep breath. The immediate response in the lungs finishes within an hour, to be followed some six hours later by the late-phase response.

Some 6–8 hours after the immediate reaction has subsided, a second reaction—the **late-phase reaction**—occurs at the site of injection (see Figure 10.15, right panel). This consists of a more widespread swelling and is due to the leukotrienes, chemokines, and cytokines synthesized by mast cells after IgE-mediated activation.

Although skin tests are a good way of determining whether a patient has developed an allergen-specific IgE response, they cannot assess, for example, whether the response to a particular allergen is the cause of a patient's asthma—an allergic reaction in which the airways become inflamed, constricted, and blocked with mucus. To ascertain this, allergists make direct measurements of a person's breathing capacity in the absence and presence of inhaled allergen (Figure 10.16). On inhalation of an allergen to which a patient is sensitized, mucosal mast cells in the respiratory tract degranulate. The released mediators cause immediate constriction of the bronchial smooth muscle, which results in the expulsion of material from the lungs by coughing, and difficulty in breathing. As in the skin test, the immediate response in the lungs finishes within an hour, to be followed some six hours later by the late-phase response, which is due to leukotrienes and other mediators. In allergies to inhaled antigens, such as chronic asthma, the late-phase reaction is the more damaging. It induces the recruitment of leukocytes, particularly eosinophils and T_H2 lymphocytes, into the site and, if antigen persists, the late-phase response can easily develop into a chronic inflammatory response in which allergen-specific T_H2 cells promote eosinophilia and IgE production.

10-8 The effects of IgE-mediated allergic reactions vary with the site of mast-cell activation

When sensitized people are re-exposed to an allergen, the effect of the IgE-mediated reaction varies depending on the allergen and the tissues that it comes in contact with. Only those mast cells at the site of exposure degranulate and, once released, the preformed mediators are short-lived. Their effects on blood vessels and smooth muscles are therefore confined to the immediate vicinity of the activated mast cells. The more sustained effects of the late-phase response are also restricted to the site of allergen exposure, as the leukotrienes and other induced mediators are also short-lived. The anatomy of the site of contact also determines how quickly the inflammatory reaction subsides. The tissues most commonly

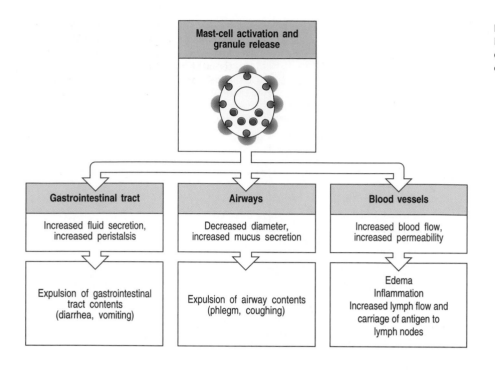

Figure 10.17 The physical effects of IgE-mediated mast-cell degranulation vary with the tissue exposed to allergen.

exposed to allergens are the mucosa of the respiratory and gastrointestinal tracts, the blood, and connective tissues. Airborne allergens irritate the respiratory tract and food-borne antigens the gastrointestinal tract; the blood and connective tissues receive allergens through insect bites and other wounds and also from absorption via the gut and respiratory mucosa (Figure 10.17).

Although direct evidence is limited, it is widely believed that the IgE response evolved as a defense against invertebrate parasites, especially worms (see Section 7-13, p. 179). Interactions between antigen, IgE, and mast cells trigger violent muscular contractions that expel material from the gastrointestinal tract and increase fluid flow to wash it out. In the lungs, the muscular spasms and increased mucus secretion that result from mast-cell activation can be seen as a way to expel organisms attached to respiratory epithelium, such as lung flukes. In allergy, this crude defense is mistakenly ranged against non-threatening particles and proteins that are often smaller than any microorganism. The violence and power of the IgE-mediated response means that the clinical manifestations of an allergic reaction can vary markedly with the amount of IgE that a sensitized person has made, the amount of allergen triggering the reaction, and the route by which it entered the body. In the next four sections we shall examine how allergic reactions vary when they occur in different tissues.

10-9 Systemic anaphylaxis is caused by allergens in the blood

When an allergen enters the bloodstream it can cause widespread activation of the connective tissue mast cells associated with blood vessels. This causes a dangerous hypersensitivity reaction called **systemic anaphylaxis**. During systemic anaphylaxis, disseminated mast-cell activation causes both an increase in vascular permeability and a widespread constriction of smooth muscle. Fluid leaving the blood causes the blood pressure to drop drastically, a condition called **anaphylactic shock**, and the connective tissues to swell. Damage is sustained by many organ systems and their function is impaired. Death is usually caused by asphyxiation due to constriction of the airways and swelling of the epiglottis (Figure 10.18). In the USA, more than 160 deaths a year are the result of anaphylaxis, which is the most extreme over-reaction of the body's defenses. Because this form

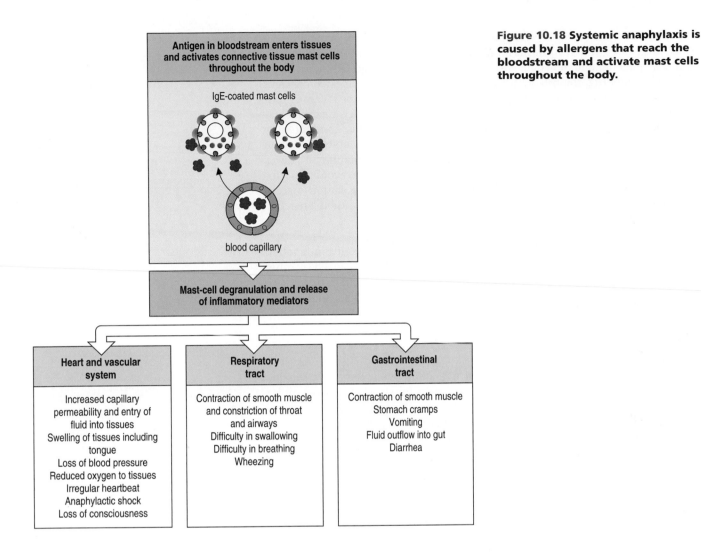

of immunity is fatal rather than protective it was called 'anaphylaxis', meaning anti-protection, to contrast with the 'prophylaxis' afforded by protective immunity.

Potential allergens are introduced directly into the blood by stings from wasps, bees, and other venomous insects; these account for about a quarter of the fatalities from anaphylaxis in the USA. Drug injections can also lead to anaphylaxis. Systemic anaphylaxis can also be provoked by food or drugs taken orally if the allergens that they contain are absorbed rapidly from the gut into the blood. Foods that can cause anaphylaxis include peanuts and brazil nuts. Anaphylactic reactions can be fatal, but treatment with an injection of epinephrine (adrenaline) will usually bring them under control. Epinephrine stimulates the reformation of tight junctions between endothelial cells. This reduces their permeability and prevents fluid loss from the blood, diminishing tissue swelling and raising blood pressure. Epinephrine also relaxes constricted bronchial smooth muscle and stimulates the heart (Figure 10.19). The danger of anaphylaxis is such that patients with known anaphylactic sensitivity to insect venoms or food are advised to carry a syringe full of epinephrine at all times.

Figure 10.19 The change in blood pressure during systemic anaphylaxis and its treatment with epinephrine. Time 0 indicates the time at which the anaphylactic reaction was reported by the patient. The arrows indicate the times at which injections of epinephrine were given.

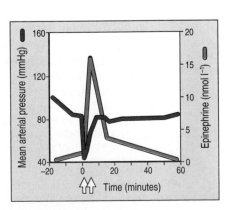

The most common cause of systemic anaphylaxis is allergy to penicillin and related antibiotics, which accounts for about 100 fatalities a year in the USA. Penicillin is a small organic molecule with a reactive β-lactam ring. On ingestion or injection of the drug, the β-lactam ring can be opened up to produce covalent conjugates with proteins of the body, creating new 'foreign' epitopes. These modified proteins can then stimulate a variety of hypersensitivity reactions, as described in later sections. In some individuals, the response is dominated by T$_H$2 cells and thus by B cells producing IgE specific for the new epitopes. When penicillin is given to a person who has been sensitized in this manner, it causes anaphylaxis and even death. For this reason, clinicians take care to avoid prescribing any drug to patients who have a history of allergy to that drug. Unfortunately, penicillin cannot be modified to remove its allergic potential because the reactive β-lactam ring is essential for its antibiotic activity.

Reactions resembling anaphylaxis can occur in the absence of specific interaction between an allergen and IgE. Such **anaphylactoid reactions** are caused by other stimuli that induce mast-cell degranulation and can be occasioned by exercise or by certain drugs and chemicals. Anaphylactoid reactions are also treated with epinephrine.

10-10 Rhinitis and asthma are caused by inhaled allergens

Allergens most commonly enter the body by inhalation. Mild allergies to inhaled antigens are common, being manifested as violent bursts of sneeezing and a runny nose, a condition called **allergic rhinitis** or hay fever. It is caused by allergens that diffuse across the mucous membrane of the nasal passages and activate mucosal mast cells beneath the nasal epithelium (Figure 10.20). Allergic rhinitis is characterized by local edema leading to obstruction of the nasal airways and a nasal discharge of mucus that is rich in eosinophils. There is also a generalized irritation of the nose due to histamine release. The reaction can extend to the ear and throat, and the accumulation of fluid in the blocked sinuses and eustachian tubes is conducive to bacterial infection. The same exposure to allergen that produces rhinitis can affect the conjuctiva of the eyes, where the reaction is called **allergic conjunctivitis**. It produces itchiness, tears, and inflammation. Although these reactions are uncomfortable and distressing, they are generally of short duration and cause no long-lasting tissue damage.

Much more serious is **allergic asthma**, a condition in which allergic reactions cause chronic difficulties in breathing, such as shortness of breath and wheezing. Asthma is triggered by allergens activating submucosal mast cells in the lower airways of the respiratory tract. Within seconds of mast-cell degranulation there is an increase in the fluid and mucus being secreted into the respiratory tract, and bronchial constriction due to contraction of the smooth muscle surrounding the airway. Chronic inflammation of the airways is a characteristic feature of asthma, involving a persistent infiltration of leukocytes, including T$_H$2 lymphocytes, eosinophils, and neutrophils (Figure 10.21). The overall effect of the asthmatic attack is to trap air in the lungs, making breathing more difficult. Patients with allergic asthma often need treatment, and asthmatic attacks can prove fatal.

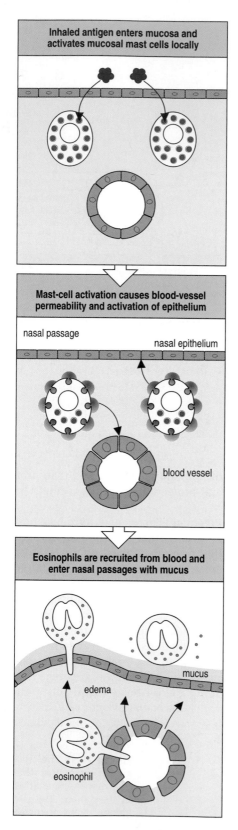

Figure 10.20 Allergic rhinitis is caused by allergens entering the respiratory tract. Histamine and other mediators released by activated mast cells increase the permeability of local capillaries and activate nasal epithelium to produce mucus. Eosinophils attracted into the tissues from the blood become activated and release their inflammatory mediators. Activated eosinophils are shed into the nasal passages.

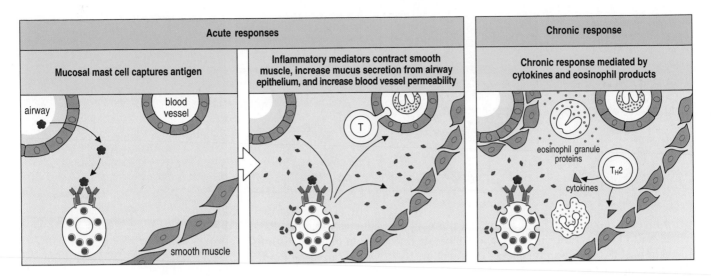

Figure 10.21 The acute response in allergic asthma leads to T$_H$2-mediated chronic inflammation of the airways. In sensitized individuals, mast cells carrying IgE specific for the allergen are present in the mucosa of the airways, as shown in the first panel. As shown in the second panel, crosslinking of specific IgE on the surface of mast cells by inhaled allergen triggers them to secrete inflammatory mediators, causing bronchial smooth muscle contraction and increased mucus secretion from mucosal epithelium, which together lead to airway obstruction. Increased blood vessel permeability also caused by inflammatory mediators leads to edema and an influx of inflammatory cells, including eosinophils and T$_H$2 lymphocytes. Activated mast cells and T$_H$2 cells secrete cytokines that also augment eosinophil activation and degranulation, which causes further tissue injury and influx of inflammatory cells, as shown in the third panel. The end result is chronic inflammation, which can then cause irreversible damage to the airways.

Although allergic asthma is initially driven by a response to a specific allergen, the chronic inflammation that subsequently develops seems to be perpetuated in the absence of further exposure to the allergen. In **chronic asthma** the airways can become almost totally occluded by mucus plugs (Figure 10.22). A generalized hypersensitivity in the airways also develops, and environmental factors other than re-exposure to specific allergen can trigger asthmatic attacks. Typically, the airways of chronic asthmatics are hyper-responsive to chemical irritants commonly present in air, such as cigarette smoke and sulfur dioxide. Disease can be exacerbated by immune responses to bacterial or viral infections of the respiratory tract, especially when they are dominated by T$_H$2 cells. For this reason, chronic asthma is classified as a type IV hypersensitivity reaction caused by T cells.

10-11 Urticaria, angioedema, and eczema are allergic reactions in the skin

Allergens that activate mast cells in the skin to release histamine cause raised itchy swellings called **urticaria** or **hives**. Urticaria means 'nettle-rash' and the word derives from *Urtica*, the Latin name for stinging nettles; the origin of the

Figure 10.22 Inflammation of the airways in chronic asthma restricts breathing. Panel a shows a light micrograph of a section through the bronchus of a patient who died of asthma; there is almost total occlusion of the airway by a mucus plug (MP). The small white circle is all that is left of the lumen of the bronchus. In panel b, a light micrograph at higher magnification gives a closer view of the bronchial wall. It shows injury to the epithelium lining the bronchus, accompanied by a dense inflammatory infiltrate that includes eosinophils, neutrophils, and lymphocytes. L, lumen of the bronchus. Photographs courtesy of T. Krausz.

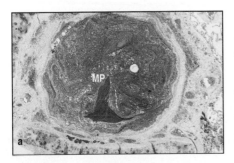

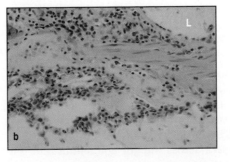

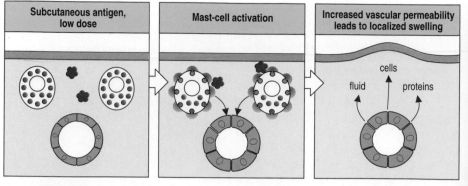

| Subcutaneous antigen, low dose | Mast-cell activation | Increased vascular permeability leads to localized swelling |

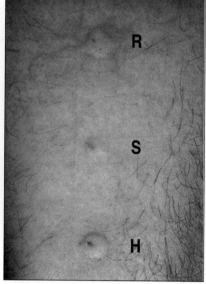

word hives is unknown. This reaction is essentially the same as the immediate wheal-and-flare reaction caused by the deliberate introduction of allergens into the skin in tests to determine allergy (see Section 10-7), which can also be produced by the injection of histamine alone (Figure 10.23). Activation of mast cells in deeper subcutaneous tissue leads to a similar but more diffuse swelling called **angioedema** (or angioneurotic edema). Urticaria and angioedema can arise as a result of any allergy to a food or drug if the allergen gets carried to the skin by the bloodstream, and they are among the many reactions that occur during systemic anaphylaxis. Insect bites are a common cause of urticaria, and such local reactions usually occur without inducing a more general anaphylaxis.

A more prolonged allergic response in the skin is observed in some atopic children. This condition is called **atopic dermatitis** or **eczema**. The word eczema is of Greek derivation and means to 'break out' or 'boil over'. The condition is characterized by an inflammatory response that causes a chronic and itching skin rash with associated skin eruptions and fluid discharge. This response has similarities to that occurring in the bronchial walls of asthmatics. Eczema frequently presents in families with a history of asthma and allergic rhinitis, and is often associated with high IgE levels. However, the severity of the dermatitis is not readily correlated with exposure to particular allergens or to the levels of allergen-specific IgE. Thus the etiology of eczema remains poorly understood. Neither is it understood why ezcema usually clears in adolescence, whereas rhinitis and asthma more often persist throughout life.

10-12 Food allergies cause systemic effects as well as gut reactions

Humans probably eat a greater variety of foodstuffs than any other living thing. All human food is derived from plants and animals and contains a large number of different proteins, many of which are potentially immunogenic. As food passes down the gastrointestinal tract, the proteins are degraded by proteases into peptides of ever-decreasing size. These are a potential source of peptides for presentation to T_H2 cells. Despite the quantity and variety of food that humans eat, IgE is made against an extremely small proportion of the proteins ingested. However, people sensitized to a particular protein will be allergic to any food containing that protein. Foods that commonly cause allergies include grains, nuts, fruits, legumes, fish, shellfish, eggs, and milk.

Once sensitized to a food allergen, any subsequent intake causes a marked and immediate reaction that can drive the person from the dining room. The allergen passes across the epithelial wall of the gut and binds to IgE on the mucosal mast cells associated with the gastrointestinal tract. The mast cells degranulate, releasing their mediators, principally histamine. The local blood vessels become permeable, and fluid leaves the blood and passes across the gut epithelium into the lumen of the gut. Meanwhile, contraction of smooth muscles of the stomach wall

Figure 10.23 Allergen-induced release of histamine by mast cells in skin causes localized swelling. As shown in the panels on the left, allergen introduced into the skin of a sensitized individual causes mast cells in the connective tissue to degranulate. The histamine released dilates local blood vessels, causing rapid swelling due to leakage of fluid and proteins into tissues. The photograph shows the raised swellings (wheals) that appear 20 minutes after intradermal injections of ragweed pollen antigen (R) or histamine (H) into a person who is allergic to ragweed. The small wheal at the site of saline injection (S) is due to the volume of fluid injected into the dermis. Photograph courtesy of R. Geha.

Figure 10.24 Ingested allergen can cause vomiting, diarrhea, and urticaria. Localized reactions are caused by histamine acting on intestinal epithelium, blood vessels and smooth muscles. Urticaria is caused by antigen that enters blood vessels and is carried to the skin.

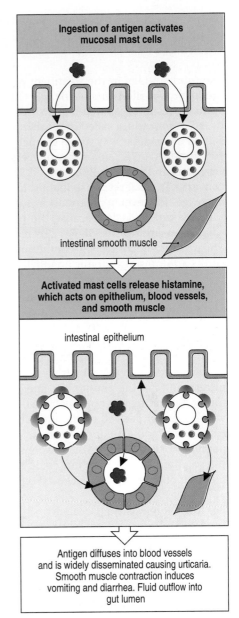

Ingestion of antigen activates mucosal mast cells

intestinal smooth muscle

Activated mast cells release histamine, which acts on epithelium, blood vessels, and smooth muscle

intestinal epithelium

Antigen diffuses into blood vessels and is widely disseminated causing urticaria. Smooth muscle contraction induces vomiting and diarrhea. Fluid outflow into gut lumen

produces cramps and vomiting, while the same reaction in the intestine produces diarrhea (Figure 10.24). These reactions probably evolved originally to expel gut parasites; in the allergic reaction they serve to expel the allergen-containing food from the human body. This goal is indeed accomplished, but at the expense of dehydration, weakness, and a waste of food.

In addition to allergic reactions localized in the gut, food allergens also produce reactions in other tissues, notably the skin. Depending on the timing and course of the gut reaction and of the uptake of allergen from the gut, allergen can enter the circulation and be transported elsewhere in the body. Mast cells in the connective tissue in the deeper layers of the skin tend to be activated by such blood-borne allergens, and their degranulation produces urticaria and angioedema. In this context, orally administered drugs behave similarly to food; they too can produce intestinal reactions, urticaria, and angioedema in sensitized individuals.

10-13 People with parasite infections and high levels of IgE rarely develop allergic disease

Almost all parasites stimulate strong immune responses, although it is difficult to distinguish which aspects of the immune response benefit the human host. A common feature of parasite infections, particularly helminth infections, is the stimulation of CD4 T_H2-like responses that produce raised levels of IgE and increased numbers of eosinophils in the blood. Only a small fraction of the IgE is parasite-specific; the remainder is highly heterogeneous and represents the product of a non-specific, polyclonal B- and T-cell activation by the parasite. In this situation, the non-specific IgE competes successfully with the parasite-specific IgE for binding to the FcεRI receptors of mast cells, basophils, and activated eosinophils. This strategy prevents the parasite from triggering IgE-mediated effector mechanisms, thereby allowing it to escape killing by activated eosinophils whose degranulation would otherwise have been focused on the parasite surface by specific IgE. Consistent with this is the observation that people with helminth infections and high levels of IgE show only transient symptoms of urticaria or wheezing when exposed to the infective form of the parasite, and rarely develop the allergic conditions experienced by uninfected people who live in areas where parasite infections are not endemic.

10-14 Allergic reactions are prevented and treated by using three complementary approaches

Three distinct strategies are used to reduce the effects of allergic disease. The first strategy is one of prevention—to modify a patient's behavior and environment so that contact with the allergen is avoided. Allergen-containing foods are avoided, houses are refurnished in ways that discourage mites, pets are kept outside, and desert vacations or sea cruises are taken during the pollen season.

The second strategy is pharmacological—to use drugs that reduce the impact of any contact with allergen. Such drugs block the effector pathways of the allergic response and limit the inflammation after IgE-induced activation of mast cells, eosinophils, and basophils. Antihistamines reduce rhinitis and urticaria by preventing histamine from binding to H1 histamine receptors on vascular endothelium and thus increasing vascular permeability. Corticosteroids, which suppress leukocyte function generally, are often administered topically or systemically to suppress the chronic inflammation of asthma, rhinitis, or eczema. Cromolyn sodium prevents the degranulation of activated mast cells and granulocytes and is inhaled by asthmatics as a prophylactic to prevent attacks. Epinephrine is used to treat anaphylactic reactions.

The third strategy in the treatment of allergy is immunological—to prevent the production of allergen-specific IgE. One way of achieving this is to modulate the antibody response so that it shifts from one dominated by IgE to one dominated by IgG. A procedure called **desensitization**, which was first described in 1911 and is used in a similar way today, can achieve this for some patients and some allergies. Patients are given a series of allergen injections in which the dose is initially very small and is gradually increased. This schedule of immunization can gradually convert a T_H2 cell response that produces IgE into a T_H1 response that ceases to make IgE. However, a potential consequence of the injections used for desensitization is anaphylaxis, because the patient is being exposed to the allergen to which they are sensitized. For this reason 'allergy shots' should always be given under carefully controlled conditions in which patients are monitored for the early symptoms of systemic anaphylaxis and, if need be, given epinephrine.

A more recent approach to desensitization is to vaccinate patients with allergen-derived peptides that are known to be presented by HLA class II molecules to CD4 T_H2 cells. The aim is to induce anergy of allergen-specific T cells *in vivo* by decreasing the expression of the CD3:T-cell receptor complex at the T_H2 cell surface. In principle, the advantage of this method over the older approach is that anaphylaxis should never be triggered by the injections, as only the native allergen protein and not the peptide vaccine can interact with allergen-specific IgE. Factors complicating this approach are the HLA class II polymorphisms determining which allergen-derived peptides can be presented by any individual. Any vaccine with general applicability to a human population must contain sufficient different peptides so that an anergizing response can be produced irrespective of HLA class II type. Alternatively, vaccines could be custom-made from peptides selected on the basis of a patient's HLA class II type.

Because allergies are a prevalent and increasing problem for humans in the richer countries in the world, there is much interest within the biotechnology and pharmaceutical industries in developing new approaches to the relief or cure of allergy. One potential set of targets for drugs are the signaling pathways that cells of the immune system use to enhance the IgE response. For example, inhibitors of the cytokines IL-4, IL-5, or IL-13 could block such pathways. Conversely, the administration of cytokines that promote T_H1 responses could shift the antibody response away from IgE and towards IgG. In experiments on mice, IFN-γ and IFN-α have both been shown to reduce IL-4-stimulated IgE synthesis. The high-affinity IgE receptor is also a potential target for drugs that bind the receptor and prevent the arming of mast cells with allergen-specific IgE.

Summary

Type I hypersensitivity reactions are caused by protein allergens binding to IgE molecules and activating mast cells. The activated mast cells release a variety of chemical mediators that orchestrate a local state of inflammation. Smooth muscles constrict, blood vessels dilate, and eosinophils and basophils enter the affected area. The effects of IgE-mediated reactions vary with the route of allergen entry to the body and the tissue affected. Inhaled allergens, for example plant pollens and animal dander, activate mast cells in the respiratory tract. Rhinitis is caused by reactions in the upper airways; asthma by reactions in the lower airways. Insect stings deliver allergens into the skin, where mast-cell mediators cause hives and urticaria. Allergens in certain foods, such as peanuts or shrimp, trigger mast cells of the gastrointestinal tract, resulting in vomiting and diarrhea. When food allergens are absorbed into the blood they become disseminated throughout the body, causing systemic mast-cell activation and resulting in wide-spread urticaria or even systemic anaphylaxis. The violent reactions activated by IgE are believed to have evolved as a defense against parasites. In allergic reactions this defensive mechanism is misguidedly aimed at environmental proteins that pose no threat.

Types II, III, and IV hypersensitivity reactions

IgG antibodies and specific T cells can cause both acute and chronic adverse hypersensitivity reactions. A variety of different effector mechanisms are involved, which cause inflammatory reactions and tissue destruction.

10-15 Type II hypersensitivity reactions are caused by antibodies specific for altered components of human cells

Occasional side-effects seen after the administration of certain drugs are hemolytic anemia caused by the destruction of red blood cells, or thrombocytopenia caused by the destruction of platelets. These are examples of type II hypersensitivity reactions and have been associated with the antibiotic penicillin, quinidine (a drug used to treat cardiac arrhythmia), and methyldopa, which is used to reduce high blood pressure. In each case chemically reactive drug molecules bind to surface components of red blood cells or platelets and create new epitopes to which the immune system is not tolerant (Figure 10.25).

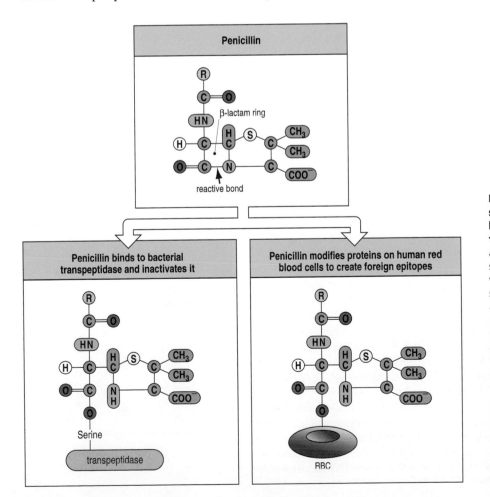

Figure 10.25 Penicillin and other small-molecule drugs can modify human cells so that they display foreign epitopes. Penicillin exerts its antibacterial action by mimicking the substrate for bacterial transpeptidase, which is required for bacterial cell wall synthesis. On binding to transpeptidase, a reactive bond in the β-lactam ring of penicillin opens and forms a covalent bond with an amino-acid residue in the active site of the transpeptidase, thereby inactivating the enzyme permanently (lower left panel). The same mechanism occasionally results in molecules of penicillin becoming covalently bonded to surface proteins of human cells (lower right panel). This modification of human proteins creates new epitopes that can act like foreign antigens. Red blood cells (RBC) are the cells most commonly modified in this way.

Figure 10.26 Penicillin–protein conjugates stimulate the production of anti-penicillin antibodies. Red cells that have been covalently bonded to penicillin (P) are phagocytosed by macrophages, which process the penicillin-modified proteins and present peptide antigens to specific CD4 T cells. These are activated to become effector T_H2 cells, which stimulate antigen-specific B cells to produce antibodies against the penicillin-modified epitope.

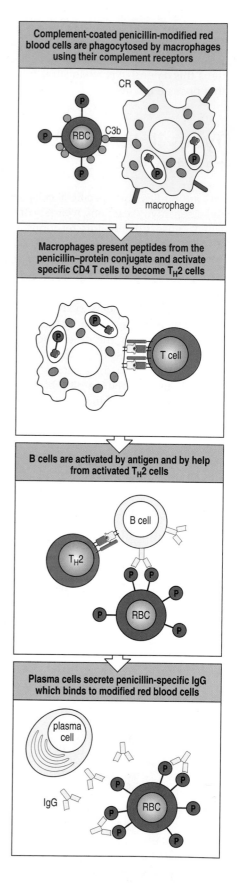

These epitopes stimulate the formation of IgM and IgG antibodies that are specific for the conjugate of drug and cell-surface component.

The penicillin-modified red cells acquire a coating of the complement component C3b as a side-effect of complement activation by the bacterial infection for which the drug has been given. This facilitates their phagocytosis by macrophages via complement receptors. These cells process the penicillin-modified protein and present peptides from it to specific CD4 T cells, which are activated to become effector T_H2 cells. These then stimulate antigen-specific B cells to produce antibodies against the penicillin-modified epitope (Figure 10.26). Binding of antibodies to the drug-conjugated cells activates complement by the classical pathway, resulting either in cell lysis by the terminal complement components or receptor-mediated phagocytosis by macrophages in the spleen (Figure 10.27).

10-16 Type III hypersensitivity reactions are caused by immune complexes formed from IgG and soluble antigens

Complexes of soluble protein antigens and their high-affinity IgG antibodies are generated in almost all immune responses and in most situations they are cleared without causing tissue damage. Immune complexes vary greatly in size, from a simple complex of one antigen and one antibody molecule through to large aggregates containing millions of antigen and antibody molecules. The larger aggregates fix complement efficiently and are readily taken up by phagocytes and removed from the circulation. Smaller immune complexes are less efficient at fixing complement; they tend to circulate in the blood and become deposited in blood vessel walls. When these complexes accumulate at such sites, they become capable of fixing complement and initiating tissue-damaging inflammatory reactions through their interactions with the Fc receptors and complement receptors on circulating leukocytes. This type of hypersensitivity is called type III hypersensitivity. Complement activation produces C3a and C5a. The former stimulates mast cells to release histamine, causing urticaria; the latter recruits inflammatory cells into the tissue. Platelets accumulate around the site of immune-complex deposition, and the clots that they form cause the blood vessels to burst, producing hemorrhage in the skin.

The size of the immune complexes formed at a particular time and place is strongly influenced by the relative concentrations of soluble antigen and antibody (Figure 10.28). Whether large immune complexes can be formed also depends on the size and complexity of the antigen; most antigens that will be encountered in normal circumstances contain multiple epitopes and thus can, in principle, form extensive immune complexes by crosslinking antibodies. When antigen is in excess, as occurs early in the immune response, each antigen-binding site on an antibody binds an antigen molecule, producing small immune complexes that often contain a single antibody molecule and two antigen molecules. Late in the immune response, antibodies are in excess; in these circumstances each antigen molecule binds to several antibody molecules. At intermediate times, when the amounts of antigen and antibody are more evenly balanced, larger immune complexes are formed in which several antibody and antigen molecules are

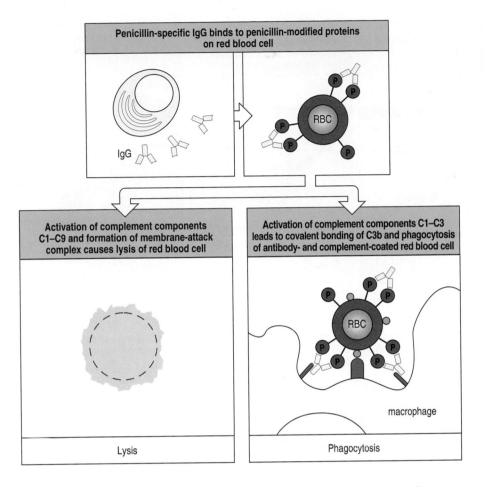

Figure 10.27 Binding of antibodies to penicillin-modified red cells makes them susceptible to complement-mediated lysis or to phagocytosis via Fc receptors and complement receptors.

crosslinked. A minimum of two IgG molecules per complex is needed to fix complement, so it is at the beginning of the immune response, when soluble immune complexes fix complement poorly, that they are most likely to circulate in the blood and become deposited in blood vessel walls.

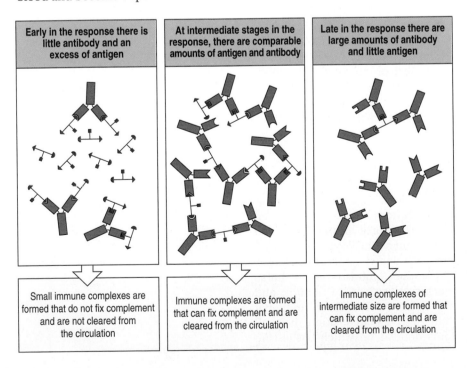

Figure 10.28 Immune complexes of different sizes and stoichiometries are formed during the course of an immune reponse.

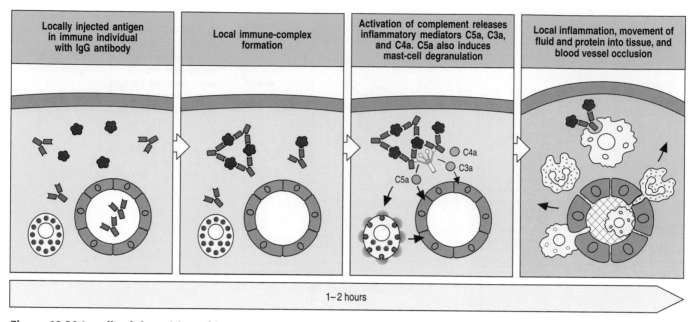

Figure 10.29 Localized deposition of immune complexes within a tissue causes a type III hypersensitivity reaction. In sensitized individuals, the introduction of allergen into a tissue leads to the formation of immune complexes with IgG in the extracellular fluid. The immune complexes activate complement and recruit inflammatory cells to the site, causing a hard swelling. Platelets accumulate in the capillary, leading to occlusion and rupture of the vessel, causing erythema.

In people who have made IgG against a soluble protein, a type III hypersensitivity reaction can be experimentally induced in the skin by subcutaneous injection of the antigen. Specific IgG diffuses from the blood into the connective tissue at the site of injection and combines with antigen to form immune complexes. The complexes activate complement, creating an inflammatory reaction that draws leukocytes and antibodies into the site of injection. Here the Fc and complement receptors of leukocytes engage the immune complexes, activating the cells and further propagating the local inflammatory reaction. This type of reaction was first described by Nicholas-Maurice Arthus and is called an **Arthus reaction** (Figure 10.29). In humans, Arthus reactions usually appear as localized areas of erythema and hard swelling (induration) that subside within a day. Such reactions can often be seen at the site of injections used to desensitize IgE-mediated allergies. The dependence of the Arthus reaction on immune-complex interactions with Fc receptors is demonstrated by the failure to produce Arthus reactions in mice lacking the γ chain common to all Fc receptors.

10-17 Systemic disease caused by immune complexes can follow the administration of large quantities of soluble antigens

During the late nineteenth century and the first half of the twentieth century, diphtheria, scarlet fever, tetanus, and other life-threatening bacterial infections were treated by injecting patients with serum taken from horses that had been immunized with these bacteria or their toxins. The horse antibodies helped human patients to control and clear the infection, but could also produce a systemic type III hypersensitivity reaction that became known as **serum sickness**. This condition occurred some 7–10 days after the administration of horse serum and was characterized by chills, fevers, rash, arthritis, vasculitis, and sometimes glomerulonephritis. The cause of serum sickness is the formation of antibodies against the foreign horse proteins and the deposition of small immune complexes in tissues; the symptoms depend on which tissues are affected (Figure 10.30).

Route	Resulting disease	Site of immune-complex deposition
Intravenous (high dose)	Vasculitis	Blood vessel walls
	Nephritis	Renal glomeruli
	Arthritis	Joint spaces
Subcutaneous	Arthus reaction	Perivascular area
Inhaled	Farmer's lung	Alveolar/capillary interface

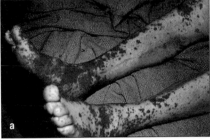

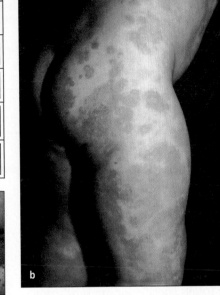

Figure 10.30 The pathology of type III hypersensitivity reactions is determined by the sites of immune-complex deposition. The table shows the types of reaction that result from different routes of antigen entry to the body. Serum sickness follows the intravenous administration of large amounts of foreign antigen. Photographs show hemorrhage in the skin (panel a) and urticarial rash (panel b) resulting from serum sickness. Photographs courtesy of R. Geha.

Therapeutic administration of serum from immunized horses is rarely used today. One remaining application is the use of horses to prepare anti-venom antisera for neutralizing the effects of snake-bite. However, the symptoms of serum sickness are now seen in other circumstances in which patients have received an infusion of large amounts of a foreign protein. Transplant patients given mouse monoclonal antibodies specific for human T cells to prevent rejection can get serum sickness. It also occurs occasionally in patients who have had a myocardial infarction (heart attack) and are treated with the bacterial enzyme streptokinase to degrade their blood clots. Serum sickness can also result from the intravenous administration of large amounts of a drug such as penicillin, which binds to host proteins on, for example, erythrocytes (see Section 10-15) and provokes an IgG response. This type of reaction can occur in people with no history of allergy to penicillin. Drug-induced serum sickness is now the most common example of this condition.

The onset of serum sickness coincides with the synthesis of antibodies, which form immune complexes with the antigenic proteins. Because the serum is loaded with antigen, large quantities of immune complexes are formed and dispersed throughout the body. The complexes fix complement and activate leukocytes bearing Fc receptors or complement receptors. These activated cells create an inflammatory response that causes widespread damage (Figure 10.31).

The formation of immune complexes induces the clearance of the antigenic proteins by the normal phagocytic pathways; consequently, serum sickness is of limited duration unless additional injections of the foreign antigen are given. If a second dose of antigen is given after the effects of the first dose have subsided, a secondary response will follow, with disease symptoms being manifested within a day or two of the second injection.

A disease similar to serum sickness can be seen in certain infections in which the immune system fails to clear the pathogen, and both the infection and the immune response persist. For example, in subacute bacterial endocarditis, or in

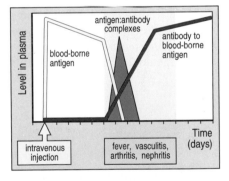

Figure 10.31 Serum sickness is a classic example of a transient immune-complex mediated syndrome. An injection of a large amount of a foreign antigen into the circulation leads to an antibody response. These antibodies form immune complexes with the circulating foreign antigens. The complexes are deposited in small blood vessels and activate complement and phagocytes, inducing fever and the symptoms of vasculitis, nephritis, and arthritis. All these effects are transient and resolve once the foreign antigen has been cleared.

chronic viral hepatitis, the multiplying pathogens continue to produce antigens, and plasma cells continue to make antibodies. Immune complexes are continually being generated, deposited, and cleared, processes that can cause injury to small blood vessels and nerves in many organs, including the skin and kidneys.

Some common inhaled antigens tend to provoke an IgG rather than an IgE response and cause type III hypersensitivity reactions. Continued exposure to the antigen leads to the formation of immune complexes and their deposition in the walls of the alveoli in the lungs. Such deposits stimulate an inflammatory response. The resulting accumulation of fluid, antigen, and cells impedes the lungs' normal function of gas exchange, and the patient experiences difficulty in breathing. Occupations in which workers are exposed daily to quantities of the same airborne antigens can lead to this condition. The immune systems of farm workers exposed to hay dust and mold spores are often provoked in this manner, giving rise to the occupational disease called **farmer's lung**. Without a change in work habits the continuing deposition of immune complexes in the alveolar membranes leads to irreversible lung damage.

10-18 Type IV hypersensitivity reactions are mediated by antigen-specific effector T cells

Hypersensitivity reactions caused by effector T cells specific for the sensitizing antigen are known as type IV hypersensitivity or **delayed-type hypersensitivity reactions** (**DTH**), because they occur 1–3 days after contact with antigen. This time course contrasts with those of antibody-mediated hypersensitivity, which are generally apparent within a few minutes. The amount of antigen required to elicit a type IV hypersensitivity reaction is a hundred to a thousand times greater than that required to produce antibody-mediated hypersensitivity reactions. This difference reflects the intrinsic inefficiency in generating peptide epitopes from protein antigens for presentation by HLA molecules. Antigens that cause common type IV hypersensitivity reactions are shown in Figure 10.32.

The best studied example of a type IV hypersensitivity reaction is the **tuberculin test**, the clinical test used to determine whether a person has been infected with *Mycobacterium tuberculosis*. In the tuberculin test a small amount of protein extracted from *M. tuberculosis* antigen should be injected intradermally or

Type IV hypersensitivity reactions are mediated by antigen-specific effector T cells		
Syndrome	**Antigen**	**Consequence**
Delayed-type hypersensitivity	Proteins: Insect venom Mycobacterial proteins (tuberculin, lepromin)	Local skin swelling: Erythema Induration Cellular infiltrate Dermatitis
Contact hypersensitivity	Haptens: Pentadecacatechol (poison ivy) Small metal ions: Nickel Chromate	Local epidermal reaction: Erythema Cellular infiltrate Contact dermatitis
Gluten-sensitive enteropathy (celiac disease)	Gliadin	Villous atrophy in small bowel Malabsorption

Figure 10.32 Examples of common type IV hypersensitivity reactions. Hapten is the name given to any small molecule which when covalently bonded to a protein stimulates an immune response. Gliadin is a proline-rich protein that is a component of cereal gluten.

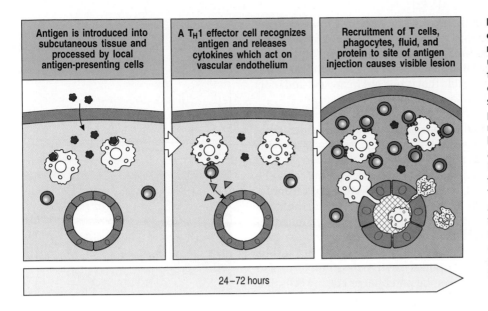

| Antigen is introduced into subcutaneous tissue and processed by local antigen-presenting cells | A T$_H$1 effector cell recognizes antigen and releases cytokines which act on vascular endothelium | Recruitment of T cells, phagocytes, fluid, and protein to site of antigen injection causes visible lesion |

24–72 hours

Figure 10.33 The stages and time course of a type IV hypersensitivity reaction. The first phase involves uptake, processing, and presentation of the antigen by local antigen-presenting cells. In the second phase, tuberculin-specific memory T cells produced during previous exposure to the antigen migrate into the site of injection and become activated. Because these antigen-specific cells are rare, and there is no inflammation to attract them into the site, it can take several hours for a T cell of the correct specificity to arrive. Activated T$_H$1 cells release mediators that activate local endothelial cells, recruiting an inflammatory cell infiltrate dominated by macrophages and causing accumulation of fluid and protein. At this point, the lesion becomes apparent.

intracutaneously. People with immunity to *M. tuberculosis* develop an inflammatory reaction around the site of injection 24–72 hours later. The response is mediated by T$_H$1 cells that recognize peptides derived from the *M. tuberculosis* protein that are presented by HLA class II molecules. The peptides are presented by macrophages and dendritic cells in the vicinity of the injection and stimulate tuberculin-specific memory T cells as they leave the blood and enter the tissue. After activation, the T$_H$1 cells initiate further inflammatory reactions that recruit fluid, proteins, and other leukocytes to the site (Figure 10.33). Each of these phases takes several hours, accounting for the time taken before the response is seen or felt. The activated T$_H$1 cells produce cytokines that mediate these effects (Figure 10.34).

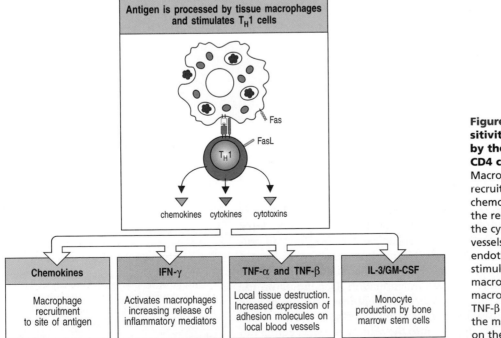

Antigen is processed by tissue macrophages and stimulates T$_H$1 cells

Fas

FasL

T$_H$1

chemokines   cytokines   cytotoxins

Chemokines	IFN-γ	TNF-α and TNF-β	IL-3/GM-CSF
Macrophage recruitment to site of antigen	Activates macrophages increasing release of inflammatory mediators	Local tissue destruction. Increased expression of adhesion molecules on local blood vessels	Monocyte production by bone marrow stem cells

Figure 10.34 Most type IV hypersensitivity reactions are orchestrated by the cytokines released by T$_H$1 CD4 cells in response to antigen. Macrophages or tissue dendritic cells recruited to the site of inflammation by chemokines present antigen and amplify the response. The release of TNF-α and the cytotoxin TNF-β affect local blood vessels through their effects on endothelial cells. IL-3 and GM-CSF stimulate the production of macrophages. IFN-γ and TNF-α activate macrophages. Macrophages are killed by TNF-β and by the interaction of Fas on the macrophage with Fas ligand (FasL) on the T cell.

Type IV hypersensitivity responses can be developed to a variety of antigens in the environment. For example, the dermatitis caused by contact with the North American plant poison ivy (*Rhus toxicodendron*) is due to a reaction of this type and involves both CD4 and CD8 T cells. The reaction is caused by pentadecacatechol, a small highly reactive lipid-like molecule that is present in the leaf and roots of the plant and is easily transferred to human skin (Figure 10.35).

When a person contacts poison ivy, pentadecacatechol penetrates the outer layers of the skin and indiscriminately forms covalent bonds with extracellular proteins and skin cell surface proteins. On degradation of the chemically modified proteins by skin macrophages and Langerhans' cells, antigenic peptides that carry the pentadecacatechol adduct are generated and presented by HLA class II molecules to T_H1 cells. The cytokines secreted by the T_H1 cells activate macrophages and produce inflammation (see Figure 10.34). When pentadecacatechol penetrates the skin it also crosses the plasma membranes of cells and chemically modifies intracellular proteins. The processing of such modified proteins results in the presentation of chemically modified peptides by HLA class I molecules to CD8 T cells. On activation, the CD8 T cells have the potential to kill any cells that contacted the chemical and present modified peptides on their surfaces.

On first contact with poison ivy, a person might experience only a minor or undetectable reaction. During this primary or sensitizing response, Langerhans' cells or dendritic cells carry pentadecacatechol-modified proteins to the draining lymph node, where the T-cell response is initiated. Once immunological memory has been developed, subsequent contacts with the plant yield unpleasant reactions of the type shown in Figure 10.35. The red, raised, weeping skin lesions are due to heavy infiltration of the contact sites with blood cells, combined with the localized destruction of skin cells and the extracellular matrix that holds the skin together. Because of the delayed nature of the reaction there is plenty of time for a person to transfer pentadecacatechol from the initial site of contact to other parts of the body, a process that often exacerbates the extent of the reaction.

The reaction to poison ivy is an example of **contact sensitivity**, so-called because contact with the skin is necessary to initiate the allergic response. Contact sensitivity can also develop to coins, jewelry, and other metallic objects containing nickel. In this case the bivalent nickel ions are chelated by histidine in human proteins; processing of these proteins forms T-cell epitopes to which the immune system responds.

Summary

Type II hypersensitivity reactions are mediated by IgG antibodies directed against cell-surface or matrix antigens and are commonly caused by drugs that are being given as treatment for other diseases. Because of their chemical reactivity, drugs bind to surface proteins of human cells, for example erythrocytes or platelets, creating new epitopes that stimulate an antibody response. On binding to the drug-modified cells, the antibodies activate complement, which leads to cell destruction. Type III hypersensitivity reactions are caused by soluble immune complexes formed by IgG binding to the soluble antigens against which they were made. They can be seen when non-human proteins, such as mouse monoclonal antibodies, are given therapeutically. Antibodies specific for the non-human proteins are made, and early in the response small immune complexes are formed that are not efficiently cleared from the circulation. These tend to deposit in blood vessels, where they activate complement and inflammation. Depending on the sites of activation, these reactions can cause vasculitis, nephritis, arthritis, or lung disease. Type IV hypersensitivity reactions are caused by effector T cells, often responding to reactive chemicals transferred into the skin by physical contact to

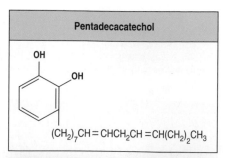

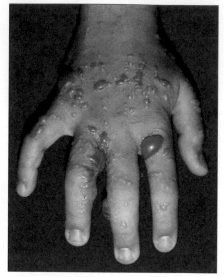

Figure 10.35 Physical contact with poison ivy transfers pentadecacatechol, which causes dermatitis. The chemical structure of pentadecacatechol is shown in the top panel. The photograph shows characteristic blistering skin lesions on the hand of a patient with dermatitis caused by contact with poison ivy. Photograph courtesy of R. Geha.

cause a contact dermatitis. In some conditions, tissue damage is caused by the activation of macrophages by T_H1 cells and the actions of cytotoxic T cells; in others, such as chronic asthma, cytokines produced by T_H2 cells specific for an inhaled allergen activate eosinophils and other inflammatory cells.

Summary to Chapter 10

The immune system provides the body with powerful defenses against infection. These include adaptive immune responses that have the potential to respond to any structure that is not a normal part of the body. Inevitably, all humans develop adaptive immunity to some foreign substances, such as animal and plant proteins in foods, that are not associated with infection. In general, such immune responses are harmless, but this is not always so, and various allergies or hyper-sensitivities are caused by the immune system's over-reactions to non-threatening environmental antigens. A first exposure to an allergen is rarely noticeable, hypersensitivity reactions being brought on by subsequent exposures in which the allergen interacts with previously formed antibodies or stimulates memory lymphocytes. Four types of hypersensitivity reaction are conventionally defined on the basis of the effector mechanisms that cause them. Hypersensitivity reactions of types I, II, and III are triggered by antibodies, whereas type IV hypersensitivity reactions are triggered by effector T cells. In all four types of reaction, recognition of the allergen triggers an unwanted inflammatory response of varying severity and duration.

Disruption of Healthy Tissue by the Immune Response

11

Chapter 10 described how strong immune responses to harmless environmental antigens lead to chronic conditions of hypersensitivity and disease. In this chapter another kind of chronic disease is considered, disease caused by adaptive immunity that becomes misdirected at healthy cells and tissues of the body. Such diseases are known as **autoimmune diseases**, and many different types have been described. They are not uncommon; 2–3% of the population of developed countries have one or more of these conditions. Autoimmune diseases are chronic inflammatory conditions in which the capacity of the body to repair damaged tissue cannot compete with the rate of tissue destruction by the immune system. When the targeted tissue is involved in essential day-to-day functions of the body, the autoimmune disease can become a threat to life.

Autoimmune diseases represent failures of the mechanisms that maintain self-tolerance in the populations of circulating B and T cells. Although much is known of the effects of autoimmune disease, less is understood of the events that break tolerance and cause an autoimmune response. The first part of the chapter describes some of the more common autoimmune diseases; the second part discusses some of the predisposing factors for autoimmune disease and how these might lead to a break in self-tolerance.

Autoimmune diseases

There are many chronic diseases in which the immune system is active, causing the affected tissues to be inflamed and abnormally infiltrated by lymphocytes and other leukocytes, but in which there seems to be no associated active infection. These diseases are caused by the immune system itself, which attacks cells and tissues of the body as though they were infected, causing chronic impairment of tissue and organ function. Chronic diseases of this kind are collectively known as autoimmune diseases because they are caused by immune responses directed towards autologous (self) components of the body. An immune response that causes an autoimmune disease is called an **autoimmune response** and produces a state of **autoimmunity**.

Autoimmune diseases vary widely in the tissues they attack and the symptoms they cause. For most autoimmune diseases the incidence differs between females and males, with females being more commonly affected. A defining characteristic of autoimmune diseases is the presence of antibodies and T cells specific for antigens expressed by the targeted tissue. These antigens are called autoantigens; the effectors of adaptive immunity that recognize them are known as autoantibodies and autoimmune T cells. The mechanisms of antigen recognition and effector function in autoimmunity are the same as those used in responding to pathogens and environmental antigens. Autoimmune disease can be caused by a variety of immune effector mechanisms; this part of the chapter describes a selection of the more common autoimmune diseases.

11-1 The effector mechanisms of autoimmunity resemble those causing some types of hypersensitivity reaction

Autoimmune diseases can be classified by the type of immunological effector mechanism causing the disease. Three kinds are distinguished, their effector mechanisms corresponding to three of the four types of hypersensitivity reactions described in Chapter 10. The first kind of autoimmune disease, corresponding to type II hypersensitivity (see Section 10-15, p. 289), is caused by antibodies directed against components of cell surfaces or the extracellular matrix; the second kind, corresponding to type III hypersensitivity (see Section 10-16, p. 290), is caused by the formation of soluble immune complexes that are deposited in tissues; the third kind, corresponding to type IV hypersensitivity (see Section 10-18, p. 294), is caused by effector T cells. Autoimmune diseases of all three kinds are described in Figure 11.1. Autoimmune diseases are never caused by IgE, the source of type I hypersensitivity reactions.

Autoimmunity corresponding to the type II hypersensitivity reaction is frequently directed at cells of the blood. In **autoimmune hemolytic anemia**, IgG and IgM antibodies bind to components of the erythrocyte surface, where they activate

Autoimmune disease	Autoantigen	Consequence
Antibody against cell-surface or matrix antigens (Type II)		
Autoimmune hemolytic anemia	Rh blood group antigens, I antigen	Destruction of red blood cells by complement and phagocytes, anemia
Autoimmune thrombocytopenia purpura	Platelet integrin gpIIb:IIIa	Abnormal bleeding
Goodpasture's syndrome	Non-collagenous domain of basement membrane collagen type IV	Glomerulonephritis, pulmonary hemorrhage
Pemphigus vulgaris	Epidermal cadherin	Blistering of skin
Acute rheumatic fever	Streptococcal cell wall antigens. Antibodies cross-react with cardiac muscle	Arthritis, myocarditis, late scarring of heart valves
Graves' disease	Thyroid-stimulating hormone receptor	Hyperthyroidism
Myasthenia gravis	Acetylcholine receptor	Progressive weakness
Insulin-resistant diabetes	Insulin receptor (antagonist)	Hyperglycemia, ketoacidosis
Hypoglycemia	Insulin receptor (agonist)	Hypoglycemia
Immune-complex disease (Type III)		
Subacute bacterial endocarditis	Bacterial antigen	Glomerulonephritis
Mixed essential cryoglobulinemia	Rheumatoid factor IgG complexes (with or without hepatitis C antigens)	Systemic vasculitis
Systemic lupus erythematosus	DNA, histones, ribosomes, snRNP, scRNP	Glomerulonephritis, vasculitis, arthritis
T-cell mediated disease (Type IV)		
Insulin-dependent diabetes mellitus	Pancreatic β-cell antigen	β-cell destruction
Rheumatoid arthritis	Unknown synovial joint antigen	Joint inflammation and destruction
Multiple sclerosis	Myelin basic protein, proteolipid protein	Brain degeneration. Paralysis

Figure 11.1 A selection of autoimmune diseases, the symptoms they cause, and the autoantigens associated with the immune response. The autoimmune diseases are classified as types II, III, and IV because their tissue-damaging effects are similar to those of hypersensitivity reactions types II, III, and IV respectively (see Chapter 10). snRNP, small nuclear ribonucleoprotein; scRNP, small cytoplasmic ribonucleoprotein.

complement by the classical pathway, triggering the destruction of red cells. Complete activation of the classical pathway leads to formation of the membrane-attack complex and hemolysis—lysis of red cells. Alternatively, erythrocytes coated with antibody and C3b are cleared from the circulation, principally by Fc

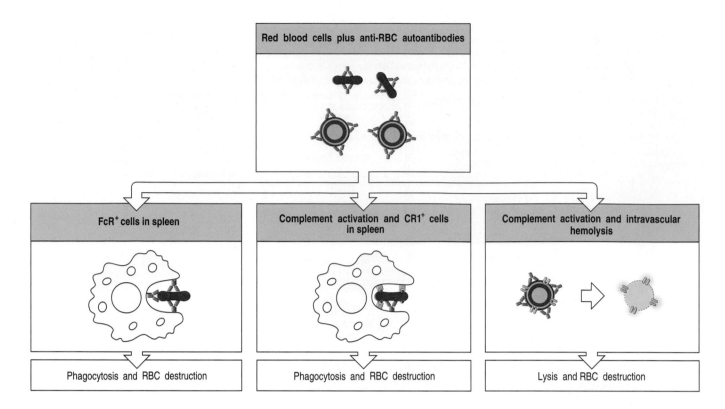

Figure 11.2 Three mechanisms destroy red blood cells in autoimmune hemolytic anemia. RBC, red blood cell.

and complement receptors on phagocytes in the spleen (Figure 11.2). All mechanisms lead to a reduction in the red-cell count. This condition is called anemia, a term derived from Greek words meaning 'without blood'.

White blood cells can also be a target for autoantibodies and complement activation. Because nucleated cells are less susceptible than erythrocytes to complement-mediated lysis, the major effect of complement activation is to facilitate phagocyte-mediated clearance of white blood cells in the spleen. Patients who develop antibodies against neutrophil surface antigens suffer from reduced numbers of circulating neutrophils, a state called **neutropenia**. Blood cells that have bound antibody and complement are, however, still able to function, so one treatment used for patients with chronic autoimmunity to blood cells is splenectomy, which reduces the rate at which the opsonized cells are cleared from the blood.

Antibody responses to components of the extracellular matrix are uncommon, but can be very damaging when they occur. In **Goodpasture's syndrome**, IgG is formed against the α_3 chain of type IV collagen, the collagen found in basement membranes throughout the body. The antibodies deposit along the basement membranes of renal glomeruli and renal tubules, eliciting an inflammatory response in the renal tissue (Figure 11.3). These basement membranes are an essential part of the blood-filtering mechanism of the kidney and, as IgG and inflammatory cells accumulate, kidney function becomes progressively impaired, leading to kidney failure and death if not treated. Treatment involves plasma exchange to remove existing antibodies and immunosuppressive drugs to prevent the formation of new ones. Although autoantibodies are deposited in the basement membranes of other organs, the high-pressure filtering of blood by renal glomeruli is the vital function most sensitive to their effects. Indeed, kidney damage by the immune system is responsible for one-quarter of all cases of end-stage renal failure.

Glomerulonephritis can also be a consequence of autoimmune diseases that produce soluble immune complexes, as in type III hypersensitivity reactions. In

Figure 11.3 Autoantibodies specific for type IV collagen react with the basement membranes of kidney glomeruli, causing Goodpasture's syndrome. The panels show sections of a renal corpuscle in serial biopsies taken from a patient with Goodpasture's syndrome. Panel a, glomerulus stained for IgG deposition by immunofluoresence. Antibody against glomerular basement membrane is deposited in a linear fashion (green staining) along the glomerular basement membrane. The autoantibody causes the local activation of cells bearing Fc receptors, complement activation, and the influx of neutrophils. Panel b, hematoxylin and eosin staining of a section through a renal corpuscle shows that the glomerulus is compressed by the formation of a crescent (C) of proliferating mononuclear cells within the Bowman's capsule (B) and there is influx of neutrophils (N) into the glomerular tuft. Photographs courtesy of M. Thompson and D. Evans.

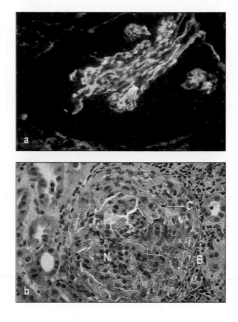

the disease **systemic lupus erythematosus** (**SLE**), IgG made against common intracellular macromolecules forms soluble complexes with these autoantigens when they are released from damaged cells. Glomerulonephritis caused by deposition of the immune complexes in the blood capillaries of the renal glomeruli can lead to death. T-cell mediated immunity of the kind seen in type IV hypersensitivity reactions is responsible for the damage caused to the pancreas in insulin-dependent diabetes mellitus, to joints in rheumatoid arthritis, and to the brain in multiple sclerosis. All three of these diseases, as well as SLE, will be described in more detail in later sections.

11-2 Endocrine glands contain specialized cells which are targets for organ-specific autoimmunity

The characteristic properties of endocrine glands make them particularly prone to involvement in autoimmune disease. Firstly, the function of an endocrine gland, the synthesis and secretion of a particular hormone, involves tissue-specific proteins not expressed in other cells or tissues. Secondly, because they secrete their hormones into the blood, endocrine cells are well vascularized, which facilitates interactions with the cells and molecules of the immune system. Loss of endocrine function has drastic systemic effects, causing pronounced disease and in some cases death. Each autoimmune disease of endocrine tissue is due to an impaired function of a single type of epithelial cell within an endocrine gland (Figure 11.4). For this reason such diseases are called **organ-specific autoimmune diseases**. To illustrate the principle we shall consider autoimmune diseases of the thyroid gland and of the islets of Langerhans in the pancreas in the next three sections.

11-3 Autoimmune diseases of the thyroid can cause either underproduction or overproduction of thyroid hormones

The thyroid gland regulates the basal metabolic rate of the body through the secretion of two related hormones, tri-iodothyronine and tetra-iodothyronine (thyroxine). These small iodinated derivatives of the amino acid tyrosine are produced in an elaborate way. The epithelial cells of the thyroid make a large glycoprotein called thyroglobulin, which is stored within follicles formed by the spherical arrangement of thyroid cells. Thyroid cells are uniquely specialized to take up iodide, which they use to iodinate and crosslink the tyrosine residues of stored thyroglobulin. When increased cellular metabolism is required, for example when the outside temperature drops, signals from the nervous system induce the pituitary, another endocrine gland, to secrete thyroid-stimulating hormone (TSH). Surface receptors on thyroid cells bind TSH, inducing the endocytosis of iodinated thyroglobulin and the release of thyroid hormones by proteolytic

Autoimmune diseases of endocrine glands	
Thyroid gland	Hashimoto's thyroiditis Graves' disease Subacute thyroiditis Idiopathic hypothyroidism
Islets of Langerhans (pancreas)	Insulin-dependent diabetes Insulin-resistant diabetes
Adrenal gland	Addison's disease

Figure 11.4 Autoimmune diseases of endocrine glands.

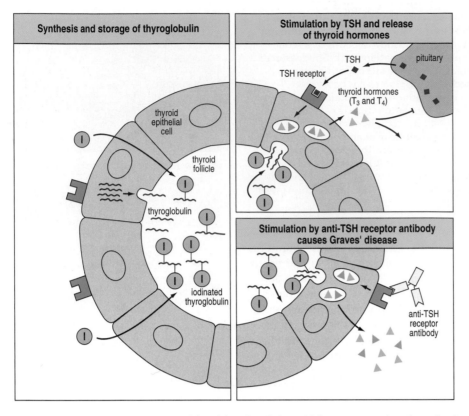

Figure 11.5 Autoantibodies against the TSH receptor cause the overproduction of thyroid hormones in Graves' disease. The diagram shows a thyroid follicle surrounded by epithelial cells. The left half of the figure shows uptake of iodide (green circles), the iodination of thyroglobulin and its storage in the follicle at times when thyroid hormones are not required. The upper right shows what occurs when thyroid hormones are needed. Thyroid-stimulating hormone (TSH) from the pituitary gland induces the endocytosis and breakdown of iodinated thyroglobulin, releasing the thyroid hormones tri-iodothyronine (T_3) and thyroxine (T_4). In addition to their effects on metabolism, T_3 and T_4 signal the pituitary to stop releasing TSH. On the lower right is shown the situation in Graves' disease. Autoantibodies, which are continually present, bind to the TSH receptor of thyroid cells, mimicking the action of TSH and inducing the continuous synthesis and release of thyroid hormones. In patients with Graves' disease, the production of thyroid hormones becomes independent of the presence of TSH.

degradation of the protein. As blood levels of thyroid hormones rise they feed back on the pituitary, inhibiting the further release of TSH (Figure 11.5).

The proteins thyroglobulin, thyroid peroxidase, TSH receptor and the thyroid iodide transporter are uniquely expressed in thyroid cells. Immune responses to all these autoantigens have been detected in autoimmune thyroid disease. However, the disease symptoms caused by different types of autoimmunity to thyroid antigens are very different. Some conditions cause a loss of thyroid hormones; others increase their production.

In **chronic thyroiditis**, also called **Hashimoto's disease**, the thyroid loses the capacity to make thyroid hormones. Hashimoto's disease seems to involve a CD4 T_H1 cell response, and both antibodies and effector T cells specific for thyroid antigens are produced. Lymphocytes infiltrate the thyroid, causing a progressive destruction of the normal thyroid tissue. The formation of organized lymphoid tissue within the thyroid, including germinal centers, gives the impression that the thyroid is being converted into a lymph node. Patients with Hashimoto's disease become **hypothyroid**, and eventually are unable to make thyroid hormone.

In **Graves' disease**, by contrast, the autoimmune response is focused on antibody production and the symptoms are caused by antibodies that bind to the TSH receptor. By mimicking the natural ligand, the bound antibodies cause a chronic overproduction of thyroid hormones that is independent of regulation by TSH and insensitive to the metabolic needs of the body (see Figure 11.5). This **hyperthyroid** condition causes heat intolerance, nervousness, irritability, warm moist skin, weight loss, and enlargement of the thyroid. Other aspects of Graves' disease are outwardly bulging eyes and a characteristic stare, symptoms caused by the binding of autoantibodies that react with the muscles of the eye. The autoimmune response in Graves' disease is biased towards a CD4 T_H2 response and, in comparison with Hashimoto's disease, there are fewer infiltrating lymphocytes and there is relatively little tissue destruction.

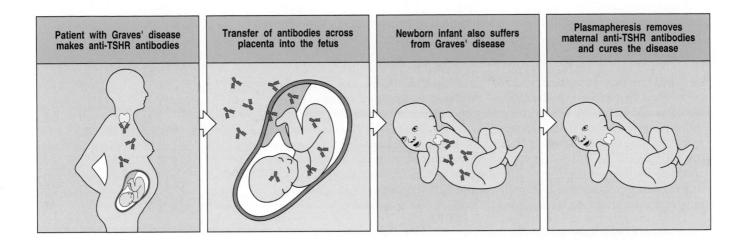

Treatment for Hashimoto's disease is replacement therapy with synthetic thyroid hormones taken orally on a daily basis. For Graves' disease the short-term treatments are drugs that inhibit thyroid function. The long-term treatments are the removal of the thyroid by surgery, or its destruction by uptake of the radioisotope ^{131}I, followed by daily doses of thyroid hormones.

Figure 11.6 Temporary symptoms of antibody-mediated autoimmune diseases can be passed from affected mothers to their newborn babies. TSHR, thyroid-stimulating hormone receptor.

11-4 The cause of autoimmune disease can be revealed by the transfer of immune effectors to neonatal infants or animals

A central goal in the study of autoimmune disease is the identification of autoantigens and effector mechanisms that cause the disease. This is no trivial task. One problem is that various types of autoimmunity are demonstrable in healthy people; a second is that once cell and tissue destruction has begun, it will often initiate further autoimmune responses that are consequences, not causes, of the disease.

The causes of autoimmune diseases involving antibodies are more easily identified than those due to effector T cells. For the former, the transfer of autoantibody from the patient to another human or to an animal has been shown to induce disease. Pregnant women who have an antibody-mediated autoimmune disease transport IgG molecules, but not lymphocytes, across the placenta to the fetal circulation. When the mother has an antibody-mediated disease such as Graves' disease, the baby is born with the symptoms of disease. As the baby grows and maternal IgG degrades, disease symptoms gradually go away. For diseases that could affect the baby's growth, treatment involves removal of antibody from the circulation by total exchange of blood plasma (Figure 11.6).

Rats injected with serum from Graves' disease patients start to overproduce thyroid hormones (Figure 11.7). Such laboratory assays have been used in the past in the immunological diagnosis of Graves' disease. This whole-animal approach has now been superseded by assays with a cultured line of rat thyroid cells. IgG fractions prepared from patients' plasma are added to thyroid-cell cultures and their capacity to cause cell activation and proliferation is assessed by measurement of the production of cyclic AMP and the synthesis of DNA respectively. A complementary assay determines whether antibodies in a patient's plasma compete with TSH for binding to the TSH receptor of thyroid membranes isolated from

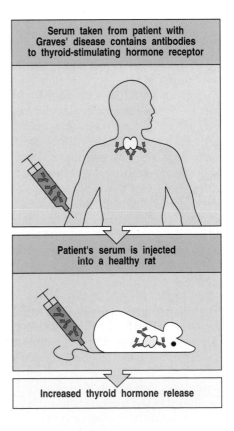

Figure 11.7 Symptoms of disease can be transferred to experimental animals with serum from patients with Graves' disease.

pigs. Symptoms of Graves' disease can also be induced in normal rats when they are injected with serum from patients, but not when injected with serum from normal individuals.

Because lymphocytes cannot pass from the maternal to the fetal circulation, babies born to mothers who have T-cell mediated autoimmune diseases do not show disease symptoms. Observations on human pregnancy therefore provide no positive information on the role of effector T cells in human autoimmune disease; neither do experiments in which T cells from patients are transferred into experimental animals. Such experiments fail to work because human T cells cannot recognize antigens presented by the MHC molecules of other species. Only in models of autoimmune disease in inbred strains of rodents has it been possible to transfer disease with T cells and show that they cause a disease.

11-5 Insulin-dependent diabetes mellitus is caused by the selective destruction of insulin-producing cells in the pancreas

Insulin is secreted from the pancreas in response to the increased blood glucose level arising after a meal. By binding to surface receptors, insulin stimulates the body's cells to take up glucose and incorporate it into carbohydrates and fats. **Insulin-dependent diabetes mellitus (IDDM)** is an autoimmune disease caused by the selective destruction of the insulin-producing cells of the pancreas. Because insulin is a major regulator of cellular metabolism it is essential for childrens' normal growth and development. Symptoms of disease are usually manifested in childhood or adolescence and they rapidly progress to coma and death in the absence of treatment. IDDM principally affects populations of European origin, in which one person in 300 is a sufferer. This distribution, and the impact of the disease on young children, has made IDDM a major target for research in the countries of western Europe, North America, and Australia.

Scattered within the exocrine tissue of the pancreas are the **islets of Langerhans**, small clumps of endocrine cells that make the hormones insulin, glucagon, and somatostatin. The pancreas contains about half a million islets, each consisting of a few hundred cells. Each islet cell is programmed to make a single hormone: α cells make glucagon, β cells make insulin, and δ cells make somatostatin.

In patients with IDDM, antibody and T-cell responses are made against insulin, glutamic acid decarboxylase, and other specialized proteins of the pancreatic β cell. Which of these responses cause disease remains unclear. CD8 T cells specific for some protein unique to β cells are believed to mediate β-cell destruction, gradually decreasing the number of insulin-secreting cells. Individual islets become successively infiltrated with lymphocytes, a process called **insulitis**. β cells comprise about two-thirds of the islet cells; as they die, the architecture of the islet degenerates. A healthy person has about 10^8 β cells, providing an insulin-making capacity much greater than that needed by the body. This excess, and the slow rate of β-cell destruction, means that disease symptoms do not become manifest until years after the start of the autoimmune response. Disease commences when there are insufficient β cells to provide the insulin necessary to control the level of glucose in the blood (Figure 11.8).

The usual treatment for patients with IDDM is daily injection with insulin purified from the pancreata of pigs or cattle. Because of amino-acid sequence differences between human insulin and animal insulins, some patients develop an immune response to animal insulin. The antibodies they make against insulin can have two effects: they reduce the activity of the insulin and they form soluble immune complexes that can lead to further tissue damage. Recombinant human insulin produced in the laboratory from the cloned insulin gene is prescribed to patients who make antibodies against animal insulins.

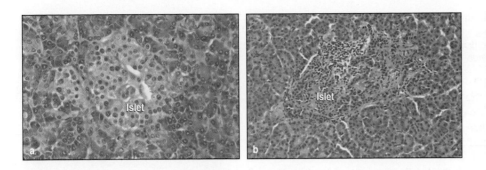

Figure 11.8 Comparison of histological sections of a pancreas from a healthy person and a patient with insulin-dependent diabetes mellitus (IDDM). Panel a shows a micrograph of a tissue section through a healthy human pancreas, showing a single islet. The islet is the discrete light-staining area in the center of the photograph. It is composed of hormone-producing cells, including the β cells that produce insulin. Panel b shows a micrograph of an islet from the pancreas of an IDDM patient with acute onset of disease. The islet shows insulitis characterized by a mainly lymphocytic infiltration at the islet periphery. Both tissue sections are stained with hematoxylin and eosin; magnification × 250. Photographs courtesy of G. Klöppel.

Good animal models are available for human IDDM. Most extensively studied is the non-obese diabetic (NOD) strain of inbred mouse. In these animals diabetes arises spontaneously and has clinical and immunological characteristics that parallel those seen in the human disease. There is also a similar bias towards females in the incidence of disease. Diabetes can be given to healthy mice by transfer of T cells from diseased animals.

11-6 Autoantibodies against common components of human cells can cause systemic autoimmune disease

So far, this chapter has concentrated on diseases in which a single type of cell is targeted for autoimmune attack. At the other end of the spectrum is systemic lupus erthythematosus (SLE), a disease in which the autoimmune response is directed at autoantigens present in almost every cell of the body. SLE is an example of a **systemic autoimmune disease**.

Characteristic of SLE are circulating IgG antibodies specific for constituents of cell surfaces, cytoplasm, and nucleus, including nucleic acids and nucleoprotein particles. The binding of autoantibodies against cell-surface components initiates inflammatory reactions that cause cell and tissue destruction. In turn, these processes release soluble cellular antigens that form soluble immune complexes. On being deposited in blood vessels, kidneys, joints, and other tissues, such complexes can initiate further inflammatory reactions (Figure 11.9). All this provides a disrupted environment in which the immune system is increasingly stimulated to respond to common self components. SLE is thus a chronic inflammatory disease that can affect all tissues of the body. The disease commonly follows a course in which outbreaks of intense inflammation alternate with periods of relative calm.

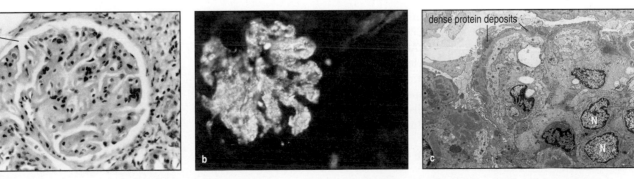

Figure 11.9 Deposition of immune complexes in the kidney glomeruli in systemic lupus erythematosus (SLE). Panel a shows a section through a glomerulus of a patient with SLE. Deposition of immune complexes causes thickening of the basement membrane (B). In panel b a similar kidney section is stained with fluorescent anti-immunoglobulin antibodies, revealing the presence of immunoglobulin in the basement membrane deposits. Panel c is an electron micrograph of part of a glomerulus. Dense protein deposits are seen between the glomerular basement membrane and the renal epithelial cells. Neutrophils (N) are also present, attracted by the deposited immune complexes. Photographs courtesy of M. Kashgarian.

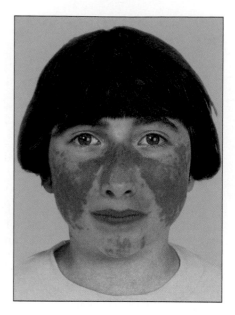

Figure 11.10 The characteristic facial rash of systemic lupus erythematosus. Although this butterfly-shaped rash was first used to recognize the disease, it is only seen in a proportion of patients who have the disease when defined immunologically. Photograph courtesy of M. Walport.

For individual patients the course of the disease is highly variable, both in its severity and in the organs and tissues involved. Many patients with SLE eventually die of the disease, because of failure of vital organs such as the kidneys or brain.

Systemic lupus erythematosus was first described as 'lupus erythematosus' on the basis of the butterfly-shaped skin rash (erythema) that can occur on the face and gives the face an appearance like a wolf's head (*lupus* is Latin for wolf) (Figure 11.10). The rash is caused by the deposition of immune complexes in skin and is similar to that caused by a type III hypersensitivity reaction (see Figure 10.30, p. 293). Further study of the disease revealed its systemic nature and so the word 'systemic' was added to the name. SLE is particularly common in women of African or Asian origin, of whom 1 in 500 has the disease. The factors that trigger the autoimmunity that leads to SLE remain unknown. However, they are likely to be simple in comparison with the complex autoimmunity and inflammation that develop during the course of disease. SLE is a disease in which the unwanted reactions of autoimmunity stimulate further autoimmunity that can send the immune system along a path of uncontrolled destruction.

11-7 Most rheumatological diseases are caused by autoimmunity

Common sites of disease in SLE are joints, where the deposition of immune complexes causes inflammation and arthritis. Over 90% of patients with SLE suffer from arthritis and it is often the first symptom to be noticed. SLE is one of a number of rheumatic diseases, almost all of which are autoimmune in nature (Figure 11.11)

Rheumatoid arthritis is the most common of the rheumatic diseases, affecting 1–3% of the US population, with women outnumbering men by three to one. The disease involves chronic and episodic inflammation of the joints, usually starting at an age of 20–40 years. A feature in 80% of patients with rheumatoid arthritis is the stimulation of B cells that make IgM, IgG, and IgA antibodies specific for the Fc region of human IgG. **Rheumatoid factor** is the name given to these anti-immunoglobulin autoantibodies.

In the joints affected by rheumatoid arthritis there is a leukocyte infiltrate into the joint synovium consisting of CD4 and CD8 T cells, B cells, lymphoblasts, plasma cells, neutrophils, and macrophages. Among these are plasma cells making rheumatoid factor. Prostaglandins and leukotrienes produced by inflammatory cells are major mediators of the inflammation. In addition, neutrophils release lysosomal enzymes into the synovial space, causing damage to the tissue and proliferation of the synovium. CD4 T cells activate macrophages, which accumulate in the inflamed synovium and add to the tissue destruction. Proteinases and collagenases produced by inflammatory cells within the joint can also damage cartilage and supporting structures such as ligaments and tendons.

The formation of immune complexes by rheumatoid factor is not considered to initiate joint inflammation but to exacerbate the inflammation once it has begun. Evidence for this view is that no rheumatoid factor is seen in 20% of patients with typical rheumatoid arthritis. Conversely, rheumatoid factor is found in other diseases, for example in 30% of patients with SLE.

Rheumatic diseases caused by autoimmunity
Systemic lupus erythematosus (SLE)
Rheumatoid arthritis
Juvenile arthritis
Sjögren's syndrome
Scleroderma (progressive systemic sclerosis)
Polymyositis–dermatomyositis
Behcet's disease
Ankylosing spondylitis
Reiter's syndrome
Psoriatic arthritis

Figure 11.11 Rheumatic diseases are autoimmune in nature.

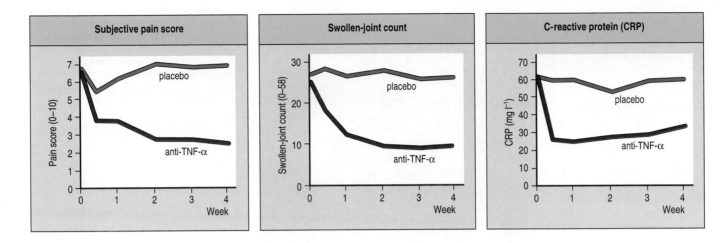

Rheumatoid arthritis is treated by a combination of physiotherapy and various anti-inflammatory and immunosuppressive drugs. In recent clinical trials, treatment with antibody against tumor necrosis factor-α (TNF-α) alleviated the symptoms of this disease (Figure 11.12).

Figure 11.12 The effects of treatment of rheumatoid arthritis with anti-TNF-α. For each parameter measured (pain, swollen joints, and level of C-reactive protein), the values for patients given a placebo treatment are depicted by the blue curve and the values for patients treated with anti-TNF-α antibody are depicted by the red curve.

11-8 Multiple sclerosis and myasthenia gravis are autoimmune diseases of the nervous system

In **multiple sclerosis**, an autoimmune response against the myelin sheath of nerve cells causes sclerotic plaques of demyelinated tissue in the white matter of the central nervous sytem. The plaques are readily revealed by magnetic resonance imaging. Symptoms of the disease include motor weakness, impaired vision, lack of coordination, and spasticity. The effects of multiple sclerosis are highly variable. It can take a slow progressive course, or involve acute attacks of exacerbating disease followed by periods of gradual recovery. The incidence of multiple sclerosis reaches 1 in 1000 in some populations, with onset commonly occurring at an age in the late 20s or early 30s. In extreme cases it leads to severe disability or death within a few years. In contrast, some patients with mild disease function well, with little neurological impediment.

Activated T_H1 CD4 cells secreting interferon-γ (IFN-γ) are the effectors implicated in multiple sclerosis. These cells are enriched in the blood and cerebrospinal fluid and activated macrophages are present in sclerotic plaques. Macrophages release proteases and cytokines, which are the direct cause of demyelination. Tragically, the role of IFN-γ was most clearly demonstrated by a clinical trial in which disease worsened in the patients treated with IFN-γ.

In 90% of patients the sclerotic plaques contain plasma cells that secrete oligoclonal IgG into the cerebrospinal fluid. The autoantigens in multiple sclerosis are thought to be structural proteins of myelin, such as myelin basic protein, proteolipid protein, or myelin oligodendrocyte glycoprotein. Regular subcutaneous injection of IFN-$β_1$ reduces the incidence of disease attacks and the appearance of plaques. High doses of immunosuppressive drugs are used to reduce the severity of disease episodes.

An animal model for multiple sclerosis is **experimental allergic encephalomyelitis (EAE)**. This condition can be induced in certain inbred strains of mouse by immunization with myelin basic protein isolated from the same inbred strain. An autoimmune response to myelin basic protein is made in which leukocytes infiltrate the central nervous system, causing progressive paralysis (Figure 11.13).

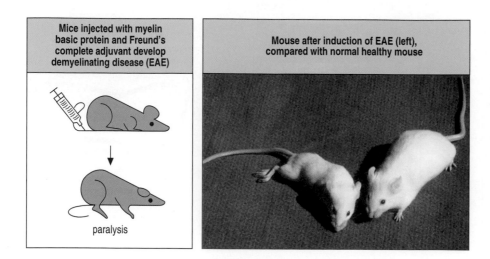

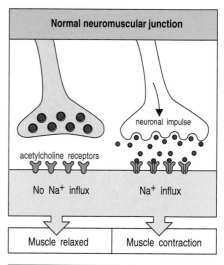

Figure 11.13 Immunization of mice with myelin basic protein causes a demyelinating disease called experimental autoimmune encephalomyelitis (EAE). The photograph compares a mouse with EAE on the left with a healthy mouse on the right. The first symptoms of paralysis are seen in the hind limbs and the tail.

Myasthenia gravis is an autoimmune disease in which signaling from nerve to muscle across the neuromuscular junction is impaired (Figure 11.14). Autoantibodies bind to the acetylcholine receptors on muscle cells, inducing their endocytosis and intracellular degradation in lysosomes. The loss of cell-surface acetylcholine receptors makes the muscle less sensitive to neuronal stimulation. Consequently, patients with myasthenia gravis suffer progressive muscle weakening as levels of autoantibody rise; the name of the disease means severe ('gravis') muscle ('myo') weakness ('asthenia'). Early symptoms of the disease are droopy eyelids and double vision. With time, other facial muscles weaken and similar effects on chest muscles impair breathing. This makes patients susceptible to respiratory infections, and can even cause death. One treatment for myasthenia gravis is the drug pyridostigmine, an inhibitor of the enzyme cholinesterase, which degrades acetylcholine. By preventing acetylcholine degradation, pyridostigmine increases the capacity of acetylcholine to compete with the autoantibodies for the receptors. A second treatment for myasthenia gravis is the immunosuppressive drug azathioprine, which inhibits the production of autoantibodies.

Autoantibodies against cell-surface receptors can act either to stimulate or to inhibit the receptor. The autoantibodies in myasthenia gravis prevent the function of the acetylcholine receptor and are called receptor **antagonists**. In contrast, the autoantibodies in Graves' disease facilitate receptor function (see Section 11-3) and are called receptor **agonists**. Autoantibodies of both types can be made against the insulin receptor; they lead to different symptoms. The cells of patients with antagonistic autoantibodies are unable to take up glucose, which accumulates in the blood, causing hyperglycemia and a form of diabetes mellitus resistant to treatment with insulin. In contrast, in patients with agonistic antibodies, cells deplete blood glucose to abnormally low levels. This state of hypoglycemia deprives the brain of glucose, causing light-headedness (Figure 11.15).

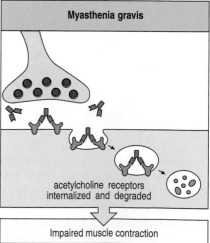

Figure 11.14 Autoantibodies against the acetylcholine receptor cause myasthenia gravis. In a healthy neuromuscular junction, signals generated in nerves cause the release of acetylcholine, which binds to the acetylcholine receptors of the muscle cells, causing an inflow of sodium ions that indirectly causes muscle contraction (upper panel). In patients with myasthenia gravis, autoantibodies specific for the acetylcholine receptor reduce the number of receptors on the muscle-cell surface by binding to the receptors and causing their endocytosis and degradation (lower panel). Consequently, the efficiency of the neuromuscular junction is reduced, which is manifested as muscle weakening.

Diseases mediated by antibodies against cell-surface receptors			
Syndrome	**Antigen**	**Antibody**	**Consequence**
Graves' disease	Thyroid-stimulating hormone receptor	Agonist	Hyperthyroidism
Myasthenia gravis	Acetylcholine receptor	Antagonist	Progressive muscle weakness
Insulin-resistant diabetes	Insulin receptor	Antagonist	Hyperglycemia, ketoacidosis
Hypoglycemia	Insulin receptor	Agonist	Hypoglycemia

Figure 11.15 Diseases mediated by antibodies against cell-surface receptors. Antibodies act as agonists when they stimulate a receptor on binding it, and as antagonists when they block a receptor's function on binding it.

Summary

In autoimmunity the mechanisms of antigen recognition and effector function are identical to those used in the response to pathogens. In contrast, the symptoms of autoimmune disease are highly variable, depending on the triggering autoantigen and the target tissue. Autoimmunity diseases can be classified into three main types corresponding to the type II, type III, and type IV categories of hypersensitivity reactions, with which they share common effector mechanisms. Some autoimmune diseases, such as insulin-dependent diabetes or Graves' disease of the thyroid, are directed against antigens of one particular organ or tissue, and are known as organ- or tissue-specific autoimmune diseases. In contrast, systemic lupus erythematosus is a systemic autoimmune disease, which is directed against components common to all cells. Autoimmune diseases caused by antibodies have been more readily studied than those caused by effector T cells, because antibodies can exert an effect when transferred from human patients to experimental animals. Understanding how effector T cells might function in human autoimmune disease remains largely dependent on extrapolation from mouse models of autoimmune disease. Autoimmune diseases tend to be chronic conditions in which episodes of acute disease are interspersed with periods of recovery. Once a tissue becomes inflamed and damaged by an autoimmune response, the resulting increase in the processing and presentation of self-antigens frequently leads to expansion and diversification of the autoimmune response. In addition, infections can exacerbate autoimmunity in a non-specific manner by inducing the production of inflammatory cytokines within the infected tissues.

Genetic and environmental factors which predispose to autoimmune disease

All autoimmune diseases involve a breach of the normal mechanisms of self-tolerance, the means by which the immune system is prevented from attacking the cells of the body. Self-tolerance is achieved in several different ways and can therefore be broken by a variety of mechanisms. Both genetic and environmental factors contribute to the loss of self-tolerance that leads to autoimmune disease.

11-9 HLA type and other genetic factors affect susceptibility to autoimmune disease

Susceptibility to autoimmune disease runs in families and varies between populations. Such clinical observations first raised the possibility that genetic factors confer differential susceptibility to disease. This turned out to be true, with the

Figure 11.16 Family studies reveal that HLA type correlates with susceptibility to insulin-dependent diabetes mellitus (IDDM). The top panel shows the frequency with which two siblings share HLA haplotypes in the population as a whole. The percentages are those expected from a simple mendelian segregation of the two maternal and two paternal HLA haplotypes. In the bottom panel, the analysis has been confined to pairs of siblings who both have IDDM. In these sibling pairs, the frequency distribution differs greatly from that expected from simple mendelian segregation. Pairs of siblings with the disease are much more likely to have the same HLA type than are pairs of siblings who are healthy.

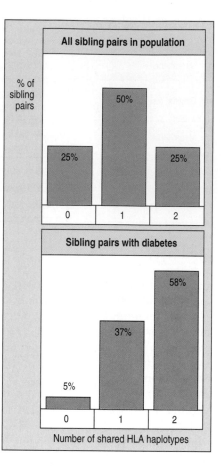

most important genetic factors being HLA class I and class II genes. The dominance of HLA over other genes in this respect is seen from comparison of the HLA types of siblings who both suffer from the same autoimmune disease. Pairs of affected siblings almost always share one or both HLA haplotypes (Figure 11.16).

Comparison of HLA type in patients and healthy controls shows that autoimmune diseases, including those described in preceding sections of this chapter, are more likely to afflict people expressing particular HLA allotypes. For example, in comparison with people who do not express HLA-DR2, people who do express HLA-DR2 are almost 16 times more likely to develop Goodpasture's syndrome. Although most disease associations are with HLA class II type, the strongest ones are with HLA class I type (Figure 11.17).

As methods for HLA typing have been refined, so have the associations with disease. For example, multiple sclerosis was initially correlated with HLA-B8, a class I serotype, before being shown to correlate more strongly with HLA-DR2, which is

Associations of HLA serotype with susceptibility to autoimmune disease			
Disease	HLA serotype	Relative risk	Sex ratio (♀:♂)
Ankylosing spondylitis	B27	87.4	0.3
Acute anterior uveitis	B27	10.04	<0.5
Goodpasture's syndrome	DR2	15.9	~1
Multiple sclerosis	DR2	4.8	10
Graves' disease	DR3	3.7	4–5
Myasthenia gravis	DR3	2.5	~1
Systemic lupus erythematosus	DR3	5.8	10–20
Insulin-dependent diabetes mellitus	DR3 and DR4	3.2	~1
Rheumatoid arthritis	DR4	4.2	3
Pemphigus vulgaris	DR4	14.4	~1
Hashimoto's thyroiditis	DR5	3.2	4–5

Figure 11.17 Associations of HLA serotype with autoimmune disease.

often found in the same haplotype as HLA-B8. In IDDM, both HLA-DR3 and HLA-DR4 were correlated with disease susceptibility and their combination conferred a susceptibility greater than the sum of the individual susceptibilities. In contrast, HLA-DR2 confers a resistance to IDDM that in heterozygotes is dominant over the susceptibilities due to HLA-DR3 or -DR4 (Figure 11.18). The molecular basis for these correlations has subsequently been shown to reside in polymorphisms in the HLA-DQ class II β-chain gene which are linked to HLA-DR3 and -DR4 respectively.

The associations of HLA type with autoimmune disease are not simple like the relation between a mutant allele and a genetic disease, as in cystic fibrosis for example. There, homozygosity or heterozygosity for mutant alleles at one gene determines absolutely whether the disease will occur or not. In contrast, the vast majority of people who have a particular HLA type will never suffer from the autoimmune diseases associated with that type. Conversely, not all the patients who develop a disease have HLA types that correlate with disease susceptibility. Even for the strongest association of HLA type with disease, that of HLA-B27 with ankylosing spondylitis, only 2% of people who have HLA-B27 develop ankylosing spondylitis and, conversely, 2% of patients with clinically defined disease do not express HLA-B27. Thus, factors other than HLA type also affect susceptibility to autoimmune disease.

The contribution of other genetic factors to autoimmunity is shown by the differential susceptibility of males and females to most autoimmune diseases (see Figure 11.17). Further evidence for the importance of other genes comes from family studies. For example, although 2% of people expressing HLA-B27 in the population as a whole develop ankylosing spondylitis, in families with a history of the disease, 20% of HLA-B27-expressing individuals develop disease. For the diseases associated with HLA class II allotypes, the concordance of disease in monozygotic twins is around 20%, compared with 5% for dizygotic twins. From whole-genome comparisons of affected families it has been estimated that HLA genes provide 50% of the genetic contribution to IDDM susceptibility, while many other loci make up the remaining 50%.

Inherited homozygous deficiency of the early complement proteins (C1, C4, or C2), for example, strongly predisposes to the development of SLE. Although the underlying mechanism has yet to be determined, this effect indicates that subtle polymorphisms in the complement genes might affect disease susceptibility.

11-10 Environmental factors influence the course of autoimmune diseases

The preceding section shows how the interaction of many different functional gene polymorphisms determines genetic predisposition to autoimmune disease. The fact that predisposed individuals develop disease with a maximum frequency of about 20% also emphasizes the importance of environmental factors in determining who actually gets ill. A simple illustration of the influence of an environmental factor is provided by Goodpasture's syndrome, the disease caused by autoantibodies against type IV collagen of basement membranes (see Section 11-1). Whereas all patients suffer with glomerulonephritis, only 40% of them develop pulmonary hemorrhage as well. The patients who develop lung damage are those who habitually smoke cigarettes. In non-smokers the basement membranes of lung alveoli are inaccessible to antibodies and there is neither antibody deposition nor disruption of the tissue. In smokers' lungs, by contrast, the alveoli are already damaged from their continual exposure to cigarette smoke. This lack of tissue integrity gives circulating antibodies access to the basement membranes, where deposition and complement activation burst blood vessels,

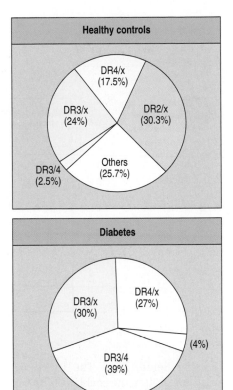

Figure 11.18 Population studies show association of susceptibility to insulin-dependent diabetes mellitus (IDDM) with HLA genotype. The HLA genotypes (determined by serotyping) of diabetic patients (bottom panel) are not representative of those found in the population (top panel). Almost all diabetic patients express HLA-DR3 and/or HLA-DR4, and HLA-DR3/DR4 heterozygosity is greatly over-represented among diabetics compared with controls. These alleles are linked tightly to HLA-DQ alleles that confer susceptibility to IDDM. In contrast, HLA-DR2 protects against the development of IDDM and is extremely rare in diabetic patients. The small letter x represents any allele other than DR2, DR3, or DR4.

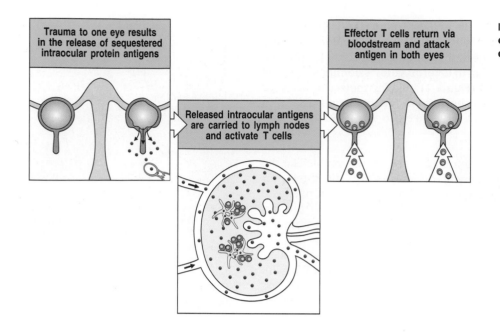

Figure 11.19 Physical trauma to one eye initiates autoimmunity that can destroy vision in both eyes.

causing hemorrhage. The inaccessibility of basement membranes in other anatomical sites, the cochlea of the ear for example, prevents them from becoming affected in Goodpasture's syndrome. The heterogeneity in pulmonary hemorrhage observed in Goodpasture's syndrome shows that circulating autoantibodies are not sufficient to cause disease; they must also have access to autoantigens.

Physical trauma can also give the immune system access to anatomical sites and autoantigens to which it is not normally exposed. One such immunologically privileged site is the anterior chamber of the eye, which contains specialized proteins involved in vision. When an eye is ruptured by a blow, eye proteins can drain to the local lymph node and there induce autoimmunity. On occasion, this response will cause blindness in the damaged eye. Surprisingly, the immune response also gains access to the undamaged eye and can cause blindness there unless treatment is given. This condition is called **sympathetic ophthalmia** (Figure 11.19). The fact that both eyes are attacked shows that immunological privilege is due to mechanisms that prevent the induction of an immune response, rather than a lack of access for effector cells and molecules once they have been activated. To preserve a patient's sight in the undamaged eye it is necessary to remove the damaged eye, which prevents the further induction of autoimmunity. Patients are also given immunosuppressive drugs to stop the ongoing autoimmune response. Sympathetic ophthalmia illustrates that healthy tissues need not be involved in initiating autoimmunity to be vulnerable to attack when a response of the right specificity is started elsewhere.

Infection with a pathogen is one environmental factor that can affect the course of an autoimmune disease, as seen in **Wegener's granulomatosis**. This disease is characterized by the severe inflammation and destruction of small blood vessels, particularly in the lungs. Autoantibodies that bind to proteinase-3, a component of neutrophil granules, are believed to cause the condition (Figure 11.20). In resting neutrophils, the autoantigen is inaccessible to antibody, which explains why individuals with high levels of circulating autoantibody can remain asymptomatic. However, once neutrophils are activated in response to an infection, proteinase-3 translocates to the cell surface and becomes accessible to autoantibody. Binding of the antibody triggers neutrophil degranulation and the release of destructive free radicals. Degranulation of activated neutrophils within small blood vessels

Figure 11.20 Serum from patients with Wegener's granulomatosis contains autoantibodies reactive with neutrophil cytoplasmic granules. Normal neutrophils with permeabilized cell membranes have been incubated with serum from a patient with Wegener's granulomatosis. IgG antibodies in the serum that react with cytoplasmic granules are detected by the addition of fluorescein-conjugated antibodies against IgG (green staining). N, nucleus; C, cytoplasm. Photograph courtesy of C. Pusey.

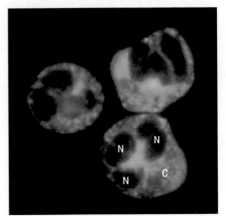

damages the vessels and causes disease. Because the accessibility of the neutrophil autoantigen is dependent on cellular activation, the occurrence of infectious disease can induce a flare-up or exacerbation of an existing autoimmune disease. This effect does not depend on infection with any particular pathogen. A similar phenomenon is seen in multiple sclerosis, in which relatively mild viral infections can excacerbate disease symptoms.

11-11 Autoimmune disease can be a consequence of the immune response to infection

As well as non-specific exacerbation of an existing autoimmune disease, an infection can also be the trigger that starts an autoimmune response. For almost every autoimmune disease there are clinical observations suggesting that autoimmunity might start with an infection and the immune response it provoked. For certain diseases the evidence is compelling; for others it is correlative, circumstantial, or anecdotal. Some autoimmune conditions that are well correlated with an infection are listed in Figure 11.21.

The best evidence comes from diseases caused by antibodies. Rheumatic fever is an autoimmune disease involving inflammation of the heart, joints, and kidneys, which can follow 2–3 weeks after a throat infection with certain strains of *Streptococcus pyogenes*. In defeating the bacterial infection, antibodies specific for cell-wall components of *S. pyogenes* are made. Some of these antibodies happen to cross-react with epitopes present on human heart, joint, and kidney tissue. On binding to the heart they activate complement and generate an acute and widespread inflammation—rheumatic fever—which sometimes causes heart failure. Rheumatic fever is a transient autoimmune disease; because of a lack of T-cell help the autoantigens cannot continue to stimulate antibody synthesis. This situation

Associations of infection with immune-mediated tissue damage		
Infection	HLA association	Consequence
Group A *Streptococcus*	?	Rheumatic fever (carditis, polyarthritis)
Chlamydia trachomatis	HLA-B27	Reiter's syndrome (arthritis)
Shigella flexneri, Salmonella typhimurium, Salmonella enteritidis, Yersinia enterocolitica, Campylobacter jejuni	HLA-B27	Reactive arthritis
Borrelia burgdorferi	HLA-DR2, DR4	Chronic arthritis in Lyme disease

Figure 11.21 Infections associated with the start of autoimmunity.

Figure 11.22 Induction of experimental allergic encephalomyelitis (EAE) requires microbial products as well as autoantigen. To induce EAE, immunologists immunize mice with homogenate of spinal cord that has been mixed with Freund's complete adjuvant. An adjuvant is a preparation that non-specifically promotes an immune response. Freund's complete adjuvant consists of mineral oil and killed mycobacteria and is the most potent adjuvant known. Far less effective is Freund's incomplete adjuvant, which lacks the mycobacteria.

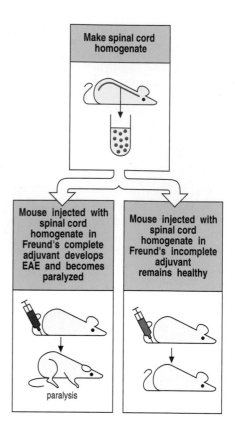

arises because the helper T cells that served in the anti-bacterial response are not stimulated by the autoantigens. The incidence of rheumatic fever has greatly diminished since streptococcal infections began to be treated with antibiotics. Bacterial infections have also been implicated in Graves' disease and viral infections in Goodpasture's syndrome.

Gastrointestinal bacterial infections are implicated in the initiation of Reiter's syndrome and reactive arthritis, two of the arthritic diseases associated with HLA-B27. When a cohort of people are all simultaneously infected with the same strain of food-poisoning bacteria, the frequency of autoimmune disease among those expressing HLA-B27 in the infected group is higher than for the population as a whole. Such evidence for the involvement of gastrointestinal infection has come from outbreaks of food poisoning among passengers on cruise ships, and among police on crowd-control duty. Coxsackie B virus, which infects the β cells of the pancreas, has been similarly implicated in triggering IDDM.

Indirect evidence that infection is necessary for inducing autoimmunity comes from the experience of immunologists trying to induce autoimmune disease experimentally in animals. Injection of tissues, tissue extracts, or purified autoantigens does not produce an autoimmune response. But when such self-components are mixed with microbial products that induce inflammation at the site of injection, then autoimmunity is reliably produced. For example, experimental autoimmune encephalomyelitis (EAE) is induced in mice after immunization with spinal cord homogenate mixed with Freund's complete adjuvant (mineral oil and dead mycobacteria). When mice are immunized with spinal cord homogenate in Freund's incomplete adjuvant (mineral oil alone) they do not succumb to EAE (Figure 11.22). Further evidence for the importance of infection is the change in the course and type of experimental autoimmunity seen when animals are kept under pathogen-free conditions.

11-12 HLA allotypes correlating with disease susceptibility may determine the T-cell responses that initiate autoimmunity

Autoimmune diseases are caused either by high-affinity antibodies or by effector T cells. Central to the development of either type of effector mechanism are antigen-specific T-cell responses, which require the presentation of antigenic peptides by HLA class I or II molecules. The fact that particular HLA class I or II allotypes confer susceptibility to autoimmune diseases suggests that peptide presentation by these allotypes is the trigger for the initiation of autoimmunity. Although such a mechanism has yet to be definitively established for any human autoimmune disease, it has been demonstrated in animal models and correlative evidence supports its validity for human disease.

One approach to this question has been to examine the peripheral blood of HLA-typed patients with autoimmune diseases for T cells that respond to potential autoantigens presented by the susceptibility HLA allotype, and to see

Figure 11.23 Polymorphism at residue 57 of the HLA-DQβ chain is correlated with susceptibility and resistance to IDDM. The top panel shows a ribbon diagram of the peptide-binding site of the HLA-DQ molecule; the α chain is shaded in light gray, the β chain is shaded blue. Position 57 on the β chain is shown as a red circle.

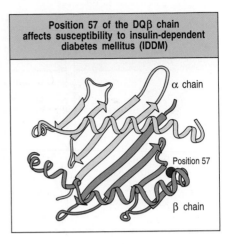

Position 57 of the DQβ chain affects susceptibility to insulin-dependent diabetes mellitus (IDDM)

α chain

Position 57

β chain

whether such T cells are absent from healthy controls. So, for example, T cells specific for myelin basic protein can be found in patients with multiple sclerosis. However, such T cells are also found in healthy controls. An important lesson from such studies is that the presence of specific T cells is not in itself an indication of disease; it is necessary but not sufficient. The discovery that an element of autoimmunity seems to be ever present in healthy bodies has led some immunologists to theorize that controlled autoimmunity is part of the normal mechanisms that maintain and renew the body's tissues.

Another approach is to identify the individual amino-acid residues in the HLA class II molecule that correlate with changes in disease susceptibility. Susceptibility to IDDM, for example, is associated with certain polymorphisms of the HLA class II DQβ chains. Comparison of the amino-acid sequences of allotypes that confer resistance with those that confer susceptibility revealed a strong but not absolute correlation with substitutions at position 57 in the β chain (Figure 11.23). Whereas aspartic acid is the most frequent residue at this position, patients with IDDM tend to have allotypes with other residues: valine, serine, or alanine. Similarly, the occurrence of diabetes in NOD mice is dependent on the expression of an allotype of the MHC class II molecule called I-A, the mouse equivalent of HLA-DQ. The I-A β chain of the NOD mouse has serine at position 57, whereas other strains of mice have aspartic acid at this position. Residue 57 is within the peptide-binding site of the MHC class II molecule and substitutions at that position modify the site's architecture.

In patients with SLE, HLA class II polymorphism influences the specificity of autoantibodies made, presumably by its effect upon the helper T-cell response. The greatest susceptibility to SLE is associated with HLA-DR3; patients with this HLA type tend to make antibodies against proteins of a small cytoplasmic ribonucleoprotein complex. In contrast, patients of HLA-DR2 type make antibodies against double-stranded DNA, and patients of HLA-DR5 type make antibodies against the nuclear ribonucleoprotein complex known as the spliceosome.

Pemphigus vulgaris is an autoimmune disease that is characterized by extensive blistering of the skin. It is due to autoantibodies that attack a component of desmosomes, structures that bind the epidermal cells of the skin tightly together. In some ethnic groups it is closely associated with a particular allele of HLA-DR4, which binds a peptide antigen derived from desmosomes.

11-13 Autoimmune disease can be induced by the transfer of T-cell clones in animal models

A major challenge in the study of autoimmune diseases is to distinguish the autoimmune response that starts the disease from the autoimmunity that develops as a consequence of inflammation and tissue damage. The definitive demonstration that a particular antibody or a T cell is the cause of disease is to purify it from patients and show that it replicates the disease on transfer to experimental animals.

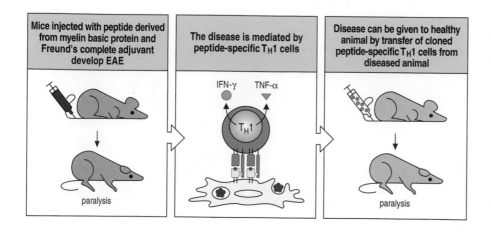

Figure 11.24 Experimental allergic encephalomyelitis (EAE) can be caused by a single clone of CD4 T$_H$1 cells.

This approach is not usually feasible for T cells because of the constraints placed on the T-cell recognition of antigen by MHC polymorphism (see Section 3-18, p. 77). Even within the same species, transfer of T-cell activity can be accomplished only when donor and recipient share MHC allotypes. Our understanding of autoimmune diseases that are thought to be caused by effector T cells is therefore much less secure than that of antibody-mediated autoimmunity. In this area, animal models of human diseases have been instrumental in showing how effector T cells can cause disease. The mouse model of multiple sclerosis, EAE, and the diabetic NOD mouse have been particularly valuable.

The disease-causing mechanism in EAE has been defined in detail. The system was first developed by immunizing mice with homogenates of spinal cord. Then it was found that immunization with myelin basic protein (MBP) alone was sufficient to cause the disease. Clones of T cells specific for MBP could be cultured from the affected animals and the epitopes recognized by these T cells were then defined by using synthetic peptides derived from the amino-acid sequence of MBP. The transfer of a single clone of MBP-specific CD4 T$_H$1 cells into a healthy unimmunized animal is sufficient to cause EAE (Figure 11.24). In this type of experiment the transfer of disease is absolutely dependent on the presence of a particular MHC allotype and a particular autoantigenic peptide.

The model was simplified further by showing that immunization with one type of antigenic peptide could induce EAE. Susceptibility to EAE varies between mouse strains and is dependent on the peptides of MBP that their MHC class II molecules present. Although the extent of the similarities between EAE and multiple sclerosis is debatable, this animal model has provided valuable pointers as to how MHC polymorphism and the T-cell response might contribute to human autoimmunity.

Similar studies have been performed on autoimmune diabetes in the NOD mouse. Diabetes can be transferred from one mouse to another by T cells, showing that effector T cells are the cause of disease. However, the autoimmune T-cell response is more complicated than that in EAE and involves both CD4 and CD8 T cells. Multiple sclerosis, rheumatoid arthritis, and IDDM are all believed to be caused by effector T cells. Current research on these diseases aims at defining their underlying molecular mechanisms with the precision accomplished for EAE.

If just one or a small number of T-cell clones is responsible for disease-causing autoimmunity in humans, this would be consistent with the influence of HLA type on disease susceptibility. The small size of the response might also contribute to the slow rate at which tissue is destroyed. In general, attempts to correlate susceptibility to infectious diseases with HLA type have been disappointing, probably because the complexity of the response dilutes out the bias imposed by any particular peptide:HLA complex. By contrast, in an autoimmune response, where only one or a few T-cell clones are active, each peptide:MHC complex becomes highly influential and leads to a correlation of the autoimmune response with HLA type.

11-14 Autoimmune disease results from perturbation of the mechanisms that normally maintain self-tolerance

Autoimmune diseases affect a minority of the population, in whom a small number of lymphocytes escape the normal mechanisms that maintain tolerance. Various mechanisms are used to maintain self-tolerance; the mechanisms leading to autoimmunity are likely to be equally varied. During T-cell development, negative selection removes T cells that respond to peptides presented by the MHC molecules of thymic cells. In this way T-cell tolerance is produced to most of the peptides normally presented by cells of the body. Because high-affinity antibody responses require T-cell responses to the same antigen, T-cell tolerance also provides B-cell tolerance to many self-components. Autoantigens are therefore likely to be tissue-specific proteins that are not expressed in the thymus, or proteins expressed in other tissues at higher levels than in the thymus. In normal circumstances, mechanisms of peripheral tolerance prevent lymphocytes from being activated by these candidate autoantigens.

The activation of naive T cells requires antigen presentation by a professional antigen-presenting cell, which expresses B7 co-stimulatory molecules as well as the specific peptide:MHC complex (see Section 6-3, p. 134). Most of the cells making tissue-specific proteins do not express co-stimulatory molecules and cannot present tissue-specific peptides in a way that activates naive T cells. On the contrary, this manner of presentation can lead to T-cell anergy. T cells could, however, be activated by tissue-specific antigens in situations where professional antigen-presenting cells pick up and present tissue-specific peptides derived from other cell types. This could happen in the course of an infection, when dead or infected tissue cells are taken up by macrophages. Once autoimmunity is stimulated in this manner, activated effector T cells would then directly attack any cells making the tissue-specific proteins.

When tissues become inflamed, some cells are induced to increase the expression of MHC class I and class II molecules. These changes, which increase the number of different peptide antigens presented by a cell and their density on the cell surface, can lead to interactions with clones of T cells that were not sensitive to the lower levels of antigen presentation. The greatest changes occur when the inflammatory cytokine IFN-γ induces MHC class II expression on cells that normally do not express these MHC molecules. Such cells, which include thyroid cells, pancreatic β cells, astrocytes, and microglia, are conspicuously present in tissues targeted by T-cell mediated autoimmunity. Although induced expression of MHC class II in the absence of co-stimulatory molecules is insufficient to activate naive T cells, it provides potential new targets for T cells activated by other means (Figure 11.25).

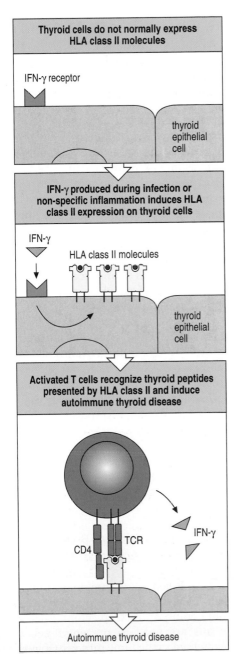

Figure 11.25 Induction of MHC class II expression on tissue cells facilitates autoimmunity. Thyroid epithelial cells do not normally express HLA class II molecules (top panel). They are induced to do so by IFN-γ (center panel). They are then able to present thyroid peptides to activated specific T cells, which induce autoimmune thyroid disease (bottom panel). IFN-γ produced by activated T cells feeds back to amplify HLA expression and antigen presentation.

Tolerance to certain sites in the body is maintained by anatomical barriers that seal them off from surveillance by circulating naive lymphocytes, as already noted in the discussion of sympathetic ophthalmia (see Section 11-10). Such anatomical barriers can be breached by wounding or infection, which exposes the sequestered autoantigens to circulating lymphocytes. A striking demonstration that autoreactive T cells can circulate in healthy individuals without being disruptive comes from the study of genetically engineered mice in which every T cell expresses the same T-cell receptor. The receptor originated in a clone of CD4 T_H1 cells that causes EAE and is specific for a peptide of MBP presented by MHC class II. When left alone, the mice remain healthy and their T cells are quiescent because naive T cells never travel through the central nervous system, where MBP is expressed. However, when the mice are immunized in another anatomical site with MBP, the T cells become activated and migrate to the central nervous system, where their actions cause paralysis.

11-15 Autoreactive CD4 T_H2 cells can suppress tissue destruction by autoreactive CD4 T_H1 and CD8 effector T cells

Animal models of infectious disease have revealed situations in which the induction of CD4 T cells to make either a T_H1 or a T_H2 response makes the difference between a beneficial or a detrimental immune response (see Section 6-8, p. 142). Similar effects are seen in animal models of autoimmune disease. If mice are immunized with MBP in the absence of killed mycobacteria they do not get EAE (see Figure 11.22). Moreover, such immunization protects the mice from EAE when they are subsequently immunized with MBP in the presence of mycobacteria. A similar outcome occurs when the mice are first fed with MBP (Figure 11.26). This oral route of immunization stimulates the production of CD4 T_H2 cells, which produce transforming growth factor-β (TGF-β) and IL-4, and does not stimulate the production of the IFN-γ-producing T_H1 cells that cause EAE. The CD4 T_H2 cells migrate to the brain, where they secrete cytokines but do not produce disease. The effect of feeding with MBP is not antigen-specific, because mice thus fed are resistant to EAE induced by other myelin components. The effect of feeding with MBP is to produce an immune response in the brain that is governed by the T_H2 type of cytokines and that through the action of TGF-β actively and nonspecifically suppresses CD4 T_H1 cells.

Similar effects have been observed in NOD mice. Feeding insulin to NOD mice prevents diabetes by activating clones of CD4 T cells that migrate to the pancreas and secrete TGF-β. Moreover, clones of CD4 T cells have been isolated that on transfer to other NOD mice prevent the destruction of β cells by cytotoxic CD8 T cells. Such regulatory T cells may well be generated in the course of human IDDM and could explain why it takes many years for the immune response to kill just 10^8 pancreatic β cells. In total, these observations suggest there are active mechanisms for controlling the autoimmune response that involve competition between CD4 T_H1 and CD4 T_H2 cells. Multiple sclerosis, rheumatoid arthritis, and IDDM are often chronic and relapsing diseases. Such a course would be consistent with ongoing competition and a changing balance between CD4 T_H1 and CD4 T_H2 cells.

11-16 Autoreactive B cells are held in check by T-cell tolerance and other mechanisms

During B-cell maturation in the bone marrow, clonal deletion and inactivation of self-reactive B cells prevent the emergence of cells with antigen receptors that bind common molecules of human cell surfaces or plasma. Additional mechanisms are used to prevent B-cell activation by tissue-specific antigens not expressed in the bone marrow. Of most importance is the control exerted by T-cell tolerance, which results in a lack of T-cell help. When a B cell is stimulated by an

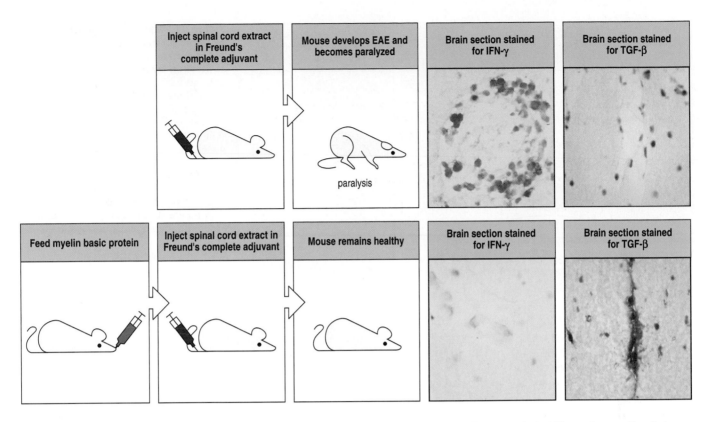

Figure 11.26 Oral administration of myelin basic protein protects against EAE on subsequent immunization with myelin basic protein in Freund's complete adjuvant. Experimental allergic encephalomyelitis (EAE) is induced in mice by immunization with spinal cord homogenate in Freund's complete adjuvant (upper panels); the disease is mediated by T_H1 cells specific for myelin antigens. These cells produce the cytokine IFN-γ (top left photograph, where the brown staining reveals the presence of IFN-γ), but not TGF-β (top right photograph). These T cells are presumably responsible for the damage that results in paralysis. When mice are first fed with myelin basic protein (MBP), later immunization with spinal cord or MBP fails to induce the disease (lower panels). In these orally tolerized mice, IFN-γ-producing cells are absent (lower left photograph) whereas TGF-β-producing T cells (lower right photograph, brown staining) are found in the brain in place of the autoaggressive T_H1 cells and presumably protect the brain from autoimmune attack. Photographs courtesy of S. Khoury, W. Hancock, and H. Weiner.

autoantigen it migrates to the T-cell area of a secondary lymphoid tissue. If antigen-activated helper T cells are not available, the antigen-activated B cells die by apoptosis.

The importance of T-cell tolerance in preventing the production of autoantibodies in normal healthy people can be inferred from what occurs in patients with SLE, a disease in which many different autoantibodies are made against the components of intracellular nucleoprotein particles. B cells with a cell-surface receptor that binds one component of a nucleoprotein particle endocytose the whole particle and process and present peptides derived from all the particle's components. Because of this property, T cells specific for one peptide derived from the particle can provide help to the many different B cells that are specific for an epitope on any one of the particle's components. This mechanism even allows T cells responding to peptide epitopes to activate B cells making high-affinity antibodies against nucleic acids, macromolecules that cannot be recognized by T cells. The wide range of autoantibodies that can be produced once T-cell help is available suggests that autoimmune diseases arise mostly from a breach of T-cell tolerance (Figure 11.27).

A B cell encountering an autoantigen can also become anergic, which involves a decrease in IgM expression at the cell surface and partial inhibition of the pathways of signal transduction. Another mechanism of B-cell inactivation is one that

Figure 11.27 Study of systemic lupus erythematosus (SLE) shows how autoreactive B cells can be controlled by lack of T-cell help. In patients with SLE, the emergence of a single clone of autoreactive T cells can lead to a diverse antibody response against the components of an intracellular particle. The top and middle panels show how a T cell specific for a peptide of histone 1 can provide help to B cells specific for histone 1 and also to B cells specific for other components of nucleosomes, for example DNA as shown here. This patient, who is making autoantibodies against nucleosomes, does not make antibodies against ribosomes because there is no autoreactive T cell specific for a ribosomal peptide (bottom panel).

is thought to eliminate B cells that become autoreactive during the process of somatic hypermutation in germinal centers. Experiments performed on mice suggest that autoreactive B cells die by apoptosis if and when they encounter their specific autoantigens during proliferation in the germinal center.

11-17 Infections can trigger autoimmunity through molecular mimicry

Rheumatic fever is caused by antibodies which are raised in response to bacterial surface antigens and cross-react with constituents of healthy human heart tissue (Figure 11.28). Such a resemblance of pathogen and host antigens is called

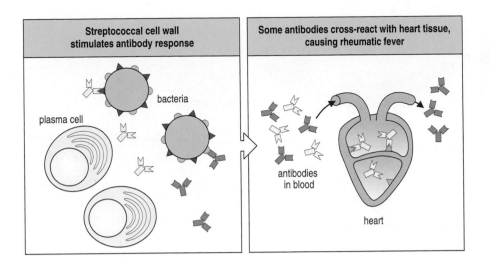

Figure 11.28 Antibodies against streptococcal cell-wall antigens cross-react with antigens on heart tissue. The immune response to the bacteria produces antibodies against various epitopes of the bacterial cell surface. Some of these antibodies cross-react with the heart (yellow) whereas others do not (blue). An epitope in the heart (orange) is structurally similar but not identical to a bacterial epitope (red).

molecular mimicry. This simple example illustrates how health-benefiting immune responses that successfully control infections can inadvertently become life-threatening autoimmunities. Other autoimmune diseases might also be due to chance antigenic similarities between pathogen and human macromolecules.

Studies with monoclonal antibodies have proved that cross-reaction with structurally similar antigens is a general property of antibodies. Cross-reactions usually involve antigens of very similar structures, as when autoantibodies from patients with Graves' disease bind to human, rat, or pig receptors for thyroid-stimulating hormone. Cross-reactions of this type are familiar not simply because they are more common, but also because they are the cross-reactions that most interest immunologists. There are, however, cross-reactivities involving very different molecules; these appear unexpectedly in experiments and are usually dismissed as artifacts. Without the clinical evidence, the discovery of anti-streptococcal antibodies that react with human heart tissue would almost certainly have been dismissed in this way. Antibody cross-reactions involving dissimilar types of antigenic macromolecule occur if the antigens share a localized region of structural similarity which interacts with the antibody, or if different epitopes on the two molecules are able to interact with the antibody's CDR loops in different ways.

T-cell receptors also cross-react with different antigens. T-cell cross-reactivities are frequently found in experiments in which peptides with related sequences are bound to the same MHC allotype, or in which the same peptide is bound to very similar MHC allotypes. As with antibodies, these are the cross-reactions that researchers tend to seek. However, cross-reactivities have also been found involving peptide:MHC complexes in which neither the peptides nor the MHC allotypes are particularly closely related. Because all antigens recognized by T cells conform to the common structure of a peptide bound to an MHC molecule, T cells are intrinsically more cross-reactive than B cells, whose antigens are under no such structural limitation. Indeed, the developmental processes of positive selection, negative selection, and T-cell activation are all based on the inherent cross-reactivity of T-cell receptors.

The need for a T-cell response in persistent autoimmunity, and the inherent cross-reactivity of T-cell receptors, have led to the idea that chronic autoimmune diseases are caused by autoreactive T cells that arise in the course of combating infection. During an infection, each clone of T cells is activated by a pathogen-derived peptide bound to the MHC allotype that positively selected the T-cell clone. As naive cells, all these clones were tolerant of the complexes of self-peptides and MHC allotypes to which they were exposed. Once activated, however, effector T cells

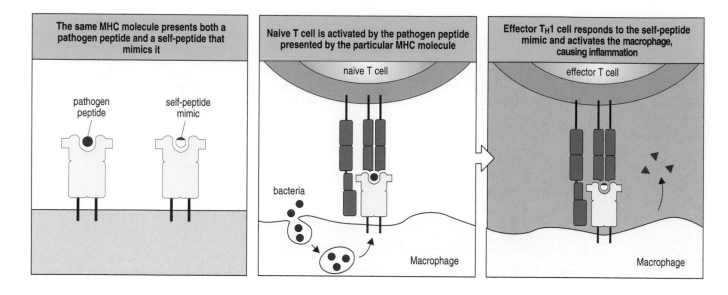

The same MHC molecule presents both a pathogen peptide and a self-peptide that mimics it	Naive T cell is activated by the pathogen peptide presented by the particular MHC molecule	Effector T$_H$1 cell responds to the self-peptide mimic and activates the macrophage, causing inflammation

Figure 11.29 Autoimmunity may be caused by self-peptides that mimic pathogen-derived peptides and stimulate a T-cell response. The first panel shows the same MHC allotype presenting two different peptides: one from the pathogen, and a self-peptide that mimics it. The second panel shows a naive CD4 T cell being activated by the pathogen peptide. This activated CD4 T$_H$1 cell can then activate a macrophage presenting the self-peptide mimic, as shown in the third panel, thus initiating an inflammatory reaction.

can respond to other peptide:MHC complexes that bind with lower affinity than is required for activation. So T cells activated by a pathogen-derived peptide have the potential to attack cells presenting self-peptide:MHC complexes that are antigenically cross-reactive (Figure 11.29). This is the model of molecular mimicry applied to T-cell antigens. The sequences of human proteins and pathogen proteins reveal many potential T-cell epitopes that could provide such mimics.

Activated T cells gain access to most of the body's tissues, whereas naive T cells are restricted to the blood, secondary lymphoid tissues, and lymph. Within tissues, T cells activated by pathogens will encounter tissue-specific self-peptide:MHC complexes that did not participate in positive selection, negative selection, or the induction of peripheral tolerance. Some of these complexes might bind T-cell receptors with affinities even higher than that of the activating antigen. Similarly, the induction of MHC class II expression on cells within infected tissues creates a new array of tissue-specific antigens, which can become the target for cross-reactive attack by effector T cells.

Summary

Susceptibility to certain autoimmune diseases is associated with particular HLA class I and II allotypes. Although HLA is the most highly correlated genetic factor, it accounts for only about half the genetic predisposition. Even for HLA the correlations are weak in absolute terms; most people having a predisposing HLA type do not develop the relevant autoimmune disease, and not all people who develop the disease have a predisposing HLA type. The same is true for the environmental factors that affect autoimmune disease, which include trauma, exposure to chemicals, and infection. No known infection inevitably leads to autoimmunity, and different pathogens can trigger the same autoimmune disease in people of different HLA types.

Immunological tolerance is generally well maintained and most people never suffer from autoimmune disease. Of principal importance is T-cell tolerance, because all the effector mechanisms of autoimmunity—CD4 T cells, CD8 T cells, and high-affinity antibody—depend on a T-cell response. Many autoimmune conditions seem to be started by one or a few clones of autoreactive T cells. How they are activated remains poorly understood, but circumstantial evidence implicates infections during which some of the T cells activated by a pathogen then react with peptides derived from a human protein. After the infection has been cleared, the autoreactive effector T cells become chronically stimulated by cells of the body. Identifying the infections that cause autoimmune disease is difficult, firstly because the infections are successfully resolved and rarely recorded, and secondly because the symptoms of autoimmune disease begin only long after the start of the immune response.

Summary to Chapter 11

Central to the success of adaptive immunity is the provision of strong responses to foreign antigens while sustaining tolerance toward human macromolecules. Among the mechanisms that prevent autoreactivity are clonal deletion and inactivation of potentially autoreactive lymphocytes, anatomical barriers that prevent lymphocytes from reaching certain tissues, and lymphocyte activation requirements that depend on the presence of infection. Failure of any of these mechanisms can lead to autoimmune responses that attack the body's tissues and cause disease. Autoimmune diseases are complicated multifactorial diseases for which no single entity can be identified as the necessary and sufficient cause of a particular disease. They arise as a consequence of unlucky combinations of genetic and environmental factors, each of which by itself is weakly correlated with disease. Autoimmune diseases do not arise from obvious genetic or functional deficiencies in the immune system but are the occasional and perhaps inevitable side effects of successfully fighting infections. A major challenge in treating autoimmune diseases is that patients usually present when the destructive effects of autoimmunity are at an advanced stage, and the initiating autoimmune response has become expanded and diversified. With better knowledge of the mechanisms operating in human autoimmune disease it should be possible to identify patients at earlier stages of disease. At such times tissue damage is limited, the autoimmune response is more narrowly defined, and it should be easier to control without resorting to non-specific immunosuppression.

Manipulation of the Immune Response

12

The preceding four chapters of this book describe the various ways in which the immune system responds helpfully or otherwise to environmental antigens. Given the spectrum of disease that results from inadequate or inappropriate immunity, clinical immunologists are driven to search for ways of manipulating the immune response to the benefit of their patients. This chapter examines three areas of medicine where this quest is being pursued—vaccination, organ transplantation, and cancer. In vaccination, the adaptive immune system is manipulated in an antigen-specific manner to stimulate protective immunity, and vaccination has protected billions of people against certain infectious diseases. In transplantation, non-specific immunosuppression allows a diseased organ to be replaced by a healthy one from another individual, procedures that have extended the lives of many thousands of patients. In contrast, manipulation of the immune response in cancer patients is still at the stage of research and development. The immune response to cancer provides too little too late; the challenge in this case is to enhance a patient's response in ways that prevent the proliferation and spread of malignant disease.

Prevention of infectious disease by vaccination

The modern era of vaccination began in the 1780s with Edward Jenner's use of cowpox as a vaccine against smallpox. Although Jenner's procedure was widely embraced, vaccines for other diseases did not emerge until well into the nineteenth century. That was the period when microorganisms were first isolated, grown in culture, and shown to cause disease. Largely by a process of trial and error, methods of inactivating microbes or impairing their ability to cause disease were developed and vaccines were produced. Vaccines were eventually developed against most of the epidemic diseases that had plagued the populations of western Europe and North America. Indeed, by the middle of the twentieth century it began to seem that the combination of vaccines and antibiotics would solve the problem of infection once and for all. Such optimism was soon tempered by the difficulty of developing effective vaccines for some diseases, and by the emergence of new diseases and antibiotic-resistant forms of old ones. This part of the chapter first surveys the vaccines in current use and then turns to the challenge of pathogens that have yet to be tamed by vaccination.

12-1 Viral vaccines are made from whole viruses or viral components

The first medically prescribed vaccine was against smallpox, a disease characterized by a rash of spots that develop into scarring pustules. The very first smallpox vaccines were made from dried pustules taken from people who seemed to have less severe symptoms of the disease. This early vaccine thus contained the smallpox virus itself. Small amounts of this material were given to healthy people, either intranasally or intradermally through a scratch on the arm—a procedure known as **variolation**, from the word **variola**, the Latin name given to both the pustule and to the disease itself. Although successful in many cases, the drawback of variolation was the frequency with which it produced full-blown smallpox, resulting in the death of around one in a hundred of those vaccinated. However, despite the risk, variolation was widely used in the eighteenth century because the threat from smallpox was so much greater. At that time smallpox killed one in four of those infected and there were regular epidemics. In London, for example, more than a tenth of all deaths were due to smallpox.

Jenner's innovation towards the end of the eighteenth century was to use the related cowpox virus as a vaccine for smallpox. The cowpox virus, called **vaccinia**, causes only very mild infections in humans, but the immunity produced gives effective protection against smallpox as well as cowpox because the two viruses have some antigens in common (Figure 12.1). In the nineteenth century, Jenner's vaccine replaced variolation and was eventually responsible for the eradication of smallpox in the twentieth century. The words vaccinia and **vaccination** derive from *vaccus*, the Latin word for cow. Although the term vaccination was once used specifically in the context of smallpox it now refers to any deliberate immunization that induces protective immunity against a disease.

Jenner's strategy for vaccination against smallpox is not possible for most pathogenic viruses because very few have a natural 'safe' counterpart. Most vaccines in use today are composed of preparations of the disease-causing virus for which the ability to cause disease has been destroyed or weakened. One type of vaccine consists of virus particles that have been chemically treated with formalin or physically treated with heat or irradiation so that they are no longer able to replicate. These are called **killed** or **inactivated virus vaccines**. The vaccines for influenza and rabies, and the Salk polio vaccine, are of this type. Only viruses whose nucleic acid

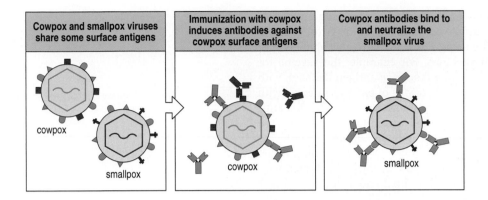

Figure 12.1 Vaccination with cowpox virus elicits neutralizing antibodies that react with antigenic determinants shared with smallpox virus. Shared antigenic determinants of cowpox also elicit protective T-cell immunity against smallpox (not shown here).

can be reliably inactivated make suitable killed virus vaccines. Such vaccines also have the drawback that large amounts of pathogenic virus must be produced during their manufacture.

A second type of vaccine consists of live virus that has mutated so that it has a reduced ability to grow in human cells and is no longer pathogenic to humans. These vaccines, called **live-attenuated virus vaccines**, are usually more potent at eliciting protective immunity than killed virus vaccines because the attenuated virus can usually replicate to a limited extent and thus mimics a real infection. Most of the viral vaccines currently used to protect humans are live-attenuated vaccines. Attenuated virus is produced by growing the virus in cells of non-human animal species. Such conditions select for variant viruses that grow well in the non-human host and are consequently less fit for growth in humans (Figure 12.2). The measles, mumps, polio (Sabin vaccine), and yellow fever vaccines consist of live-attenuated viruses. Attenuated viral strains can also arise naturally. As a viral infection passes though the human population the virus generally diversifies by mutation, sometimes producing a strain with reduced pathogenicity. These strains represent natural forms of live-attenuated virus and also provide candidates for vaccines. One of the three live-attenuated polio virus strains that make up the oral polio vaccine is of this type (the Sabin strain 2).

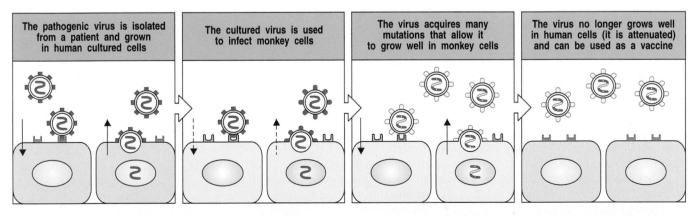

Figure 12.2 Attenuated viruses are selected by growing human viruses in non-human cells. To produce an attenuated virus, the virus must first be isolated by growing it in cultured human cells. This in itself can cause some attenuation; the rubella vaccine, for example, was made in this way. In general, however, the virus is then grown in cells of a different species, such as a monkey, until it has become fully adapted to those cells and grows only poorly in human cells. The adaptation is a result of mutation, usually a combination of several point mutations. It is usually hard to tell which of the mutations in the genome of an attenuated viral stock are critical to attenuation. An attenuated virus grows poorly in the human host, where it produces immunity but not disease.

In the human immune response to viruses, protective neutralizing antibodies tend to be directed towards particular surface components of the virus. Vaccines against some viruses have been made by using just these antigenic viral components and they are known as **subunit vaccines**. For example, the vaccine for hepatitis B virus (HBV) consists of the hepatitis B surface antigen (HBsAg), with which most anti-HBV antibodies react. The hepatocytes of patients infected with HBV secrete the surface antigen into the blood as minute particles. The first anti-hepatitis B vaccine was made from HBsAg purified from the blood of infected people, a procedure requiring the careful removal of infective virions. The vaccine now used is made from HBsAg produced by recombinant DNA technology, which avoids this problem. This vaccine provides about 85% of those vaccinated with protective immunity against HBV infection. People who do not respond may lack an effective CD4 T-cell response because their HLA class II allotypes fail to present HBsAg-derived peptides.

There remain, however, viruses for which no vaccine has yet been developed, despite considerable effort. Some are those, such as HIV, that naturally persist in the body for long periods, evading the efforts of the immune system to dislodge them. Others, like the viruses that cause the common cold, naturally provoke rather weak immune responses and also exist in such a large number of different strains that it is virtually impossible to develop a vaccine that will protect against all of them.

12-2 Bacterial vaccines are made from whole bacteria, their secreted toxins, or capsular polysaccharides

Vaccines against bacterial diseases have been developed by using the same approaches as outlined above for viruses. One difference in the result of these efforts, is that the number of live-attenuated bacterial vaccines remains small. The oldest and most commonly used is the BCG vaccine, which uses the Bacille Calmette-Guérin, derived from a bovine strain of *Mycobacterium tuberculosis*, to provide protection against tuberculosis. The efficacy of this vaccine varies in different populations and although routinely given to children in Europe, BCG is not used in the USA. More recently, live-attenuated vaccines against species of *Salmonella* have been introduced for both medical and veterinary use. An attenuated strain of *Salmonella typhi*, the bacterium that causes typhoid fever, was made by mutagenesis and selection for loss of a lipopolysaccharide necessary for pathogenesis. In this strain, which is now used as a vaccine, an enzyme necessary for lipopolysaccharide synthesis is defective.

Many bacterial diseases result from the effects of toxic proteins secreted by the bacteria. Examples are diphtheria, caused by *Corynebacterium diphtheriae*, and tetanus, caused by *Clostridium tetani*. Vaccines against these pathogens are made by purifying the respective toxin—**diphtheria toxin** or **tetanus toxin**—and treating it with formalin to destroy its toxic activity. The inactivated proteins, called **toxoids**, retain sufficient antigenic activity to provide protection against disease. Diphtheria and tetanus vaccines are thus comparable to the viral subunit vaccines. In medical practice, tetanus and diphtheria toxoids are combined in a single vaccine with a killed preparation of the bacterium *Bordetella pertussis*, the agent that causes whooping cough. This **combination vaccine** is called DTP, signifiying *D*iphtheria, *T*etanus, *P*ertussis. The immune response to the diphtheria and tetanus toxoids is enhanced by the adjuvant effect of the whole pertussis bacteria, which produces a strong inflammatory reaction at the site of injection. In some countries, Japan was the first, the DTP vaccine is being superseded by DTaP vaccines in which the whole pertussis bacteria is replaced by an acellular pertussis component (aP). This acellular component contains a modified inactivated form of pertussis toxin—pertussis toxoid—and one or more of the other pertussis antigens, such as filamentous hemagglutinin, pertactin, and fimbrial antigen.

Current immunization schedule for children (USA)							
Vaccine given	1–2 months	2 months	4 months	6 months	15 months	16–18 months	4–6 years
Diphtheria–tetanus–pertussis (DTP/DTaP)	☐	▓	▓	▓	▓	☐	▓
Trivalent oral polio (TVOP)	☐	▓	▓	☐	▓	☐	▓
Measles/mumps/rubella (MMR)	☐	☐	☐	☐	▓	☐	▓
Haemophilus B conjugate (HiBC)*	☐	▓	▓	▓	▓	☐	☐
Hepatitis B†	▓	☐	▓	☐	☐	▓	☐

Figure 12.3 Recommended current vaccination schedules for children in the USA. Vaccinations are indicated by the red shaded boxes. DTP, diphtheria, tetanus, and whole-cell pertussis vaccine; DTaP, diphtheria, tetanus, and an acellular pertussis vaccine. *An alternative option is to immunize at 2, 4, and 12 months. †An alternative option is to immunize at birth, 1–2 months, and 16–18 months.

Many pathogenic bacteria have an outer capsule composed of polysaccharides that define species-specific and strain-specific antigens. For these **encapsulated bacteria**, which include the pneumococcus (*Streptococcus pneumoniae*), salmonellae, the meningococcus (*Neisseria meningitidis*), *Haemophilus influenzae*, *Escherichia coli*, *Klebsiella pneumoniae*, and *Bacteroides fragilis*, the capsule determines both the pathogenicity and the antigenicity of the organism. In particular, the capsule prevents fixation of complement by the alternative pathway. Only when antibodies bind to the capsule does complement fixation lead to bacterial clearance. Consequently, the aim of vaccination against such bacteria is to produce complement-fixing antibodies that bind to the capsule.

Subunit vaccines composed of purified capsular polysaccharides have been developed for some of these bacteria and they are effective in adults, in whom they provoke a T-independent antibody response (see Section 7-2, p. 162). In contrast, these vaccines are ineffective in children under 18 months, probably because children do not develop good T-independent responses to polysaccharide antigens until a few years after birth (see Section 7-8, p. 174), and in the elderly. Both these groups are particularly susceptible to infection by encapsulated bacteria. The solution has been to convert the bacterial polysaccharide into a T-dependent antigen. This is done by covalently coupling the polysaccharide to a carrier protein, for example tetanus or diphtheria toxoid, which provides antigenic peptides for stimulating a CD4 T-cell response. On vaccination, T cells responding to the protein carrier can then stimulate polysaccharide-specific B cells to make antibodies. Vaccines of this type are called **conjugate vaccines**. A conjugate vaccine against *H. influenzae* (see Figure 6.31, p. 156) has reduced the incidence and severity of childhood meningitis caused by this bacterium.

A current schedule for recommended childhood immunizations in the USA is given in Figure 12.3. Multiple immunizations with most vaccines are recommended because they are needed to develop the memory B cells and T cells necessary for long-term protection.

12-3 Adjuvants non-specifically enhance the immune response

A prerequisite for a good immune response is a state of inflammation. During an infection this is initiated by microbial products that activate macrophages and recruit inflammatory cells (see Chapter 8). To work effectively, vaccination must also create a state of inflammation at the site in the body where the antigens are injected. In general, immunization with purified proteins leads to a poor immune

Adjuvants that enhance immune responses		
Adjuvant name	**Composition**	**Mechanism of action**
Freund's incomplete adjuvant	Oil-in-water emulsion	Delayed release of antigen; enhanced uptake by macrophages
Freund's complete adjuvant	Oil-in-water emulsion with dead mycobacteria	Delayed release of antigen; enhanced uptake by macrophages; induction of co-stimulators in macrophages
Freund's adjuvant with MDP	Oil-in-water emulsion with muramyldipeptide (MDP), a constituent of mycobacteria	Similar to Freund's complete adjuvant
Alum (aluminum hydroxide)	Aluminum hydroxide gel	Delayed release of antigen; enhanced macrophage uptake
Alum plus *Bordetella pertussis*	Aluminum hydroxide gel with killed *B. pertussis*	Delayed release of antigen; enhanced uptake by macrophages; induction of co-stimulators
Immune stimulatory complexes (ISCOMs)	Matrix of lipid micelles containing viral proteins	Delivers antigen to cytosol; allows induction of cytotoxic T cells

Figure 12.4 Some adjuvants used in experimental immunology.

response. The response can be enhanced, however, by substances that induce inflammation by antigen-independent mechanisms. Such enhancing substances are called adjuvants, a word meaning helpers (Figure 12.4). In experimental immunology the most effective adjuvant is **Freund's complete adjuvant**, an emulsion of killed mycobacteria and mineral oil into which antigens are vigorously mixed. The active components contributed by the mycobacteria are a dipeptide, N-acetylmuramyl-L-alanine-D-isoglutamine (MDP), and elements of the bacterial cell-wall skeleton. In addition to stimulating inflammation, adjuvants cause soluble protein antigens to aggregate and precipitate forming particles, which facilitates their efficient uptake by antigen-presenting cells. The particulate nature of the antigen also reduces the rate at which antigen is cleared from the system.

Although many substances and preparations are known to be adjuvants, the only adjuvant approved for use in human vaccines is alum, a form of aluminum hydroxide. Because vaccines are given to large numbers of healthy people, the safety standards for vaccines are high and they do not permit the side-effects of the most potent adjuvants. Although alum is a safe adjuvant, it is not particularly powerful, especially in stimulating immunity mediated by CD4 T_H1 cells and CD8 T cells. Adjuvant function can also be provided by the bacterial components of vaccines, such as the *B. pertussis* component of the DTP vaccine. ISCOMs (immune stimulatory complexes) are lipid carriers that act as adjuvants but have minimal toxicity. They seem to load peptides and proteins into the cell cytoplasm, allowing class I-restricted T-cell responses to peptides (Figure 12.5). These carriers are being considered for use in human immunization.

Another important practical aspect of vaccination is the route by which a vaccine is introduced into the human body. Today, most vaccines are given by injection or scarification. These procedures are disliked because of the pain that they cause, and for most pathogens they do not mimic the normal route of infection, which is

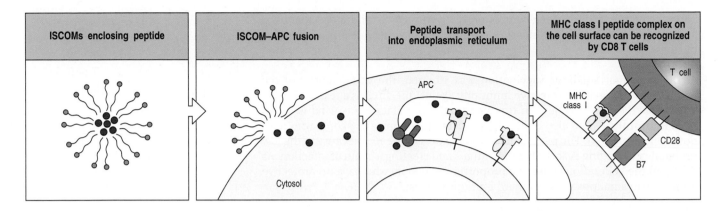

| ISCOMs enclosing peptide | ISCOM–APC fusion | Peptide transport into endoplasmic reticulum | MHC class I peptide complex on the cell surface can be recognized by CD8 T cells |

via mucosal surfaces. Vaccination by the oral or nasal routes would be less traumatic and potentially more effective in stimulating protective immunity. It would also be cheaper, because less skill and time is required of the person administering the vaccine. That oral vaccines can be effective is shown by the live-attenuated polio vaccine. Moreover, just as the disease-causing polio virus can be transmitted by the orofecal route, so can the vaccine, for example by fecal contamination of swimming pools.

12-4 Vaccination can inadvertently cause disease

Live-attenuated viruses have made the best vaccines because they challenge the immune system in ways most like the natural pathogen. Consequently, the immune system is best prepared for a real infection after vaccination with a live virus vaccine. However, because of their similarity to the pathogen, attenuated viruses can revert to becoming pathogenic. For example, the Sabin polio vaccine, which markedly reduced the incidence of polio in North America and Europe, induces polio and paralysis in three people per million vaccinated. The Sabin vaccine (trivalent oral polio vaccine, TVOP) consists of three different live-attenuated polio strains, all of which are needed to give protection against the natural variants of polio virus. Of these, strain 3 is the one responsible for causing disease after vaccination.

Strain 3 differs from a natural strain of polio by ten nucleotide substitutions. Unfortunately, mutation back to the original nucleotide at just one of the substituted positions in strain 3 is sufficient to cause reversion to pathogenicity, and this happens at a low frequency either during preparation of the vaccine (which is carefully monitored) or after vaccination. In contrast, strain 1 differs from natural strains by 57 or more nucleotide substitutions, and reversion to pathogenicity is much less likely because it requires more than one back-mutation. As the incidence of natural polio infection decreases in a population, fear of the side-effects of vaccination can become greater than fear of the disease itself. For this reason there is considerable social pressure to improve the polio vaccine and some parents refuse to have their children vaccinated. A modified protocol for vaccination helps to overcome the problem; the killed polio vaccine is first given to induce some immunity; this is then followed by the live vaccine.

12-5 The need for a vaccine and the demands placed on it change with the prevalence of the disease

In eighteenth-century Europe, the high probability of death or permanent scarring from smallpox made the risk of variolation acceptable to those who could afford it. Later, the rare side-effects of the cowpox vaccine were tolerated while smallpox still posed a threat. In the late twentieth century, smallpox has been eradicated

Figure 12.5 ISCOMs (immune stimulatory complexes) can be used to deliver peptides to the MHC class I processing pathway. ISCOMs are lipid micelles that will fuse with cell membranes. Peptides trapped in ISCOMs can be delivered to the cytosol of an antigen-presenting cell (APC), allowing the peptide to be transported into the endoplasmic reticulum, where it can be bound by newly synthesized MHC class I molecules and hence transported to the cell surface as peptide:MHC class I complexes. This is a possible method of delivering vaccine peptides to activate CD8 cytotoxic T cells. ISCOMs can also be used to deliver proteins to the cytosol of other types of cell, where they can be processed and presented as though they were a protein produced by the cell.

and vaccination has now been completely discontinued. Smallpox vaccination is an example of a preventive measure that was so successful it put itself out of business.

At the time that smallpox vaccination was stopped, the vast majority of the world's population had protective immunity against the virus as a result of previous vaccination or infection. In this situation, a population has what is called **herd immunity**, which also indirectly protects the minority of people who have not been vaccinated. The pathogen cannot create a epidemic, because of the low probability of finding susceptible individuals and creating a chain of infection. As time and the generations pass, the proportion of people who have no protective immunity to smallpox rises and herd immunity is lost.

Concern for the safety of a vaccine can sometimes lead to a resurgence of disease, as seen for whooping cough in the 1970s. At the beginning of the twentieth century 1 in 20 children in the USA died from whooping cough. The DTP vaccine containing whole killed *B. pertussis* bacteria was introduced in the 1940s and was routinely given to infants at three months of age. The vaccination program produced a hundredfold decline in the annual incidence of whooping cough, from 2000 cases per million to 20 cases per million.

But as people's fear of the disease abated, their concern for the side-effects of the vaccine increased. All children vaccinated with pertussis vaccine develop inflammation at the injection site, and some develop a fever that induces persistent crying. Very rarely, the vaccinated child suffers fits and either a short-lived sleepiness or a transient floppy unresponsive state, all of which cause parental anxiety. In the 1970s, awareness of these established neurological side-effects was heightened by anecdotal reports of vaccination causing encephalitis and permanent brain damage, a connection that has never been conclusively proved. Distrust of pertussis vaccination grew, most notably in Japan.

In Japan, DTP vaccination was introduced in 1947. By 1974 the incidence of pertussis had been reduced by more than 99% and in that year no deaths were ascribed to the disease. In the following year, two children died soon after vaccination, raising fears that the vaccine was the direct cause of the deaths. During the next five years, the number of Japanese children being vaccinated fell from 85% to 15% and, as a consequence, the incidence of whooping cough increased by about twentyfold, as did the number of deaths from the disease. Japanese companies then developed vaccines containing antigenic components of pertussis instead of whole bacteria; acellular pertussis vaccines replaced the whole-cell vaccines in 1981 in Japan. By 1989 the incidence of pertussis was again at the very low levels of 1974. Acellular vaccines are now being increasingly used in other countries. These vaccines have a reduced incidence of the common side-effects such as inflammation, pain, and fever.

12-6 Vaccines have yet to be found for many chronic pathogens

The diseases for which we already have vaccines are ones in which the infection is acute and resolves in a matter of weeks, either by elimination of the pathogen or by the death of the patient. In the absence of vaccination, many of those infected would clear the infection, showing that the human immune system can defeat the invader. Even at its worst, for example, smallpox killed only a third of those infected. For such diseases, a vaccine that mimics the pathogen and provokes an immune response similar to that raised against the pathogen itself is likely to provide protective immunity. The diseases for which vaccines are available are listed in Figure 12.6.

Available vaccines for infectious diseases in humans	
Bacterial diseases	**Types of vaccine**
Diphtheria	Toxoid
Tetanus	Toxoid
Pertussis (*Bordetella pertussis*)	Killed bacteria Subunit vaccine composed of pertussis toxoid and other bacterial antigens
Paratyphoid fever (*Salmonella paratyphi*)	Killed bacteria
Typhus fever (*Rickettsia prowazekii*)	Killed bacteria
Cholera (*Vibrio cholerae*)	Killed bacteria or cell extract
Plague (*Yersinia pestis*)	Killed bacteria or cell extract
Tuberculosis	Attenuated strain of bovine *Mycobacterium tuberculosis* (BCG)
Typhoid fever (*Salmonella typhi*)	Vi polysaccharide subunit vaccines. Live-attenuated oral vaccine
Meningitis (*Neisseria meningitidis*)	Purified capsular polysaccharide
Bacterial pneumonia (*Streptococcus pneumoniae*)	Purified capsular polysaccharide
Meningitis (*Haemophilus influenzae*)	*H. influenzae* polysaccharide conjugated to protein
Viral diseases	
Yellow fever	Attenuated virus
Measles	Attenuated virus
Mumps	Attenuated virus
Rubella	Attenuated virus
Polio	Attenuated virus (Sabin) or killed virus (Salk)
Varicella (chickenpox)	Attenuated virus
Rotaviral diarrhea	Attenuated virus
Influenza	Inactivated virus
Rabies	Inactivated virus (human). Attenuated virus (dogs and other animals). Recombinant live vaccinia-rabies (animals)
Hepatitis A	Subunit vaccine (recombinant hepatitis antigen)
Hepatitis B	Subunit vaccine (recombinant hepatitis antigen)

Figure 12.6 Diseases for which vaccines are available. Note that not all of these vaccines are equally effective, and not all are in routine use.

Some diseases for which effective vaccines are not yet available		
Disease	Annual mortality	Annual incidence
Malaria	856,000	213,743,000
Schistosomiasis	8000	No numbers available
Worm infestation	22,000	No numbers available
Tuberculosis	1,960,000	6,346,000
Diarrhea*	2,946,000	4,073,920 000
Respiratory disease	4,299,000	362,424,000
AIDS	138,000	411,000
Measles†	1,458,000	44,334,000
Hepatitis C	No numbers available	~170,000,000 ‡

Figure 12.7 Diseases for which better vaccines are needed. *A vaccine has recently been developed against the rotavirus, a common cause of childhood diarrhea. †The measles vaccines currently used are effective but they are heat sensitive and require reconstitution before use, and carefully controlled refrigeration. In some tropical countries this reduces their usefulness. ‡This figure is the number of people worldwide estimated by the World Health Organization to be chronically infected with the hepatitis C virus. Annual mortality is difficult to estimate because hepatitis C is associated with chronic conditions such as cirrhosis of the liver and liver cancer, from which death eventually results. Data courtesy of C.J.L. Murray and A.D. Lopez.

In contrast, most of the diseases for which vaccines have been difficult to find are due to chronic infections (Figure 12.7). Such pathogens are adept at evading and subverting the immune system, and so can live for years in a human host (see Chapter 9). Despite considerable research and investment, the approaches to vaccine design that have worked so well for acute bacterial and viral infections have failed to find vaccines for the chronic infections that plague humanity, especially infections with parasites. In most chronic infections there is little evidence that the immune system can clear the infection unaided, and as a group, pathogens that cause chronic infections divert the immune system into making responses that do not clear the infection. Consequently, successful vaccines against these diseases must stimulate different immune responses from those resulting from most natural exposures to the pathogen.

For many chronic infections, a minority of people exposed to the pathogen become resistant to long-term infection. A study of their resistance should reveal examples of successful immune responses that might be emulated by vaccination. For example, of people infected with the hepatitis C virus, less than 30% clear the infection quickly, whereas the majority develop a chronic infection in which the liver goes through cycles of destruction and regeneration (Figure 12.8). Vaccine design for hepatitis C virus could be helped by a comparison of the unsuccessful immunity in patients with chronic hepatitis with the successful immunity in people who clear hepatitis C infection. This, however, is more easily said than done. Studying people who make unsuccessful responses is relatively easy, because they become sick and seek medical help. In contrast, people who successfully repel the virus are healthy, do not show up in a doctor's office and are consequently difficult to identify. As a result, knowledge of the immune response against pathogens causing chronic infections encompasses mainly the failures of the immune system and not its successes.

Figure 12.8 A minority of people exposed to hepatitis C virus resist the virus, whereas the majority develop chronic infection.

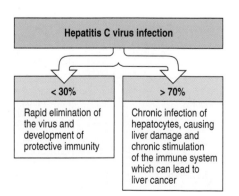

12-7 Increasing knowledge and new technologies enable new approaches to vaccine design

In the 1980s, the first viral genomes were completely sequenced; most pathogenic human viruses have now been characterized in this way. In the 1990s, the genomic sequences for pathogenic bacteria are being similarly catalogued. The complete knowledge of the genetic blueprint of a pathogen provides a foundation for defining the biology of its pathogenesis and its interactions with the human immune system. In turn, this knowledge permits the direct genetic manipulation of a pathogen's genes to produce attenuated strains with the properties required in a vaccine (Figure 12.9). For example, such approaches should enable strain 3 of the Sabin polio vaccine to be mutated in ways that reduce its reversion frequency.

Cloned genes from pathogens can be safely expressed in bacterial or other cell cultures to produce subunit vaccines, thus avoiding the handling of large amounts of infectious agents or material purified from infected blood, as was used for the first hepatitis B vaccine. The ease with which genes can be moved from one organism to another opens up the possibility of using current bacterial and viral vaccine strains as vehicles for antigens that could stimulate protective immunity against other pathogens. Natural pathogens have acquired many mechanisms for evading and escaping the immune system; the removal of these functions could also generate improved strains for vaccination. In a complementary fashion, microbial genes encoding adjuvants could be incorporated into vaccine strains of other organisms.

The ideal starting point for vaccine design would be a comprehensive knowledge of how the human immune system responds to the particular infection and which mechanisms enable the pathogen to be quickly cleared from the body. With this understanding, candidate vaccines could be designed to stimulate the clones of B and T cells that make up a successful immune response. Until recently, most vaccines have seemed to work by the induction of protective antibodies and have been evaluated purely in those terms. An appreciation of the importance of T-cell responses in protective immunity has stimulated the study of vaccines containing peptide epitopes that bind MHC molecules and thus can stimulate T cells specific for the pathogen. Although this approach has been shown to work, its application is complicated by the diversity of human HLA types and the peptides that they are able to present.

Another approach to vaccination is to use DNA encoding the antigen of a pathogen as the vaccine. The principle has been demonstrated in mice but has yet to be applied to humans. The mechanism by which injected DNA generates an immune response has yet to be worked out.

One factor critical to the success of vaccines against chronic infections will be the prevention of unsuccessful immune responses. In animal models of infectious disease, the choice between CD4 T_H1 or T_H2 responses can determine whether an infection resolves promptly or degenerates into chronic disease (see Section 8-16, p. 226). The inclusion of cytokines in vaccines might help to drive immunity in the desired direction. For example, the addition of IL-12 to a vaccine against the

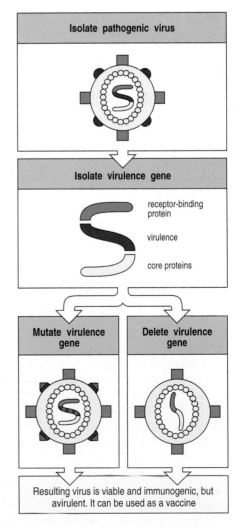

Figure 12.9 Production of live-attenuated viral strains by recombinant DNA techniques. If a viral gene that is necessary for virulence but not for growth or immunogenicity can be identified, this gene can be either mutated (left lower panel) or deleted (right lower panel) by using recombinant DNA techniques. This procedure creates an avirulent (non-pathogenic) virus that can be used as a vaccine. The mutations made in the virulence gene are usually many, so that the virus cannot easily revert to the wild type.

Isolate pathogenic virus

Isolate virulence gene

receptor-binding protein

virulence

core proteins

Mutate virulence gene

Delete virulence gene

Resulting virus is viable and immunogenic, but avirulent. It can be used as a vaccine protozoan parasite *Leishmania* changes the induced immune response in mice from one that is ineffective and dominated by CD4 T_H2 cells to a protective response dominated by CD4 T_H1 cells. A complementary strategy that has also worked in mice is to use antibodies to inactivate cytokines that could produce the unwanted immune response.

Summary

Vaccination involves the deliberate immunization of healthy people with some form of a pathogen or its component antigens. It induces a protective immunity that prevents subsequent infection with the same pathogen from causing disease. Vaccines can consist of killed whole pathogens, live-attenuated strains, non-pathogenic species related to the pathogen, or secreted and surface macromolecules of the pathogen. Vaccination has saved millions of lives and reduced the incidence of many common infectious diseases, particularly in the industrialized countries. By reducing the incidence of disease, successful vaccination programs inevitably lead to decreasing public awareness of the effects of disease and increasing concern with the safety of vaccination. Vaccine development has largely been a process of trial and error, one in which knowledge of immunological mechanisms played little part and the guiding principle was for the vaccine to resemble the natural pathogen as closely as possible. Although this approach has worked for pathogens causing acute infections, it has failed to produce vaccines against pathogens that establish chronic infections and cause chronic disease. These pathogens are adept at fooling the immune system; most people infected make immune responses that fail to clear the infection. Successful vaccines against these pathogens might need to push the immune system to respond in ways that are different from those seen in most natural infections with the pathogen.

Transplantation of tissues and organs

The replacement of diseased, damaged or worn-out tissue was for a long time a dream of the medical profession. Achieving the reality required the solution of three basic problems. First, transplants must be introduced in ways that allow them to perform their normal functions. Second, the health of both the recipient and the transplant must be maintained during the surgery and other procedures used in transplantation. Third, the immune system of the patient must be prevented from developing adaptive immune responses to antigens on the grafted tissue—responses that can result in rejection of the transplant and other complications.

During the past 50 years, solutions to these problems have been found, and organ transplantation has progressed from being an experimental procedure to the treatment of choice for a variety of conditions. In clinical practice, selective suppression of the response to the transplanted tissue has yet to be achieved and so non-specific suppression is accomplished by using a variety of drugs and antibodies. Thus, in contrast to vaccination, which selectively stimulates immunity to a particular pathogen, transplantation involves manipulations which cause widespread inactivation of the immune response.

12-8 Transplant rejection and graft-versus-host reaction are immune responses caused by genetic differences between transplant donors and recipients

Immune responses against transplanted tissues or organs are caused by genetic differences between donor and recipient, most importantly differences in the highly polymorphic HLA antigens. Antigens such as these, which vary between members of the same species, are known as alloantigens; the immune responses

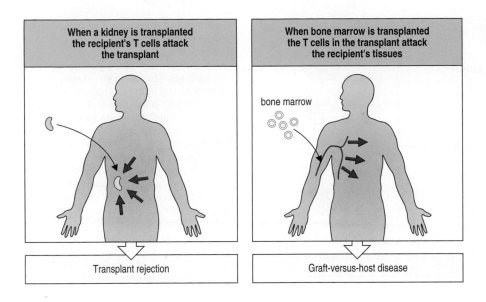

When a kidney is transplanted the recipient's T cells attack the transplant	When bone marrow is transplanted the T cells in the transplant attack the recipient's tissues

Transplant rejection

Graft-versus-host disease

Figure 12.10 Alloreactions in transplant rejection and graft-versus-host reaction. As shown in the left panel, rejection of a transplanted organ occurs when the recipient makes an immune response against it. Graft-versus-host disease occurs when cells in transplanted bone marrow make an immune response against the cells of the host, as shown in the right panel.

that they provoke are known as alloreactions (see Section 3-17, p. 76). A subfield of immunology, called **immunogenetics**, is devoted to the genetics of alloantigens. After transplantation of a solid organ, alloreactions developed by the recipient's immune system are directed at the cells of the graft and can kill them, a process called **transplant rejection**. In contrast, in bone marrow transplantation the recipient's immune system has been destroyed, and the major type of alloreaction arises from mature T cells in the grafted bone marrow that attack the recipient's tissues. This type of alloreactive response by donor lymphocytes is called a **graft-versus-host (GVH) reaction**. It causes **graft-versus-host disease** (**GVHD**), which, with varying degrees of severity, affects almost all patients who have a bone marrow transplant (Figure 12.10).

Because people do not normally make an immune response to their own tissues, tissues transplanted from one site to another on the same person are not rejected. This type of transplant, called an **autograft**, is used to treat patients who have suffered burns. Skin from unaffected parts of the body is grafted into the burnt areas, where it facilitates wound healing. Immunogenetic differences are also avoided when tissue is transplanted between identical twins. The first successful kidney transplant, in 1954, involved the donation of a kidney by a healthy twin to his brother, who was suffering from kidney failure. A transplant between genetically identical individuals is called a **syngeneic** transplant or an **isograft**. A transplant made between two genetically different individuals is called an **allograft** or an **allogeneic** transplant.

12-9 In blood transfusion, donors and recipients are matched for the A,B,O system of blood group antigens

The barriers to transplanting blood are fewer than for other tissues and, in 1812, blood transfusion was the first transplant to save a life. Today, blood transfusion remains the most common clinical transplantation procedure. For example, it is used when trauma or surgery causes blood loss and patients need the immediate replacement of fluid, serum proteins, red cells, and platelets. Today, donated blood is commonly separated into its component cells and plasma, and transfusions are made of whatever components a patient needs. Transfused blood components are usually only needed in the short term, because within a few weeks the patient's bone marrow makes up the loss. This demand is less stringent than that placed on transplanted organs such as hearts or kidneys, which need to function for years.

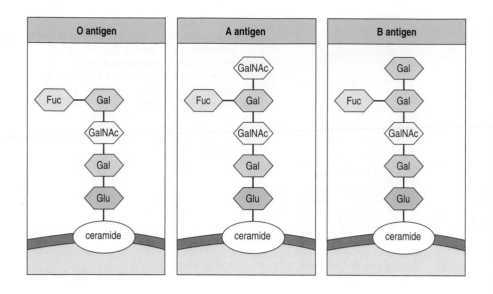

Figure 12.11 Structures of the A,B,O blood group antigens. The A,B,O antigens arise from a family of glycolipids present at the surface of erythrocytes. Their core structure consists of the lipid ceramide, to which is attached an oligosaccharide consisting of glucose (Glu), galactose (Gal), *N*-acetyl galactosamine (GalNAc), galactose, and fucose (Fuc). In people of blood group O this is the only glycolipid made. People of blood group A have an enzyme that can add an additional *N*-acetyl galactosamine to the oligosaccharide, forming the A antigen. People of blood group B have an enzyme that can add an additional galactose to the core structure, forming the B antigen. The erythrocytes of people of blood groups A and B also express the core structure alone, which is why alloantibodies against O are not made.

However, blood transfusion can give rise to life-threatening alloreactions. Human erythrocytes do not express HLA class I and II molecules, and so these highly polymorphic cell-surface molecules are of little consequence in blood transfusion. The major immunogenetic barrier to transfusion with red blood cells arises from structural polymorphisms in the carbohydrates on glycolipids of the erythrocyte surface. The resulting antigenic differences in these carbohydrates are the basis for the A,B,O system of blood group antigens (Figure 12.11). Because of structural similarities between blood group antigens and bacterial cell-surface carbohydrates, many people possess antibodies, made during the course of common infections, that react with blood group antigens different from those that they express themselves.

For example, people of blood group O invariably have antibodies against antigens A and B. If such a person is transfused with group A or B blood, the antibodies bind to the transfused erythrocytes, causing complement fixation and rapid clearance of the cells from the circulation. Besides thwarting the purpose of the blood transfusion, such hemolytic reactions can cause fever, chills, shock, renal failure, and even death. This hemolytic reaction in blood transfusion is analogous to the type II hypersensitivity reactions caused by antibodies made against drug-modified erythrocytes or thrombocytes described in Section 10-15, p. 289.

To prevent hemolysis and other transfusion reactions, patients needing transfusion are given only blood of compatible A,B,O type. Four blood types, O, A, B, and AB, account for the vast majority of people, which makes matching for this system of alloantigens much simpler than for HLA (Figure 12.12). Some twenty other polymorphic systems of blood group antigens have been defined, but apart from the Rhesus (Rh) blood group antigens, which are also matched, they are less important for matching in blood transfusion than the A,B,O system. To eliminate unexpected transfusion reactions, a recipient's serum is, however, tested directly for its reactivity with the red cells selected for transfusion on the basis of A,B,O typing. This direct assessment of compatibility is called a **cross-match test**. Cross-match tests are not performed between a recipient's cells and serum from the blood to be transfused, because antibodies in transfused blood that bind to a recipient's erythrocytes have no detrimental effect. The quantity of antibody transfused is insufficient to produce the density of antibody on the red-cell surface required to trigger hemolysis.

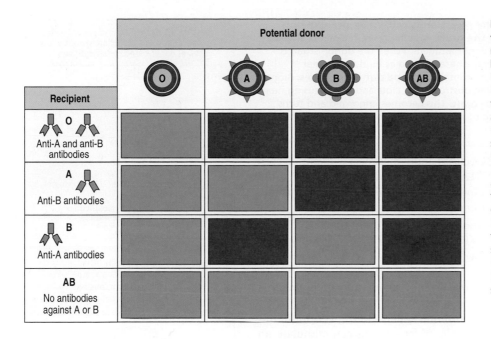

Figure 12.12 Donors and recipients for blood transfusion must be matched for the A,B,O system of blood group antigens. Common gut bacteria bear antigens that are similar or identical to blood group antigens, and these stimulate the formation of antibodies to these antigens in individuals who do not bear the corresponding antigen on their own red blood cells (left column); thus, type O individuals, who lack A and B, have both anti-A and anti-B antibodies, whereas type AB individuals have neither. The combinations of donor and recipient blood groups that allow blood transfusion are indicated by green squares; the combinations that would result in an immune reaction, and must be avoided, are indicated by the red squares.

12-10 Antibodies against A,B,O or HLA antigens cause hyperacute rejection of transplanted organs

A,B,O antigens are also expressed on the endothelial cells of blood vessels, an important factor in the transplantation of solid organs such as kidneys. For example, were a type O recipient to receive a kidney graft from a type A donor, then anti-A antibodies in the recipient's circulation would quickly bind to blood vessels throughout the graft. By fixing complement throughout the vasculature of the graft, the antibodies would produce a very rapid rejection of the graft (Figure 12.13). This type of rejection, called **hyperacute rejection**, can occur even before a transplanted patient has left the operating room. Hyperacute rejection is the most devastating form of rejection of organ grafts and is directly comparable to type III hypersensitivity reactions (see Section 10-16, p. 290) in which immune-complex deposition causes complement activation within blood vessel walls. To avoid hyperacute rejection, transplant donors and recipients are typed and cross-matched for the A,B,O blood group antigens.

As HLA class I molecules are expressed constitutively on vascular endothelium, pre-existing antibodies against HLA class I polymorphisms can also cause hyperacute rejection. It is therefore essential that transplant recipients do not have antibodies that bind to the HLA class I allotypes of the transplanted organ. To a smaller extent, antibodies against HLA class II can also contribute to hyperacute rejection. HLA class II molecules are not normally expressed on endothelium but can be induced by infection, inflammation, or trauma, all of which can occur during transplantation.

No reliable method of reversing hyperacute rejection has been found, so this type of rejection is avoided by choosing compatible transplant donors and recipients. Compatibility is assessed with a cross-match test in which blood serum from the prospective recipient is assessed for the presence of antibodies that bind to the white blood cells of the prospective donor. The traditional method for cross-match reveals antibodies in the patient's serum that can trigger complement-mediated lysis of the donor's lymphocytes. The assay is usually performed on separated B cells and T cells so that reactivities due to antibodies against HLA class I and II

Figure 12.13 Hyperacute rejection is caused by pre-existing antibodies binding to the graft. In some cases, recipients already have antibodies against donor antigens. When the donor organ is grafted into such recipients, these antibodies bind to vascular endothelium, initiating the complement and clotting cascades. Blood vessels in the graft become obstructed by clots and leak, causing hemorrhage of blood into the graft. This becomes engorged and turns purple from the presence of deoxygenated blood. The graft dies.

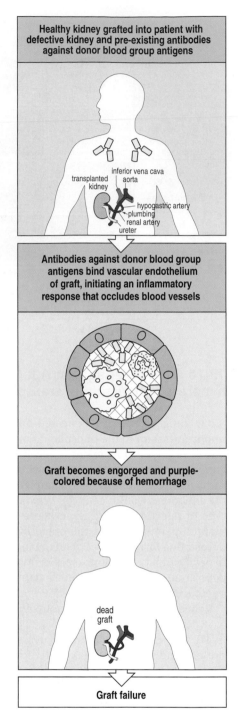

Healthy kidney grafted into patient with defective kidney and pre-existing antibodies against donor blood group antigens

transplanted kidney — inferior vena cava — aorta — hypogastric artery — plumbing — renal artery — ureter

Antibodies against donor blood group antigens bind vascular endothelium of graft, initiating an inflammatory response that occludes blood vessels

Graft becomes engorged and purple-colored because of hemorrhage

dead graft

Graft failure

molecules can be distinguished: anti-HLA class I antibodies react with both B cells and T cells, whereas antibodies against HLA class II react only with B cells. A more sensitive cross-match assay examines binding of the patient's antibodies to the donor's lymphocytes by using a flow cytometer (see Figure 5.4, p. 110).

12-11 Anti-HLA antibodies can arise from pregnancy, blood transfusion, or previous transplants

Circulating anti-HLA class I and II antibodies in prospective transplant patients can have arisen for several reasons. The most common is a previous pregnancy. The fetus is protected from the mother's immune system during gestation, but during the trauma associated with birth, cells of fetal origin can stimulate an immune response in the mother. Anti-HLA antibodies can be formed if the fetus expresses paternal HLA allotypes different from those of the mother. With successive pregnancies, increasing levels of anti-HLA antibodies can develop, and multiparous women are the main source of the anti-HLA sera used in HLA typing.

The fetus can be considered as an allograft within the mother's body (Figure 12.14) and any anti-HLA antibodies in a mother's circulation have the potential to bind to fetal cells and cause hyperacute rejection. That they never do so shows how well the fetus is protected from the maternal immune system.

Blood transfusions can also lead to the production of anti-HLA antibodies. In routine blood transfusion, no assessment of HLA type is performed and so although the donor and recipient are matched for A,B,O type they are not matched for HLA type. The infusion of HLA-incompatible leukocytes and platelets in a blood transfusion can therefore generate antibodies specific for the donor HLA allotypes. Patients who have had multiple blood transfusions have been stimulated by many HLA allotypes and can develop antibodies that react with the cells of most other people in the population. The degree to which a patient seeking a transplant has been sensitized to potential donors is assessed by testing their sera against a representative panel of individuals from the population and expressing the number of positive reactions as a percentage **panel reactive antibody** (**PRA**). The higher the value of a patient's PRA, the more difficult it is to find a suitable transplant donor.

A third way in which patients develop anti-HLA antibodies is from a previous transplant. Now that transplantation has been in routine practice for more than 30 years, many patients have had more than one transplant. As with blood transfusions, the more transplants that a person has had, the greater their percentage PRA tends to be.

12-12 Acute rejection is caused by effector T cells responding to HLA differences between donor and recipient

The alloreactive responses that cause the rejection of a transplant have been systematically studied with inbred strains of mice. In these studies skin grafting was the preferred model for transplantation, in part because rejection was easily visible.

Figure 12.14 The fetus is an allograft that is protected from rejection. In human families the mother and father almost always have different HLA types. When the mother becomes pregnant she carries for nine months a fetus that expresses one HLA haplotype of maternal origin (pink) and one HLA haplotype of paternal origin (blue). Although the paternal HLA class I and II molecules expressed by the fetus are alloantigens against which the mother's immune system has the potential to respond, the fetus does not provoke such a response during pregnancy and is protected from pre-existing alloreactive antibodies or T cells. The trauma associated with childbirth can lead to fetal stimulation of the mother's immune system.

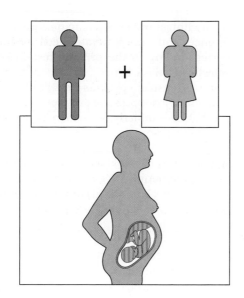

Skin grafts made between animals of different MHC type are rejected 11–15 days after transplantation. (Such 'naive' animals have no preformed anti-MHC antibody against the other strain that could cause hyperacute rejection.) The kinetics of rejection follow those of a primary immune response; the effector cells are CD4 T_H1 cells and CD8 T cells that recognize the graft's MHC class II and class I molecules, respectively. When mice that have already rejected one graft are given a second graft from the same strain, the graft is rejected in 6–8 days, a more rapid response that corresponds to a secondary immune response (Figure 12.15). In the secondary response, antibodies made against the MHC class I and class II molecules also contribute to rejection. The secondary response can be shown to be antigen specific; when mice that have rejected a skin graft from one strain are given a graft from another strain of different MHC type, the graft is rejected with the kinetics of a primary response.

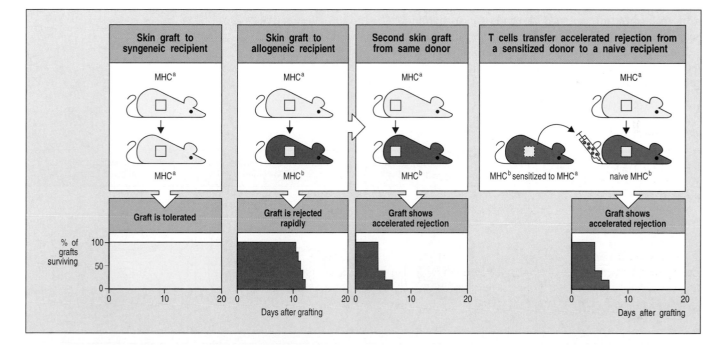

Figure 12.15 The rejection of skin grafts by mice is mediated by effector T cells. The three vertical panels on the left show that syngeneic skin grafts are accepted and allogeneic skin grafts are rejected. A first allogeneic skin graft stimulates a primary rejection response, whereas a second allogeneic skin graft of the same type stimulates a faster secondary rejection response. The panels on the right depict the type of experiment used to demonstrate that skin-graft rejection is mediated by T cells. When T cells from a mouse that has rejected an allogeneic skin graft are transferred to a naive animal, they confer the ability to reject the same type of allogeneic skin graft with the kinetics of a secondary response.

Most clinical transplants are performed across some HLA class I and/or II difference, and thus generate T-cell mediated alloreactions that cause rejection corresponding to the primary response described above. This is called **acute rejection**. Unlike hyperacute rejection, it takes days to develop and can be prevented. Transplant patients are carefully monitored for early signs of acute rejection and immunosuppressive drugs or anti-T-cell antibodies are used to prevent it. The mechanisms underlying acute rejection are just like those causing type IV hypersensitivity reactions (see Section 10-18, p. 294).

Alloreactions of the kind that cause acute transplant rejection can be studied by mixing the peripheral blood lymphocytes of two HLA-incompatible persons in a tissue culture. This mixed lymphocyte reaction causes a burst of T-cell proliferation followed by the differentiation of effector T cells. The reaction can be simplified by having T cells from one person respond to antigen-presenting cells of the second person, that have been treated so they cannot proliferate (see Figure 5.15, p. 123). Because different HLA allotypes bind different sets of self-peptides, each allotype selects a different repertoire of T-cell receptors. Consequently, the T-cell repertoire in a person selected by one HLA type contains numerous T-cell clones that can respond to the HLA:self-peptide complexes presented by cells of a different HLA type. For this reason, alloreactive T-cell responses stimulated by HLA differences are much stronger than T-cell responses to other antigens or to pathogens. Two pathways by which HLA molecules stimulate T-cell responses have been distinguished. In the direct pathway, the recipient's T cells recognize ligands composed of donor HLA molecules binding donor self-peptides. In the indirect pathway, peptides derived from donor HLA molecules are processed by the recipient's antigen-presenting cells and presented by the recipient's HLA molecules. The latter is a special case of the normal conventional pathways of antigen processing and presentation.

The surgery performed during organ transplantion causes both tissue damage and inflammation in and around the transplant. The direct alloreactive response is started by donor dendritic cells in the transplant, which carry donor self-antigens and travel to draining lymph nodes or spleen, and stimulate circulating T and B cells. After proliferation and differentiation, the effector T cells migrate to the grafted organ, where they attack the allogeneic cells (Figure 12.16).

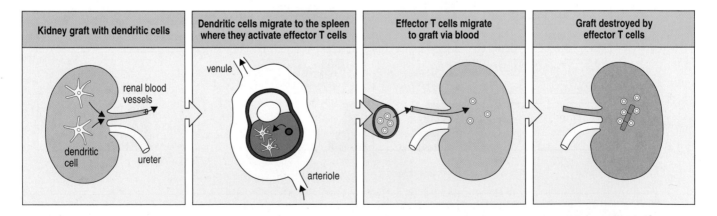

Figure 12.16 The direct pathway for stimulating acute rejection of a kidney graft. Donor antigen-presenting cells in the graft (in this case a kidney) are carried to a secondary lymphoid organ (the spleen is illustrated here). Here, they encounter recipient T lymphocytes whose receptors have specificity for the allogeneic MHC class I or class II molecule in combination with a donor peptide. These alloreactive T cells are activated by the donor antigen-presenting cells. After activation, the effector T cells travel back to the grafted organ, where they attack cells bearing that specific HLA:peptide complex.

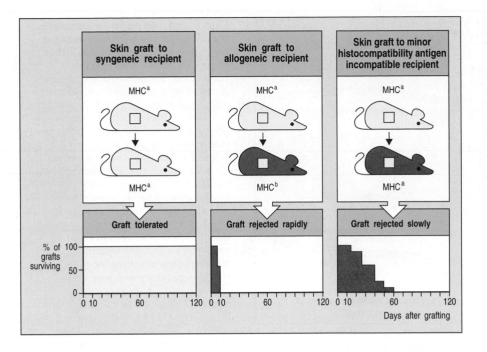

12-13 When donor and recipient are MHC identical, alloreactive T cells can still respond to minor histocompatibility antigens

Skin grafts made between mouse strains that are identical at the MHC but differ elsewhere in the genome are usually rejected. However, the rejection is much slower than for grafts mismatched at the MHC, typically taking around 60 days (Figure 12.17). This type of rejection is mediated by effector T cells responding to peptides derived from polymorphic non-MHC proteins for which the donor and recipient strains have different allotypes. Peptides derived uniquely from the allotype carried by the graft are presented to recipient T cells in secondary lymphoid tissues. The polymorphic peptides are called **minor histocompatibility antigens**, and the genes encoding them **minor histocompatibility loci**. The human genome encodes many polymorphic non-HLA proteins that can furnish minor histocompatibility antigens. Most provide peptides that are presented by donor HLA class I allotypes to CD8 T cells (Figure 12.18).

Studies of alloreactions after bone marrow transplantation between HLA-identical siblings have defined a number of human minor histocompatibility antigens. When boys receive a bone marrow transplant from their HLA-identical sisters, a cytotoxic T-cell response is sometimes made by the graft to an antigen encoded on the Y chromosome called H-Y (H for histocompatibility). The protein that provides the peptides recognized by H-Y-specific T cells is the human homolog of a mouse protein called SMCY. (The acronym stands for *s*elected *m*ouse *c*DNA on *Y* chromosome, which describes how it was found, not what it does.) Intracellular degradation of human SMCY protein produces peptides that can be presented by various HLA class I allotypes, so that T-cell responses to H-Y can be made by females of various HLA class I types.

In addition to hyperacute and acute rejection, transplanted human organs can be rejected by a third mechanism called **chronic rejection**. This phenomenon, which can occur months or years after transplantation, is characterized by reactions in the vasculature of the graft that cause thickening of the vessel walls and a narrowing of their lumina. Gradually the blood supply to the graft becomes inadequate, a state called **ischemia**, the function of the graft is lost, and it eventually dies. The mechanisms of chronic rejection are poorly understood.

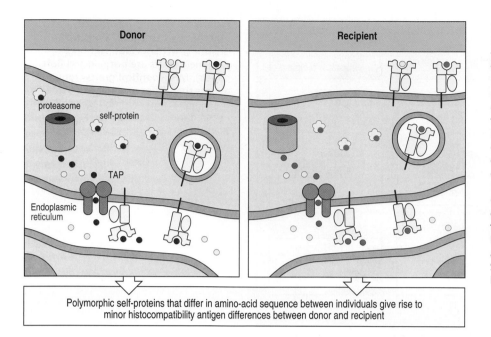

Polymorphic self-proteins that differ in amino-acid sequence between individuals give rise to minor histocompatibility antigen differences between donor and recipient

Figure 12.18 Minor histocompatibility antigens are peptides derived from polymorphic proteins other than HLA class I and class II molecules. Self-proteins are routinely digested by proteasomes within the cell's cytosol, and peptides derived from them are delivered to the endoplasmic reticulum, where they can bind to MHC class I molecules and be delivered to the cell surface. If a polymorphic protein differs between the graft donor (shown in red on the left) and the recipient (shown in blue on the right), it can give rise to an antigenic peptide (red on the donor cell) that can be recognized by the recipient's T cells as non-self and elicit an immune response. Such antigens are the minor histocompatibility antigens.

12-14 Matching donor and recipient for HLA class I and class II allotypes improves the outcome of transplantation

Studies in mice indicated that matching for HLA class I and II should reduce alloreactions after clinical transplantation in humans and improve both the function of engrafted organs and the long-term health of transplant recipients. That the first successful kidney transplant in 1954 was achieved between identical twins was consistent with this indication. Since then, the continued refinement of two complementary approaches has made kidney transplantation available to a range of patients with kidney disease. The first approach was to develop methods for determining HLA type and assessing the **histocompatibility**, meaning tissue compatibility, between prospective donors and recipients. The second approach was to find immunosuppressive drugs that could be used to prevent the alloreactive responses arising from transplantation across immunogenetic differences.

Clinical transplantation was pioneered with the kidney for two principal reasons. First, patients whose kidneys had failed could be sustained by the well-established procedure of dialysis. This meant that graft failure or rejection did not inevitably lead to the patient's death. Second was the simple fact that everyone has two kidneys but can manage with one, so healthy relatives could donate a kidney to a needy patient. Immunogenetic differences within a family are much smaller than within the population at large, and so the probability of finding an HLA-matched person within the family is higher. Analysis of the fate of kidney transplants performed between HLA-identical and non-identical family members was instrumental in showing that the better the HLA class I and II match, the better the clinical outcome. Of the various HLA loci, matching for HLA-A, -B and -DR is the most important. Clinical HLA typing is currently performed by a combination of serological and DNA methods.

After the success of transplantation between living relatives, methods were developed to transplant kidneys from unrelated donors who had been killed in accidents. Over 100,000 kidney transplants have been performed worldwide and a statistical analysis of these data demonstrates that both graft performance and long-term health of the recipient increase with the degree of HLA match (Figure 12.19).

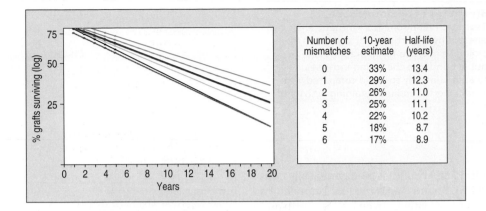

Number of mismatches	10-year estimate	Half-life (years)
0	33%	13.4
1	29%	12.3
2	26%	11.0
3	25%	11.1
4	22%	10.2
5	18%	8.7
6	17%	8.9

Figure 12.19 HLA matching improves the survival of transplanted kidneys. The colored lines in the left panel represent the actual (to 5 years) and projected survival rates of kidney grafts in patients with no (blue), 1 (orange), 2 (red), 3 (dark blue), 4 (green), 5 (black), and 6 (brown) HLA mismatches, plotted on a semi-log scale. Data courtesy of G. Opelz, T. Wujciak and B. Döhler.

12-15 Allogeneic transplantation is made possible by the use of immunosuppressive drugs

The limited supply of donated organs and the diversity of HLA type in the population mean that a majority of patients receive organs that are mismatched at one or more HLA loci. Drugs are used to suppress the alloreactions that could lead to transplant rejection. The **immunosuppressive** drugs used in clinical transplantation are of three kinds. The **corticosteroids** are steroids with anti-inflammatory properties that are very similar to natural glucocorticosteroid hormones produced by the adrenal cortex. The second group consists of cytotoxic drugs that, by interfering with DNA replication, kill proliferating lymphocytes activated by graft alloantigens. The third category of immunosuppressive drugs are microbial products that inhibit the signaling pathways of T-cell activation.

Any drug powerful enough to inhibit alloreactions also inhibits the normal immune responses to infecting pathogens. Consequently, administration of these drugs, which is greatest during the period immediately before and after transplantation, renders transplant patients highly susceptible to infection. Patients are initially cared for in conditions under which their exposure to pathogens is reduced. As their immune systems accommodate to the graft, the dose of immunosuppressive drugs is gradually reduced to 'maintenance levels' that prevent rejection while sustaining active defenses against infection. Aggressive early treatment has reduced the incidence of acute rejection in clinical transplantation. However, with the reduction of immunosuppression and the restoration of a patient's immunocompetence, the likelihood of chronic rejection increases.

All immunosuppressive drugs are also toxic to other tissues in varying degrees. Because these 'side-effects' vary for the different drugs, immunosuppressive drugs are generally used in combination so that their immunosuppressive effects are additive whereas their toxic effects are not. Certain side-effects emerge only after patients have taken immunosuppressive drugs for long periods of time. These include a higher incidence of certain types of malignant disease, particularly carcinomas of the skin and the genital tract, lymphoma, and Kaposi's sarcoma. The incidence of cancer in transplant recipients is on average three times that of similarly aged people who have not received a transplant.

12-16 Corticosteroids change patterns of gene expression

Hydrocortisone, also called cortisol, is the principal steroid made by the adrenal cortex; for more than 50 years it has been used clinically to reduce inflammation. The steroid commonly given to transplant patients is **prednisone**, a synthetic derivative of hydrocortisone that is about four times more potent in reducing

Figure 12.20 Chemical structures of hydrocortisone, prednisone, and prednisolone. Prednisone is a synthetic analog of the natural adrenocorticosteroid hydrocortisone, or cortisol. It is converted *in vivo* to the biologically active form, prednisolone. Introduction of the 1,2 double bond into the A ring increases anti-inflammatory potency approximately fourfold compared with hydrocortisone, without modifying the sodium-retaining activity of the compound.

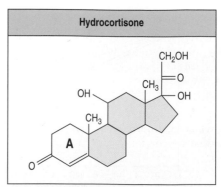

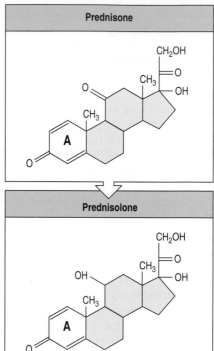

inflammation. Prednisone has no biological activity until it is enzymatically converted *in vivo* to **prednisolone** (Figure 12.20). Prednisone is thus an example of a **pro-drug**, a name given to drugs that are given to patients in an inactive form and which become chemically or enzymatically converted to the active form within the body. By itself, prednisone is insufficiently immunosuppressive to prevent graft rejection, but it works well in combination with a cytotoxic drug.

Corticosteroids have wide-ranging physiological effects and affect all white blood cells, not just lymphocytes, as well as other cells of the body. Unlike many other biologically active molecules, steroid hormones do not act at cell-surface receptors but diffuse across the plasma membrane and bind to specific receptors in the cytoplasm. Before steroid binding, the receptors are associated with another cytoplasmic polypeptide called Hsp90 (*heat-shock protein* of 90 kDa molecular weight). Steroid binding induces a conformational change in the receptor, which then dissociates from Hsp90 and enters the nucleus. There, the complex of receptor and steroid binds selectively to certain genes, activating their transcription (Figure 12.21). The transcription of about 1% of cellular genes can be influenced by corticosteroids.

In the context of their anti-inflammatory action, an important effect of corticosteroids is inhibition of the function of NFκB, a transcription factor important for cellular activation and cytokine production in the immune response. In quiescent cells, NFκB is held in the cytoplasm through its association with a protein called IκBα. On cellular activation, IκBα becomes phosphorylated, allowing NFκB to dissociate and enter the nucleus, where it initiates cytokine gene transcription.

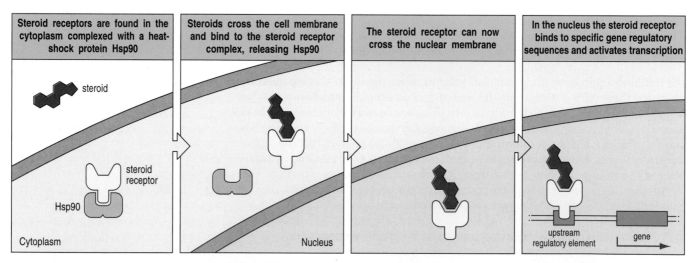

Figure 12.21 Steroids act at intracellular receptors.
Corticosteroids are lipid-soluble compounds that diffuse across the plasma membrane and bind to their receptors in the cytosol. The binding of corticosteroid to the receptor displaces a dimer of a heat-shock protein named Hsp90, exposing the DNA-binding region of the receptor, which then enters the nucleus and binds to specific DNA sequences in the promoter regions of steroid-responsive genes. Corticosteroids exert their effects by modulating the transcription of a wide variety of genes.

Corticosteroids increase the production of IκBα, thereby preventing NFκB from gaining access to the nucleus. By this mechanism they suppress the production of cytokines, such as IL-1 by monocytes, which stimulate inflammation and immune responses (Figure 12.22).

Owing to their multifarious effects on gene expression and cellular metabolism, corticosteroid drugs have many adverse side-effects, including fluid retention, weight gain, diabetes, loss of bone mineral, and thinning of the skin. Because of their mechanism of action, corticosteroids are most effective as immunosuppressive drugs when they are first administered before transplantation. With this approach, the patterns of cytokine gene expression are already changed in the recipient's cells at the time of alloantigenic challenge. They are often used as an acute immunosuppressive agent during episodes of rejection that are frequently caused by infection, but their continued use is to be avoided wherever possible.

12-17 Cytotoxic drugs kill proliferating cells

A cytotoxic drug commonly used in solid organ transplantation is **azathioprine**, a pro-drug that is first converted *in vivo* to 6-mercaptopurine and then to 6-thioinosinic acid (Figure 12.23). The latter compound inhibits the production of inosinic acid, an intermediate in the biosynthesis of adenylic and guanylic acids, which are essential components of DNA. The principal effect of azathioprine is therefore to inhibit DNA replication. Azathioprine has no effect on cells until they

Corticosteroid therapy	
Activity	**Effect**
↓ IL-1, TNF-α, GM-CSF ↓ IL-3, IL-4, IL-5, IL-8	↓ Inflammation ↓ caused by cytokines
↓ NOS	↓ NO
↓ Phospholipase A₂ ↓ Cyclo-oxygenase type 2 ↑ Lipocortin-1	↓ Prostaglandins ↓ Leukotrienes
↓ Adhesion molecules	Reduced emigration of leukocytes from vessels
Induction of endonucleases	Induction of apoptosis in lymphocytes and eosinophils

Figure 12.22 Effects of corticosteroids on the immune system. Corticosteroids regulate the expression of many genes, with a net anti-inflammatory effect. First, they reduce the production of inflammatory mediators, including some cytokines, prostaglandins, and nitric oxide (NO). As well as the cytokines listed here, corticosteroids also indirectly cause a decrease in IL-2 synthesis by activated lymphocytes, by their effects on other cytokines. Second, they inhibit inflammatory cell migration to sites of inflammation by inhibiting the expression of adhesion molecules. Third, corticosteroids promote the death by apoptosis of leukocytes and lymphocytes. NOS, nitric oxide synthase.

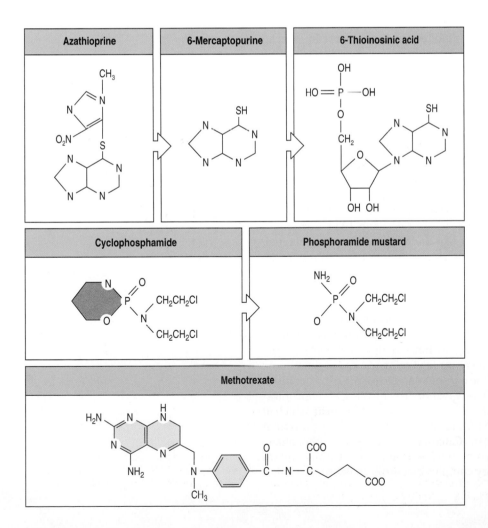

Figure 12.23 The chemical structures and metabolism of cytotoxic drugs. Azathioprine was developed as a modification of the anti-cancer drug 6-mercaptopurine; by blocking the reactive thiol group, the metabolism of this drug is slowed down. It is slowly converted *in vivo* to 6-mercaptopurine, which is then metabolized to 6-thioinosinic acid, which blocks the pathway of purine biosynthesis. Cyclophosphamide was similarly developed as a stable pro-drug, which is activated enzymatically in the body to phosphoramide mustard, a powerful and unstable DNA alkylating agent. Methotrexate blocks DNA synthesis by interfering with thymidine synthesis.

attempt to replicate their DNA, whereupon they die. While helpfully inhibiting the proliferation of alloantigen-activated lymphocytes, azathioprine, like other cytotoxic drugs, damages all the tissues of the body that are normally active in cell division. Principally affected are the bone marrow, the intestinal epithelium, and hair follicles, which can lead to anemia, leukopenia, thrombocytopenia, intestinal damage, and loss of hair. When pregnant women have to take cytotoxic drugs, fetal development can be adversely affected. Because azathioprine cannot act until a patient's immune system has been stimulated by alloantigen, it need only be administered after transplantation.

Cyclophosphamide is one of the nitrogen mustard compounds that were developed as chemical weapons and saw heavy use during World War I. It is a pro-drug that is converted in the body to phosphoramide mustard; this alkylates and crosslinks DNA molecules (see Figure 12.23). These covalent modifications render cells incapable of normal division and also affect transcription. Consequently, cyclophosphamide is equally immunosuppressive when given before or after antigenic stimulation.

Cyclophosphamide has many toxic effects that limit its clinical application. In addition to side-effects shared with other cytotoxic drugs, cyclophosphamide specifically damages the bladder, sometimes causing cancer or a condition called hemorrhagic cystitis. Unlike azathioprine, cyclophosphamide is not particularly toxic to the liver, and for patients who have sustained liver damage or become otherwise sensitized to azathioprine, it is a useful alternative. Cyclophosphamide is most effective when used in short courses of treatment.

Methotrexate was one of the first cytotoxic drugs shown to be effective in treating cancer cells. It prevents DNA replication by inhibiting dihydrofolate reductase, an enzyme essential for the cellular synthesis of thymidine. Methotrexate is the drug of choice for inhibiting the graft-versus-host reactions in bone marrow transplant recipients (see Figure 12.23).

12-18 Cyclosporin A, tacrolimus, and rapamycin selectively inhibit T-cell activation

During the 1960s and 1970s, transplant physicians depended on combinations of corticosteroids and cytotoxic drugs to prevent the rejection of transplanted organs. Towards the end of the 1970s, new kinds of immunosuppressive drug were introduced that selectively inhibited T-cell activation. These drugs have had a marked impact on clinical transplantation in the 1980s and 1990s, leading to improved graft survival, a wider range of tissues and organs being transplanted, and an increased number of diseases for which transplantation is recommended. This period has become known as the cyclosporin era, because cyclosporin A was the first of these drugs to be introduced.

Cyclosporin A (also known as **cyclosporine**) is a cyclic decapeptide derived from a soil fungus, *Tolypocladium inflatum*, originally isolated in Norway. It inhibits the antigen activation of T cells by disrupting the transduction of signals from the T-cell receptor. Signals from the T-cell receptor lead to hydrolysis of membrane lipids to give inositol triphosphate and the consequent release of Ca^{2+} from intracellular stores (see Section 6-5, p. 138). The elevation in cytosolic Ca^{2+} concentration activates a cytoplasmic serine/threonine phosphatase, **calcineurin**, which in turn activates the transcription factor NFAT. In resting T cells, NFAT is present in the cytoplasm in a phosphorylated form. Calcineurin removes the phosphate, permitting NFAT to enter the nucleus, where it binds to the AP-1 transcription factor to form a transcriptional regulatory complex that turns on the transcription of the IL-2 gene.

Cyclosporin interferes with calcineurin activity. It diffuses across the plasma membrane into the cytosol, where it binds peptidyl-prolyl isomerase enzymes, which in this context are called **cyclophilins**. The complex of cyclosporin A and cyclophilin binds to calcineurin, inhibiting its phosphatase activity and preventing it from activating NFAT. In the presence of cyclosporin therefore, IL-2 cannot be made and the program of T-cell activation, proliferation, and differentiation is shut down at a very early stage (Figure 12.24).

Tacrolimus, also called **FK506**, was isolated from the soil actinomycete *Streptomyces tsukabaensis*. It is a macrolide, a class of compound with structures based on a many-membered lactone ring that is attached to one or more deoxy sugars.

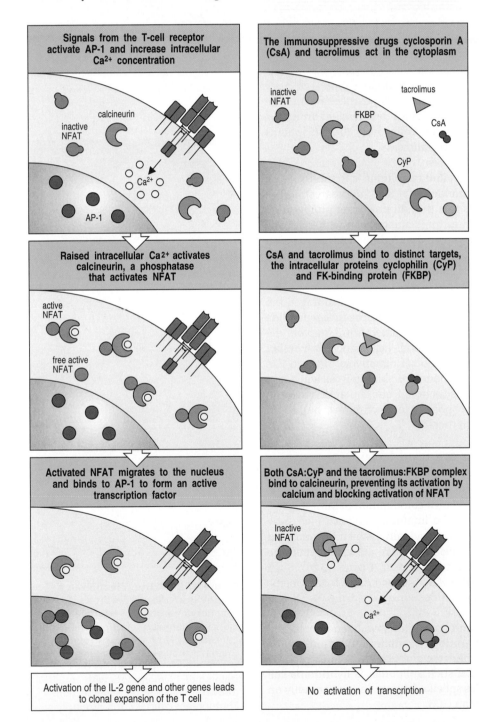

Figure 12.24 Cyclosporin A and tacrolimus inhibit T-cell activation by interfering with the serine/ threonine phosphatase calcineurin. Signaling via T-cell receptor-associated tyrosine kinases leads to the activation and increased synthesis of the transcription factor AP-1, as well as increasing the concentration of Ca^{2+} in the cytoplasm (left panels). The Ca^{2+} binds to calcineurin and thereby activates it to dephosphorylate the cytoplasmic form of the nuclear factor of activated T cells (NFAT). Once dephosphorylated, the active NFAT migrates to the nucleus to form a complex with AP-1; the NFAT:AP-1 complex can then induce the transcription of genes required for T-cell activation, including the IL-2 gene. When cyclosporin A (CsA) or tacrolimus are present, they form complexes with their immunophilin targets, cyclophilin (CyP) and FK-binding protein (FKBP), respectively (right panels). The complex of cyclophilin with cyclosporin A can bind to calcineurin, blocking its ability to activate NFAT. The complex of tacrolimus with FKBP binds to calcineurin at the same site, also blocking its activity.

Cell type	Effects
T lymphocyte	Reduced expression of IL-2, IL-3, IL-4, GM-CSF, TNF-α Reduced cell division because of decreased IL-2 Reduced Ca^{2+}-dependent exocytosis of cytotoxic granules Inhibition of antigen-driven apoptosis
B lymphocyte	Inhibition of cell division because T-cell cytokines are absent Inhibition of antigen-driven cell division Induction of apoptosis following B-cell activation
Granulocyte	Reduced Ca^{2+}-dependent exocytosis of granules

Figure 12.25 Immunological effects of cyclosporin A and tacrolimus.

Although structurally distinct from cyclosporin A, tacrolimus suppresses T-cell activation by the inhibition of calcineurin through a similar mechanism. The peptidyl-prolyl isomerases to which tacrolimus binds are distinct from the cyclophilins and are known as FK-binding proteins. Collectively, the cyclophilins and **FK-binding proteins** are known as **immunophilins**.

Although the principal effect of cyclosporin A and tacrolimus is to inhibit T-cell activation, the activation of B cells and granulocytes is also suppressed (Figure 12.25). A major advantage of these drugs is that they do not target proliferating cells, so the reduced hematopoiesis and intestinal damage seen with cytotoxic drugs does not occur. A side-effect associated with the continued administration of cyclosporin A or tacrolimus is nephrotoxicity, and some patients can no longer tolerate the drug—they are said to have become sensitized to it.

The success of cyclosporin A and tacrolimus in transplantation encouraged the search for similarly selective drugs. One find is an immunosuppressive macrolide called **rapamycin** (also called **sirolimus**), which was isolated from *Streptomyces hygroscopicus*, a soil bacterium found on Easter Island. The island's Polynesian name, 'Rapa ui', was used to name the drug. Although rapamycin binds to FK-binding proteins, it does not interfere with calcineurin but blocks T-cell activation at a later stage by preventing signal transduction from the IL-2 receptor. Rapamycin is more toxic than either cyclosporin A or tacrolimus, but could become a useful component of combination therapy.

12-19 Antibodies specific for T-cells are used to control acute rejection

After transplantation, patients are maintained on a combination of immunosuppressive drugs. Because of their toxicity and the immunodeficiency that these drugs cause, physicians seek to lower the dosage gradually to the minimum that will maintain tolerance of the transplant. Inevitably, there are times when the balance is upset and early symptoms of rejection appear (Figure 12.26). Such episodes can be treated with a 5–15-day course of daily injections of T-cell specific antibodies, as well as an increased dose of immunosuppressive drugs.

Anti-T-cell antibodies are made in sheep and goats that have been immunized with human thymocytes or lymphocytes. Antibody-containing fractions called **antithymocyte globulin** (ATG) or **antilymphocyte globulin** (ALG) are prepared from the animals' blood. A second source of such antibodies is hybridoma cell lines making mouse monoclonal antibodies specific for proteins present only on the human T-cell surface, for example human CD3.

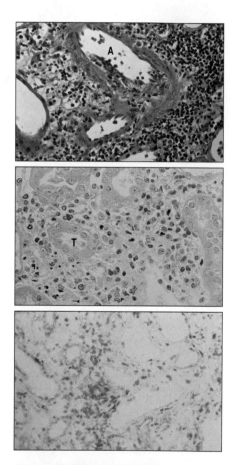

Figure 12.26 Acute rejection in a kidney graft. Top panel shows lymphocytes around an arteriole (A) in a kidney undergoing rejection. The middle panel shows lymphocytes surrounding the renal tubules (T) of the same kidney, and the bottom panel shows the staining of T lymphocytes with anti-CD3 (brown staining) in the same section. Photographs courtesy of F. Rosen.

Immunosuppressive antibodies work in one of two ways. ALG and ATG cause the destruction of the lymphocytes to which they bind, through complement fixation and phagocytosis. In contrast, monoclonal anti-CD3 interferes with the function of the T cells to which it binds, causing reduced expression of the CD3:T-cell receptor complex on the cell surface. Because the immunosuppressive antibodies come from a non-human species, they tend to stimulate an antibody response, which reduces their immunosuppressive activity when used on subsequent occasions. In such situations the patient's antibodies form immune complexes with the immunosuppressive antibodies, clearing them from the circulation before they can bind to T cells. Such reactions can also lead to serum sickness (see Section 10-17, p. 292). For this reason, physicians generally use each immunosuppressive antibody to counter just one episode of acute rejection per patient.

12-20 An increasing number of diseases are being treated by transplantation

Kidney transplantation procedures have developed to a point at which it is now possible to transplant cadaveric kidneys across considerable HLA mismatches. This progress helped the development of heart transplantation, for which only cadaveric donors could be considered. Heart transplantation is inherently more difficult than kidney transplantation, principally because the failure of a grafted heart is fatal, whereas patients with failed kidney grafts can go back to dialysis. The use of cyclosporin A and tacrolimus has increased the success of heart transplantation, mainly by preventing death from acute rejection or infection during the first few months after transplantation. As a consequence, the number of heart transplants increased considerably after 1979. In the USA some 4000 patients each year receive a heart transplant (Figure 12.27) and more than half of them are projected to be alive 10 years after the operation.

Liver transplantation has similarly progressed in the cyclosporin era, from being a relatively risky procedure to one offering considerable benefit. In 1979 only 30–40% of patients survived a liver transplant for more than a year; today 70–90% of patients are surviving after 1 year and 60% are alive after 5 years. A similar improvement has been seen for lung transplantation.

Bone marrow transplantation, which permanently replaces an individual's entire hematopoietic system, including their immune system, is used as a treatment for an increasing number of diseases. In a bone marrow transplant, the important cells are the pluripotent stem cells that reconstitute the patient's immune system as well as their red cells and platelets. In 2–3 weeks after a successful transplant,

Tissue transplanted	Number of transplants		Patients waiting for a transplant (August 1999)
	Oct 1987 – Dec 1998	1998	
Kidney	110,290	11,990	42,875
Kidney–Pancreas	6906	965	1929
Pancreas	1209	253	502
Liver	32,482	4450	13,698
Heart	23,626	2340	4287
Heart–Lung	638	45	218
Lung	5948	849	3343

Figure 12.27 Organs commonly transplanted in medicine. The numbers of organs transplanted in the USA during the period 1987–1998 and in 1998 are shown. That the availability of organs and other factors are limiting transplantation is shown by the numbers of patients who could benefit from a transplant. Data courtesy of United Network for Organ Sharing.

new circulating blood cells begin to be produced from the transplanted marrow. This is a sign that the puripotent stem cells have colonized the bones, the process known as **engraftment**.

The diseases treatable by bone marrow transplantation fall into two distinct groups. The first consists of genetic diseases, in which one or more of the cell types produced by hematopoiesis is defective. These include red-cell deficiencies, such as thalassemia major, sickle-cell anemia, and Fanconi's anemia, as well as the severe combined immune deficiencies (SCID) (see Section 9-14, p. 257). For these diseases the transplant replaces a defective hematopoietic system with one that is normal.

The second group of diseases treatable by bone marrow transplantation consists of malignancies for which the radiation and chemotherapy necessary to eliminate the tumor cells also compromises bone marrow function. For these diseases, which include various types of leukemia and lymphoma, the transplant replaces a hematopoietic system that has been damaged in the course of treating the primary disease (Figure 12.28).

The outcome of bone marrow transplantation is more sensitive to HLA mis-matching than is solid organ transplantation. This is because the procedure involves the transplantation of lymphocytes. If these are of a different HLA type from that of the recipient, they can attack almost all tissues of the body. In addition, a functional immune system cannot be reconstituted in the transplant recipient unless the donor and recipient share some HLA class I and II allotypes (see Section 9-14, p. 257).

There are, however, large numbers of potential bone marrow donors, because bone marrow, like blood, can be donated by healthy individuals without compromising their immunological or hematological functions. Marrow is usually aspirated from the iliac crests of the pelvis under anesthesia. For patients without HLA-identical siblings who will donate bone marrow, there are national and international registries of HLA-typed people that are searched to identify histocompatible donors. Worldwide, some 5 million potential donors have been HLA typed for this purpose.

Some cancer patients can be treated with an **autologous bone marrow transplant**. Here, samples of the patient's own bone marrow are taken before the remainder is destroyed by the treatments given for the cancer. After separation of the bone marrow stem cells from any tumor cells, the stem cells are reinfused into the patient. Autologous transplants avoid the problems of histoincompatibility and graft-versus-host disease and thus patients do not require immunosuppression. However, their use is limited by the rate of relapse of malignant disease, which is greater than with allogeneic transplants. This difference has led to the hypothesis that some alloreactions help by eradicating residual tumor cells.

Autologous stem cells for transplantation can now be obtained by less invasive procedures than bone marrow aspiration. Pluripotent hematopoietic stem cells are first mobilized from the patient's bone marrow into the peripheral blood by treatment with granulocyte colony-stimulating factor (G-CSF) and granulo-cyte–macrophage colony-stimulating factor (GM-CSF), which are sometimes combined with cyclophosphamide chemotherapy. Leukocytes are then selectively removed from the blood by connecting the patient's circulation to a machine that performs apheresis, and a cell fraction containing the stem cells is isolated from the removed leukocytes. The CD34 cell-surface glycoprotein expressed by stem cells is used to determine their yield. Between a quarter and half a billion CD34-positive cells are needed to ensure prompt engraftment after transplantation.

Genetic diseases treatable by bone marrow transplantation
SCID Wiskott–Aldrich syndrome Fanconi's anemia Kostmann's syndrome Chronic granulomatous disease Osteopetrosis Ataxia telangiectasia Diamond–Blackfan syndrome Mucocutaneous candidiasis Chédiak–Higashi syndrome Cartilage-hair hypoplasia Mucopolysaccharidosis Gaucher's disease Thalassemia major Sickle cell anemia

Malignant diseases treatable by bone marrow transplantation	
Allogeneic/syngeneic transplant	Autologous transplant

Breast cancer
Aplastic anemia
Leukemia
  AML
  ALL
  CML
Myelodysplasia
Multiple myeloma
Non-Hodgkin's
  lymphoma
Hodgkin's disease
	Leukemia
	  AML
	  ALL
	Multiple myeloma
	Non-Hodgkin's
	  lymphoma
	Hodgkin's disease
	Solid tumors
	  Breast
	  Ovarian
	  Testicular
	  Neuroblastoma

Figure 12.28 Diseases for which bone marrow transplantation is a therapy. In North America, bone marrow transplants are most commonly used for treating breast cancer, but their use in this condition is controversial.

Another source of stem cells is umbilical cord blood, obtained from placentas after birth. Umbilical cord blood is rich in stem cells and it also contains fewer cells that contribute to the graft-versus-host reaction. Because stem cells for transplantation are now being obtained from sources other than bone marrow, the term **hematopoietic stem cell transplantation** is beginning to be used instead of bone marrow transplantation.

12-21 The nature of the alloreactions after transplantation depends on the type of tissue or organ transplanted

Each tissue has a unique anatomy and vasculature, and these properties affect both the strength of antigenic stimulation that it gives upon transplantation and the nature of the alloreactive response. An extreme case is the cornea of the eye, which is not vascularized and can be transplanted successfully without immuno-suppression. Preceding sections of this chapter have described the alloreactions stimulated by allogeneic kidney grafts. Some of the effects specific to the transplantation of other organs will be examined in this section.

The outcome of liver transplantation is not improved by matching the donor and recipient for HLA class I or II antigens. Indeed, it has been claimed that transplant outcome is inversely correlated with the degree of HLA match. As a consequence, HLA type or cross-match are not assessed before liver transplantation; A,B,O type is the only genetic factor affecting donor selection. Clinical experience indicates that the liver is relatively refractory to either acute or hyperacute rejection, yet the use of cyclosporin A and tacrolimus has markedly improved the success of liver transplants. The liver has a specialized architecture and vasculature, and hepato-cytes express very low levels of HLA class I and no HLA class II. These properties, and the daily exposure of liver cells to the digestion products of a myriad of foreign proteins in the intestines, could all contribute to the distinct immunobiology of the transplanted allogeneic liver.

The major cause of morbidity and mortality after bone marrow transplantation is acute graft-versus-host disease (GVHD). In solid organ transplantation the alloreactions are localized to a single organ, whereas those produced by allogeneic bone marrow transplantation can attack almost every tissue of the body. In essence, GVHD is an acute autoimmune disease that can prove fatal. The severity of GVHD varies, four grades being defined in clinical diagnosis (Figure 12.29). The principal tissues targeted are the skin, the intestines, and the liver. The charac-teristic skin rash of GVHD tends to develop with the kinetics of a primary

Tissue reactions in the four grades of graft-versus-host disease			
Grade	Skin	Liver	Gastrointestinal tract
1	Maculopapular rash on <25% of body surface	Serum bilirubin 2–3 mg dl^{-1}	>500 ml diarrhea day^{-1}
2	Maculopapular rash on <25–50% of body surface	Serum bilirubin 3–6 mg dl^{-1}	>1000 ml diarrhea day^{-1}
3	Generalized erythroderma	Serum bilirubin 6–15 mg dl^{-1}	>1500 ml diarrhea day^{-1}
4	Generalized erythroderma with bullous formation and desquamation	Serum bilirubin 15 mg dl^{-1}	Severe abdominal pain with or without ileus

Figure 12.29 Characteristics of the four grades of graft-versus-host disease.

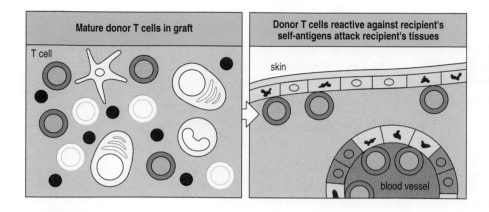

Figure 12.30 Graft-versus-host disease is due to donor T cells in the graft that attack the recipient's tissues. After bone marrow transplantation, any mature donor T cells present in the graft that are reactive for antigens on the recipient's tissues become activated and mount an attack on the recipient's cells.

immune response during the 10–28 days after transplantation. The fine, diffuse erythematous rash begins on the palms of the hands, on the soles of the feet, and on the head and then spreads to the trunk. The intestinal reaction causes cramps and diarrhea, and inflammation of the bile ducts in the liver causes hyperbilirubinemia and a rise in the levels of liver enzymes in the blood. Methotrexate in combination with cyclosporin A is used to reduce the incidence and severity of GVHD.

The chronic rejection reactions seen in kidney transplant patients have their analogs in the chronic GVHD experienced by 25–45% of bone marrow transplant patients who survive longer than 6 months after transplant. The course of this condition resembles that of an autoimmune disease and the cumulative effect is to produce severe immunodeficiency, leading to recurrent life-threatening infections.

Acute graft-versus-host disease is due to mature T cells in the transplanted bone marrow that react against the recipient's tissues (Figure 12.30). Bone marrow can be depleted of mature T cells by treatment with the appropriate monoclonal antibodies before being given to the patient. Although markedly reducing GVHD, this procedure leads to higher rates of graft rejection by residual immune function in some patients, and a relapse of the malignant disease for which the patient is being treated. In leukemia patients, these results and others suggest that the alloreactions of GVHD involve T cells that help to rid the body of residual leukemia cells. This phenomenon is called the **graft-versus-leukemia** (**GVL**) effect. The goals of current research are to define the cells responsible for the GVL effect and to determine whether the cell populations used for hematopoietic stem cell transplantation can be purged of cells that cause GVHD while preserving the cells that perform GVL reactions.

12-22 Xenotransplantation could enable more patients to receive an organ transplant than is now possible

The very success of solid organ transplantation has created its own problem, namely that there are many more patients who could benefit from a kidney, heart, or liver transplant than there are organs available from live and cadaveric donors (see Figure 12.27). Patients are therefore placed on waiting lists and chosen for transplantation on the basis of various criteria, including the severity of disease and the HLA match with available organs. To increase the organ supply, some countries have instituted a policy whereby cadaveric organs from accident victims become automatically available for clinical transplantation unless the person has deliberately opted out. Other countries retain the policy that organs can be used for transplantation only if an accident victim has deliberately opted in by previously signing a consent form. Even then, relatives can overrule the victim's wishes. The increasing demand for kidneys has inevitably led to an unregulated international trade in human kidneys, generally involving donors in poorer countries selling one of their kidneys for transplantation to rich patients.

A possible way of overcoming the limitations of organ supply would be to use organs from animals. This type of transplantion, in which donor and recipient are of different species, is called **xenotransplantation**, and the grafted tissue a **xenograft**. At present pigs are considered the most suitable donor species for humans: first because their organs are of a similar size to those of humans, and second because they are already farmed and consumed by humans in large numbers.

The first major immunological problem facing xenotransplantation is hyperacute rejection. This is because most humans happen to have circulating antibodies that are specific for carbohydrate antigens on pig endothelial cells. In this context the carbohydrate antigens of pig endothelium are called **xenoantigens**, and the human antibodies that react with them **xenoantibodies**. As with the alloantibodies that humans make against A,B,O blood group alloantigens, the xenoantibodies are probably induced by infections with common bacteria whose surface carbohydrates resemble those on pig cells. Adding to the severity of hyperacute rejection in xenotransplantation, the cell-surface proteins in pigs that prevent complement activation from damaging their own cells (e.g. the pig versions of CD59, DAF, and MCP; see Section 7-23, p. 194), do not inhibit human complement. The binding of human xenoantibodies to pig endothelium therefore triggers a reaction far more violent than the hyperacute rejection of an allogeneic transplant. Until these problems are overcome it is impossible to assess whether the acute rejection reactions that are also to be expected will be manageable.

One approach to improving the compatibility of pig organs with the human immune system is to make the pigs transgenic for certain human genes. As a first step, pigs have been made transgenic for human decay-accelerating factor (DAF). In experiments with monkeys as the recipients, the expression of human DAF has prolonged the survival of pig kidney grafts. This approach can be extended by adding other human transgenes and by removing the genes for pig xenoantigens, for example MHC classes I and II.

T-cell-mediated acute rejection of pig organs might be less severe than antibody-mediated hyperacute rejection. In this part of the xenoreactive response, species differences could prevent human T cells from recognizing antigens presented by pig antigen-presenting cells. Such impairment is seen in mice made transgenic for human HLA class I molecules, where CD8 T-cell responses are very poor because mouse CD8 cannot bind to human HLA class I molecules.

Some virologists have urged caution with regard to xenotransplantation, because it provides a potential route for endogenous pig viruses to infect humans. During the period immediately after transplantation, patients are immunodeficient and susceptible to all manner of infections. Of particular concern is the possibility of a pig retrovirus having effects like those of HIV after transfer to humans. As has occurred on many previous occasions, the development of procedures to save lives by tissue transplantation raises complicated ethical questions.

Summary

Many human diseases involve the malfunction of a single organ or tissue, which can cause incapacitation or death. With transplantation, diseased tissues are replaced by healthy ones, leading to improved health and longer life. Of the many problems faced by clinical transplantation, the most challenging has been control of the alloreactive immune response, which rejects transplanted organs or causes graft-versus-host disease (GVHD) in bone marrow transplantation. These alloreactions are due to genetic differences between transplant donor and recipient, which produce epitopes recognized by host or donor lymphocytes. Two complementary strategies are used to reduce the effects of alloreactions: the first is to match donor and recipient for the most critical genetic factors; the second is the administration of drugs that suppress the immune system.

A transplanted tissue can be rejected at various times after transplantation. Hyperacute rejection, which occurs immediately, is caused by preformed antibodies in the recipient that react with cells of the transplant, particularly the vascular endothelium. Common antibodies of this type are directed against A,B,O antigens or HLA class I molecules. Hyperacute rejection cannot be treated and it is therefore avoided by matching for A,B,O and performing a cross-match test. Acute rejection is caused by the recipient's mature CD4 and CD8 T cells, responding to HLA class I and II differences between transplant donor and recipient. Acute GVHD is similarly caused by mature immunocompetent T cells in the graft specific for antigens on the recipient's tissues. Acute rejection can be prevented by matching transplant donors and recipients for HLA class I and II allotypes and by treatment with immunosuppressive drugs and antibodies. After transplantation there is a gradual accommodation between the grafted tissue and the recipient that allows the doses of immunosuppressive drugs to be reduced. During this period, episodes of rejection can occur, which are treated with additional immunosuppressive drugs and anti-T-cell antibodies. Chronic rejection and chronic GVHD are caused by incomplete suppression therapy, allowing reactions to HLA mismatches or to minor histocompatibility antigens. With time, many patients can be maintained on a level of immunosuppression that allows the immune system to recover and provide defense against infection.

Cancer and its interactions with the immune system

Cancer is a diverse collection of life-threatening diseases that are caused by abnormal and invasive cell proliferation. It accounts for about 20% of deaths in the industrialized countries; worldwide there are some 6 million new cases of cancer each year and half of these people will die from the disease. Cancer is largely a disease of older people, and in the industrialized societies where both life expectancy and the average age of the population are increasing, so is fear of cancer.

Cancer cells are very similar to normal cells and the immune system seems unable to attack them effectively. In treating cancer, physicians resort to surgery, radiation, and cytotoxic drugs, sometimes referred to as stagies of slash, burn, and poison. Although these treatments give remission or cure to some patients, more often they are limited by the incomplete elimination of cancer cells and the deleterious side-effects of the treatment. For more than a century, cancer immunologists have sought to harness a patient's immune system to augment conventional therapies. Although ideas for boosting cancer immunity have been regularly demonstrated to work to some extent in animal models, useful routine immunotherapies for human cancer have yet to emerge. In this part of the chapter we first consider the processes by which cancers emerge in the body. Then we turn to what is known of the antigens that distinguish cancer cells and the types of immune responses that can be made. Lastly, we describe the relatively modest contribution that immunological therapies have made so far to the treatment of cancer.

12-23 Cancer results from mutations that cause uncontrolled cell growth

Maintaining the human body involves continuing cell division, to replace worn out cells, repair damaged tissue, and mount immune responses against invading pathogens. Of the order of 10^{16} cell divisions are estimated to occur in a human body during a lifetime. An essential preparation for cell division is DNA replication, which makes two identical copies of the diploid genome. Although the enzymes that replicate DNA are extremely accurate and use proof-reading mechanisms to correct mistakes, rare errors are still made. Additional changes in DNA arise from chemical damage that escapes the DNA repair machinery.

Changes in DNA are called **mutations**. They include the substitution, insertion, and deletion of nucleotides, recombination between different members of a gene family, and chromosomal rearrangements. Mutations in the germline—the eggs and sperm—provide the variation that allows the human species to diversify and evolve. In contrast, mutations in somatic cells affect only the individual in which they arise. Most somatic mutations are never noticed because their effects, if any, are manifest in a single cell. There are, however, certain mutations that abolish the normal controls on cell division and cell survival. When that happens, the mutant cell proliferates to form an expanding population of mutant cells that eventually disrupts the body's physiology and organ function, causing the diseases that are collectively called cancer.

Tumor, which means swelling, and **neoplasm**, which means new growth, are words used synonymously to describe tissue in which cells are multiplying abnormally. The branch of medicine that deals with tumors is called **oncology**, the prefix *onco* deriving from the Greek word for swelling, *ogkos*. Not all tumors are malignant: **benign tumors**, such as warts, are encapsulated, localized, and limited in size; the **malignant tumors**, in contrast, can continously increase their size by breaking through basal laminae and invading adjacent tissues (Figure 12.31).

The word **cancer**, meaning an evil spreading in the manner of a crab (Latin, *cancer*), is used to describe the diseases caused by malignant tumors. In addition to spreading out locally from their site of origin, cancer cells can be carried by lymph or blood to distant sites, where they initiate new foci of cancerous growth. This mode of spreading is called **metastasis**, the site of origin being called the primary tumor, and the sites of spreading are called secondary tumors. Cancers arise most

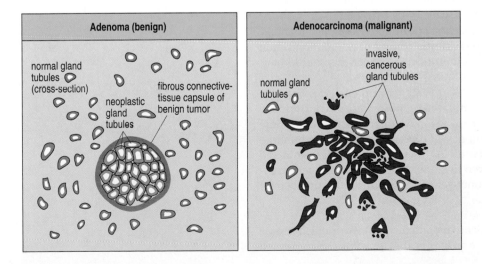

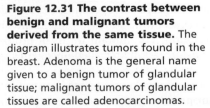

Figure 12.31 The contrast between benign and malignant tumors derived from the same tissue. The diagram illustrates tumors found in the breast. Adenoma is the general name given to a benign tumor of glandular tissue; malignant tumors of glandular tissues are called adenocarcinomas.

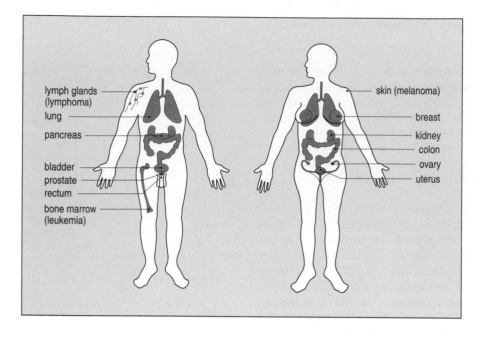

Figure 12.32 The tissues that give rise to the most common cancers in humans in the USA. For simplicity, tissues common to males and females are not all shown on both figures.

commonly in tissues that are actively undergoing cell division and are thus more likely to accumulate mutations due to errors in DNA replication. These tissues include the epithelial linings of the gastrointestinal tract, urogenital tract, and the mammary glands (Figure 12.32). Cancers of epithelial cells are known as **carcinomas**, cancers of other cell types as **sarcomas**. Cancers of immune system cells are known as **leukemias** when they involve circulating cells, **lymphomas** when they involve solid lymphoid tumors, and **myelomas** when they involve bone marrow (see Sections 4-10, p. 101 and 5-12, p. 125).

12-24 A cancer arises from a single cell that has accumulated multiple mutations

The best defenses against cancer lie within each cell of the human body and not with the specialized cells of the immune system. The integrity of the body is so dependent on well-controlled cell division that many mechanisms have evolved to ensure this. These include the repair of certain types of DNA damage as well as mechanisms that prevent the survival and division of cells with badly damaged DNA. Consequently, the control of cell division never depends on the function of just one protein, and a cell cannot become cancerous by mutation in just one gene. For a cell to give rise to a cancer, it must first accumulate multiple mutations, and these must occur in genes concerned with the control of cell multiplication and cell survival. When a cell becomes able to form a cancer it is said to have undergone **malignant transformation**.

There are two main classes of gene that if mutated or misexpressed, contribute to malignant transformation. **Proto-oncogenes** are genes that normally contribute positively to the initiation and execution of cell division. More than a hundred human proto-oncogenes have been identified; they encode growth factors and their receptors, and also proteins involved in signal transduction and gene transcription. The mutant forms of proto-oncogenes that contribute to malignant transformation are called **oncogenes**.

The second class of genes involved in cellular transformation are called **tumor suppressor genes** because they encode proteins that prevent the unwanted

Figure 12.33 A typical series of mutations acquired during the development of a cancer. This illustration shows the development of colorectal cancer by successive mutations in different genes. The morphological changes accompanying each change are indicated. *RAS* is an oncogene; *APC*, *DCC*, and *p53* are tumor suppressor genes. Both copies of a tumor suppressor gene must be mutated to contribute to malignant transformation, so the number of mutations that has occurred here totals seven. The sequence of events shown here usually takes 10–20 years or more. After a cell has undergone the initial series of mutations that render it malignant, the tumor cells then rapidly accumulate more mutations. Some of these contribute to making the cancerous cells more invasive, but others are thought to be simply a consequence of the cell's becoming cancerous rather than a cause. Once a tumor has become genetically heterogeneous, selection acts to favor those cells that divide more rapidly and are more invasive.

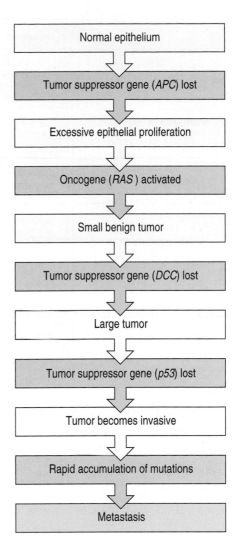

proliferation of mutant cells. A typical tumor suppressor gene is *p53*. This is expressed in response to DNA damage and encodes a protein that causes the damaged cell to die by apoptosis. Loss of the *p53* gene or mutations that interfere with its protective function are the most frequent mutations present in human cancers. Over 50% of cases of human cancer have a mutation in *p53*, revealing its importance in protecting the body from cancer. Indeed, *p53* might well have evolved as an internal cellular defense against cancer.

It has been estimated that a cell must accumulate at least five or six independent mutations before it can become cancerous. The actual number depends on the cell type and the particular genes in which the mutations occur (Figure 12.33). Because of the random manner in which mutations occur, the combination of genetic lesions in each cancer is unique.

The low frequency of mutation together with the requirement for multiple mutations means that each cancer inevitably arises from a single cell that has undergone malignant transformation. This explains why in paired organs such as the lungs, cancer initially affects only one of them. In the course of a lifetime, a person's cells accumulate mutations, and the probability that a cell somewhere in the body will have the right combination of mutations to cause a cancer increases in a non-linear fashion. For this reason, the incidence of cancer increases with age, and cancers are largely diseases of older people.

12-25 Exposure to chemicals, radiation, and viruses can facilitate the progression to cancer

Cancer is not, however, an inevitable outcome of aging. Most people never suffer from cancer, even in old age. This reveals the existence of genetic and environmental factors that influence the risk of cancer's development. Genetic factors include the possession of a germline mutation in one copy of a tumor suppressor gene. Such a mutation in the *p53* gene is the underlying cause of Li–Fraumeni syndrome, a condition in which there is an unusually strong predisposition to multiple cancers that arise at a relatively early age. This is because only one new mutation is needed, in the good copy of *p53*, to abolish all p53 function. In other people, at least two new mutations, in both copies of *p53*, would be needed.

Of greater importance to the population at large are environmental insults that increase the number of mutations suffered by the body. Chemical and physical agents that damage DNA in such a way as to cause an increased rate of mutation are called **mutagens**. Many known mutagens can increase the risk of cancer, and are known as **carcinogens**. People who have had heavy or prolonged exposure to

Viruses associated with human cancers		
Virus	**Associated tumors**	**Areas of high incidence**
DNA viruses		
Papillomavirus (many distinct strains)	Warts (benign) Carcinoma of uterine cervix	Worldwide Worldwide
Hepatitis B virus	Liver cancer (hepatocellular carcinoma)	Southeast Asia Tropical Africa
Epstein–Barr virus	Burkitt's lymphoma (cancer of B lymphocytes). Nasopharyngeal carcinoma B-cell lymphoproliferative disease	West Africa Papua New Guinea Southern China Greenland (Inuit) Immunosuppressed or immunodeficient patients
RNA viruses		
Human T-cell leukemia virus type I (HTLV-1). Human immunodeficiency virus (HIV-1)	Adult T-cell leukemia/lymphoma. Kaposi's sarcoma	Japan (Kyushi) West Indies Central Africa

Figure 12.34 Viruses associated with human cancers. For all the viruses listed here, the number of people infected is much larger than the number who develop cancer; the viruses must act in conjunction with other factors. Some of the viruses probably contribute to cancer only indirectly; for example, HIV-1, by obliterating cell-mediated immune defenses, might allow endothelial cells transformed by some other agent to thrive as a tumor instead of being destroyed by the immune system.

carcinogenic agents, which include certain chemicals, ultraviolet light, and other forms of radiation, are more at risk of developing cancer than those who have not been exposed. Most marked is the increased frequency of lung cancer among people who smoke.

Chemical carcinogens tend to cause mutations due to single nucleotide substitutions in DNA. Radiation, in contrast, tends to produce grosser forms of damage such as DNA breaks, cross-linked nucleotides, abnormal recombination, and chromosome translocations. The radiation released by the atomic bombs exploded at Hiroshima and Nagasaki in 1945 caused an increased incidence of leukemia in those who survived the acute effects of the blasts. Less marked, but more pervasive, is the increasing incidence of skin cancer caused by overexposure to ultraviolet radiation from the sun in the pursuit of work, fun and fashion. However, despite the fear of cancer in developed societies and the well-known risk factors, people in those same societies still purposefully engage in behavior that they know increases their chances of developing cancer in later years.

Certain viruses also have the potential to transform cells, and viruses are associated with some 15% of human cancers (Figure 12.34). Such viruses are called **oncogenic viruses**. The known oncogenic viruses that affect humans are DNA viruses, except for the RNA retrovirus HTLV-1, which is associated with adult T-cell leukemia.

Human oncogenic viruses typically set up chronic infections in a cell, producing novel virally encoded proteins that override or interfere with the cell's normal mechanisms for regulating cell division. Infected cells therefore start to proliferate. For example, the Epstein–Barr virus induces infected B cells to divide repeatedly (see Section 9-4, p. 246), and if the infected cells are not cleared rapidly, a proportion of them go on to become transformed. Certain strains of papilloma virus predispose to cancer of the cervix. They encode proteins that prevent the normal tumor suppressor mechanisms from acting within the infected cells. Viral proteins bind to p53 protein and to another tumor suppressor protein called Rb, blocking their functions and enabling the virally infected epithelial

cells to proliferate. Other chronic infections can lead to cancer because the tissue damage that they cause necessitates high rates of cell renewal and, as a consequence, mutations accumulate more rapidly. The liver cancers arising from hepatitis B and C virus infections might be of this type, as might the stomach cancers associated with ulcers caused by infection with the bacterium *Helicobacter pylori*.

12-26 Tumor cells are easily recognized and killed by allogeneic CD8 T cells

In the latter part of the nineteenth century, cell biologists began to investigate the unusual growth properties of tumors. To expand the scope of their studies they tried to propagate tumors by transplanting them from the mouse in which they arose to other mice. They found that in outbred colonies of mice the transplanted tumors would never grow, but when partly inbred mice were used there was some measure of success. These observations led eventually to the discovery of the major histocompatibility complex and to the development of the completely inbred strains of mice that are the subject of much immunological research today. In the early experiments in transplantation with outbred mice, the transplanted tumors were rejected by alloreactive CD8 T cells that recognized the allogeneic MHC class I molecules expressed by the tumor. In fact, the rules governing the rejection and acceptance of transplanted tumors are the same as those governing the transplantation of skin or any other tissue.

When a transplanted tumor expresses MHC class I or II allotypes not expressed by the recipient mouse, it is acutely rejected by a potent T-cell response (Figure 12.35). Because most tumors, like most normal cells, express MHC class I but not MHC class II molecules, the alloreactive response to transplanted tumors is usually directed towards MHC class I polymorphisms. The high polymorphism of HLA class genes in human populations is therefore a barrier that prevents any tumor cells that are passed from one person to another from growing. Situations in which such transfer could potentially occur are the same as those that serve to spread HIV: intimate contact, blood transfusion, sharing of syringes and needles in the non-medical use of drugs. Thus cancer cells themselves are not infectious, although oncogenic viruses can be.

12-27 Cancers studied in the laboratory can behave differently from those that cause human disease

If a tumor that arose independently in one mouse, known as a spontaneous tumor, is transplanted to MHC-identical mice of the same inbred strain, the rate of successful establishment of a tumor is low, and the resulting tumor grows only slowly. However, when some tumors are serially passaged among mice of the same strain, they do adapt to this mode of existence and successive passages become increasingly efficient. Because these transplantable mouse tumors are convenient to work with, much more is known of their interactions with the immune system in comparison with spontaneously arising tumors. However, the medical relevance of this knowledge is uncertain, because all human cancer is due to newly arising tumors that have grown in a single host. A further important difference is that humans live some 20 times longer than mice and have the potential to accumulate larger numbers and combinations of mutations.

12-28 Tumor-cell surfaces have antigens that can be recognized by cytotoxic CD8 T cells

The fact that allogeneic tumor cells can be killed by CD8 T cells showed that tumors are not inherently resistant to the effector mechanisms of cytotoxic T cells. That cancers are rarely terminated by the immune system is largely due to failures in immune recognition and/or activation. Cells that become malignantly

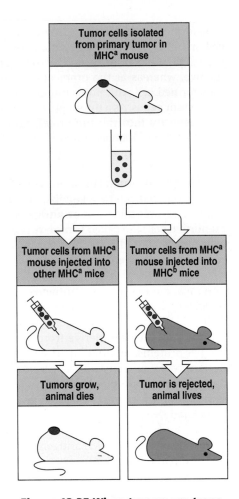

Figure 12.35 When tumors are transplanted between MHC-incompatible mice they are rejected by the alloreactive response to MHC differences. The left panels show the transplantation of a tumor between two mice of the same MHC type. The tumor grows in the recipient. In the right panels, the tumor is transplanted to a mouse of a different MHC type and is rejected.

transformed have genomic differences that distinguish them from every other cell in the body, but at the beginning of the transformation process these are only very slight. Cells transformed by oncogenic viruses are genetically the most distinctive, whereas at the other end of the spectrum are tumors arising from a transformed cell with point mutations in some half-dozen oncogenes or tumor suppressor genes. As a tumor grows, further mutations occur, introducing genetic heterogeneity into the tumor-cell population. Selection for variant cells with faster growth or increased metastatic potential causes the genomes of the tumor and host to diverge even further.

Some of the mutations in tumor cells produce antigenic changes on the tumor-cell surface that can be recognized by the immune system. The new antigens on tumor cells are called **tumor antigens**. Antigens present on tumor cells but not on normal cells are called **tumor-specific antigens**; antigens expressed on tumor cells but also found on certain normal cells, often in smaller amounts, are called **tumor-associated antigens**.

The most common tumor antigens seem to be peptides bound by HLA class I molecules and thus are presented to cytotoxic CD8 T cells. Within this category, tumor-specific peptide antigens have amino-acid sequences not found in any normal cell; they can derive from viral proteins, the mutated parts of mutant cellular proteins, or amino-acid sequences encoded by tumor-specific recombination between genes. In contrast, tumor-associated peptide antigens are derived from unmutated protein sequences encoded by the host genome. The same mutant protein could thus give rise to both tumor-specific and tumor-associated peptide antigens.

Tumor-associated peptide antigens come from a variety of sources. Some derive from proteins produced in greater quantities in the tumor than in normal cells, for example proteins involved in cell division. Others are parts of proteins that are

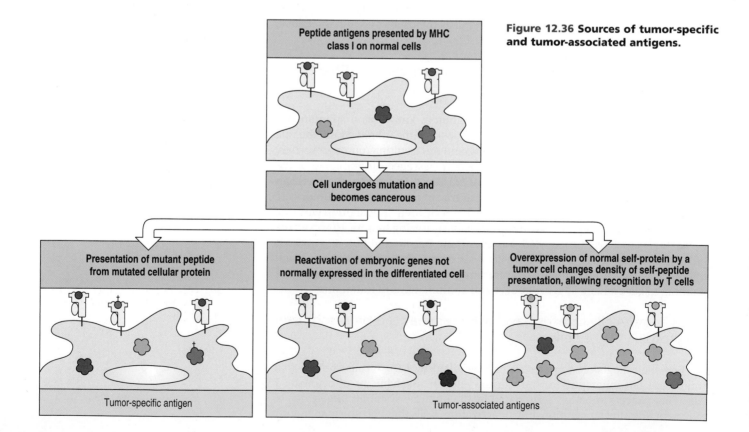

Figure 12.36 Sources of tumor-specific and tumor-associated antigens.

normally expressed on embryonic cells but have become re-expressed in the tumor cell (Figure 12.36). Another source is proteins constitutively expressed by tumor cells, but expressed transiently or by very small subsets of normal cells.

12-29 Tumor-cell surfaces have antigens that can be recognized by antibodies

Changes at the tumor-cell surface also produce tumor antigens recognized by B cells and antibodies. These antigens arise by mutation within a cell-surface protein, by alterations in glycosylation of glycoproteins and glycolipids, or by increased levels of expression of cell-surface components. One category of tumor-associated antigens comprises the cell-surface immunoglobulins and T-cell receptors of B- and T-cell lymphomas. In the healthy human body, each immunoglobulin and T-cell receptor is expressed by a minute subset of the lymphocyte population. In contrast, in a patient with lymphoma the vast population of tumor cells expresses an identical immunoglobulin or T-cell receptor. Under these circumstances, the antigen receptor is for practical purposes a tumor-specific antigen.

One of the first cell-surface molecules designated as a tumor antigen is a 200 kDa cell-surface glycoprotein. It was initially shown to be expressed by cells of human colon cancer and by fetal gut cells, but not by healthy adult colon cells. On this basis the glycoprotein was called **carcinoembryonic antigen** (**CEA**). Subsequently, when more sensitive methods of detection were used, CEA was found to be expressed by fetal liver and pancreas, and at low levels on healthy cells of the adult colon, liver, pancreas, lung, and lactating breast. In addition, the heightened CEA expression in adults was not restricted to malignant cells but was also seen at sites of chronic inflammation, such as occur in the colons of patients with inflammatory bowel disease, the lungs of cigarette smokers, and the pancreata of patients with pancreatitis. The expression of CEA is in fact correlated more closely with the proliferation of epithelial cells than with their malignant transformation. This story illustrates the problems that bedevil the search for tumor-specific antigens. Cancer cells differ from normal cells in only subtle ways, most of which involve changes in the expression of components that are also present on healthy cells.

12-30 Protective immune responses to tumors can be demonstrated in experimental situations

The experimental system for studying immune responses to tumor antigens that most closely parallels the human situation is when numerous tumors are induced independently within a strain of inbred mice and all subsequent manipulations are performed with mice of the same strain. Such studies have been made of sarcomas induced by exposing mice to the chemical methylcholanthrene. If mice are injected with sarcoma cells from one tumor (irradiated so that they cannot grow), subsequent attempts to establish live cells of the same tumor in the injected mice frequently fail. The mice have developed immunity against the tumor. However, the immunity developed against one methylcholanthrene-induced sarcoma provides no immunity against any other of the sarcomas induced in the same way (Figure 12.37).

These observations indicated that each sarcoma expresses tumor-specific antigens that can be recognized by cytotoxic CD8 T cells. That every sarcoma has distinct tumor-specific antigens is consistent with the stochastic manner by which cells accumulate mutations on the way to malignant transformation. Each

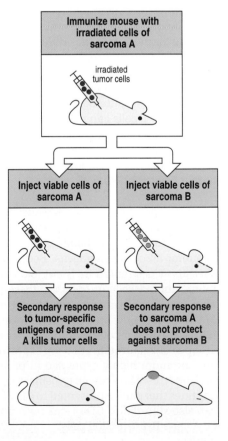

Figure 12.37 Immunity to one chemically induced sarcoma does not protect against other chemically induced sarcomas.

Class of antigen	Antigen	Nature of antigen	Tumor type
Embryonic	MAGE-1 MAGE-3	Normal testicular proteins	Melanoma Breast Glioma
Abnormal post-translational modification	MUC-1	Underglycosylated mucin	Breast Pancreas
Differentiation	Tyrosinase	Enzyme in pathway of melanin synthesis	Melanoma
	Surface immunoglobulin	Specific immunoglobulin after gene rearrangements in B-cell clone	Lymphoma
Mutated oncogene or tumor suppressor	Cyclin-dependent kinase 4	Cell-cycle regulator	Melanoma
	β-catenin	Relay in signal transduction pathway	Melanoma
	Caspase-8	Regulator of apoptosis	Squamous cell carcinoma
Oncoviral protein	HPV type 16, E6 and E7 proteins	Viral transforming gene products	Cervical carcinoma

Figure 12.38 Antigens recognized by tumor-specific cytotoxic CD8 T cells. The antigens listed here have all been found to be recognized by cytotoxic T cells raised from patients with the tumor type listed. MAGE, melanoma antigen encoding; MUC, mucin; HPV, human papilloma virus.

cell that becomes transformed does so because of a different set of mutations. The tumor-specific antigens recognized by cytotoxic CD8 T cells are peptides whose sequences include amino acids determined by the mutations. Like the mutations, these peptides differ between sarcomas and they therefore activate different clones of T cells.

Cytotoxic CD8 T-cell responses to tumor antigens can be demonstrated in many cancer patients, although they have failed to eliminate the cancer or prevent its spread. The T cells can be isolated from tumor tissue, the draining lymph node, or peripheral blood and can then be cultured and analyzed in the laboratory. By using T-cell killing of target cells as an assay, individual peptides that function as tumor antigens can be identified, as can the cellular proteins from which they derive (Figure 12.38).

Two of the peptide antigens to which melanoma patients respond come from proteins that are normally expressed only in the testicles (see Figure 12.38). These proteins, called MAGE-1 and MAGE-3 (*m*elanoma *a*ntigen *e*ncoding), are examples of tumor-associated antigens. Because testicles are immunologically privileged sites, the T-cell repertoire is not tolerant to MAGE-1 or MAGE-3. When expressed by melanoma cells in the skin, these proteins can therefore stimulate a T-cell response. Other cancers, including glioma and breast cancer, also express the MAGE genes. How and why certain cancers express these 'testicle-specific' proteins is not understood. Another source of melanoma-associated peptide antigens are specialized proteins that are expressed only by cells of the melanocyte lineage and are more highly expressed in proliferating melanoma cells than in normal melanocytes. The enzyme tyrosinase is one such protein. In addition to providing peptides that can be presented to CD8 T cells by HLA class I molecules, tyrosinase can also stimulate a CD4 T-cell response after being processed and presented by HLA class II-expressing cells.

12-31 Tumors can escape from the immune response

As a tumor grows, cells within the tumor acquire different mutations and so the tumor becomes genetically heterogeneous. This variation can be acted on by selection. As well as selection for faster growth and metastasis, there can also be selection by the immune response. A CD8 T-cell response will kill cells presenting tumor antigens, thereby selecting for variant tumor cells that do not present the antigen. Between one-third and one-half of human tumors have defects in expression of one or more of their HLA class I allotypes (Figure 12.39), which indicates that they have been subject to such selection.

Loss of HLA class I expression could make a tumor susceptible to attack by NK cells. NK cells do not attack normal healthy cells because of an inhibitory interaction between their cell-surface receptors and polymorphic epitopes of HLA class I allotypes (see Section 8-12, p. 222). Conversely, NK cells are able to recognize and kill tumor cells with certain defects in HLA class I expression (Figure 12.40). For example, cells that have completely lost the expression of HLA class I,

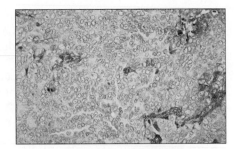

Figure 12.39 Loss of HLA class I expression in a cancer of the prostate gland. The section is of a human prostate cancer that has been stained with a monoclonal antibody specific for HLA class I molecules. The antibody is conjugated to horseradish peroxidase, which produces a brown stain wherever the antibody binds. The stain and HLA class I molecules are not seen on the tumor mass, but are restricted to lymphocytes infiltrating the tumor and tissue stromal cells. Photograph courtesy of G. Stamp.

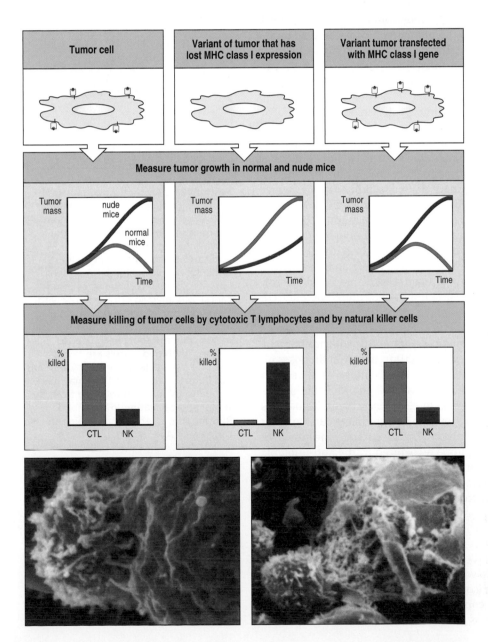

Figure 12.40 Cytotoxic T cells and NK cells can have complementary functions in killing tumor cells. In this experiment the killing of three types of tumor cell by cytotoxic CD8 T cells and NK cells is compared in normal and nude mice. Nude mice are a mutant strain of mouse that lack a thymus and thus has low levels of T cells and high levels of NK cells. Tumors that express MHC class I molecules are killed more effectively by CD8 T cells, whereas tumors lacking MHC class I are more effectively killed by NK cells. The photographs show scanning electron micrographs of NK cells attacking leukemic cells. The left photograph shows the NK cell, the smaller cell at the left, shortly after it has bound to a leukemic cell. The right photograph shows the interaction 1 hour after mixing, when much damage has been done to the leukemic cell. Photographs courtesy of J.C. Hiserodt.

or of all HLA-C allotypes, are susceptible to attack by NK cells. In contrast, for most individuals the loss of a single HLA-A, -B or -C allotype, a common occurrence in human tumors, would not render the tumor cells susceptible to NK cells.

Another strategy used by tumors to escape the immune response is to release cytokines that create an immunosuppressive environment around the tumor. An example of such a cytokine is transforming growth factor-β (TGF-β), which suppresses T_H1 cell responses and cell-mediated immunity.

12-32 Monoclonal antibodies against cell-surface tumor antigens can be used for diagnosis and immunotherapy

Changes in cell-surface antigens on tumor cells can be recognized by B cells, which produce antibodies against these tumor antigens. Monoclonal antibodies that recognize tumor antigens have been used in both tumor analysis and therapy. The location of tumor cells within the body can be revealed by giving patients purified monoclonal antibody specific for a tumor antigen and covalently coupled to a radioactive isotope, for example ^{131}I. The binding of the antibodies to tumor cells concentrates the radioactivity at sites in the body containing tumors. These can then be detected by exposing the patient to photographic film or to an imaging device (Figure 12.41). In this manner the size and location of a primary tumor can be determined as well as the extent of its metastasis. Such information can help to determine the type of therapy.

Monoclonal antibodies that are used to reveal tumors can also be used to destroy the tumor cells to which they bind. Two approaches are being explored. The first relies on the natural effector functions of the antibody to direct complement activation at the tumor-cell surface and thus harness the destructive forces of either phagocytes or the terminal complement components against the tumor cell. This approach has had some success in the treatment of B-cell lymphoma. The tumor antigen targeted in these studies is the cell-surface immunoglobulin. Within a particular patient, all the lymphoma cells carry the same immunoglobulin, whereas only very few healthy B cells express that same immunoglobulin. Thus the B-cell lymphoma immunoglobulin can be considered to be a tumor-specific antigen. Mouse monoclonal antibodies specific for the unique antigen-binding site of the lymphoma immunoglobulin can be made. These are known as anti-idiotypic antibodies, and are directed against epitopes determined by the CDR loops of the antigen-binding site.

The advantage of targeting the surface immunoglobulin of B-cell lymphoma is that the destruction of all cells expressing a particular immunoglobulin poses no threat to the patient's immunocompetence. The disadvantage is that each patient's therapy requires a custom-made monoclonal antibody. The administration of anti-idiotypic antibody invariably leads to clearance of lymphoma cells and a remission that, for a few patients, has proved to be a complete cure. In other patients, however, remission is followed by relapse. This is caused by the growth of lymphoma cells that do not possess the epitope recognized by the anti-idiotypic antibody and have been selected by the antibody therapy. Tumors of hematopoietic cells are best suited to this therapeutic approach because the tumor cells are accessible to the antibody, which is not true of solid tumors derived from other tissues.

The second use of monoclonal antibodies in cancer therapy does not rely on the natural effector functions of the antibodies. Instead, the antibodies are used to deliver a toxic agent that kills the cells to which the antibody binds. Various types of toxic agent are being tried: radioactive isotopes such as iodine-131 or yttrium-99, chemical toxins such as those used in conventional chemotherapy, plant toxins such as ricin, and bacterial toxins such as *Pseudomonas* toxin. The advantage of

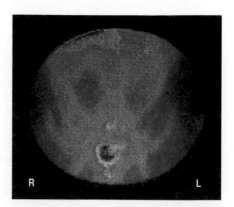

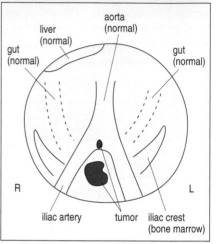

Figure 12.41 Detection of colon cancer by using a radiolabeled antibody against carcinoembryonic antigen. Anterior projection of the pelvis showing uptake of anti-carcinoembryonic antigen (CEA) antibody in a pelvic tumor. A patient with a suspected recurrence of colon cancer was injected intravenously with ^{111}In (indium)-labeled monoclonal antibody against CEA. The tumor is seen as two dark red spots in the pelvic region. The blood vessels are faintly outlined by circulating antibody that has not bound to the tumor, but may have bound to soluble CEA released from the tumor and circulating in the blood. R, right side; L, left side. Photograph courtesy of A.M. Peters.

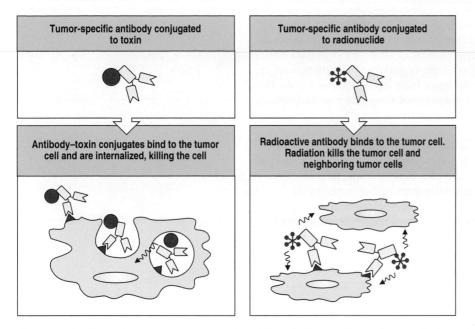

Tumor-specific antibody conjugated to toxin	Tumor-specific antibody conjugated to radionuclide
Antibody–toxin conjugates bind to the tumor cell and are internalized, killing the cell	Radioactive antibody binds to the tumor cell. Radiation kills the tumor cell and neighboring tumor cells

Figure 12.42 Antibodies can target toxins or radioactive isotopes to the tumor-cell surface. The tumor-specific antibody is coupled to either a toxin (red ball in left panels) or a radionuclide (red star in right panels). Toxins such as ricin exert a toxic effect on the cell only after internalization followed by proteolytic cleavage, which releases the toxic part of the molecule from the antibody.

these antibody conjugates over conventional irradiation or chemotherapy is that the destructive power of the toxin is more specifically targeted toward the tumor cells and away from proliferating normal tissues (Figure 12.42).

The conjugates of biological toxins and antibodies are called **immunotoxins**. In addition to the selectivity imposed by the binding of the monoclonal antibody to the tumor antigen, immunotoxins are activated only after their entry into the cell and thus should not cause damage to healthy non-cancerous cells. On binding to tumor cells, immunotoxins are endocytosed. Once within the endosomes, the toxin is cleaved from the antibody, allowing it to exert its toxic effect. A related approach uses a pro-drug that is administered in a conventional manner but can be activated only by an enzyme administered separately as a conjugate linked to a tumor-specific antibody.

A drawback to all therapeutic applications of mouse monoclonal antibodies is that they stimulate the human immune system to produce antibodies that bind the mouse antibody and thus inactivate it as a therapeutic agent. In general this limits their application to a period of 2 weeks, the time taken to generate a high-affinity antibody response. One way of reducing this effect is to use genetic engineering to transfer the antigenic specificity of the mouse antibody to a human antibody. This can usually be accomplished by simple replacement of the CDR loops of the heavy and light chains. Although it eliminates many of the mouse epitopes that stimulate the human immune system, this humanization of mouse antibodies does not eliminate the idiotypic determinants. With the large quantities of antibodies used in therapy, the stimulation of anti-idiotypic responses is still a limiting factor.

12-33 Boosting the T-cell response is another approach to immunotherapy

The immune response to an infectious disease is started by inflammation produced at the initial site of infection. This enables adaptive immune responses to be made within two weeks of the start of infection. In comparison, the malignant transformation of a single cell that marks the start of a cancer goes unnoticed by the body. Indeed, for many years the growth of a cancer might cause no damage of the sort that would trigger inflammation and an immune response (Figure 12.43). By the time that the cancer interferes with bodily functions, the tumor load is usually so great that the immune system is likely to be overwhelmed.

Growth of breast cancer

Diameter of tumor (mm)

death of patient (10^{12} cells)

tumor first palpable (10^9 cells)

tumor first visible on X-ray (10^8 cells)

Tumor cell population doublings

Figure 12.43 The growth and lifespan of a typical human tumor such as a tumor of the breast. The diameter of the tumor is plotted on a logarithmic scale. Years can elapse before the tumor becomes noticeable.

One way of boosting the immune response is to isolate a patient's tumor-specific T cells, stimulate them with antigen and cytokines so that they proliferate in tissue culture, and then give the expanded population of effector T cells back to the patient. In this way, many more effector cells are targeted at the tumor than are produced by the natural response. Remissions have been achieved with this approach in patients with melanoma, a cancer that does not respond well to conventional therapies.

There is a history of anecdotal observations suggesting that stimulation of the immune system by other agents can help it to respond to cancer. For example, tumor regression has been seen in the wake of a bacterial infection or conventional anti-tumor therapy. More established is the treatment of superficial bladder cancer by introducing the BCG vaccine into the bladder, where it elicits a chronic inflammation. However, this seems to be a rather specific application, because BCG has no effect on other cancers.

New approaches being explored are based on vaccination, the procedure that has been so successful in boosting immunity against infectious disease. As we saw in the first part of this chapter, the best vaccines are modified versions of the disease-causing pathogen. In the context of cancer, this would ideally involve removing tumor cells from patients, modifying them in ways that improve their presentation of antigens and stimulation of immunity, and reintroducing them as a vaccine into the patient. For example, vaccination with tumor cells transfected with the gene encoding the B7 co-stimulatory molecule improved the ability of mice to eliminate both B7-transfected and unmodified tumor cells. Alternatively, transfection of tumor cells with cytokine genes is aimed at attracting cells of the immune system to the tumor. In experiments in mice, most success has been obtained with GM-CSF, which causes hematopoietic precursor cells to differentiate into dendritic cells in close proximity to the tumor cells. The dendritic cells process and present tumor antigens, which they carry to the draining lymphoid tissue, where an immune response is generated (Figure 12.44).

Another way of boosting the presentation of tumor antigens is to isolate antigen-presenting dendritic cells from a patient, load them with tumor peptide epitopes *in vitro* and then give them back to the patient. This approach has had some success in producing temporary remissions in patients with advanced melanoma.

Summary

Unlike infection, cancer is a disease that arises from within and is caused by cells almost identical to normal human cells. Each case of cancer derives from one cell that has undergone malignant transformation, a process requiring mutations in several genes important for cell survival and cell division. Cancer cells become divorced from the mechanisms that maintain tissue integrity and they gradually compete with normal cells for food and space within the human body. In time, the effects of that competition become manifest as disease and ultimately death. Every cancer is unique and dies with the person in which it arose.

The principal defenses against cancer are present in every cell of the body. These are the normal cell processes that monitor DNA integrity, and control DNA replication and cell division; they ensure that multiple insults are required for malignant transformation. Although tumor-specific cytotoxic CD8 T cells and antibodies can be detected in patients with cancer, these immune responses do not enable them to control or eliminate the disease. In some cases the infusion of patients with tumor-specific antibodies or CD8 T cells that have been derived *in vitro* has ameliorated or even cured disease. Other approaches include boosting the tumor cells' ability to activate cytotoxic CD8 T cells. None of these procedures has yet entered routine medical practice.

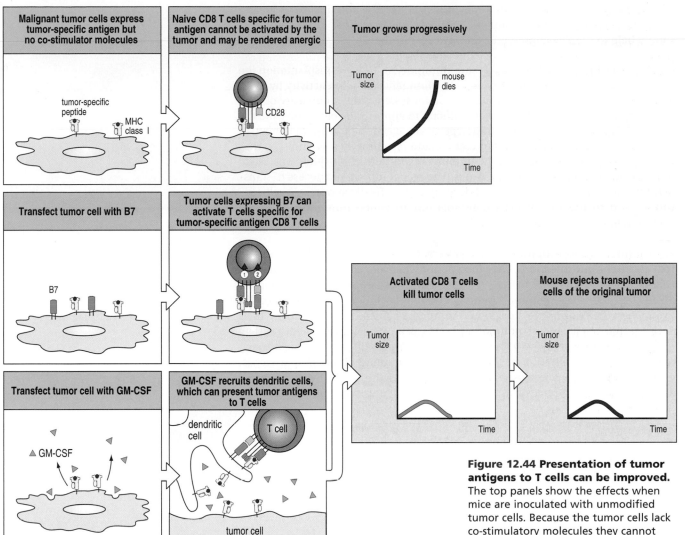

| Malignant tumor cells express tumor-specific antigen but no co-stimulator molecules | Naive CD8 T cells specific for tumor antigen cannot be activated by the tumor and may be rendered anergic | Tumor grows progressively |

tumor-specific peptide — MHC class I — CD28 — Tumor size — mouse dies — Time

| Transfect tumor cell with B7 | Tumor cells expressing B7 can activate T cells specific for tumor-specific antigen CD8 T cells |

B7

| Transfect tumor cell with GM-CSF | GM-CSF recruits dendritic cells, which can present tumor antigens to T cells |

GM-CSF — dendritic cell — T cell — tumor cell

| Activated CD8 T cells kill tumor cells | Mouse rejects transplanted cells of the original tumor |

Tumor size — Time — Tumor size — Time

Figure 12.44 Presentation of tumor antigens to T cells can be improved. The top panels show the effects when mice are inoculated with unmodified tumor cells. Because the tumor cells lack co-stimulatory molecules they cannot activate CD8 T cells and may indeed make them anergic. As a consequence the tumor grows and the mice die. The middle and bottom panels show the effect of transfecting the tumor cells with genes for the co-stimulatory molecule B7 and the cytokine GM-CSF respectively, before giving them to mice. Tumor cells expressing B7 can now activate CD8 T cells specific for the tumor antigens. The GM-CSF secreted by the tumor recruits dendritic cells which then present the tumor antigens to CD8 T cells. As shown in the panels on the right, in both of these cases, the tumor is eliminated by activated cytotoxic CD8 T cells. Their role is confirmed by challenging the same mouse with transplanted cells of the original unmodified tumor. These are rejected, showing that active immunity against that particular tumor has been established.

Summary to Chapter 12

The prevention of infectious diseases by vaccination illustrates superbly how manipulation of the immune response can benefit public health. The pathogens for which effective vaccines have been found are those that cause acute infections and are not highly mutable. When these pathogens caused epidemic disease, many people survived and developed immunity, showing that the human immune system could respond to the infection in a productive way. Death, when it occurred, was because the immune system was just too slow. What vaccination achieved was to start the immune response ahead of the infection, giving the immune system an edge on the pathogen.

The pathogens that have foiled the vaccinologists are those that foil the immune system. Their infections are protracted and they cause chronic disease. For such diseases, we know little about what constitutes a successful immune response. Against some of these pathogens the human immune response is uniformly unsuccessful; for others a minority of people seem to be able to clear the infection soon after infection; and for yet others, some immunity does develop but what it consists of, and when and how it is produced, is still unclear. A greater understanding of the successful immune responses against such pathogens should help in the development of vaccines against them.

Whereas vaccination provides a boost to immunity, in tissue transplantation the goal is immune suppression. Although the elimination of alloreactivity by exact MHC matching can be demonstrated in model systems, the complexity of HLA polymorphism has made this concept difficult to apply in the clinic. Instead, the practical approach has been to avoid hyperacute rejection by cross-matching, to reduce the T-cell alloreactions that cause acute rejection by matching for HLA where possible, and to suppress the immune response with the judicious use of non-specific immunosuppressive drugs. The recent increased success of transplantation is due to the discovery of drugs that specifically suppress T-cell activation, and to the use of drug combinations in which immunosuppressive activities, but not toxicities, are additive.

In their interactions with the immune system, cancers are akin to the pathogens that cause chronic infections. Over long periods, cancer cells selfishly exploit the cellular society of the human body to expand their own populations. They largely avoid stimulating the immune system by being almost identical to normal cells. Even when the lymphocytes respond, the tumors can escape by changing their antigens. Although cancer immunotherapy is still an experimental field, a variety of approaches are showing some promise. In the past 50 years, clinical transplantation has benefited from the strategy of combining different approaches, each with its own advantages and limitations. Over the next 50 years it seems likely that a similar strategy will lead to advances in cancer immunotherapy.

Glossary

α:β T cells T cells that express an antigen receptor made up of α and β chains. This majority population of T cells includes all T cells that recognize peptide antigen presented by MHC class I and II molecules.

α:β T-cell receptor *see* T-cell receptor.

Acquired immune deficiency syndrome (AIDS) the disease caused by infection with the human immunodeficiency virus (HIV). It involves a gradual destruction of the CD4 T-cell population and increasing susceptibility to infection.

Acute lymphoblastic leukemia (ALL) a type of cancer that results from uncontrolled growth of a progenitor cell thought to give rise to both the B and T lineages of lymphoid cells.

Acute rejection rejection of transplanted allogeneic cells, tissues, or organs that is due to the primary adaptive immune reponse stimulated by the transplant.

Acute-phase proteins plasma proteins made by the liver whose synthesis is quickly increased in response to infection. They include mannose-binding protein (MBP), C-reactive protein (CRP), and fibrinogen.

Acute-phase response a response of innate immunity that occurs soon after the start of an infection and involves the synthesis of acute-phase proteins by the liver and their secretion into the blood.

ADA *see* adenosine deaminase.

Adaptive immune response the response of antigen-specific B and T lymphocytes to antigen, including the development of immunological memory.

Adaptive immunity the state of resistance to infection that is produced by the adaptive immune response.

ADCC *see* antibody-dependent cell-mediated cytotoxicity.

Adenoids mucosa-associated lymphoid tissues located in the nasal cavity.

Adenosine deaminase (ADA) enzyme involved in purine breakdown. Its absence leads to the accumulation of toxic purine nucleosides and nucleotides, resulting in the death of most developing lymphocytes within the thymus. The disease that results from a genetically determined lack of adenosine deaminase is called **adenosine deaminase deficiency**. Its symptoms are severe combined immunodeficiency.

Adjuvants substances used to enhance the adaptive immune response to any antigen. To exert its effect an adjuvant must be mixed with the antigen before injection or vaccination.

Afferent lymphatic vessels the several vessels that bring lymph draining from connective tissue into a lymph node en route to the blood.

Affinity a measure of the strength with which one molecule binds to another when using a single binding site.

Affinity maturation the increase in affinity of the antigen-binding site of an antibody for the antigen that occurs during the course of a humoral immune response.

Agglutination the clumping together of particles, usually caused by antibody or some other multivalent binding molecule interacting with antigens on the surfaces of adjacent particles. Such particles are said to be **agglutinated**. When the particles are red blood cells, the phenomenon is called hemagglutination. When they are white blood cells it is called leukoagglutination.

Agonist any molecule that binds to a receptor and causes it to function.

Agonist peptides peptide antigens that activate their specific T cells, inducing them to make cytokines and to proliferate.

ALG *see* antilymphocyte globulin.

ALL *see* acute lymphoblastic leukemia.

Alleles the natural variants of a single gene.

Allelic exclusion refers to the fact that in a heterozygous individual, only one of the alternative C-region alleles of the immunoglobulin heavy- or light-chain loci is ever expressed in each B cell. In the B cell population roughly half of the cells express one allotype and the remaining cells express the other allotype.

Allergens antigens that elicit hypersensitivity or allergic reactions. They are frequently innocuous proteins that do not inherently threaten the integrity of the body.

Allergic asthma disease involving constriction of the bronchial tree and difficulty in breathing which is caused by an allergic reaction to inhaled antigen.

Allergic conjunctivitis allergic reactions in the conjunctiva of the eye, usually caused by airborne antigens.

Allergic reactions the result of secondary immune responses to innocuous environmental antigens or allergens. They can be caused by either antibodies or effector T cells.

Allergic rhinitis an allergic reaction in the nasal mucosa that causes runny nose, sneezing, and tears. It is also known as hay fever.

Allergy a state of hypersensitivity to a normally innocuous environmental antigen. It results from the interaction between the antigen and antibodies or T cells produced by earlier exposure to the same antigen.

Alloantibodies antibodies that are made by immunization of one member of a species with antigen derived from another member of the same species. Alloantibodies recognize antigens that are the result of allelic variation at polymorphic genes. Common types of alloantibodies are those recognizing blood group antigens and HLA class I and class II molecules.

Alloantigens antigens that differ between members of the same species, such as HLA molecules and blood group antigens. They are due to allelic differences at polymorphic genes.

Allogeneic describes two members of the same species who are genetically different.

Allograft a tissue graft made between genetically non-identical members of the same species.

Alloreactivity stimulation of the immune system by the antigenic differences between two members of the same species. These antigenic differences are called alloantigens. The immune response to an alloantigen is also called an alloreaction.

Allotypes natural protein variants that are produced by the alleles of a gene.

Alternative pathway one of three pathways of complement activation. It does not involve antibody, but is triggered by the binding of complement protein C3b to the surface of a pathogen. *See also* **classical pathway** and **lectin-mediated pathway**.

Anaphylactic shock IgE-mediated allergic reaction to systemically administered antigen that causes circulatory collapse and suffocation due to tracheal swelling. Also called systemic anaphylaxis.

Anaphylactoid reactions reactions that are clinically similar to anaphylactic shock but do not involve IgE.

Anaphylatoxins complement fragments C5a, C4a, and C3a, which are produced during complement activation. They recruit fluid and inflammatory cells to sites of antigen deposition.

Anchor residues residues in MHC-binding peptides that interact with pockets of the MHC-binding groove and show limited variation among the peptides that bind to a given MHC allotype.

Anergy a state of non-responsiveness to antigen. People are said to be anergic when they cannot mount delayed-type hypersensitivity reactions on challenge with an antigen. T and B cells are said to be anergic when they cannot respond to their specific antigen.

Angioedema swelling of the skin as a result of IgE-mediated allergic reactions, which cause increased permeability of subcutaneous blood vessels.

Antagonist any molecule that binds to a receptor and prevents its function.

Antagonist peptides peptides that inhibit the response of T cells to their specific peptide antigen and are usually closely related in sequence to it. In this context the antigenic peptide is called an agonist peptide.

Antibody the secreted form of the immunoglobulin made by a B cell.

Antibody humanization the process by which the antigen-binding site of a mouse monoclonal antibody is transferred to a human antibody.

Antibody repertoire the total variety of antibodies that a person can make.

Antibody-dependent cell-mediated cytotoxicity (ADCC) the killing of antibody-coated target cells by NK cells having the FcγRIII receptor (CD16) that recognizes the Fc region of the bound antibody.

Antigen any molecule that can bind specifically to an antibody. The name was given because they are **anti**body **gen**erators.

Antigen-binding site the site on an immunoglobulin or T-cell receptor molecule that binds specific antigen.

Antigen presentation the display of antigen as peptide fragments bound to MHC molecules on the surface of cells. This is the form of antigen recognized by most T cells.

Antigen-presenting cells cells that express either MHC class I and/or MHC class II molecules and thus display complexes of MHC molecule and peptide on their surfaces.

Antigen processing the degradation of proteins into peptides that bind to MHC molecules for presentation to T cells.

Antigen receptors for a B cell, the antigen receptor is its cell-surface immunoglobulin; for a T cell the antigen receptor is called the T-cell receptor. Each individual lymphocyte bears receptors of a single antigen specificity.

Antigenic determinant *see* **epitope**.

Antigenic drift a process by which point mutations in influenza virus genes cause differences in the structure of viral surface antigens. This causes year-to-year differences in influenza strains.

Antigenic shift a process by which influenza viruses reassort their segmented genomes and change their surface antigens radically. New viruses arising by antigenic shift are the usual cause of influenza pandemics.

Antilymphocyte globulin (ALG) a preparation of antibodies made by immunizing animals with human lymphocytes. It is used clinically to prevent rejection in organ transplantation.

Antiserum (plural **antisera**) the fluid component of clotted blood from an immune individual that contains antibodies against an antigen. An antiserum contain a heterogeneous collection of antibodies that bind the antigen.

Antithymocyte globulin (ATG) a preparation of antibodies made by immunizing animals with human thymocytes. It is used clinically to prevent rejection in organ transplantation.

AP-1 a family of transcription factors, some of which participate in lymphocyte activation.

Apoptosis a mechanism of cell death in which the cells to be killed are induced to degrade themselves from within, in a tidy manner. Also called programmed cell death.

Appendix a gut-associated lymphoid tissue located at the beginning of the colon.

Arthus reaction an immune reaction in the skin caused by the injection of antigen into the dermis. The antigen reacts with specific IgG antibodies in the extracellular spaces, activating complement and phagocytic cells to produce a local inflammatory response.

ATG *see* **antithymocyte globulin.**

Atopic allergy, or **atopy** the genetically determined tendency of some people to produce IgE-mediated hypersensitivity reactions against innocuous substances.

Atopic dermatitis *see* **eczema.**

Attenuated pathogen a strain of a pathogen that will grow in the host and induce immunity without causing disease.

Autocrine term applied to a cytokine or other secreted molecule that acts on the same type of cell that secreted it.

Autograft a graft of tissue made from one anatomical site to another on the same individual.

Autoimmune diseases diseases in which the pathology is caused by immune responses to normal components of healthy tissue.

Autoimmune hemolytic anemia a disease characterized by reduced numbers of red blood cells (anemia), which is caused by autoantibodies that bind to surface antigens of red blood cells.

Autoimmune response an adaptive immune response directed at self-antigens.

Autoimmunity adaptive immunity specific for self-antigens.

Autologous bone marrow transplant a bone marrow transplant in which the donor and recipient are the same person.

Azathioprine an immunosuppressive drug that kills dividing cells. It is used in transplantation.

β_2-microglobulin (β_2m) the invariant polypeptide that is common to all MHC class I molecules. Also called the light chain of MHC class I molecules.

B cells, or **B lymphocytes** lymphocytes that are dedicated to making immunoglobulins and antibodies.

B-cell antigen receptor, or **B-cell receptor** each B cell is programmed to make a single type of immunoglobulin. The cell-surface form of this immunoglobulin serves as the B-cell receptor for specific antigen. Associated in the membrane with the immunoglobulin are the signal transduction molecules Igα and Igβ.

B-cell co-receptor a complex of the CD19, TAPA-1, and CR2 polypeptides which associates with the B-cell receptor and augments its response to specific antigen.

B-cell mitogens substances that cause resting B cells to undergo cell division and expand their numbers.

B-1 cells the minority population of B cells. They express the CD5 glycoprotein and make antibodies of broad specificities. They are also known as CD5 B cells.

B-2 cells the majority population of B cells. They do not express the CD5 glycoprotein and make antibodies of narrow specificities.

B7 molecules the B7.1 and B7.2 co-stimulatory molecules.

B lymphocyte *see* **B cells.**

Bacteria diverse prokaryotic microorganisms that are responsible for many infectious diseases of humans and other animals.

Bare lymphocyte syndromes genetic diseases in which MHC class I or II molecules are not expressed on cells. They can be caused by various different regulatory gene defects and their effect is severe immunodeficiency.

Basophils rare white blood cells that are one of the three types of granulocyte. They contain granules that stain with basic dyes.

Benign tumor a growth due to abnormal proliferation of cells that is localized and contained by epithelial barriers.

Bone marrow the major site of generation of all the cellular elements of blood (hematopoiesis).

Brambell receptor (FcRB) an Fc receptor that transports IgG across epithelia and has a structure like an MHC class I molecule.

Bronchial-associated lymphoid tissues (BALT) the lymphoid cells and organized lymphoid tissues in the respiratory tract.

Bronchiectasis chronic inflammation of the bronchioles of the lung.

C domain *see* **constant domains.**

C region *see* **constant region.**

C-reactive protein (CRP) an acute-phase protein that binds to phosphatidylcholine, a surface constituent of various bacteria. C-reactive protein binds bacteria, opsonizing them for uptake by phagocytes.

C1 inhibitor (C1INH) a regulatory protein in plasma that inhibits the activity of activated complement component C1. C1INH deficiency causes the disease hereditary angioneurotic edema, in which spontaneous complement activation causes episodes of epiglottal swelling and suffocation.

C3 convertase proteolytic enzyme that is formed during complement activation and cleaves complement component C3, thereby enabling it to bond covalently to antigens.

C3 convertase of the alternative pathway (C3b,Bb) C3-cleaving enzyme made up of the complement fragments C3b and Bb.

C3 convertase of the classical pathway (C4b,2b) C3-cleaving enzyme made up of the complement fragments C4b and C2b.

C4-binding protein (C4BP) a regulatory protein in plasma that inactivates the C3 convertase of the classical pathway by binding to C4b and displacing C2b.

C5 convertase of the alternative pathway (C3b$_2$Bb) C5-cleaving enzyme consisting of two molecules of C3b and one of Bb.

C5 convertase of the classical pathway (C4b,2b,3b) C5-cleaving enzyme consisting of the complement fragments C4b, C2b, and C3b.

C5 convertases proteolytic enzymes that are formed during complement activation and cleave complement component C5.

Calcineurin a cytosolic serine/threonine phosphatase that contributes to T-cell activation. The immunosuppressive drugs cyclosporin A and tacrolimus act by inhibiting calcineurin.

Calnexin a membrane protein in the endoplasmic reticulum that facilitates the folding of MHC molecules and other glycoproteins.

Calreticulin a soluble protein in the endoplasmic reticulum that is structurally related to calnexin and also helps MHC molecules and other glycoproteins to fold properly.

CAMs *see* **cell adhesion molecules**.

Cancer diseases caused by abnormal and invasive cell proliferation.

Carcinoembryonic antigen (CEA) a large glycoprotein of epithelial cells which has increased expression on dividing cells.

Carcinogens chemical or physical agents that increase the risk of cancer.

Carcinomas cancers of epithelial cells.

Carriers foreign proteins to which small non-immunogenic antigens, or haptens, can be coupled to render the hapten immunogenic. *In vivo*, self proteins can also serve as carriers if they are suitably modified by the hapten; this is important in allergy to drugs.

Caseation necrosis a form of necrosis seen in the center of some large granulomas. The term comes from the white cheesy appearance of the central necrotic area.

Catalytic antibodies antibodies that bind their antigen, chemically change it, and then release it.

CD2 an adhesion molecule of T cells that binds to the LFA-3 adhesion molecule of antigen-presenting cells. Also called LFA-2.

CD3 complex a complex of signaling molecules that associates with T-cell receptors. It consists of CD3γ, δ, and ε chains, and ζ chains.

CD4 a cell-surface glycoprotein on T cells that recognize antigens presented by MHC class II molecules. CD4 binds to the MHC class II molecules on the antigen-presenting cell and acts as a co-receptor to augment the T cell's response to antigen.

CD4 T cells the subset of T cells that express the CD4 co-receptor and recognize peptide antigens presented by MHC class II molecules.

CD5 B cells *see* **B-1 cells**.

CD8 a cell-surface glycoprotein on T cells that recognize antigens presented by MHC class I molecules. CD8 binds to MHC class I molecules on the antigen-presenting cell and acts as a co-receptor to augment the T-cell's response to antigen.

CD8 T cells the subset of T cells that express the CD8 co-receptor and recognize peptide antigens presented by MHC class I molecules.

CD19 a component of the B-cell co-receptor.

CD21 another name for complement receptor 2 (CR2). A component of the B-cell receptor.

CD28 the low-affinity receptor on T cells that interacts with B7 co-stimulatory molecules to promote T-cell activation.

CD34 a vascular addressin expressed on high endothelial venules in lymph nodes.

CD40 cell-surface glycoprotein on B cells whose interaction with **CD40 ligand** on T cells triggers B-cell proliferation.

CD81 a component of the B-cell receptor that is also a cellular receptor for hepatitis C virus. Also called TAPA-1.

CDR *see* **complementarity-determining regions**.

CEA *see* **carcinoembryonic antigen**.

Cell adhesion molecules (CAMs) cell-surface proteins that enable cells to bind to each other.

Cell-mediated immunity, or **cellular immunity** any adaptive immune response in which antigen-specific effector T cells dominate. It is defined operationally as all adaptive immunity that cannot be transferred to a naive recipient with serum antibody.

Central lymphoid tissues the anatomical sites of lymphocyte development. In humans, B lymphocytes develop in bone marrow, whereas T lymphocytes develop in the thymus.

Centroblasts large dividing B cells present in germinal centers. Somatic hypermutation occurs in centroblasts and antibody-secreting and memory B cells derive from them.

Chediak–Higashi syndrome a genetic disease in which phagocytes malfunction. Their lysosomes fail to fuse properly with phagosomes and killing of ingested bacteria is impaired.

Chemokines small cytokines involved in guiding white blood cells to sites where their functions are needed. They have a central role in inflammatory responses.

Chronic asthma a disease characterized by chronic inflammation of the airways and difficulty in breathing. Although probably initiated by exposure to an allergen, chronic asthma can be perpetuated in its absence.

Chronic granulomatous disease an immunodeficiency disease in which multiple granulomas form as a result of defective elimination of bacteria by phagocytic cells. It is caused by a defect in the NADPH oxidase system of enzymes that generate the superoxide radical involved in bacterial killing.

Chronic rejection rejection of organ grafts that occurs years after transplantation and is characterized by degeneration and occlusion of the blood vessels in the graft. Its underlying cause is poorly understood.

Chronic thyroiditis an autoimmune disease that causes progressive destruction of the thyroid gland. Also called Hashimoto's thyroiditis or Hashimoto's disease.

Class I region the part of the major histocompatibility complex that contains the MHC class I heavy-chain genes.

Class II region the part of the major histocompatibility complex that contains the MHC class II α- and β-chain genes.

Class II-associated invariant chain peptide (CLIP) is a peptide of variable length cleaved from the class II invariant chain by proteases. It remains associated with the MHC class II molecule in an unstable form until it is removed by the HLA-DM protein.

Class switching *see* isotype switching.

Classes *see* isotypes.

Classical pathway one of three pathways of complement activation. It is activated by antibody bound to antigen, and involves complement components C1, C4, and C2 in the generation of the C3 and C5 convertases. *See also* alternative pathway and lectin-mediated pathway.

Clonal deletion the elimination of immature lymphocytes that bind to self-antigens. Clonal deletion is the main mechanism that produces self-tolerance.

Clonal selection the central principle of adaptive immunity by which adaptive immune responses derive from individual antigen-specific lymphocytes that proliferate in response to antigen and differentiate into antigen-specific effector cells.

Co-receptor a cell-surface protein that increases the sensitivity of an antigen receptor to its antigen. This can be accomplished by increasing adhesion interactions and/or by enhancing signal transduction.

Co-stimulator molecules the activation of naive lymphocytes requires both an antigen-binding signal and a co-stimulatory signal that is delivered by a cell-surface molecule on the antigen-presenting cell. For T cells, the **co-stimulatory** signals are delivered by the B7.1 and B7.2 proteins, which engage the T-cell molecules CD28 and CTLA-4. For B cells, CD40 ligand binding to CD40 on the B cell serves an analogous role.

Cognate interactions cell–cell interactions between B and T lymphocytes that are specific for the same antigen.

Collectins a family of calcium-dependent sugar-binding proteins or lectins containing collagen-like sequences. An example is mannose-binding protein.

Combination vaccines vaccines that contain antigens derived from more than one pathogen and are designed to provide protection against more than one disease.

Common lymphoid progenitors stem cells that give rise to all lymphocytes and are derived from pluripotent hematopoietic stem cells.

Complement a set of plasma proteins that act in a cascade of reactions to attack extracellular forms of pathogens. As a result of complement activation, pathogens become coated with complement, which can either kill the pathogen directly or cause its engulfment and destruction by phagocytes.

Complement activation the initiation by pathogens of a series of reactions involving the complement components of plasma, leading to the death and elimination of the pathogen.

Complement receptors (CR) cell-surface proteins on various cell types that recognize and bind complement proteins bound to antigens. Complement receptors on phagocytes facilitate their engulfment of pathogens coated with complement. Complement receptors include **CR1, CR2, CR3, CR4**, and the receptor for C1q.

Complement system a set of over 30 plasma and cell-surface proteins that serve to target pathogens for destruction by phagocytes or in some cases kill them directly.

Complementarity-determining regions (CDRs) the localized regions of immunoglobulin and T-cell receptor chains that determine the antigenic specificity and bind to the antigen. The CDRs are the most variable parts of the variable region and are also called hypervariable regions.

Conformational epitopes epitopes on a protein antigen that are formed from several separate regions in the primary sequence of a protein brought together by protein folding. Antibodies that bind conformational epitopes bind only to native folded proteins. Also called **discontinuous epitopes.**

Conjugate vaccines vaccines made from capsular polysaccharides bound to immunogenic proteins such as tetanus toxoid. The protein provides peptide epitopes that stimulate CD4 T cells to help B cells specific for the polysaccharide.

Constant domains (C domains) the constituent domains of constant regions of immunoglobulin and T-cell receptor polypeptides.

Constant region (C region) the part of an immunoglobulin or T-cell receptor that is of identical amino-acid sequence in molecules of the same isotype but different antigen-binding specificity. The term constant region is also used to refer to the parts of the individual polypeptide chains that are of identical amino-acid sequence in antibodies and T-cell receptors of the same isotype but different antigen-binding specificity.

Contact sensitivity a form of delayed-type hypersensitivity in which T cells respond to antigens that are introduced into the body by contact with the skin.

Corticosteroids compounds related to the steroid hormones produced in the adrenal cortex, such as cortisone. They suppress immune responses and are widely used as anti-inflammatory and immunosuppressive agents in medicine.

CR, CR1, CR2, CR3, CR4 *see* complement receptors.

Cross-match testing a test used in blood typing and histocompatibility testing to determine whether donor and recipient have antibodies against each other's cells that might interfere with successful transfusion or transplantation.

CRP *see* **C-reactive protein.**

CTLA-4 a high-affinity inhibitory receptor on T cells that interacts with B7 co-stimulatory molecules.

Cutaneous T-cell lymphoma a malignant growth of T cells that home to the skin. Also known as mycosis fungoides.

Cyclophilins cytoplasmic proteins that bind to the immunosuppressive drugs cyclosporin A and tacrolimus. The complex of drug and cyclophilin binds calcineurin and this prevents T-cell activation.

Cyclophosphamide an alkylating agent used as an immunosuppressive drug. It acts by killing rapidly dividing cells, including lymphocytes proliferating in response to antigen.

Cyclosporin A an immunosuppressive drug that specifically prevents T-cell activation and effector function. Also called **cyclosporine.**

Cytokines proteins made by cells that affect the behavior of other cells. Cytokines made by lymphocytes are often called lymphokines or interleukins (abbreviated IL). Cytokines bind to specific receptors on their target cells.

Cytotoxic CD8 T cells the subset of T cells that express the CD8 co-receptor and recognize peptide antigen presented by MHC class I molecules.

Cytotoxic T cells T cells that can kill other cells. Almost all cytotoxic T cells are CD8 T cells. Cytotoxic T cells are important in host defense against viruses and other cytosolic pathogens.

Cytotoxins proteins made by cytotoxic T cells that participate in the destruction of target cells. Perforins and granzymes or fragmentins are examples of cytotoxins.

D gene segments, or **diversity gene segments** short DNA sequences that connect the V and J gene segments in rearranged immunoglobulin heavy-chain genes and in T-cell receptor β and δ chain genes.

Dark zone the part of a germinal center in secondary lymphoid tissue that contains dividing centroblasts.

Decay-accelerating factor (DAF) a cell-surface protein that prevents complement activation on human cells. DAF binds to C3 convertases of both the alternative and classical pathway and, by displacing Bb and C2b respectively, prevents their action.

Degenerate binding specificity the type of antigen-binding specificity exhibited by MHC class I and II molecules. Each MHC allotype can bind numerous peptides of different amino acid sequences.

Delayed-type hypersensitivity (DTH) a form of cell-mediated immunity elicited by antigen in the skin and mediated by CD4 T_H1 cells. It is called delayed-type hypersensitivity because the reaction appears hours to days after antigen is injected.

Dendritic cells cells with a branched, dendritic morphology that are the most potent stimulators of T-cell responses. Also known as interdigitating reticular cells, they are distinct from the follicular dendritic cell that presents antigen to B cells.

Dendritic epidermal T cells (dETC) a specialized class of γ:δ T cells found in the skin of mice and some other species, but not humans. All dEC have the same γ:δ T-cell receptor; their function is unknown.

Desensitization a therapeutic procedure in which an allergic individual is exposed to increasing doses of allergen with the goal of inhibiting their allergic reactions. It probably works by shifting the balance between CD4 T_H1 and T_H2 cells and thus changing the antibody produced from IgE to IgG.

Diapedesis the movement of white blood cells from the blood across blood vessel walls into tissues.

DiGeorge syndrome a recessive genetic immunodeficiency disease in which thymic epithelium fails to develop.

Diphtheria toxin the toxin secreted by *Corynebacterium diphtheriae* which causes the disease symptoms. The diphtheria vaccine consists of an inactive form of the toxin called diphtheria toxoid.

Discontinuous epitopes *see* **conformational epitopes.**

Diversity gene segment *see* **D gene segments.**

Double-negative thymocytes immature T cells within the thymus that express neither CD4 nor CD8.

Double-positive thymocytes T cells at an intermediate stage of development in the thymus. They express both CD4 and CD8.

Draining lymph node the lymph node to which fluid collected at a site of infection first travels on its return to the blood.

DTH *see* **delayed-type hypensensitivity.**

E-selectin *see* **selectins.**

EAE *see* **experimental allergic encephalomyelitis.**

Early pro-B cell *see* **pro-B cell.**

Eczema a common skin disease of children; its etiology is poorly understood. Also called atopic dermatitis.

Edema abnormal accumulation of fluid in connective tissue, leading to swelling.

Effector cells lymphocytes that can mediate the removal of pathogens from the body without the need for further differentiation.

Efferent lymphatic vessel the single vessel in which lymph and lymphocytes leave a lymph node en route to the blood.

Encapsulated bacteria bacteria that possess thick carbohydrate coats that protect them from phagocytosis. Encapsulated bacteria cause extracellular infections and can be dealt with by phagocytes only if they are first coated with antibody and complement.

Endocytic vesicle membrane vesicles that are pinched off from the plasma membrane and take extracellular material into cells.

Endocytosis the uptake of extracellular material into cells by endocytic vesicles that form by pinching off pieces of plasma membrane.

Engraftment the time at which a bone marrow transplant is making new blood cells.

Eosinophils white blood cells that are one of the three types of granulocyte. They contain granules that stain with eosin. Eosinophils contribute chiefly to defense against parasitic infections.

Eotaxin-1 and **eotaxin-2** chemokines that act specifically on eosinophils.

Epidemic an outbreak of infectious disease that affects many individuals within a population.

Epitope the portion of an antigenic molecule that is bound by an antibody or gives rise to the MHC-binding peptide that is recognized by a T-cell receptor. Also called an antigenic determinant.

Erythroid progenitor an immature cell that gives rise to erythrocytes and megakaryocytes but no other type of blood cell.

Experimental allergic encephalomyelitis (EAE) an inflammatory disease of the central nervous system that develops after mice are immunized with neural antigens in a strong adjuvant.

Extravasation the movement of cells or fluid from within blood vessels to the surrounding tissues.

F(ab')₂ a proteolytic fragment of IgG that consists of the two Fab arms held together by a disulfide bond. Produced by digesting IgG with pepsin.

Fab fragment a proteolytic fragment of IgG that consists of the light chain and the amino-terminal half of the heavy chain held together by an interchain disulfide bond. It is called Fab because it is the Fragment with antigen binding specificity. In the intact IgG molecule the parts corresponding to the Fab fragment are often called Fab or Fab arms.

Factor B plasma protein that binds to C3(OH) or C3b and is cleaved to form part of the C3 convertases (C3(OH),Bb and C3b,Bb) in the alternative pathway of complement activation.

Factor D plasma protease that cleaves factor B to Bb and Ba in the alternative pathway of complement activation.

Factor H a complement regulatory protein of plasma that inactivates the C3 convertase of the alternative pathway and C5 convertases by binding to C3b and rendering it susceptible to cleavage by factor I to produce inactive iC3b.

Factor I a regulatory protease of complement that can cleave C3b and C4b into inactive forms.

Factor P *see* **properdin**.

Farmer's lung a hypersensitivity disease caused by the interaction of IgG antibodies with large particles of inhaled allergen in the alveolar wall of the lung. The resulting inflammation compromises gas exchange by the lungs. The disease is caused by chronic exposure to large quantities of the allergen.

Fas a member of the TNF receptor family that is expressed on certain cells and makes them susceptible to being killed by cells expressing **Fas ligand**, a cell-surface member of the TNF family of proteins. Binding of Fas ligand to Fas triggers apoptosis in the Fas-bearing cell.

Fc fragment a proteolytic cleavage fragment of IgG consisting of the carboxy-terminal halves of the two heavy chains disulfide-bonded to each other by the residual hinge region. It is called Fc because it is the Fragment that was most readily crystallized. In the intact IgG molecule the part corresponding to the Fc fragment is called **Fc**, **Fc region**, or **Fc piece**.

Fc receptors cell-surface receptors for the Fc portion of some immunoglobulin isotypes. They include the Fcγ and Fcε receptors.

Fc region *see* **Fc fragment**.

Fcε receptor (FcεRI) a receptor present on the surface of mast cells, basophils and activated eosinophils that binds free IgE with very high affinity. When antigen binds to IgE and crosslinks FcεRI it causes cellular activation and degranulation.

FDC *see* **follicular dendritic cells**.

Fever a rise of body temperature above the normal. Is caused by cytokine production in response to infection.

FK-binding proteins another name for the cyclophilins that bind tacrolimus (FK506). *See also* **cyclophilins**.

FK506 *see* **tacrolimus**.

Follicular center cell lymphoma a malignancy of mature B cells.

Follicular dendritic cells (FDCs) characteristic non-lymphoid cells of follicles in secondary lymphoid tissues. They have long branching processes that make intimate contact with B cells and have Fc and complement receptors that hold antigen:antibody:complement complexes on their surfaces for long periods. These cells are crucial in selecting antigen-binding B cells during antibody responses.

Framework regions relatively invariant regions within the variable domains of immunoglobulins and T-cell receptors that provide a protein scaffold for the hypervariable regions.

Freund's complete adjuvant an adjuvant consisting of an emulsion of mineral oil and water containing killed mycobacteria.

Fungi single-celled and multicellular eukaryotic organisms, including the yeasts and molds, that can cause a variety of diseases. Immunity to fungi involves both humoral and cell-mediated responses.

γ:δ T cells a minority population of T cells that express receptors made up of γ and δ chains. The antigen specificities and functions of these cells are uncertain.

GALT *see* **gut-associated lymphoid tissues**.

Gene conversion a process of recombination between two homologous genes in which a localized segment from one gene is replaced by the homologous segment of the second gene. In birds and rabbits, immunoglobulin receptor diversity is generated mainly by gene conversion, in which homologous inactive V gene segments exchange short sequences with an active, rearranged variable-region gene.

Gene knock-out laboratory jargon for a cell or animal in which the function of a particular gene has been deliberately eliminated by replacement of the normal gene with an inactive mutant gene.

Gene segments multiple short DNA sequences in the immunoglobulin and T-cell receptor genes. These can be rearranged in many different combinations to produce the vast diversity of immunoglobulin or T-cell receptor polypeptide chains. *See also* **D gene segments, J gene segments, V gene segments**.

Genetic polymorphism variation in a population due to the existence of two or more alleles of a gene.

Germinal centers areas in secondary lymphoid tissues that are sites of intense B-cell proliferation, selection, maturation, and death. Germinal centers form around follicular dendritic cell networks when activated B cells migrate into lymphoid follicles.

Germline configuration the organization of the immunoglobulin and T-cell receptor genes in the DNA of germ cells and in the vast majority of somatic cells that do not undergo somatic recombination. Also called **germline form**.

GlyCAM-1 a vascular addressin found on the high endothelial venules of lymphoid tissues. It is an important ligand for the L-selectin molecule of naive lymphocytes and directs these cells to leave the blood and enter the lymphoid tissues.

GM-CSF *see* **granulocyte–macrophage colony-stimulating factor.**

Goodpasture's syndrome an autoimmune disease in which autoantibodies against basement membrane of type IV collagen cause extensive vasculitis.

Graft-versus-host (GVH) reaction the response of mature donor-derived T cells in transplanted bone marrow against the alloantigens of the recipient's tissues. These alloreactive responses cause **graft-versus-host disease (GVHD)**.

Graft-versus-leukemia (GVL) in bone marrow transplantation as therapy for leukemia, some degree of genetic incompatibility between donor and recipient is thought to help T cells from the transplant eliminate residual leukemia cells.

Granulocyte–macrophage colony-stimulating factor (GM-CSF) a cytokine involved in the growth and differentiation of cells in the granulocyte and monocyte lineages.

Granulocytes white blood cells with irregularly shaped, multilobed nuclei and cytoplasmic granules. There are three types of granulocyte: neutrophils, eosinophils, and basophils. They are also known as polymorphonuclear leukocytes.

Granuloma A site of chronic inflammation usually triggered by persistent infectious agents such as mycobacteria, or by a non-degradable foreign body. Granulomas have a central area of macrophages, often fused into multinucleate giant cells, surrounded by T lymphocytes.

Granzymes serine esterase enzymes present in the granules of cytotoxic T cells and NK cells. On entering the cytosol of a target cell granzymes induce apoptosis of the target. Also called fragmentins.

Graves' disease an autoimmune disease in which antibodies against the thyroid-stimulating hormone receptor cause the over-production of thyroid hormone and the symptoms of hyperthyroidism.

Gut-associated lymphoid tissues (GALT) lymphoid tissues closely associated with the gastrointestinal tract, including the palatine tonsils, Peyer's patches, and layers of intraepithelial lymphocytes.

H chain *see* **heavy chain.**

HANE *see* **hereditary angioneurotic edema.**

Haplotype for a linked set of polymorphic genes, the set of alleles carried on a single chromosome is called a haplotype. Every person inherits one haplotype from each parent. The term is used mainly in connection with the genes of the major histocompatibility complex.

Hashimoto's disease *see* **Hashimoto's thyroiditis.**

Hashimoto's thyroiditis an autoimmune disease characterized by persistent high levels of antibodies against thyroid-specific antigens. These antibodies recruit NK cells to the thyroid, leading to damage and inflammation.

Heavy chain (H chain) the larger of the two component polypeptides of an immunoglobulin molecule. Heavy chains come in a variety of heavy-chain classes or isotypes, each of which confers a distinctive effector function on the antibody molecule.

Helper CD4 T cells CD4 T cells are sometimes generally known as helper cells because their function is to help other cell types perform their functions. The term helper T cell sometimes refers to T_H2 cells only, the cells that help B cells to produce antibody.

Hematopoiesis the generation of the cellular elements of blood, including the red blood cells, white blood cells and platelets. These cells all originate from pluripotent **hematopoietic stem cells** whose differentiated progeny divide under the influence of various **hematopoietic growth factors**.

Hematopoietic stem cell transplantation any form of transplantation in which the role of the graft is to replace the hematopoietic system. This includes bone marrow transplantation.

Hemolytic anemia of the newborn a potentially fatal disease caused by maternal IgG antibodies directed towards paternal antigens expressed on fetal red blood cells. The usual target of this response is the Rh blood group antigen. Maternal anti-Rh IgG antibodies cross the placenta to attack the fetal red blood cells. Also called erythroblastosis fetalis.

Herd immunity the phenomenon whereby those people in a population who have no protective immunity against a pathogen are largely protected from infection when the majority of the population is resistant to the pathogen.

Hereditary angioneurotic edema (HANE) a genetic disease due to deficiency of the C1 inhibitor of the complement system. In the absence of C1 inhibitor, spontaneous activation of the complement system causes diffuse fluid leakage from blood vessels, the most serious consequence of which is epiglottal swelling leading to suffocation.

Highly polymorphic genes that have many alleles and for which most individuals in a population are heterozygotes.

Histamine a vasoactive amine stored in mast cell granules. Histamine is released when antigen binds to IgE molecules on mast cells and causes dilation of local blood vessels and contraction of smooth muscle, producing some of the symptoms of immediate hypersensitivity reactions. Anti-histamines are drugs that counter histamine action.

Histocompatibility literally, the ability of tissues (Greek *histo*) to get along with each other. The term is used in immunology to describe the genetic systems that determine the rejection of tissue and organ grafts resulting from the immunological recognition of alloantigens (known in this context as **histocompatibility antigens**).

HIV *see* **human immunodeficiency virus.**

Hives itchy red swellings in the skin caused by IgE-mediated reactions. Also called urticaria or nettle rash.

HLA the acronym for Human Leukocyte Antigen. It is the genetic designation for the human MHC. Individual loci are designated by capital letters, as in HLA-A, and alleles are designated by numbers, as in HLA-A*0201.

HLA-A, HLA-B, and **HLA-C** the highly polymorphic human MHC class I genes.

HLA-DM an invariant MHC class II molecule that is involved in the intracellular loading of MHC class II molecules with peptides.

HLA-DP, HLA-DQ, and **HLA-DR** the highly polymorphic human MHC class II molecules. Each class II molecule is made from α and β chains encoded by A and B genes respectively. For example, the HLA-DPα and HLA-DPβ chains are encoded by the HLA-DPA and HLA-DPB genes respectively. All the genes are in the MHC.

HLA complex the human major histocompatibility complex.

HLA type the combination of HLA class I and class II allotypes that a person expresses.

Hodgkin's disease a malignant disease caused by transformed antigen-presenting cells resembling dendritic cells. **Hodgkin's lymphoma** is a form of Hodgkin's disease in which lymphocytes predominate. It has a much better prognosis than the nodular sclerosis form of this disease in which the predominant cell type is non-lymphoid.

Human immunodeficiency virus (HIV) the causative agent of the acquired immune deficiency syndrome (AIDS). HIV is a retrovirus of the lentivirus family that infects CD4 T cells, leading to their slow depletion, which eventually results in immunodeficiency.

Humoral immune response the adaptive immune response that leads to the production of antibodies.

Humoral immunity immunity that is mediated by antibodies and can therefore be transferred to a non-immune recipient by serum.

HV regions *see* **hypervariable regions.**

Hybridomas hybrid cell lines that make monoclonal antibodies of defined specificity. They are formed by fusing a specific antibody-producing B lymphocyte with a myeloma cell that grows in tissue culture and does not make any immunoglobulin chains of its own.

Hyper-IgM syndrome a genetically determined immunodeficiency disease in which B cells cannot switch their immunoglobulin heavy-chain isotype. It arises from the absence of CD40 ligand.

Hyperacute graft rejection rejection of an allogeneic tissue graft as a result of preformed antibodies that react against A,B,O blood group antigens or HLA class I antigens on the graft. The antibodies bind to endothelium and trigger the blood clotting cascade, leading to ischemia and death of the transplanted organ or tissue.

Hypereosinophila abnormally high numbers of eosinophils in the blood.

Hypersensitivity reactions immune responses to innocuous antigens that lead to symptomatic reactions on re-exposure. These can cause **hypersensitivity diseases** if they occur repetitively. This state of heightened reactivity to an antigen is called **hypersensitivity.** Hypersensitivity reactions are classified by mechanism: type I hypersensitivity reactions involve the triggering of mast cells by IgE antibodies; type II hypersensitivity reactions involve IgG antibodies against cell-surface or matrix antigens; type III hypersensitivity reactions involve antigen:antibody complexes; and type IV hypersensitivity reactions are mediated by effector T cells.

Hyperthyroid refers to an abnormally high production of thyroid hormones by the thyroid gland.

Hypervariable regions (HV regions) small regions of high amino-acid sequence diversity within the variable regions of immunoglobulin and T-cell receptors. They correspond to the complementarity determining regions.

Hypothyroid refers to an abnormally low production of thyroid hormone by the thyroid gland.

ICAM-1, ICAM-2, and **ICAM-3** the three different types of ICAM. *See also* **intercellular adhesion molecules.**

Iccosomes immune-complex coated bodies. Small fragments of membrane coated with immune complexes that bud off from the processes of follicular dendritic cells in lymphoid follicles.

IDDM *see* **insulin-dependent diabetes mellitus.**

IFN-α, IFN-β, IFN-γ *see* **interferons.**

Ig domain *see* **immunoglobulin domains.**

Igα, Igβ *see* **B-cell antigen receptor.**

IgA the class of immunoglobulin having α heavy chains. IgA antibodies in dimeric form are the antibodies present in mucosal secretions. IgA in monomeric form is present in the blood.

IgD the class of immunoglobulin having δ heavy chains. It appears as surface immunoglobulin on mature naive B cells but its function is unknown.

IgE the class of immunoglobulin having ε heavy chains. It is involved in allergic reactions.

IgG the class of immunoglobulin having γ heavy chains. It is the most abundant class of immunoglobulin in plasma.

IgM the class of immunoglobulin having μ heavy chains. It is the first immunoglobulin to appear on the surface of B cells and the first antibody secreted during an immune response. It is secreted in pentameric form.

Ii *see* **invariant chain.**

IL-1 interleukin-1, a cytokine released by macrophages which with IL-6 and TNF-α induces a wide range of responses at early times in infection.

IL-2 interleukin-2, a cytokine produced by activated T cells which is essential for the development of adaptive immune responses.

IL-3 interleukin-3, a growth factor for hematopoietic progenitor cells.

IL-4 interleukin-4, a cytokine secreted by CD4 T_H2 cells which helps to initiate the proliferation and clonal expansion of B cells.

IL-6 interleukin-6, a cytokine released by macrophages which with IL-1 and TNF-α induces a wide range of responses at early times in infection.

IL-8 interleukin-8, a chemokine released by macrophages which is a chemotactic factor for neutrophils.

IL-10 interleukin-10, a cytokine released by CD4 T_H2 cells.

IL-12 interleukin-12, a cytokine released by macrophages which activates NK cells.

IL-13 interleukin-13, a cytokine secreted by CD4 T_H2 cells.

Immature B cells B cells that have rearranged a heavy- and a light-chain variable-region gene and express surface IgM but not IgD.

Immediate reactions reactions occuring within minutes of exposure to antigen. They are mediated by pre-existing antibodies in the circulation. Also called immediate hypersensitivity reactions.

Immune resistant to infection.

Immune complex the protein complex formed from the binding of antibodies to soluble antigens. The size of the immune complexes formed depends on the relative concentrations of antigen and antibody. Large immune complexes are cleared by phagocytes bearing Fc and complement receptors. Small soluble immune complexes tend to be deposited on the walls of small blood vessels, where they can activate complement and cause damage.

Immune system the tissues, cells, and molecules involved in host defense mechanisms.

Immunity the ability to resist infection.

Immunization the deliberate provocation of an adaptive immune response by introducing antigen into the body.

Immunodeficiency diseases a group of inherited or acquired disorders in which some part or parts of host defense are either absent or defective.

Immunogenetics a subfield of immunology that was originally concerned with the analysis of genetic traits by means of antibodies against genetically polymorphic molecules such as blood group antigens and MHC proteins. Immunogenetics now encompasses the genetic analysis, by any technique, of molecules that are of

specific importance to the immune system.

Immunoglobulin A *see* **IgA.**

Immunoglobulin D *see* **IgD.**

Immunoglobulin domains (Ig domains) components of protein structure consisting of ~100 amino acids that fold into a sandwich of two β sheets held together by a disulfide bond. Immunoglobulin heavy and light chains are made up of a series of immunoglobulin domains. Similar domains are present in many other proteins.

Immunoglobulin E *see* **IgE.**

Immunoglobulin G *see* **IgG.**

Immunoglobulin M *see* **IgM.**

Immunoglobulin superfamily, or **Ig superfamily** the name given to all the proteins that contain one or more immunoglobulin or immunoglobulin-like domains.

Immunoglobulins (Ig) the antigen-binding molecules of B cells.

Immunological memory the capacity of the immune system to make quicker and stronger adaptive immune responses to successive encounters with an antigen. Immunological memory is specific for a particular antigen and is long-lived.

Immunophilins intracellular proteins with peptidyl–prolyl *cis–trans* isomerase activity that bind the immunosuppressive drugs cyclosporin A, tacrolimus, and sirolimus (rapamycin).

Immunoreceptor tyrosine-based activation motifs (ITAMs) sequences in the cytoplasmic domains of membrane receptors that are sites of tyrosine phosphorylation and of association with tyrosine kinases and phosphotyrosine-binding proteins involved in signal transduction. Related motifs with opposing effects are **immunoreceptor tyrosine-based inhibitory motifs (ITIMs)**, which recruit phosphatases that remove the phosphate groups added by tyrosine kinases.

Immunosuppressive drugs compounds that inhibit adaptive immune responses.

Immunotoxins conjugates made of a specific antibody that is chemically coupled to a toxic protein usually derived from a plant or microbe. The antibody delivers the toxin moiety to the surface of the target cells.

Inactivated virus vaccines *see* **killed virus vaccines.**

Inflammation a general term for the local accumulation of fluid, plasma proteins, and white blood cells that is initiated by physical injury, infection, or a local immune response. This is also known as an **inflammatory response.** The cells that invade tissues undergoing inflammatory responses are often called **inflammatory cells** or an **inflammatory infiltrate.**

Inflammatory mediators a variety of substances released by various cell types that contribute to the production of inflammation at a site of infection or trauma.

Influenza hemagglutinin a glycoprotein of the influenza virus coat that binds to certain carbohydrates on human cells, the first step in viral infection. Changes in the hemagglutinin are the major source of antigenic shift.

Innate immunity the host defense mechanisms that act from the start of an infection and do not adapt to a particular pathogen. Also called the **innate immune response**.

Insulin-dependent diabetes mellitus (IDDM) an autoimmune disease in which insulin-secreting β cells of the pancreatic islets of Langerhans are gradually destroyed.

Insulitis an infiltration of the pancreatic islets of Langerhans with lymphocytes and other leukocytes. This is a symptom of incipient diabetes.

Integrins a class of cell-surface glycoproteins that mediate adhesive interactions between cells and the extracellular matrix.

Interallelic conversion a mechanism of genetic recombination between two alleles of a locus in which a segment of one allele is replaced with the homologous segment from the other. This mechanism is used to generate new HLA class I and II alleles.

Intercellular adhesion molecules (ICAMs) ligands for the leukocyte integrins that enable lymphocytes and other leukocytes to bind to other cells, including antigen-presenting cells and endothelial cells.

Interdigitating reticular cells *see* **dendritic cells.**

Interferons cytokines that help cells to resist viral infection. **Interferon-α (IFN-α)** and **interferon-β (IFN-β)** are produced by leukocytes and fibroblasts respectively, as well as by other cells, whereas **interferon-γ (IFN-γ)** is a product of CD4 T_H1 cells, CD8 T cells, and NK cells. IFN-γ acts principally to activate macrophages.

Interleukin (IL) a generic term used for many of the cytokines produced by leukocytes. *See also* **IL-1, IL-2, IL-3,** etc.

Invariant chain (Ii) polypeptide that associates with major histocompatibility complex (MHC) class II proteins in the endoplasmic reticulum and prevents them from binding peptides there. It guides the MHC class II molecules to endosomes where Ii is degraded, enabling MHC class II molecules to bind peptides present in the endosomes.

Ischemia deficient blood supply due to obstructed blood vessels.

Islets of Langerhans the endocrine hormone-producing tissue of the pancreas, which includes the β cells that produce insulin.

Isoforms the different forms of a protein that are encoded by the alleles of a gene or by different but closely related genes.

Isograft a tissue or organ graft from one genetically identical individual to another.

Isotype switching the process by which a B cell changes the class of immunoglobulin made while preserving the antigenic specificity of the immunoglobulin. Isotype switching involves somatic recombination that attaches a different heavy-chain constant-region gene to the variable-region exon.

Isotypes classes of immunoglobulin—IgM, IgG, IgD, IgA, and IgE—each of which has a distinct heavy-chain constant region encoded by a different constant-region gene. The heavy-chain constant region determines the effector properties of each antibody class.

ITAMs *see* **immunoreceptor tyrosine-based activation motifs.**

ITIMs immunoreceptor tyrosine-based inhibitory motifs. *See also* **immunoreceptor tyrosine-based activation motifs.**

J gene segments, or **joining gene segments** one of the types of gene segment in immunoglobulin and T-cell receptor genes that is rearranged to make functional variable-region exons.

Janus kinases (JAKs) a family of tyrosine kinases that transduce activating signals from cytokine receptors.

Joining gene segments *see* **J gene segments.**

Junctional diversity diversity present in immunoglobulin and T-cell receptor polypeptides that is created during the process of gene rearrangement by the addition of nucleotides into the junctions between gene segments.

kappa (κ) one of the two types of immunoglobulin light chain.

Killed virus vaccines vaccines that contain viral particles that have been deliberately killed by heat, chemicals or radiation.

Killer cell immunoglobulin-like receptors (KIRs) receptors on NK cells that bind to MHC class I molecules and can send either activating or inhibitory signals to the NK cell.

Kupffer cells Phagocytic cells of the liver that line the hepatic sinusoids.

λ5 *see* **pre-B-cell receptor.**

L chain *see* **light chain.**

L-selectin an adhesion molecule of the selectin family found on lymphocytes. It binds to CD34 and GlyCAM-1 on high endothelial venules to initiate the migration of naive lymphocytes into secondary lymphoid tissue.

lambda (λ) one of the two types of immunoglobulin light chain.

Langerhans' cells phagocytic dendritic cells found in the epidermis. They can migrate in lymph from the epidermis to lymph nodes, where they differentiate into dendritic cells.

Large pre-B cells immature B cells that have a cell-surface pre-B-cell receptor.

Late pro-B cell *see* **pro-B cell.**

Late-phase reaction that part of type 1 immediate hypersensitivity reactions that occurs 7–12 hours after contact with antigen and is resistant to treatment with anti-histamine.

Latency a state adopted by some viruses in which they have entered cells but do not replicate.

Lck protein tyrosine kinase associated with the CD4 and CD8 co-receptor of T cells.

Lectin-mediated pathway one of the three pathways of complement activation. It is activated by the binding of a mannose-binding protein present in blood plasma to mannose-containing proteoglycans on bacterial surfaces.

Lentiviruses group of slow retroviruses that includes the human immunodeficiency virus (HIV). They have a long incubation period and disease can take years to become apparent.

Leukemia the unrestrained proliferation of a malignant white blood cell characterized by abnormally high numbers of white cells in the blood. A leukemia can be lymphocytic, myelocytic, or monocytic depending on the type of white blood cell that became malignant.

Leukocyte a general term for a white blood cell. Lymphocytes, granulocytes, and monocytes are all leukocytes.

Leukocyte adhesion deficiency an immunodeficiency disease in which the common β chain of the leukocyte integrins is absent. This mainly affects the ability of leukocytes to enter sites infected with extracellular pathogens, so such infections cannot be effectively eradicated.

Leukocytosis increased numbers of leukocytes in the blood. It is commonly seen in acute infection.

LFA-1 *see* **lymphocyte function-associated antigen-1**.

LFA-3 *see* **lymphocyte function-associated antigen-3**.

Light chain (L chain) the smaller of the two types of polypeptide chain that make up immunoglobulins. It consists of one variable and one constant domain, and in the immunoglobulin molecule it is disulfide-bonded to a heavy chain. There are two classes of light chain, known as κ and λ.

Light zone the part of a germinal center in secondary lymphoid tissue that contains non-dividing centrocytes interacting with follicular dendritic cells.

Linear epitope epitope of a protein recognized by antibody that consists of a linear sequence of amino acids within the protein's primary structure.

Live attenuated virus vaccines vaccines composed of live viruses that have an accumulation of mutations that impedes their growth in human cells and their capacity to cause disease.

LPS-binding protein plasma protein that binds bacterial lipopolysaccharide (LPS) and delivers it to LPS receptors on neutrophils and macrophages.

LT *see* **lymphotoxin**.

Lymph mixture of extracellular fluid and cells that is carried by the lymphatic system.

Lymph nodes a type of secondary lymphoid tissue found at many sites in the body where lymphatic vessels converge. Antigens are delivered by the lymph and presented to lymphocytes within the lymph node where adaptive immune responses are initiated.

Lymphatic vessels, or **lymphatics** thin-walled vessels that carry lymph from tissues to secondary lymphoid tissues (with the exception of the spleen) and from secondary lymphoid tissues to the thoracic duct.

Lymphocyte function-associated antigen-1 (LFA-1) one of the leukocyte integrins that mediate the adhesion of lymphocytes, especially T cells, to endothelial cells and antigen-presenting cells.

Lymphocyte function-associated antigen-3 (LFA-3) cell adhesion molecule of the immunoglobulin superfamily that is expressed on antigen-presenting cells (and other types of cell) and mediates their adhesion to T cells.

Lymphocytes a class of white blood cells that consist of small and large lymphocytes. The small lymphocytes bear variable cell-surface receptors for antigen and are responsible for adaptive immune responses. There are two main classes of small lymphocyte—B lymphocytes (B cells) and T lymphocytes (T cells). Large lymphocytes are natural killer (NK) cells.

Lymphoid containing lymphocytes, or pertaining to lymphocytes.

Lymphoid lineage all types of lymphocyte, and the bone marrow cells that give rise to them.

Lymphoid organs, or **lymphoid tissues** organized tissues that contain very large numbers of lymphocytes held in a non-lymphoid stroma. The primary lymphoid organs, where lymphocytes are generated, are the thymus and bone marrow. The main secondary lymphoid tissues, in which adaptive immune responses are initiated, are the lymph nodes, spleen, and mucosa-associated lymphoid tissues such as tonsils and Peyer's patches.

Lymphokines cytokines produced by lymphocytes.

Lymphomas tumors of lymphocytes that grow in lymphoid and other tissues but do not enter the blood in large numbers.

Lymphotoxin (LT) tumor necrosis factor-β (TNF-β), a cytotoxic cytokine secreted by some CD4 T cells.

Lytic granules intracellular storage granules of cytotoxic T cells and NK cells that contain perforin and granzymes.

M cells specialized cells in intestinal epithelium through which antigens and pathogens enter gut-associated lymphoid tissue from the intestines.

Macrophage activation stimulation of macrophages which increases their phagocytic, antigen-presenting and bacterial killing functions. It occurs in the course of infection.

Macrophage chemoattractant protein (MCP) chemokine produced by activated T cells that attracts macrophages into a site of infection.

Macrophages large mononuclear phagocytic cells resident in most tissues. They are derived from blood monocytes and contribute to innate immunity and early non-adaptive phases of host defense. They function as professional antigen-presenting cells and as effector cells in humoral and cell-mediated immunity.

MadCAM-1 mucosal cell adhesion molecule-1. A mucosal addressin that is recognized by the lymphocyte surface proteins L-selectin and VLA-4. This interaction mediates specific homing of lymphocytes to mucosal tissues.

Major basic protein constituent of eosinophil granules that is released on eosinophil activation. It acts on mast cells to cause their degranulation.

Major histocompatibility complex (MHC) cluster of genes on the short arm of human chromosome 6 that encodes a set of polymorphic membrane glycoproteins called the MHC molecules, which are involved in presenting peptide antigens to T cells.

Major histocompatibility complex (MHC) molecules *see* **MHC molecules.**

Malignant transformation the changes that occur in a cell to make it cancerous.

Malignant tumors tumors that are capable of uncontrolled and invasive growth.

MALT *see* **mucosa-associated lymphoid tissues.**

Mannose-binding protein (MBP) an acute-phase protein in the blood that binds to mannose residues on pathogen surfaces and when bound activates the complement system. It is also known as **mannan-binding lectin.**

Mannose-binding protein pathway *see* **lectin-mediated pathway.**

Mantle zone in lymphoid follicles, a rim of B lymphocytes that surrounds a follicle.

Marginating refers to the population of neutrophils that is loosely attached to blood vessel walls.

Mast cells large bone marrow derived cells found resident in connective tissues throughout the body. They contain large granules that store a variety of chemical mediators including histamine. Mast cells have high-affinity Fcε receptors (FcεRI) that bind free IgE. Antigen binding to mast cell associated IgE triggers mast cell activation and degranulation, producing a local or systemic immediate hypersensitivity reaction. Mast cells have a crucial role in allergic reactions.

Mature B cells B cells that have IgM and IgD on their surface and are able to respond to antigen.

MCP *see* (1) **macrophage chemoattractant protein;** (2) **membrane co-factor protein.**

Membrane co-factor protein (MCP) complement regulatory protein on human cells that promotes the inactivation of C3b and C4b by factor I.

Membrane-attack complex the complex of terminal complement components that forms a pore in the membrane of the target cell, damaging the membrane and leading to cell lysis.

Memory B cells long-lived antigen-specific B cells that are produced from activated naive B cells during the primary immune response to an antigen. On subsequent exposure to their specific antigen they are reactivated to differentiate into plasma cells as part of the secondary and subsequent immune responses.

Memory cells general term for lymphocytes that are responsible for the phenomenon of immunological memory and protective immunity.

Memory T cells long-lived antigen-specific T cells that are activated in secondary and subsequent immune responses to an antigen.

Metastasis the invasion of other tissues by a malignant tumor and which establishes secondary tumors distant from the site of origin of the primary tumor.

Methotrexate cytotoxic drug used to inhibit graft-versus-host reactions in bone marrow transplant recipients.

MHC *see* **major histocompatibility complex.**

MHC class I molecules the class of MHC molecules that present peptides generated in the cytosol to CD8 T cells. They consist of a heterodimer of a class I heavy chain associated with β_2-microglobulin.

MHC class II molecules the class of MHC molecules that present peptides generated in intracellular vesicles to CD4 T cells. They consist of a heterodimer of class II α and β chains.

MHC class II transactivator (CIITA) a transcriptional activator of MHC class II genes. When defective it causes a type of bare lymphocyte syndrome.

MHC molecules major histocompatibility complex molecules. Highly polymorphic glycoproteins encoded by the major histocompatibility complex (MHC). They form complexes with peptides and present peptide antigens to T cells. There are two classes—MHC class I and MHC class II molecules—with different roles in the immune response. Also known as major histocompatibility antigens because they are the main alloantigens involved in the rejection of transplanted tissues.

MHC restriction the fact that a given T-cell receptor will recognize its peptide antigen only when the peptide is bound to a particular form of MHC molecule.

Minor histocompatibility antigens, or **minor H antigens** peptides of polymorphic cellular proteins that can lead to graft rejection when they are bound by MHC molecules and recognized by T cells.

Minor histocompatibility loci the genes encoding proteins that can act as minor histocompatibility antigens.

Mixed lymphocyte reaction technique for detecting MHC differences between two individuals. The T cells from one individual proliferate in response to allogeneic MHC molecules on the cells of the other individual.

Molecular mimicry antigenic similarity between a pathogen antigen and a cellular antigen, which results in the induction of antibodies or T cells that act against the pathogen but also cross-react with the self antigen.

Monoclonal antibodies antibodies produced by a single clone of B lymphocytes and which are therefore identical in structure and antigen specificity.

Monocytes white blood cells with a bean-shaped nucleus. They are the precursors of tissue macrophages.

Monomorphic having only one form. The term is used, for example, for genes that have only one allele.

Mucosa the mucus-secreting epithelia that line the respiratory, intestinal, and urogenital tracts. The conjunctiva of the eye and the mammary glands are also in this category.

Mucosa-associated lymphoid tissue (**MALT**) aggregations of lymphoid cells in mucosal epithelia and in the lamina propria beneath. The main mucosa-associated lymphoid tissues are the gut-associated lymphoid tissues (GALT) and the bronchial-associated lymphoid tissues (BALT).

Mucosal surfaces *see* **mucosa**.

Mucus slimy protective secretion produced by many internal epithelia.

Multiple sclerosis a chronic progressive neurological disease characterized by patches of demyelination in the central nervous system and lymphocyte infiltration into the brain. It is believed to be an autoimmune disease.

Multivalent having more than one binding site for the same or different ligands.

Mutagen any agent, such as a chemical or radiation, that can cause a mutation.

Mutation an alteration in the DNA sequence of a gene.

Myasthenia gravis an autoimmune disease in which autoantibodies against the acetylcholine receptor on skeletal muscle cells cause a block of signal transmission from nerve to muscle at neuromuscular junctions, leading to progressive muscle weakness and eventually to death.

Mycosis fungoides *see* **cutaneous T-cell lymphoma**.

Myeloid lineage a subset of bone-marrow derived cells comprising granulocytes, monocytes, and macrophages.

Myeloid progenitors stem cells in the bone marrow that give rise to granulocytes, monocytes, and macrophages.

Myelomas tumors of plasma cells resident in bone marrow.

Myelopoiesis the production of monocytes and granulocytes in the bone marrow.

N-nucleotides nucleotides added at the junctions between gene segments of T-cell receptor and immunoglobulin heavy-chain variable-region sequences during somatic recombination, which increase the diversity of these molecules. They are not encoded in the gene segments, but are inserted by the enzyme terminal deoxynucleotidyltransferase (TdT).

Naive B cell a mature B cell that has left the bone marrow but has not yet encountered its specific antigen.

Naive T cell a mature T cell that has left the thymus but not yet encountered its specific antigen.

Natural killer cells (**NK cells**) large, granular, cytotoxic lymphocytes that circulate in the blood. NK cells are important in innate immunity to viruses and other intracellular pathogens and also kill certain tumor cells. They are the cytotoxic cells in antibody-dependent cell-mediated cytotoxicity (ADCC).

Necrosis the death of cells by lysis that results from chemical or physical injury. It leaves extensive cellular debris that must be removed by phagocytes. Neighboring tissue is also damaged by the molecules released from necrotic cells.

Negative selection process in the thymus whereby developing T cells that recognize self-antigens are induced to die by apoptosis.

Neoplasm a tumor, which may be benign or malignant

Neutralization the mechanism by which antibodies binding to sites on pathogens prevent growth of the pathogen and/or its entry into cells. The toxicity of bacterial toxins can similarly be **neutralized** by bound antibody.

Neutropenia a deficiency of neutrophils in the blood.

Neutrophils Phagocytic white blood cells that enter infected tissues in large numbers. They contain granules that stain with neutral dyes. After their entry into infected tissues, neutrophils engulf and kill extracellular pathogens in large numbers. They are also known as neutrophilic polymorphonuclear leukocytes and are a type of granulocyte. Once called microphages.

NFAT nuclear factor of activated T cells, a transcription factor which is activated as a result of signaling from the T-cell receptor. It functions as a complex of the NFAT protein with the dimer of Fos and Jun proteins known as AP-1.

NFκB nuclear factor κB, a transcription factor that is involved in turning on the expression of many genes of the immune system. It was first discovered in the κ light-chain gene.

NK cells *see* **natural killer cells**.

Oncogenes genes involved in the control of cell growth. When these genes are defective in either structure or expression, they can cause cells to proliferate abnormally and form a tumor.

Oncogenic viruses viruses that are involved in causing cancer.

Oncology the clinical discipline that deals with the study, diagnosis, and treatment of cancer.

Opportunistic pathogen a microorganism that causes disease only in individuals whose immune systems are in some way compromised.

Opsonins antibodies and complement components that bind to pathogens and facilitate their phagocytosis by neutrophils or macrophages.

Opsonization the coating of the surface of a pathogen or other particle with any molecule that makes it more readily ingested by phagocytes. Antibody and complement opsonize extracellular bacteria for phagocytosis by neutrophils and macrophages because the phagocytic cells carry receptors for these molecules.

Organ-specific autoimmune diseases autoimmune diseases targeted at a particular organ, such as the thyroid in Graves' disease.

Original antigenic sin a bias seen in successive immune responses to structurally related antigens such as different strains of influenza virus. When infected for the second time with influenza the antibody response is restricted to epitopes the second strain shares with the first strain to which the person was exposed. Other highly immunogenic epitopes on the second and subsequent viruses are ignored.

P-nucleotides nucleotides added into the junctions between gene segments during the somatic recombination that generates a rearranged variable-region sequence. They are an inverse repeat (a palindrome) of the nucleotide sequence at the end of the adjacent gene segment, which gives them their name of palindromic or P-nucleotides.

P-selectin *see* **selectins.**

Pandemic outbreak of an infectious disease that spreads worldwide.

Panel reactive antibody (PRA) a measure of the likelihood that a patient seeking a transplant is sensitized to potential donors. The patient's serum is tested against tissue from a representative panel of individuals for antibodies that would cause immediate hyperacute rejection of a graft. The PRA is the percentage of individuals in the panel whose cells react with the patient's antibodies.

Paracrine term applied to a cytokine that is released from one type of cell and acts on other cells nearby.

Parasites the unicellular protozoa and multicellular worms that infect animals and humans and live within them.

Paroxysmal nocturnal hemoglobinuria (PNH) a disease in which the complement regulatory proteins CD59 and DAF are defective, so that complement activation leads to episodes of spontaneous hemolysis. The defect is in the attachment of CD59 and DAF to cell membranes via a glycolipid anchor.

Passive immunization the injection of specific antibodies to provide protection against a pathogen or toxin. The administered antibodies may derive from human blood donors, immunized animals or hybridoma cell lines.

Passive transfer of immunity transfer of immunity to a non-immune individual by the injection of specific antibody, immune serum, or T cells.

Pathogen an organism, most commonly a microorganism, that can cause disease.

Pathology the scientific study of disease or detectable damage to tissue caused by disease.

Pemphigus vulgaris an autoimmune disease involving blistering of the skin.

Pentraxins a family of acute-phase proteins to which C-reactive protein belongs. They are formed of five identical subunits.

Peptide-binding motif of an MHC isoform, the combination of anchor residues that are conserved in the amino-acid sequences of peptides that bind to the isoform.

Perforin one of the proteins released by cytotoxic T cells on contact with their target cells. It forms pores in the target cell membrane that contribute to cell killing.

Peripheral lymphoid tissues *see* **secondary lymphoid tissues.**

Peyer's patches gut-associated lymphoid tissue present in the wall of the small intestine, especially the ileum.

Phagocyte a cell specialized to perform phagocytosis. The principal phagocytic cells in mammals are neutrophils and macrophages.

Phagocytosis cellular internalization of particulate matter, such as bacteria, by means of endocytosis. Cells specialized in phagocytosis are known as phagocytes.

Phagolysosome intracellular vesicle formed by fusion of a phagosome with a lysosome, in which the phagocytosed material is broken down by degradative lysosomal enzymes.

Phagosome intracellular vesicle containing material taken up by phagocytosis.

Plasma cells terminally differentiated B lymphocytes that secrete antibody.

Plasmablast an activated B cell differentiating within a lymphoid follicle at the stage at which it shows some features of a plasma cell.

PNH *see* **paroxysmal nocturnal hemoglobinuria.**

Poly-Ig receptor a receptor present on the basolateral membrane of epithelial cells that binds polymeric immunoglobulins, especially dimeric IgA, and transports them across the epithelium by transcytosis.

Polyclonal activation the non-specific activation of multiple clones of lymphocytes of diverse antigen specificity.

Polygeny the existence in the genome of several different gene loci encoding structurally similar proteins of identical function. Often refers to families of closely linked genes.

Polymorphism existing in different forms. Genetic polymorphism is defined as the existence of two or more forms (alleles) of a given gene within the population, with the variant alleles each occurring at a frequency greater than 1%.

Polymorphonuclear leukocytes *see* **granulocytes.**

Polyspecificity the ability to bind to many different antigens, a property shown by some antibodies. It is also known as polyreactivity.

Positive selection a process in the thymus that selects immature T cells with receptors that recognize peptide antigens presented by self-MHC molecules. Only cells that are positively selected are allowed to continue their maturation.

PRA *see* **panel reactive antibody.**

Pre-B cells a stage in B-cell development at which the B cells have rearranged their heavy-chain genes but not their light-chain genes.

Pre-B-cell receptor immunoglobulin-like receptor that is expressed on the surface of pre-B cells. It consists of μ heavy chains in association with surrogate light chains. Its appearance signals the cessation of heavy-chain gene rearrangement. The receptor's ligand is unknown.

Pre-T-cell receptor receptor that is present on the surface of some immature thymocytes. It consists of a T-cell receptor β chain associated with a surrogate α chain called pTα. The receptor's ligand is unknown.

Prednisolone   the biologically active compound to which the immunosuppressive drug prednisone is converted *in vivo.*

Prednisone   a synthetic steroid drug with potent anti-inflammatory and immunosuppressive activity. Extensively prescribed to patients who have had organ transplants.

Present   cells carrying cell-surface complexes of peptide antigens and MHC molecules are said to present these antigens to T lymphocytes.

Primary immune response   the adaptive immune response that follows a person's first exposure to an antigen.

Primary lymphoid follicles   the name given to the B-cell areas of secondary lymphoid tissues in the absence of an immune response. They contain resting B lymphocytes.

Primary lymphoid tissues   anatomical sites of lymphocyte development. The bone marrow and the thymus gland.

Pro-B cells   an early stage in B-cell development at which B-cell precursors express B-cell marker proteins and rearrange their heavy-chain genes. D_H to J_H joining occurs at the **early pro-B cell** stage, followed by V_H to DJ_H joining at the **late pro-B cell** stage.

Pro-drug   biologically inactive compound that is metabolized in the body to produce an active drug.

Productive rearrangements   DNA rearrangements within immunoglobulin and T-cell receptor genes that lead to a gene that can direct the synthesis of a functional polypeptide chain.

Professional antigen-presenting cells   cells that can present antigen to naive T cells and activate them. A professional antigen-presenting cell not only displays peptide antigens bound to appropriate MHC molecules but also has co-stimulatory molecules on its surface that are needed to activate the T cell. Only dendritic cells, macrophages and B cells can be professional antigen-presenting cells.

Progenitors   the more differentiated type of stem cells that give rise to distinct subsets of mature blood cells and that lack the self-renewal capacity of pluripotent stem cells.

Programmed cell death   *see* **apoptosis.**

Properdin   a blood protein that helps to activate the alternative pathway of complement activation. It binds to and stabilizes the alternative pathway C3 and C5 convertases on the surfaces of bacterial cells. It is also known as factor P.

Proteasome   large multisubunit protease present in the cytosol of all cells that degrades cytoplasmic proteins. It generates the peptides presented by MHC class I molecules.

Protective immunity   the specific immunological resistance to a pathogen that follows either from specific vaccination or recovery from an infection with the pathogen.

Protein tyrosine kinases   enzymes that phosphorylate tyrosine residues on proteins. They are involved in signal transduction from many types of cell-surface receptor, including the antigen receptors on B cells and T cells.

Proto-oncogenes   cellular genes that regulate the control of cell growth and division. When mutated or aberrantly expressed, they contribute to the malignant transformation of cells which leads to cancer.

Provirus   the DNA form of a retrovirus when it is integrated into the host cell genome. In this state it can remain transcriptionally inactive for a long time.

pTα   *see* **pre-T-cell receptor.**

Purine nucleotide phophorylase (PNP)   enzyme involved in purine metabolism. Its deficiency results in the accumulation of purine nucleosides, which are toxic for developing T cells. This leads to severe combined immunodeficiency.

Pus   thick yellowish-white fluid that is formed in infected wounds. It is composed of dead and dying white blood cells (principally neutrophils), tissue debris, and dead microorganisms.

Pyogenic bacteria   extracellular encapsulated bacteria that cause the formation of pus at sites of infection.

RAG-1, RAG-2 *see* **recombination activation genes.**

Rapamycin   *see* **sirolimus.**

Recirculation   of lymphocytes, their continual movement from blood to secondary lymphoid tissues to lymph and back to the blood. An exception to this pattern is traffic to the spleen; lymphocytes both enter and leave the spleen in the blood.

Recombination activation genes (*RAG-1, RAG-2*)   two genes whose expression is required for immunoglobulin and T-cell receptor gene rearrangement in B cells and T cells.

Recombination signal sequences (RSSs)   short stretches of DNA that flank the gene segments that are rearranged to generate V-region exons. They are the sites at which somatic recombination occurs.

Respiratory burst   metabolic change accompanied by a transient increase in oxygen consumption that occurs in neutrophils and macrophages when they have taken up opsonized particles. It leads to the generation of toxic oxygen metabolites and other anti-bacterial substances that attack the phagocytosed material.

Respiratory scyncytial virus (RSV)   a paramyxovirus that is a common cause of severe chest infection in young children. Often associated with wheezing.

Rev   protein product of the *rev* gene of the human immunodeficiency virus (HIV). The protein promotes the passage of viral RNA from nucleus to cytoplasm during HIV replication.

Reverse transcriptase   a DNA polymerase that works from an RNA template. It is a component of the retrovirus particle and copies the RNA genome into proviral DNA that integrates into the host-cell DNA.

Rheumatoid arthritis   a common inflammatory disease of joints that is probably due to an autoimmune response.

Rheumatoid factor an IgM antibody with specificity for human IgG that is produced in some people with rheumatoid arthritis.

RSS *see* **recombination signal sequences.**

RSV *see* **respiratory syncytial virus.**

Sarcoma a tumor that arises from a cell of connective tissue.

SCID, scid *see* **severe combined immunodeficiency.**

SE *see* **staphylococcal enterotoxins.**

Secondary immune response the adaptive immune response provoked by a second exposure to an antigen. It differs from the primary response by starting sooner and building more quickly.

Secondary lymphoid follicle the name given to the B-cell areas of secondary lymphoid tissues that are responding to antigen. They contain proliferating B cells.

Secondary lymphoid tissues the lymph nodes, spleen and mucosa-associated lymphoid tissues. These are the tissues in which immune responses are initiated. The more highly organized tissues such as lymph nodes and spleen are also often known as **secondary lymphoid organs.**

Secretory component, or **secretory piece** fragment of the poly-Ig receptor left attached to dimeric IgA after its transport across epithelial cells.

Segmental exchange *see* **gene conversion.**

Selectins family of adhesion molecules present on the surfaces of leukocytes and endothelial cells. They are lectins and bind to sugar moieties on certain mucin-like glycoproteins.

Self-antigens a term used to describe all the normal constituents of the body to which the immune system would respond were it not for the mechanisms of tolerance that destroy or inactivate self-reactive B and T cells.

Self-MHC a person's own MHC molecules.

Self-peptides peptides produced from the body's own proteins. In the absence of infection these peptides occupy the peptide-binding sites of MHC molecules on cell surfaces.

Self-tolerance the normal situation whereby a person's immune system does not respond to constituents of the person's body.

Sensitization in connection with allergy, the first exposure to an allergen that elicits an IgE response. Allergic reactions occur only in individuals who have already been **sensitized.**

Sepsis the toxic effects of infection of the bloodstream. Usually caused by Gram-negative bacteria. *See also* **septic shock.**

Septic shock frequently fatal shock syndrome caused by the systemic release of the cytokine TNF-α after bacterial infection of the bloodstream, usually with Gram-negative bacteria.

Septicemia bacterial invasion of the blood.

Seroconversion the phase of an infection when antibodies against the infecting agent are first detectable in the blood.

Serotypes antigenically different strains of a bacterium or other pathogen that can be distinguished by immunological means, for example by antibody-based detection tests. Also used to describe human alloantigens such as HLA and blood group antigens.

Serum sickness the collection of symptoms that follow the injection of large amounts of a foreign molecule (such as a serum protein) into a person. It is caused by the immune complexes formed between the injected protein and the antibodies that are made against it. Characterized by fever, arthralgias and nephritis.

Severe combined immune deficiency (SCID) immune deficiency disease in which neither antibody nor T-cell responses are made. It is usually the result of genetic defects that lead to T-cell deficiencies, and is fatal in childhood if not treated.

Single-positive thymocytes a late stage of T-cell development in the thymus characterized by the expression of either the CD4 or the CD8 co-receptor on the cell surface.

Sirolimus the proprietary name for the immunosuppressive drug rapamycin. Both names are in common use.

Small lymphocytes the general name for resting T and B lymphocytes as they recirculate.

Small pre-B cells pre-B cells in which light-chain gene rearrangements are being made.

Somatic gene therapy a possible therapy for curing inherited immunodeficiency diseases. Stem cells isolated from the patient are transfected with a normal copy of the gene which is defective in the patient. The stem cells which now express a good copy of the defective gene are reinfused into the patient's circulation.

Somatic hypermutation mutation that occurs at high frequency in the rearranged variable-region DNA of immunoglobulin genes in activated B cells, resulting in the production of variant antibodies, some of which have higher affinity for the antigen.

Somatic recombination DNA recombination that occurs between gene segments in the immunoglobulin genes and T-cell receptor genes in developing B cells and T cells respectively. It generates a complete exon that encodes the variable region of an immunoglobulin or T-cell receptor polypeptide chain.

Specificity the property of antibodies and other antigen binding molecules for selective interaction with only one or a few types of molecule or cell.

Spleen organ situated adjacent to the cardiac end of the stomach. One function of the spleen is to remove old or damaged red blood cells from the circulation, the other function is to be a secondary lymphoid organ that responds to blood-borne pathogens and antigens.

Staphylococcal enterotoxins (SEs) toxins secreted by strains of staphylococcus that act on the gut and cause the symptoms of food poisoning. They are also superantigens.

STATs Signal Transducers and Activators of Transcription. Proteins that are activated to become transcription factors through the Janus kinase signaling pathway from cytokine receptors.

Stromal cells cells that provide the supporting framework of a tissue or organ.

Subunit vaccines vaccines composed only of isolated antigenic components of a pathogen and not the pathogen itself, either alive or dead.

Superantigens molecules that by binding non-specifically to MHC class II molecules and T-cell receptors, stimulate the polyclonal activation of T cells.

Surrogate light chain a protein that mimics an immunoglobulin light chain and made up of two subunits, VpreB and λ5. It is produced by pro-B cells and together with μ heavy chain forms the pre-B-cell receptor.

Switch regions, or **switch sequences** DNA sequences preceding heavy-chain constant region genes at which somatic recombination occurs when B cells switch from producing one immunoglobulin isotype to another.

Sympathetic ophthalmia an autoimmune response that sometimes follows damage to an eye and affects both the damaged eye and the healthy eye.

Syngeneic genetically identical. A syngenic tissue graft is one made between two genetically identical individuals. It does not provoke an immune response or a rejection reaction.

Systemic anaphylaxis a rapid-onset and potentially fatal form of IgE-mediated allergic reaction, in which antigen in the bloodstream triggers the activation of mast cells throughout the body, causing circulatory collapse and suffocation due to tracheal swelling.

Systemic autoimmune disease, or **systemic autoimmunity** autoimmunity that involves reactions to common components of the body and whose effects are not confined to one particular organ.

Systemic lupus erythematosus an autoimmune disease in which autoantibodies made against DNA, RNA, and nucleoprotein particles form immune complexes that damage small blood vessels.

T cells, or **T lymphocytes** lymphocytes that develop in the thymus and are responsible for cell-mediated immunity. Their cell-surface antigen receptor is called the T-cell receptor.

T-cell activation the stimulation of mature naive T cells by antigen presented to them by professional antigen-presenting cells. It leads to their proliferation and differentiation into effector T cells.

T-cell priming the activation of mature naive T cells by antigen presented to them by professional antigen-presenting cells.

T-cell receptor α chain (TCRα) one of the two polypeptide chains that make up the most common form of T-cell receptor.

T-cell receptor β chain (TCRβ) one of the two polypeptide chains that make up the most common form of T-cell receptor.

T-cell receptor complex the complex of T-cell receptor α and β chains and the invariant CD3 and ζ chains that makes up a functional antigen receptor on the T-cell surface.

T-cell receptor the highly variable antigen receptor of T lymphocytes. On most T cells it is composed of a variable α chain and β

chain and is known as the α:β T-cell receptor. On a minority of T cells, the variable chains are γ and δ chains, and this receptor is known as the γ:δ T-cell receptor. Both types of receptor are present at the cell surface in association with the complex of invariant CD3 chains and ζ chains, which have a signaling function.

T lymphocyte *see* **T cells**.

Tacrolimus immunosuppressive polypeptide drug that inactivates T cells by inhibiting signal transduction from the T-cell receptor. It is widely used to suppress transplant rejection. It is also known as FK506.

TAP *see* **transporter of antigen processing**.

Tapasin TAP-associated protein, a chaperone protein involved in the assembly of peptide–MHC class I molecule complexes in the endoplasmic reticulum.

Target cell any cell that is acted on directly by effector T cells, effector cells or molecules. For example, virus-infected cells are the targets of cytotoxic T cells, which kill them, and naive B cells are the targets of effector CD4 T cells, which help to stimulate them to produce antibodies.

Tat protein product of the *tat* gene of the human immunodeficiency virus (HIV), which enhances the rate of transcription of viral RNA.

TD antigens *see* **thymus-dependent antigens**.

Terminal deoxynucleotidyltransferase (TdT) enzyme that inserts non-templated nucleotides (N-nucleotides) into the junctions between gene segments during the rearrangement of T-cell receptor and immunoglobulin heavy-chain genes.

Tetanus toxin protein neurotoxin produced by the bacterium *Clostridium tetani*. The cause of the disease tetanus.

TGF-β *see* **transforming growth factor-β**.

T$_H$1 cells a subset of CD4 T cells that are characterized by the cytokines they produce. They are involved mainly in activating macrophages. Also called inflammatory T cells.

T$_H$2 cells a subset of CD4 T cells that are characterized by the cytokines they produce. They are involved mainly in stimulating B cells to produce antibody.

Thymectomy the surgical removal of the thymus gland.

Thymic anlage the tissue from which the thymic stroma develops during embryogenesis.

Thymic stroma the reticular epithelial cells and connective tissue of the thymus, which form the essential microenvironment for T-cell development.

Thymocytes developing T cells in the thymus.

Thymus a lymphoepithelial organ in the upper part of the middle of the chest, just behind the breastbone. It is the site of T-cell development.

Thymus-dependent antigens (TD antigens) antigens for which the elicitation of an antibody response on immunization or vaccination is dependent upon help provided by antigen-specific T cells. Most protein antigens are of this type.

Thymus-dependent lymphocyte *see* **T cells.**

Thymus-independent antigens (TI antigens) antigens that can elicit antibody production in the absence of T cells. There are two types of TI antigen: TI-1 antigens, which have intrinsic B-cell activating activity and TI-2 antigens, which have multiple identical epitopes that crosslink B-cell receptors.

TI-1 antigens, TI-2 antigens *see* **thymus-independent antigens.**

Tingible body macrophages phagocytic cells that are seen in histological sections engulfing apoptotic B cells in germinal centers.

TNF-α *see* **tumor necrosis factor-α.**

TNF-β *see* **lymphotoxin.**

Tolerance when the immune system of a person does not or cannot respond to an antigen. In such cases the individual is said to be **tolerant** of the antigen.

Tonsils large aggregates of lymphoid cells lying on either side of the pharynx.

Toxic shock syndrome a systemic toxic reaction caused by the over production of cytokines by CD4 T cells activated by the bacterial superantigen **toxic shock syndrome toxin-1 (TSST-1)**, which is secreted by *Staphylococcus aureus*.

Toxoids toxins that have been deliberately inactivated by heat or chemical treatment so that they are no longer toxic but can still provoke a protective immune response on vaccination.

Toxoplasmosis the disease caused by the protozoan parasite *Toxoplasma gondii.*

Transcytosis the transport of molecules from one side of an epithelium to the other by endocytosis into vesicles within the epithelial cells at one face of the epithelium and release of the vesicles at the other.

Transforming growth factor-β (TGF-β) a multifunctional cytokine produced by T_H2 cells and other cell types.

Translocation a chromosomal abnormality in which a piece of one chromosome has become joined to another chromosome. The rearranging genes of B and T cells are often the sites of translocation in B- and T-cell tumors.

Transplant rejection immune reaction that is directed towards transplanted tissue and leads to death of the graft.

Transplantation the grafting of organs or tissues from one individual to another.

Transporter associated with antigen processing (TAP) an ATP-binding protein in the endoplasmic reticulum membrane that transports peptides from the cytosol to the endoplasmic reticulum lumen. It is composed of two subunits, TAP-1 and TAP-2. It supplies MHC class I molecules with peptides.

Tuberculin test clinical test to detect exposure to *Mycobacterium tuberculosis*, the causal agent of tuberculosis. Purified protein derivative (PPD) of *M. tuberculosis* which is injected subcutaneously elicits a delayed-type hypersensitivity reaction in individuals who have had tuberculosis or have been immunized against it.

Tumor a growth arising from uncontrolled cell proliferation. It may be benign and self-limiting, or malignant and invasive.

Tumor antigens cell-surface components of tumor cells that can elicit an immune response in the affected individuals. *See also* **tumor-associated antigens, tumor-specific antigens.**

Tumor-associated antigens antigens characteristic of certain tumor cells that are also present on some types of normal cells.

Tumor immunology the study of the immune response to tumors and how it can be improved.

Tumor necrosis factor-α (TNF-α) a cytokine produced by macrophages and T cells that has several functions in the immune response and is the prototype of the TNF family of cytokines. These cytokines function as cell-associated or secreted proteins that interact with receptors of the tumor necrosis factor receptor (TNFR) family.

Tumor-specific antigens antigens that are characteristic of tumor cells and are not expressed by normal cells.

Tumor suppressor genes genes for cellular proteins that function to prevent cells from becoming cancerous.

Type I, type II, type III, type IV hypersensitivity reactions *see* **hypersensitivity reactions.**

Unproductive rearrangements DNA rearrangements of immunoglobulin and T-cell receptor genes that produce a gene that cannot make a functional polypeptide chain.

Urticaria the technical term for hives, which are red, itchy skin welts usually brought on by an allergic reaction.

V domain *see* **variable domain.**

V gene segment, or **variable gene segment** DNA sequence in the immunoglobulin or T-cell receptor genes that encodes the first 95 or so amino acids of the V domain. There are multiple different V gene segments in the germline genome. To produce a complete exon encoding a V domain, one V gene segment must be rearranged to join up with a J or a rearranged DJ gene segment.

V region *see* **variable region.**

Vaccination the deliberate induction of protective immunity to a pathogen by the administration of killed or non-pathogenic forms of the pathogen, or its antigens, to induce an immune response.

Vaccine any preparation made from a pathogen that is used for vaccination and provides protective immunity against infection with the pathogen.

Vaccinia the cowpox virus. It causes a limited infection in humans that leads to immunity to the human smallpox virus. Was used as a vaccine against smallpox.

Variable domain (V domain) the amino-terminal domain of immunoglobulin and T-cell receptor polypeptide chains. Paired variable domains make up the antigen-binding site.

Variable gene segment *see* **V gene segment.**

Variable region (V region) the part of an immunoglobulin or T-cell receptor, or of its constituent polypeptide chains, that varies in amino acid sequence between isoforms having different antigenic specificities. It is responsible for determining antigen specificity.

Variable surface glycoproteins (VSGs) glycoproteins that form the surface coat of African trypanosomes. The trypanosome can repeatedly change its glycoprotein coat by expressing different glycoprotein genes using a process akin to gene conversion.

Variola name given to both the smallpox virus and the disease it causes: smallpox.

Variolation historical procedure for immunization against smallpox in which a small amount of live smallpox virus was introduced by scarification of the skin.

Vascular addressins mucin-like cell adhesion molecules on endothelial cells to which some leukocyte adhesion molecules bind. They determine the selective homing of leukocytes to particular sites in the body.

Viruses submicroscopic pathogens composed of a nucleic acid genome enclosed in a protein coat. They replicate only in a living cell, as they do not possess all the metabolic machinery required for independent life. A viral particle is called a virion.

VLA-4 integrin present on the surface of T lymphocytes that binds to mucosal cell adhesion molecules.

VpreB *see* **pre-B-cell receptor.**

Wegener's granulomatosis autoimmune disease characterized by severe necrotizing vasculitis. Caused by antibodies made against a constituent of the neutrophil granule: proteinase-3.

Weibel–Palade bodies granules containing P-selectin, which are present inside endothelial cells.

Wheal-and-flare reaction the reaction observed when small amounts of allergen are injected into the dermis of an individual who is allergic to the antigen. It consists of a raised area of skin containing fluid (the wheal) with a spreading, red, itchy reaction around it (the flare).

Wiskott–Aldrich syndrome a genetic disease in which interactions of T cells and B cells are defective. Antibody responses against polysaccharide antigens are deficient, and patients are susceptible to infection with pyogenic bacteria.

X-linked agammaglobulinemia (XLA) a genetic disorder in which B-cell development is arrested at the pre-B-cell stage and neither mature B cells or antibodies are formed. The disease is due to a defect in the gene encoding the protein tyrosine kinase Btk.

Xenoantibodies antibodies produced in one species against the antigens of another species.

Xenoantigens antigens on the tissues of one species that provoke an immune response in members of another species.

Xenograft an organ from one species transplanted into another species.

Xenotransplantation the transplantation of organs from one species into another. It is being explored as a possible solution to the shortage of human organs for clinical transplantation.

ZAP-70 a cytoplasmic tyrosine kinase in T cells that forms part of the signal transduction pathway from T-cell receptors.

Zymogen the inactive form in which some enzymes are synthesized. Active enzyme is usually produced by proteolytic cleavage of the zymogen.

Index